Microbiology Monographs

Volume 18

Series Editor: Alexander Steinbüchel

Münster, Germany

Microbiology Monographs

Volumes published in the series

Inclusions in Prokaryotes
Volume Editor: Jessup M. Shively
Vol. 1, 2006

Complex Intracellular Structures in Prokaryotes
Volume Editor: Jessup M. Shively
Vol. 2, 2006

Magnetoreception and Magnetosomes in Bacteria
Volume Editor: Dirk Schüler
Vol. 3, 2007

Predatory Prokaryotes – Biology, Ecology and Evolution
Volume Editor: Edouard Jurkevitch
Vol. 4, 2007

Amino Acid Biosynthesis – Pathways, Regulation and Metabolic Engineering
Volume Editor: Volker F. Wendisch
Vol. 5, 2007

Molecular Microbiology of Heavy Metals
Volume Editors: Dietrich H. Nies and Simon Silver
Vol. 6, 2007

Microbial Linear Plasmids
Volume Editors: Friedhelm Meinhardt and Roland Klassen
Vol. 7, 2007

Prokaryotic Symbionts in Plants
Volume Editor: Katharina Pawlowski
Vol. 8, 2009

Hydrogenosomes and Mitosomes: Mitochondria of Anaerobic Eukaryotes
Volume Editor: Jan Tachezy
Vol. 9, 2008

Uncultivated Microorganisms
Volume Editor: Slava S. Epstein
Vol. 10, 2009

Microbial Megaplasmids
Volume Editor: Edward Schwartz
Vol. 11, 2009

Endosymbionts in Paramecium
Volume Editor: Masahiro Fujishima
Vol. 12, 2009

Alginates: Biology and Applications
Volume Editor: Bernd H. A. Rehm
Vol. 13, 2009

Plastics from Bacteria: Natural Functions and Applications
Volume Editor: Guo-Qiang Chen
Vol. 14, 2010

Amino-Acid Homopolymers Occurring in Nature
Volume Editor: Yoshimitsu Hamano
Vol. 15, 2010

Biology of Rhodococcus
Volume Editor: Héctor Alvarez
Vol. 16, 2010

Structures and Organelles in Pathogenic Protists
Volume Editor: W. de Souza
Vol. 17, 2010

Plant Growth and Health Promoting Bacteria
Volume Editor: Dinesh K. Maheshwari
Vol. 18, 2010

(Endo)symbiotic Methanogenic Archaea
Volume Editor: Johannes H.P. Hackstein
Vol. 19, 2010

Dinesh K. Maheshwari
Editor

Plant Growth and Health Promoting Bacteria

Springer

Editor
Prof. Dr. Dinesh K. Maheshwari
Department of Botany and Microbiology
Gurukul Kangri University
Haridwar 249404 (Uttarakhand)
India
maheshwaridk@gmail.com

Series Editor
Professor Dr. Alexander Steinbüchel
Institut für Molekulare Mikrobiologie und Biotechnology
Westfälische Wilhelms-Universität
Corrensstr. 3
48149 Münster
Germany
steinbu@uni-muenster.de

ISSN 1862-5576 e-ISSN 1862-5584
ISBN 978-3-642-13611-5 e-ISBN 978-3-642-13612-2
DOI 10.1007/978-3-642-13612-2
Springer Heidelberg Dordrecht London New York

Library of Congress Control Number: 2010935951

Cover design: SPi Publisher Services

Printed on acid-free paper

Springer is part of Springer Science+Business Media (www.springer.com)

Preface

Plants provide an excellent ecosystem for microorganisms that interact with plant cells and tissues with differing degrees of dependence. Investigation on the relationship between roots and microbiota are essential to achieve innovations in agriculture and biotechnology. Similar to other industries, one such system is adoption of biological agents in the form of Plant Growth Promoting Bacteria (PGPB). These groups of bacteria are as effective as pure chemical on plant growth enhancement and disease control besides managing abiotic and other stresses in plants. Such organisms are now alternative paradigms for commercialization. Seeing the importance of these bacteria in the protection of plant health, new biotechnological approaches are employed regulating to develop newer and much better microbial agents for management of the phytopathogens.

This volume of the Microbiology Monograph series has 18 chapters that cover various facets of current scientific knowledge on PGPB that colonize the root and rhizosphere. *Bacillus*- and *Paenibacillus*-based bioinoculant formulations have met with great success in improving plant growth. A large number of PGPB genera on one hand and rhizobia and few endophytes on the other promise benefit to crop ecosystem for sustainable agriculture. A due account is provided with respect to basic concept on plant–bacteria interaction, mineral–nutrient exchange, biofilm formation, and bacteria inhabiting in harsh and cold tropical environment and their role in ethylene regulation via ACC deaminase, as well as the mechanisms of action of PGPB-mediated antifungals. In relation to plant health, the exploitation of such beneficial bacteria may improve agriculture system with economically sound production of human food and animal feed.

This book will be useful not only for students, teachers, and researchers but also for those interested in agriculture microbiology, plant pathology, ecology, environmental science, and agronomy.

I would like to express my sincere thanks to all the contributors for their much needed cooperation, authoritative and up to date information organized in a befitting manner. I acknowledge with thanks the assistance rendered by my research students Abhinav, Rajat, Pankaj, and Dr. Sandeep. I am also thankful to Council of

Scientific and Industrial Research (CSIR), New Delhi, and Director, Uttarakhand Council of Science and Technology (UCOST), Dehradun, India, for their support in execution of my research projects on PGPB that served as a prelude to lay foundation for compilation of the volume like this. I owe my special thanks to Prof. Alexander Steinbüchel, series editor, 'Microbiology Monographs,' University of Münster, Germany, for his professional advice from time to time in multifarious manner. I extend my sincere thanks to Drs. Christina Eckey and Jutta Lindenborn from the publisher Springer for their valuable support to facilitate completion of this volume.

Haridwar, India Dinesh K. Maheshwari

Contents

Benefits of Plant Growth-Promoting Rhizobacteria and Rhizobia in Agriculture 1
Marta S. Dardanelli, S.M. Carletti, N.S. Paulucci, D.B. Medeot, E.A. Rodriguez Cáceres, F.A. Vita, M. Bueno, M.V. Fumero, and M.B. Garcia

Plant Growth Promoting Rhizobacteria: Fundamentals and Applications 21
Márcia do Vale Barreto Figueiredo, Lucy Seldin, Fabio Fernando de Araujo, and Rosa de Lima Ramos Mariano

Potential of PGPR in Agricultural Innovations 45
Haluk Caglar Kaymak

Importance of Biofilm Formation in Plant Growth Promoting Rhizobacterial Action 81
Gamini Seneviratne, M.L.M.A.W. Weerasekara, K.A.C.N. Seneviratne, J.S. Zavahir, M.L. Kecskés, and I.R. Kennedy

Plant Growth Promoting Rhizobacteria: Constraints in Bioformulation, Commercialization, and Future Strategies 97
Naveen K. Arora, Ekta Khare, and Dinesh K. Maheshwari

Antifungal Compounds of Plant Growth Promoting Rhizobacteria and Its Action Mode 117
C.S. Quan, X. Wang, and S.D. Fan

Role of Plant Growth Promoting Rhizobacteria in Biocontrol of Plant Diseases and Sustainable Agriculture 157
Mohd. Sayeed Akhtar and Zaki A. Siddiqui

Potential of Bacilli for Biocontrol and Its Exploitation in Sustainable Agriculture 197
Olga Susana Correa and Marcelo Abel Soria

Plant Growth Promoting Rhizobacteria as Biocontrol Agents Against Soil-Borne Plant Diseases 211
Nico Labuschagne, T. Pretorius, and A.H. Idris

Sustainable Approaches for Biological Control of Fusarium Wilt in Pigeon Pea (*Cajanus cajan* L. Millspaugh) 231
Piyush Pandey, Abhinav Aeron, and D.K. Maheshwari

Plant Growth Promoter Rhizobacteria in Plants Inhabiting Harsh Tropical Environments and Its Role in Agricultural Improvements 251
Suikinai Nobre Santos, Vanessa Nessner Kavamura, João Luiz da Silva, Itamar Soares de Melo, and Fernando Dini Andreote

Cold-Tolerant Agriculturally Important Microorganisms 273
Pankaj Kumar Mishra, Piyush Joshi, Shekhar Chandra Bisht, Jaideep Kumar Bisht, and Govindan Selvakumar

The Role of Plant Growth Promoting Rhizobacteria in Sustainable and Low-Input Graminaceous Crop Production 297
Stephen P. Cummings and Caroline Orr

Rice Endophytic Diazotrophic Bacteria 317
Janpen Prakamhang, Nantakorn Boonkerd, and Neung Teaumroong

***Bacillus* and *Paenibacillus* spp.: Potential PGPR for Sustainable Agriculture** 333
Venkadasamy Govindasamy, Murugesan Senthilkumar, Vellaichamy Magheshwaran, Upendra Kumar, Pranita Bose, Vikas Sharma, and Kannepalli Annapurna

The Role of ACC Deaminase Producing PGPR in Sustainable Agriculture 365
Meenu Saraf, Chaitanya Kumar Jha, and Dhara Patel

The Role of the C:N:P Stoichiometry in the Carbon Balance Dynamics of the Legume–AMF–Rhizobium Tripartite Symbiotic Association 387
Vincent M. Gray

Regulation of Nitrogen Assimilation in Foliar Fed Legume Plants at Insufficient Molybdenum Supply 417
Marieta Hristozkova, Maria Geneva, and Ira Stancheva

Index 433

Contributors

Abhinav Aeron Department of Botany and Microbiology, Gurukul Kangri University, Haridwar 249404, Uttarakhand, India, abhinavaeron@gmail.com

Mohd. Sayeed Akhtar Department of Botany, Aligarh Muslim University, Aligarh 202002, India

Kannepalli Annapurna Division of Microbiology, Indian Agricultural Research Institute, New Delhi 110 012, India, annapurna93@yahoo.co.in

Fabio Fernando de Araujo National Research and Technological Development, Brazil University of West Paulista, UNOESTE, São Paulo, SP, Brazil

Naveen K. Arora Department of Microbiology, Institute of Biosciences and Biotechnology, CSJM University, Kanpur 208024, Utter Pradesh, India, nkarora_net@rediffmail.com

Jaideep Kumar Bisht Vivekananda Institute of Hill Agriculture, (I.C.A.R.), Almora 263601, Uttarakhand, India, bishtjk@hotmail.com

Nantakorn Boonkerd School of Biotechnology, Institute of Agricultural Technology, Suranaree University of Technology, Nakhon Ratchasima 30000, Thailand

Pranita Bose Division of Microbiology, Indian Agricultural Research Institute, New Delhi 110 012, India

M. Bueno Departamento de Biología Molecular, Facultad de Ciencias Exactas, Físico-Químicas y Naturales, Universidad Nacional de Río Cuarto, Ruta Nacional N° 36, Km. 601, CP X5804BYA Río Cuarto, Córdoba, Argentina

S.M. Carletti Departamento de Ciencias Básicas, Universidad Nacional de Luján, Rutas 5 y 7, CP 6700, Luján Provincia de Buenos Aires, Argentina

Shekhar Chandra Bisht Vivekananda Institute of Hill Agriculture, (I.C.A.R.), Almora 263601, Uttarakhand, India, shekhar_cbisht@yahoo.co.in

Olga Susana Correa Microbiología Agrícola y Ambiental, Facultad de Agronomía, Universidad de Buenos Aires, Av. San Martín 4453, C1417DSE Buenos Aires, Argentina, correa@agro.uba.ar

Stephen P. Cummings School of Applied Sciences, University of Northumbria, Ellison Building, NE1 8ST, Newcastle upon Tyne, UK, stephen.cummings@northumbria.ac.uk

Marta S. Dardanelli Departamento de Biología Molecular, Facultad de Ciencias Exactas, Físico-Químicas y Naturales, Universidad Nacional de Río Cuarto, Ruta Nacional N° 36, Km. 601, CP X5804BYA Río Cuarto, Córdoba, Argentina, mdardanelli@exa.unrc.edu.ar

Fernando Dini Andreote Department of Soil Science, Surerior School of Agriculture "Luiz de Queiroz", University of Sao Paulo, Piracicaba, SP, Brazil, fdandreo@gmail.com

S.D. Fan College of Life Science, Dalian Nationalities University, Dalian 116600, China

M.V. Fumero Departamento de Biología Molecular, Facultad de Ciencias Exactas, Físico-Químicas y Naturales, Universidad Nacional de Río Cuarto, Ruta Nacional N° 36, Km. 601, CP X5804BYA Río Cuarto, Córdoba, Argentina

M.B. Garcia Departamento de Biología Molecular, Facultad de Ciencias Exactas, Físico-Químicas y Naturales, Universidad Nacional de Río Cuarto, Ruta Nacional N° 36, Km. 601, CP X5804BYA Río Cuarto, Córdoba, Argentina

Maria Geneva Acad. M. Popov, Institute of Plant Physiology, Bulgarian Academy of Sciences, Acad. G. Bonchev str., block 21, Sofia 1113, Bulgaria

Venkadasamy Govindasamy Division of Microbiology, Indian Agricultural Research Institute, New Delhi 110 012, India

Vincent M. Gray School of Molecular and Cell Biology, University of the Witwatersrand, Johannesburg, South Africa, Vincent.Gray@wits.ac.za

Marieta Hristozkova Acad. M. Popov, Institute of Plant Physiology, Bulgarian Academy of Sciences, Acad. G. Bonchev str., block 21, Sofia 1113, Bulgaria

A.H. Idris ARC-Plant Protection Research Institute, Private bag X134, Queenswood 0121, 0001, Pretoria, South Africa, HassenA@arc.agric.za

Chaitanya Kumar Jha Department of Microbiology, School of Sciences, Gujarat University, Ahmedabad 380 009, Gujarat, India

Piyush Joshi Vivekananda Institute of Hill Agriculture, (I.C.A.R.), Almora 263601, Uttarakhand, India, joshinbri@rediffmail.com

Vanessa Nessner Kavamura Superior School of Agriculture "Luiz de Queiroz", University of São Paulo, Piracicaba, SP, Brazil; Laboratory of Environmental Microbiology, Embrapa Environment, Rodovia SP 340 – km 127.5, P.O. Box 13820-000, Jaguariúna, SP, Brazil

Haluk Caglar Kaymak Faculty of Agriculture, Department of Horticulture, Atatürk University, 25240 Erzurum, Turkey, hckaymak@atauni.edu.tr

M.L. Kecskés SUNFix Centre for Nitrogen Fixation, Faculty of Agriculture, Food and Natural Resources, University of Sydney, Sydney, NSW 2006, Australia, m.kecskes@usyd.edu.au

I.R. Kennedy SUNFix Centre for Nitrogen Fixation, Faculty of Agriculture, Food and Natural Resources, University of Sydney, Sydney, NSW 2006, Australia, i.kennedy@usyd.edu.au

Ekta Khare Department of Microbiology, Institute of Biosciences and Biotechnology, CSJM University, Kanpur 208024, Utter Pradesh, India

Upendra Kumar Division of Microbiology, Indian Agricultural Research Institute, New Delhi 110 012, India

Nico Labuschagne Department of Microbiology and Plant Pathology, University of Pretoria, Pretoria, South Africa

Rosa de Lima Ramos Mariano National Research and Technological Development, Brazil Federal Rural University of Pernambuco, Recife, PE, Brazil

Vellaichamy Magheshwaran Division of Microbiology, Indian Agricultural Research Institute, New Delhi 110 012, India

Dinesh K. Maheshwari Department of Botany and Microbiology, Gurukul Kangri University, Haridwar 249404, Uttarakhand, India, maheshwaridk@gmail.com

D.B. Medeot Departamento de Biología Molecular, Facultad de Ciencias Exactas, Físico-Químicas y Naturales, Universidad Nacional de Río Cuarto, Ruta Nacional N° 36, Km. 601, CP X5804BYA Río Cuarto, Córdoba, Argentina

Itamar Soares de Melo Laboratory of Environmental Microbiology, Embrapa Environment, Rodovia SP 340 – km 127.5, P.O. Box 13820-000, Jaguariúna, SP, Brazil, itamar@cnpma.embrapa.br

Pankaj Kumar Mishra Vivekananda Institute of Hill Agriculture, (I.C.A.R.), Almora 263601, Uttarakhand, India, misrapank12@rediffmail.com

Suikinai Nobre Santos Superior School of Agriculture "Luiz de Queiroz", University of São Paulo, Piracicaba, SP, Brazil; Laboratory of Environmental Microbiology, Embrapa Environment, Rodovia SP 340 – km 127.5, P.O. Box 13820-000, Jaguariúna, SP, Brazil, suikinai@yahoo.com.br

Caroline Orr School of Applied Sciences, University of Northumbria, Ellison Building, NE1 8ST, Newcastle upon Tyne, UK

Piyush Pandey Department of Biotechnology, Division of Life Sciences, S.B.S.P.G. Institute of Biomedical Sciences and Research, Balawala, Dehradun 248161, Uttarakhand, India, piyushgkp@rediffmail.com

Dhara Patel Department of Microbiology, School of Sciences, Gujarat University, Ahmedabad 380 009, Gujarat, India

N.S. Paulucci Departamento de Biología Molecular, Facultad de Ciencias Exactas, Físico-Químicas y Naturales, Universidad Nacional de Río Cuarto, Ruta Nacional N° 36, Km. 601, CP X5804BYA Río Cuarto, Córdoba, Argentina

Janpen Prakamhang School of Biotechnology, Institute of Agricultural Technology, Suranaree University of Technology, Nakhon Ratchasima 30000, Thailand

T. Pretorius Department of Microbiology and Plant Pathology, University of Pretoria, Pretoria, South Africa

C.S. Quan College of Life Science, Dalian Nationalities University, Dalian 116600, China, mikyeken@dlnu.edu.cn

E.A. Rodriguez Cáceres Biotecnología Plantec, Avenida Cabred 1350, Open Door, CP 6708 Luján Provincia de Buenos Aires, Argentina

Meenu Saraf Department of Microbiology, School of Sciences, Gujarat University, Ahmedabad 380 009, Gujarat, India, sarafmeenu@gmail.com

Lucy Seldin National Research and Technological Development, Brazil Federal University of Rio de Janeiro, Rio de Janeiro, RJ, Brazil

Govindan Selvakumar Indian Institute of Horticultural Research, Bangalore 560089, Karnataka, India, gselva74@rediffmail.com

Gamini Seneviratne Institute of Fundamental Studies, Hantana Road, Kandy Sri Lanka, gaminis@ifs.ac.lk

K.A.C.N. Seneviratne Royal Botanic Gardens, Peradeniya, Sri Lanka

Murugesan Senthilkumar Division of Microbiology, Indian Agricultural Research Institute, New Delhi 110 012, India

Vikas Sharma Division of Microbiology, Indian Agricultural Research Institute, New Delhi 110 012, India

Zaki A. Siddiqui Department of Botany, Aligarh Muslim University, Aligarh 202002, India, zaki_63@yahoo.co.in

João Luiz da Silva Laboratory of Environmental Microbiology, Embrapa Environment, Rodovia SP 340 – km 127.5, P.O. Box 13820-000, Jaguariúna, SP, Brazil, silva@cnpma.embrapa.br

Marcelo Abel Soria Microbiología Agrícola y Ambiental, Facultad de Agronomía, Universidad de Buenos Aires, Av. San Martín 4453, C1417DSE Buenos Aires, Argentina

Ira Stancheva Acad. M. Popov, Institute of Plant Physiology, Bulgarian Academy of Sciences, Acad. G. Bonchev str., block 21, Sofia 1113, Bulgaria, ira_stancheva@abv.bg

Neung Teaumroong School of Biotechnology, Institute of Agricultural Technology, Suranaree University of Technology, Nakhon Ratchasima 30000, Thailand, neung@sut.ac.th

Márcia do Vale Barreto Figueiredo National Research and Technological Development, Brazil Agronomical Institute of Pernambuco, IPA/CARHP, 1371, Gen. San Martin Avenue, Recife, PE 50761-000, Brazil, mbarreto@elogica.com.br

F.A. Vita Departamento de Ciencias Básicas, Universidad Nacional de Luján, Rutas 5 y 7, CP 6700 Luján Provincia de Buenos Aires, Argentina

X. Wang Dalian Institute of Chemical Physics, Chinese Academy of Science, Dalian 116023, China; Graduate School of Chinese Academy of Science, Beijing 100049, China

M.L.M.A.W. Weerasekara Laboratory of Soil Microbiology, Faculty of Agriculture, Saga University, Saga, Japan, anjani6466@gmail.com

J.S. Zavahir Geocycle (SBF), Dandenong South VIC 3174, Australia, bungali@rediffmail.com

Benefits of Plant Growth-Promoting Rhizobacteria and Rhizobia in Agriculture

Marta S. Dardanelli, S.M. Carletti, N.S. Paulucci, D.B. Medeot, E.A. Rodriguez Cáceres, F.A. Vita, M. Bueno, M.V. Fumero, and M.B. Garcia

Contents

1 Introduction 2
2 Rhizosphere in Action 3
3 Role of PGPR in Agriculture 6
4 Potential Uses of PGPR and Rhizobia 8
5 PGPR Studies in Argentina 10
6 New Studies and Applications of PGPR 13
7 Perspectives and Conclusion 14
References 15

Abstract The rhizosphere is the volume of soil under the influence of plants roots, where very important and intensive microbe–plant interactions take place. These interactions can both significantly influence plant growth and crop yields and have biotechnological applications. The rhizosphere harbors a diverse community of microorganisms that interact and compete with each other and with the plant root. The activity of some of the members of this community affects the growth and the physiology of the others, as well as the physical and chemical properties of the soil. Among all these interactions, those resulting in symbiotic and non-symbiotic nitrogen fixation are considerably important. In recent years, the use of bacteria

M.S. Dardanelli (✉), N.S. Paulucci, D.B. Medeot, M. Bueno, M.V. Fumero, and M.B. Garcia
Departamento de Biología Molecular, Facultad de Ciencias Exactas, Físico-Químicas y Naturales, Universidad Nacional de Río Cuarto, Ruta Nacional N° 36, Km. 601, CP X5804BYA Río Cuarto, Córdoba, Argentina
e-mail: mdardanelli@exa.unrc.edu.ar

S.M. Carletti and F.A. Vita
Departamento de Ciencias Básicas, Universidad Nacional de Luján, Rutas 5 y 7, CP 6700 Luján, Provincia de Buenos Aires, Argentina

E.A.R. Cáceres
Biotecnología Plantec, Avenida Cabred 1350, Open Door, CP 6708 Luján, Provincia de Buenos Aires, Argentina

D.K. Maheshwari (ed.), *Plant Growth and Health Promoting Bacteria*,
Microbiology Monographs 18, DOI 10.1007/978-3-642-13612-2_1,

(rhizobacteria) to promote plant growth has increased in several regions of the world and has acquired relevant importance in developing countries that are the producers of raw materials for food. Rhizobacteria can affect plant growth by producing and releasing secondary metabolites, which either decrease or prevent the deleterious effects of phytopathogenic organisms in the rhizosphere, and/or by facilitating the availability and uptake of certain nutrients from the root environment. Significant increases in the growth and yield of agriculturally important crops in response to inoculation with rhizobacteria have been reported. This practical application of plant growth-promoting rhizobacteria is the main focus of this chapter.

1 Introduction

The roles of microbiology and biotechnology in agriculture are very important because plant sources satisfy up to 80% of humans dietary needs. The Earth's population increases by 1.4% annually and is expected to reach 8.3 billion by 2025; therefore, unprecedented increases in crop production will be needed if the current levels of N (nitrogen) are to be maintained (11 g of N per person per day) (Mannion 1998; Graham and Vance 2000). The needs for N of most crop plants are second only to their photosynthetic requirement. Because soil N deficiency is common in many areas of crop production and land areas now considered marginal, N supply, N management, and N-use efficiency are significant factors in crop production, and are important as to the availability of fossil fuel reserves for future fertilizer N production (Graham and Vance 2000). On the other hand, farmers and breeders have long known that it is often the simultaneous occurrence of several abiotic stresses, rather than a particular stress condition, that is most lethal to crops. Tolerance to a combination of different stress conditions, particularly those that mimic the field environment, should be the focus of future research programmes aimed at developing transgenic crops and plants with enhanced tolerance to naturally occurring environmental conditions (Mittler 2006).

Arable land resources are limited. Thus, meeting food needs in some regions has already led to the adoption of agricultural practices that can degrade the soil, and to the use of land that is marginal for crop production. Nutrient depletion and soil acidification are only two of the common consequences of inadequate soil management (Hungria and Vargas 2000). In this context, the presence of microorganisms in the soil is critical to the maintenance of soil function, in both natural and managed agricultural soils, because of their involvement in key processes such as soil structure formation, decomposition of organic matter, toxin removal, and the cycling of carbon, nitrogen, phosphorus, and sulphur (van Elsas and Trevors 1997). In addition, microorganisms play key roles in suppressing soil-borne plant diseases and in promoting plant growth and changes in vegetation (Doran et al. 1996). Future exploitation of interactions will be as dependent on a better understanding of the biology of plant–microbe interaction as on developments in biotechnology (Beringer 1986). The beneficial use of rhizobacteria in agriculture is discussed in this chapter.

2 Rhizosphere in Action

The associations that occur between plant roots and soil microorganisms have been known for many decades. Considerable efforts have been devoted to study ectomycorrhizal fungi, nitrogen-fixing bacteria, soil-borne pathogenic fungi, and other microorganisms. As a consequence of the many investigations of the variable response of plants to different soils, an awareness of the complexity of the interactions between roots and soil microbes has been developed (Atkinson and Watson 2000). When seeds germinate and roots grow through the soil, the loss of organic material provides the driving force for the development of active microbial populations around the root. This effect is known as "the rhizosphere effect" (Whipps 1990). The term "rhizosphere" was first defined by Lorenz Hiltner in 1904 as "the soil compartment influenced by the root" (Hiltner 1904).

Although bacteria were not proven to exist until von Leeuwenhoek in 1683 discovered microscopic "animals" under the lens of his microscope, their use to stimulate plant growth in agriculture has been exploited since ancient times. Theophrastus (372–287 BC) suggested the mixing of different soils as a means of "remedying defects and adding heart to the soil" (Tisdale and Nelson 1975). The rhizosphere of plants is a zone of intense microbial activity, and some bacteria from this zone, termed rhizobacteria, exhibit different functions. The rhizosphere contains an increased microbial biomass and activity compared with nonrhizosphere soil: the number of microorganisms in the rhizosphere is 19–32 times larger than in root-free soil (Bodelier et al. 1997). Rhizobacteria that exert beneficial effects on plant development are referred to as plant growth-promoting rhizobacteria (PGPR) because their application is often associated with increased rates of plant growth (Kloepper and Schroth 1978). The well-known PGPR include members of the genera *Arthrobacter, Azoarcus, Azospirillum, Bacillus, Burkholderia, Enterobacter, Gluconacetobacter, Herbaspirillum, Klebsiella, Paenibacillus, Pseudomonas*, and *Serratia*, among others. PGPR can affect plant growth either directly (by providing plants with a compound synthesized by the bacterium or by facilitating the uptake of certain nutrients from the environment) or indirectly (by decreasing or preventing the deleterious effects of one or more phytopathogenic organisms) (Glick 1995). In order to exert their function, PGPR must colonize the rhizosphere around the roots, the rhizoplane (root surface) or the root itself (within root tissues) (Glick 1995).

Non-pathogenic rhizobacteria can induce a systemic resistance in plants that is phenotypically similar to the pathogen-induced systemic acquired resistance (SAR). Rhizobacteria-mediated induced systemic resistance (ISR) has been demonstrated against fungi, bacteria, and viruses in bean, carnation, cucumber, radish, tobacco, and tomato under conditions in which the inducing bacteria and the challenging pathogen remained spatially separated (van Loon et al. 1998). ISR elicited by PGPR has suppressed plant diseases caused by a range of pathogens in both greenhouse and field conditions. However, fewer reports have been published on PGPR as elicitors of tolerance to abiotic stresses, such as drought, salt and nutrient deficiency or excess. Recently, Yang et al. (2009) have proposed the term "induced systemic tolerance" (IST) for PGPR-induced physical and chemical changes in plants that result in enhanced tolerance to abiotic stresses.

Beneficial bacteria that are able to establish a nitrogen-fixing symbiotic relationship with leguminous plants (collectively called rhizobia) are usually not considered as PGPR. Endosymbiotic interactions between legume plants and the genera *Azorhizobium, Bradyrhizobium, Mesorhizobium*, *Rhizobium*, and *Sinorhizobium* have been intensively studied (Vance 1998; Graham and Vance 2000; Perret et al. 2000). Rhizobia infect legumes and have a global distribution, ranging from high latitudes in Europe and North America to the equator, to tropics in Australia and South America. In equatorial and tropical areas, legumes are particularly important and are used in sylvopastoral and agroforestry systems (Dommergues and Subba Rao 2000). Intricate signaling between the host and rhizobial symbiont is required for successful symbiotic interactions, which result in the reduction of atmospheric N_2 to ammonia by the bacteroids. Recently, some of these bacteria have been shown to be plant-growth promoting on nonlegumes, through mechanisms different from nitrogen fixation. Nevertheless, these will not be further considered, as the mechanisms involved are not different from those of the well-known and better-documented PGPR (Spaepen et al. 2009). Thus, in the broadest sense, PGPR include the N_2-fixing rhizobacteria that colonize the rhizosphere and provide N to plants.

Rhizosphere interactions are based on complex exchanges that take place round plant roots. Beneficial, detrimental, and neutral relationships between plant roots and microorganisms are all regulated by complex molecular signaling. It is clear that all the biological community, rather than only the immediate micro-flora, plays a role in the interaction of the rhizosphere. The existence of both microbial responses to plants and plant responses to the presence of microbes suggests a degree of coevolution between two partners. Two of the best-studied interactions between plant hosts and bacteria include the root nodule inhabiting *Rhizobium* spp. and tumor-forming *Agrobacterium* spp. The study of these systems has led to the discovery that plants and bacteria communicate by using chemical signals, which are involved in a successful interaction (Peters et al. 1986; Bolton et al. 1986; Fisher and Long 1992; Dardanelli et al. 2008a, b, 2009).

Chemical signaling between plant roots and other soil organisms, including the roots of neighboring plants, is often based on root-derived chemicals. Forty to ninety percent of the carbon transferred to the roots is lost and is called rhizodeposition (Kennedy 1998). In the rhizosphere environment, rhizodeposition includes different fractions: root exudates, lysates, mucilage, secretions, and dead cell material (Lynch and Whipps 1990). A substantial portion of the root exudates consists of carbon and energy sources readily available for microbial growth development and the physiology of microbial cell populations (Sommers et al. 2004).

Different organic compounds, such as amino acids, sugars, vitamins, organic acids, auxins, and flavonoids, which are rapidly utilized by microorganisms, have been identified in root exudates (Sommers et al. 2004; Dardanelli et al. 2008a,b, 2009; Raaijmakers et al. 2009). The signal components largely responsible for specific host–microbe relationships belong to a class of compounds termed flavonoids (Peters et al. 1986). More than 4,000 different flavonoids have been identified in vascular plants, and a particular subset of them is involved in mediating host specificity in legumes (Perret et al. 2000). Isoflavonoids are found only in members of the legume family, and despite the great importance of chemical molecules, some problems may arise because the same chemical signals may elicit dissimilar responses from different recipients. The chemical components of root exudates may deter one organism and attract another, or two very different organisms that may cause different consequences to the plant may be attracted. For example, the secretion of isoflavones by soybean roots is able to attract both a mutualist (*Bradyrhizobium japonicum*) and a pathogen (*Phytopthora sojae*) (Morris et al. 1998).

The attraction and subsequent migration toward plant roots is probably a key factor for the initiation of several plant–bacterial interactions. Motility may increase the probability that the microbe and the plant meet in the soil environment. By means of directed movement, bacteria are able to move toward plant roots, where they can benefit from a wide range of exudate-derived nutrients, enabling them to survive in and subsequently colonize the rhizosphere (Sommers et al. 2004). A variety of compounds, such as surface proteins and polysaccharides, have been implicated in the adherence of several rhizobacteria to plant roots (Dardanelli et al. 2003; Rodríguez-Navarro et al. 2007). The importance of bacterial attachment in PGPR–plant interactions has been intensively studied in *Azospirillum* and *Pseudomonas*. It is generally believed that the main mechanism by which *Azospirillum* enhances plant growth is by the production of plant hormones (Steenhoudt and Vanderleyden 2000). These growth-promoting substances stimulate the density and length of root hairs and root surface area, improving the utilization of water and mineral nutrients. Zhu et al. (2002) have shown that *Azospirillum irakense* cells are mainly associated with rice root hairs, whereas *Azospirillum brasilense* are mainly located on root surfaces. These differences in spatial distribution are the reason why these two species do not compete for root colonization.

Plant-associated *Pseudomonas* bacteria live as saprophytes but also as pathogenic parasites on plant surfaces and inside plant tissues. In addition, some *Pseudomonas* species show plant growth-promoting activity by suppressing the growth (biocontrol) of other phytopathogenic microorganisms, synthesizing growth-stimulating plant hormones and promoting plant mechanisms involved in disease resistance. Initial attachment to biotic or abiotic surfaces leads to a global change in gene expression in *Pseudomonas putida* (Rodríguez-Navarro et al. 2007). The isolation of genes involved in the adhesion to abiotic surfaces and the attachment to plant roots suggests that initial colonization of both biotic and abiotic surfaces proceeds via similar pathways (Sauer and Camper 2001). Although agglutinin plays a major role in the adherence and colonization abilities of *P. putida* strain Corvallis to bean and cucumber, the role of agglutinins is not general for all biocontrol strains. No agglutination-dependent adherence or root colonization has been demonstrated for 30 different *Pseudomonas* isolates on tomato, potato, and grasses (Lugtenberg and Dekkers 1999).

3 Role of PGPR in Agriculture

The plant growth-promoting capacity has been related to different physiological activities that may have a profound effect on the growth and/or health of plants. Although some chapters in this book comment on different functions of PGPR and rhizobia in agriculture, in this chapter we will briefly discuss some of them.

In most agricultural ecosystems, soil-borne plant pathogens can be a major limitation in the production of marketable yields. They are also more recalcitrant to management and control as compared to pathogens that attack the above-ground portions of the plant (Bruehl 1987). In addition, they are adapted to growing and surviving in the bulk soil, but the rhizosphere is the infection court where they encounter the plant and establish a parasitic relationship (Raaijmakers et al. 2009). Estimating crop loss caused by pathogens is difficult and there are only a few well-documented studies. From 2001–2003, an average of 7–15% of major world crops (wheat, rice, potato, maize, and soybean) was lost because of fungi and bacteria (Oerke 2005). From 1996 to 1998, these pathogens caused a loss of 9.9%, although the potential loss without controls could have been 14.9% (Oerke and Dehne 2004). Losses caused by soil-borne pathogens are even more difficult to estimate, because of the difficulty of diagnosis. Some estimate that soil-borne pathogens cause 50% of the crop loss in the United States (Lewis and Papavizas 1991).

The increased use of chemical inputs causes several negative effects, i.e., development of pathogen resistance to the applied agents and their non-target environmental impacts (Gerhardson 2002). Furthermore, the growing cost of pesticides, particularly in less affluent regions of the world, and the growing consumer demand for pesticide-free food, have led to a search for substitutes for these products. There are also a number of fastidious diseases for which chemical solutions are few, ineffective, or nonexistent (Gerhardson 2002). Biological control is thus being considered as an alternative or supplemental way of reducing the use of chemicals in agriculture (Whipps 2001; Gerhardson 2002). For several years, a great diversity of rhizobacteria have been described, characterized, and tested as biocontrol agents of diseases caused by soil-borne plant pathogens. Different biocontrol activities of PGPR are mediated by the synthesis of bacterial allochemicals, including iron-chelating siderophores, antibiotics, biocidal volatiles, lytic enzymes, and detoxication enzymes, among others (Glick 1995; Compant et al. 2005). In the last few years, some studies have been carried out on bacteria applied; studies of bacteria applied to seeds and roots for the purpose of controlling bacterial diseases. One example is the application of non-pathogenic strains of *Streptomyces* to control scab of potato (*Solanum tuberosum* L.) caused by *S. scabies* (Ryan and

Kinkel 1997). Here, the biocontrol may operate through antibiosis or competition for space or nutrients in the rhizosphere.

The global market for phytosanitary products used worldwide to ensure crop yield was estimated at US$ 26.7 billion in 2005 (Thakore 2006). Synthetic pesticides dominate this market. However, irrational selection and use has led to environmental toxicity of their residues, decrease or loss of efficacy because of adaptation of pathogens, or undesirable effects on non-target organisms sharing the ecosystem (Ongena and Jacques 2007). The use of beneficial microorganisms as biopesticides is considered one of the most promising methods for more rational and safe crop-management practices. Among all biopesticides, microorganism-based products represent 30% of total sales and new products are regularly brought to the market (Thakore 2006). Biopesticides are used in field crops and greenhouses to reduce diseases on various cereals, legumes, fruits, flowers, and ornamentals caused by soil-borne, foliar, or postharvest pathogens. Most of the bacterial strains exploited as biopesticides belong to the genera *Agrobacterium*, *Bacillus*, and *Pseudomonas* (Fravel 2005). *Bacillus thuringiensis*, specifically used for insect pest control, accounts for >70% of total sales (Ongena and Jacques 2007; Sanchis and Bourguet 2008). As for the rest, *Bacillus*-based products, such as *Bacillus subtilis*, *Bacillus licheniformis*, and *Bacillus pumilus,* represent about half of the commercially available bacterial biocontrol agents (Ongena and Jacques 2007).

The *Bacillus* genus produces a wide range of biologically active molecules that are potentially inhibitory for phytopathogen growth. Among these antimicrobial compounds, cyclic lipopeptides (LPs) of the surfactin, iturin, and fengycin (or plipastatin) families have well-recognized potential uses in biotechnology and biopharmaceutical applications because of their surfactant properties (Ongena and Jacques 2007). Recent investigations indicate that these lipopeptides can also influence the ecological fitness of the producing strain in terms of root colonization (and thereby persistence in the rhizosphere) and that they have a key role in the beneficial interaction of *Bacillus* species with plants by stimulating host defence mechanisms (Ongena and Jacques 2007). The production of LPs has been demonstrated in *Bacillus* populations growing on roots, leaves, and fruits (Touré et al. 2004; Romero et al. 2007). In the rhizosphere, LPs are difficult to be estimated because of the small amounts excreted as compared to the other organic compounds present in the environment, their difficult extraction from the complex soil matrix, and the possibility that the low quantities produced are restricted from diffusing freely and can be rapidly embedded in the membrane structure of the target organism (Ongena and Jacques 2007).

Pseudomonas fluorescens strains, have been reported to control diseases caused by soil-borne pathogens and are known to survive in both rhizosphere and phyllosphere (Weller, 1988; Wilson et al. 1991). Several studies have indicated that foliar diseases could be controlled by the application of *P. fluorescens* as seed, soil, or root treatments, and it is presumed that they may produce ISR and thus protect the leaves (Wei et al. 1991). The ability of rhizosphere-associated fluorescent pseudomonads to inhibit the growth of plant pathogenic fungi has generated increased interest in their use as crop protectants (Schippers et al. 1987; Lam and Gaffney 1993;

Weller 1988; Fravel 2005). A formulation of *P. fluorescens* strains, for instance, has been reported to control the foliar pathogen *Pyricularia oryzae* that causes blast disease in rice in field trials (Vidhyasekaran et al. 1997).

4 Potential Uses of PGPR and Rhizobia

Although there are several works on the role of specific strains of PGPR and rhizobia in plant-growth promotion, N_2 fixation, biofertilizer activities, and biological control, there is a need for more attention with regard to the negative effects of environmental stresses, diseases on rhizobacteria–plant interactions (Barea et al. 1998; Kloepper et al. 1999; Jetiyanon et al. 2003; Vessey 2003; Bashan et al. 2004; Morrissey et al. 2004). For example, rhizobia are sensitive to drought stress, resulting in a significant decrease of N_2 fixation when faced with low soil-water content. In a study under drought stress, coinoculation of bean (*Phaseolus vulgaris* L.) with *Rhizobium tropici* and two strains of *Paenibacillus polymyxa* resulted in increased plant height, shoot dry weight, and nodule number (Figueiredo et al. 2008). Interestingly, the effect on IST and the increased nodule number was greater when the two strains of *P. polymyxa* were applied than individual strain, suggesting some synergistic effects from themixed strains. Recently, Dardanelli et al. (2009) have shown how biotic and abiotic stresses can alter the pattern of flavonoids exuded by Osumi soybean roots. In that work, in the presence of *Chryseobacterium balustinum* Aur9, soybean roots did not exude quercetin and naringenin, and under salt stress (50 mM NaCl), flavonoids daidzein and naringenin could not be detected. Soybean root exudates obtained under saline conditions showed a diminished capacity to induce the expression of the *nodA* gene in comparison to the exudates obtained in the absence of salt. In addition, lipochitooligosaccharides (LCOs) were either not detected or weakly detected when *Sinorhizobium fredii* SMH12 was grown in the exudates obtained under salt stress conditions or under salt stress in the presence of *C. balustinum* Aur9, respectively.

Another abiotic stress that plants face is the obtaining of adequate soil nutrients. Although soil fertilization is typically required for agricultural production, it can cause nitrate and phosphate accumulation that eventually contaminates surface and ground waters. The use of fertilizers, including chemical fertilizers and manures, to enhance soil fertility and crop productivity has often negatively affected the complex system of the biogeochemical cycles (Perrott et al. 1992; Steinshamn et al. 2004). The use of fertilizers has caused leaching and run-off of nutrients, especially N and phosphorus (P), leading to environmental degradation (Tilman 1998; Gyaneshwar et al. 2002). Important reasons for these problems are the low use efficiency of fertilizers and the continuous long-term use. Despite the negative environmental effects, the total amount of fertilizers used worldwide is projected to increase with the growing world population because of the need to produce more food through intensive agriculture (Vitousek et al. 1997; Frink et al. 1999).

The challenge, therefore, is to continue agricultural productivity in a way that minimizes harmful environmental effects of fertilizers.

Research activities aimed at achieving a better use efficiency of fertilizers, including the use of PGPR and/or arbuscular mycorrhizal fungi as supplements to fertilizers, have steadily increased in the last two decades. However, it is important to emphasize those agro-environmental problems which are not limited to the use of chemical fertilizers but also occur with manures and compost (Mitchell and Tu 2006). Both animal waste and chemical fertilizers have the potential of environmental pollution (McLaughlin and Mineau 1995; Jarecki et al. 2008). Release of greenhouse gases (Flessa et al. 2002; Jarecki et al. 2008), ozone layer depletion (Ma et al. 2007), global warming, and acid rain are reported as negative impacts of fertilizers (Vitousek et al. 1997; Frink et al. 1999). Microbial inoculants, such as PGPR, are promising components for integrated solutions to agro-environmental problems because inoculants possess the capacity to promote plant growth, enhance nutrient availability and uptake, and support the health of plants (Barea et al. 1998; Dobbelaere et al. 2001; Hodge et al. 2001; Bonfante 2003; Vessey 2003; Kloepper et al. 2004; Han and Lee 2005; Weller 2007; Adesemoye et al. 2008).

On the basis of the beneficial effects of PGPR and rhizobia, studies using inoculant mixtures are very promising (Berg 2009). Benefits to plants from plant–PGPR interactions have been shown to include increase in seed germination, root growth, yield, leaf area, chlorophyll content, nutrient uptake, protein content, hydraulic activity, tolerance to abiotic stress, shoot and root weights, biocontrol, and delayed senescence (Mahaffee and Kloepper 1994; Raaijmakers et al. 1997; Bashan et al. 2004; Mantelin and Touraine 2004; Bakker et al. 2007; Yang et al. 2009).

Amir et al. (2005) reported enhanced uptake of N and P in oil palm seedlings, following PGPR inoculation in the field nursery. Aseri et al. (2008), on the other hand, conducted field experiments in India and assessed the effectiveness of PGPR (*Azotobacter chroococcum* and *A. brasilense*) and arbuscular mycorrhizal fungi (*Glomus mosseae* and *Glomus fasciculatum*) on the growth, nutrient uptake, and biomass production of pomegranate (*Punica granatum* L.). Strains were applied individually or in combinations and the results showed that dual inoculation of PGPR and arbuscular mycorrhizal fungi led to higher biomass production and increase in the uptake of N, as well as of P, K, Ca, and Mg, in pomegranate seedling. The increase in N and P uptake was suggested to result from improved symbiotic N_2 fixation and improved phosphatase activity.

The study by Adesemoye et al. (2008) confirmed that inoculation with mixed strains was more efficient than single-strain inoculations. A proposal made by Adesemoye et al. (2009) toward solving the agro-environmental problems mentioned is integrated nutrient management (INM), which does not aim to remove fertilizer totally in the short run but to reduce the negative impacts of the overuse of fertilizers containing N, P, and other elements. The INM system promotes low chemical input but improved nutrient-use efficiency by combining natural and man-made sources of plant nutrients in an efficient and environmentally prudent manner. This will not sacrifice high crop productivity in the short term nor endanger sustainability in the long term (Gruhn et al. 2000; Adesemoye et al. 2008).

Recently, it has been demonstrated that PGPR-elicited plant-growth promotion results in enhanced N uptake by plant roots (Adesemoye et al. 2008).

Owing to the broad-range metabolic activities found in many PGPR, another interesting topic is the potential use of PGPR in rhizoremediation (microbial degradation of hazardous compounds in the rhizosphere) in contaminated zones in order to obtain a dual effect: first, the remediation of the soil and then the promoting of plant growth for agriculture purposes. It is well known that bacteria of the *Burkholderia cepacia* complex (Bcc), which include nine species or genomovars (Mahenthiralingam et al. 2005), may be found in soils (including polluted soils), rhizospheres of crop plants, water, various animal species, humans, and hospital environments (Coenye and Vandamme 2003). More recently, Caballero-Mellado et al. (2007) reported the occurrence of nitrogen-fixing *Burkholderia* species associated with tomato (*Lycopersicon esculentum*) cultivated in different locations in Mexico. These authors found that the rhizosphere of tomato is a reservoir of different known and unknown diazotrophic *Burkholderia* species that, in vitro, are able to exhibit some activities involved in bioremediation, plant-growth promotion, and biological control. Similarly, Perin et al. (2006) reported that the isolation of *Burkholderia unamae* from field-grown sugarcane in Brazil and Mexico, as well as the isolation of *Burkholderia tropica* from maize cultivated in Mexico, find probably novel diazotrophic species (the *Burkholderia* NAR group) in rhizospheric and endophytic association with both maize and sugarcane in Brazil. Manipulating biotic interactions to provide desired services and thus reduce or eliminate the need for external inputs is fundamental to the practice of ecologically sound agriculture. The challenge is how to encourage positive interactions and reduce negative ones. Shennan (2008) indicates that the potential for a greater use of ecological management approaches is high; however, owing to the nature of ecosystems as medium number systems, there is some inherent unpredictability about the responses to different management interventions, which needs to be accommodated in the development of recommendations for farm management. This requires an increased emphasis on the effective synthesis of complex and often apparently contradictory information and a greater emphasis on field-based adaptive research that includes monitoring performance as adaptations are made, along with social learning mediated by farmer/researcher collaborations.

5 PGPR Studies in Argentina

PGPR have been studied by Argentinean scientists from universities and other research laboratories for the last three decades, with the aim to assess the beneficial effects on plant growth and yield of many crops of agronomic importance. In this section, we present some of the most relevant data on this issue in Argentina.

Bacteria of the genus *Azospirillum* are free-living nitrogen-fixing rhizobacteria that are found in close association with plant roots of a large number of plants, including forage and cereal crops (Okon 1994). Early studies with these bacteria in Argentina aimed to isolate local strains. By using an improved culture medium with Congo red, the colonies were typical red scarlett and easy to isolate (Rodríguez Cáceres 1982). Several local strains were selected in greenhouse conditions by their capacity of inducing changes in root systems and were preserved at the collection of Agricultural Microbiology and Zoology Institute of INTA, Castelar, Buenos Aires. To investigate the practical use of *Azospirillum* as a plant biofertilizer, wheat field trials were carried out at different locations of country. The local isolate *A. brasilense* strain Az39, which was obtained from wheat roots in the province of Córdoba, showed a consistent positive effect on the yield of different cultivars, from 13.4 to 33% increase over the control in three growing cycles tested (Rodríguez Cáceres et al. 1996). In addition, it is known that increases in crop yield derived from *Azospirillum* inoculation are consistently obtained when water is deficient (Fig. 1) and soil nutrients are limiting (low organic matter) (Fig. 2) (Rodríguez Cáceres et al. 2008a).

Inoculation trials of corn and wheat carried out with a liquid formulation of *A. brasilense* Az39 in 2002–2003 and 2006–2007 at 110 different sites showed an average yield increase of 6%. In most of the sites, the inoculation with this liquid formulation increased root and shoot early growth and grain number of the crops (Diaz-Zorita et al. 2004). Similar increases of forage yield were obtained in foxtail millet (*Setaria italica*) inoculated with these rhizospheric bacteria (Di Ciocco and Rodríguez Cáceres 1994). All these positive results motivated the agro-industry to produce new inoculants based on *Azospirillum* and other PGPR. Coinoculation studies with PGPR and *Rhizobium* spp. have been shown to increase root and

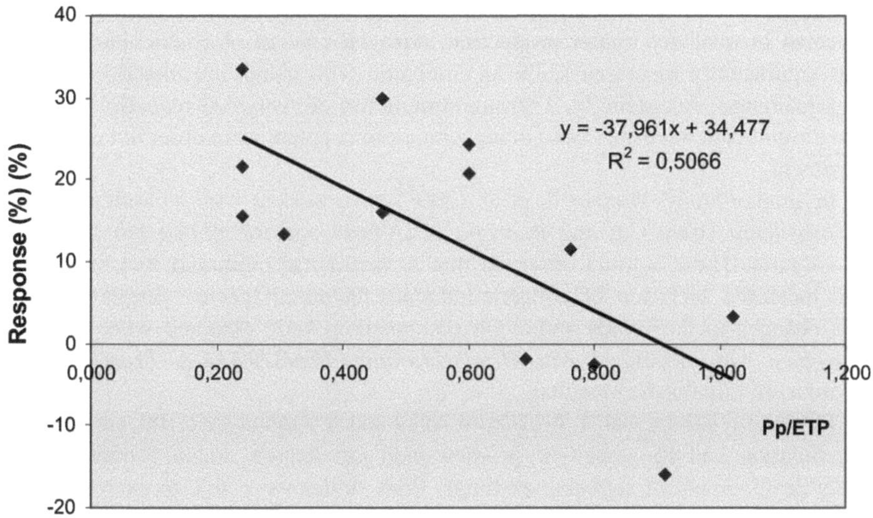

Fig. 1 Yield responses of wheat to inoculation with *Azospirillum brasilense* Az 39 conditioned to hidric conditions

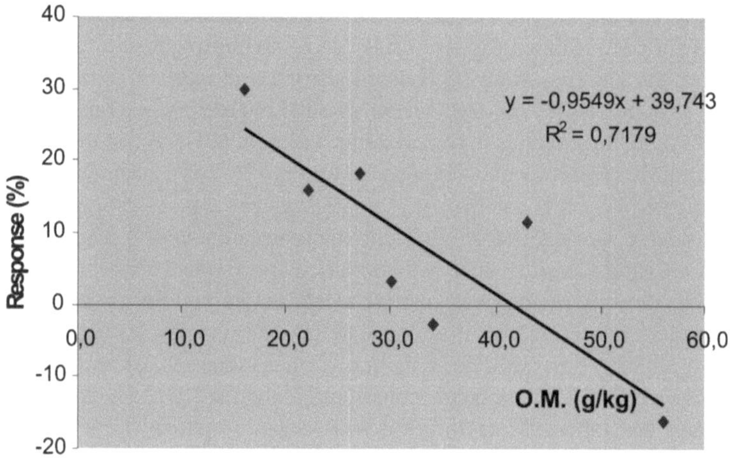

Fig. 2 Yield responses of wheat to inoculation with *Azospirillum brasilense* Az 39 conditioned to soil organic matter

shoot dry weight, plant vigour, nodulation, and nitrogen fixation in various legumes (Saxena et al. 2006). It is still a matter of controversy whether the stimulation of plant growth and N fixation by coinoculation is due either to an increase in root surface by hormonal effects or to nodulation and nutrient uptake. In relation with physiological and biochemical changes, Groppa et al. (1998) found that coinoculation with *B. japonicum* and *A. brasilense* on soybean plants showed a significantly higher proportion of nodules attached to the main root and located in the upper 3 cm of the root system. Although no significant differences were detected in total dry matter production, nitrogen content of coinoculated plants was significantly increased (23% as compared with plants inoculated only with *B. japonicum*). Accordingly, a strong stimulation of acetylene reduction activity and a significant increase (39%) in leghemoglobin content were observed using this treatment.

In another study, Dardanelli et al. (2008) have worked with a combination of *A. brasilense* strain Cd, and *R. tropici* CIAT899 and *Rhizobium etli* ISP42 in *P. vulgaris*. These authors observed that *A. brasilense* promoted root branching and increased secretion of nodgene-inducing flavonoid species. Similar results, and changes in flavonoids and sugar composition, were obtained when *Arachis hypogaea* was inoculated with *Bradyrhizobium* SEMIA6144-*A. brasilense* Cd (Dardanelli unpublished results).

It was also shown that *A. brasilense* Az39 and *B. japonicum* E109, singly or in combination, had the capacity to promote seed germination, nodule formation, and early development of soybean seedlings. Both strains were able to excrete plant-regulating substances into the culture medium, at a concentration sufficient to produce morphological and physiological changes in young seed tissues (Cassán et al. 2009). In order to obtain a positive effect on growth and nodulation of

Table 1 Effect of *Azospirillum brasilense* Az39 inoculum concentration on nodulation and acetylene reduction activity (ARA) of pouch-grown plants of soybean

Treatment	CFU ml^{-1}	Nodule number (per plant)	Nodule dry weight (mg)	ARA nmol C_2H_4 h^{-1} $plant^{-1}$
Control	10^9	144 a	0.14 a	20.75
Azospirillum	10^3	148 a	0.15 a	20.11
	10^5	226 b	0.21 b	28.00
	10^8	139 a	0.15 a	24.10

All treatments were inoculated with 1 ml (1×10^9 CFU ml^{-1}) of *B. japonicum*. Values followed by the same letter were not significantly different ($P = 0.05$). *CFU* Colony-forming unit, *ARA* acetylene reduction activity

coinoculated leguminous plants, it is necessary to establish optimal cell concentration of each biological component (Rodríguez Cáceres, unpublished data) (Table 1).

Barassi et al. (2006) also demonstrated growth-promoting effects of *A. brasilense* strain 245 on lettuce in saline conditions. These authors observed that *Azospirillum*-inoculated lettuce seeds had a better germination and vegetative growth than non-inoculated controls after exposure to NaCl. Plants grown from inoculated seeds and irrigated with saline medium displayed higher total fresh and dry weights and biomass partition to the aerial portion than non-inoculated controls.

Since in the semiarid pampas of Argentina the phosphorus distribution available is not uniform, Rosas et al. (2006) studied the possible action of phosphate-solubilizing bacteria on the leguminous–rhizobia symbiosis. The strains used were *Sinorhizobium meliloti* 3DOh13, a good solubilizer of iron and phosphorus for alfalfa, *B. japonicum* TIIIB for soybean, and two phosphorus-solubilizing strains of *P. putida* for growth-promotion treatments. Modification in the dry weights of the shoot and root systems were observed in soybean, but not in alfalfa, in the presence of the *Pseudomonas* strains (Rosas et al. 2006).

6 New Studies and Applications of PGPR

Rhizospheric bacteria such as *Azospirillum and Azotobacter* are also being applied for the induction of rooting in micropropagated plants (jojoba, photinia, ornamental grasses) (Carletti et al. 1998). It seems that PGPR can replace all or at least some synthetic plant hormones commonly used in in vitro cultures of plants (Carletti et al. 2006).

In a study by Larraburu et al. (2007), bacterial inoculation was able to induce earlier rooting of photinia (*Photinia × fraseri* Dress) shoots. *A. brasilense* Cd with an indole-3-butyric acid pulse showed a significant increase in root fresh and dry weight, root surface area, and shoot fresh and dry weight, *A. brasilense* Sp7 enhanced root fresh weight and root surface area, but no significant differences were detected with *A. chroococcum* inoculation.

On the other hand, *Bacillus* spp. and *P. polymyxa* have attracted considerable interest because of their great biotechnological potential in different industrial processes and sustainable agriculture (Lal and Tabacchioni 2009). In Argentina, for example, two sporulating bacterial strains were isolated from the rhizosphere of the legume *Cicer arietinum*. The results of DNA–DNA hybridization showed that these strains constitute a novel species of the genus *Paenibacillus*, for which the name *Paenibacillus rhizosphaerae* sp. nov. was proposed (Rivas et al. 2005).

Correa et al. (2009) showed the ability of *Bacillus amyloliquefaciens* BNM122 strain to colonize seeds and roots when applied as a coating on soybean seeds. This bacterium is a potential microbial biocontrol agent able to control the damping-off caused by *Rhizoctonia solani*, both in a plant-growth chamber and in a greenhouse. Correa et al. (2009) also observed that it had a lesser effect on soil microbial community than fungicides, because of the less environmental persistence and toxic effects of the strain.

7 Perspectives and Conclusion

The rhizosphere represents one of the most complex ecosystems on Earth with almost every root on the planet expected to have a chemically, physically, and biologically unique rhizosphere. Despite its intrinsic complexity, understanding the rhizosphere is vital if we are to solve some of the world's most impending environmental crises, such as sustainable food, fibre and energy production, preservation of water resources and biodiversity, and mitigation against climate change (Jones and Hinsinger 2008). The secretion of rhizodeposition is an important way for plants to respond to and alter their environment. Over the last several years, research and technical advances have provided a better understanding of how root exudates mediate communication between plants and other organisms. These advances could be applied to agricultural systems to enhance production by increasing defence responses against soil-borne pathogens and/or favoring the association with beneficial soil microbes. An improvement in plant–microbe symbioses should involve the reorganizations of the integrated genetic systems due to coordinated modifications in the plant and microbial genotypes. Sustainable agriculture should switch from growing plants to the cultivation of plant–microbial communities, which can reach a high productivity under minimal energy and chemical investments and with minimal pressures on the environment. However, we are only at the beginning of this process and much more efforts and cooperation between experts on plant and microbial genetics, molecular biology and ecology are required to be successful in attaining sustainable microbial-based agrotechnologies.

Acknowledgments This research was partially supported by the Secretaría de Ciencia y Técnica de la Universidad Nacional de Río Cuarto (SECyT-UNRC) and CONICET PIP 112-200801-00537. NP is fellow from CONICET. MSD is a member of the research career of CONICET, Argentina. The authors thank Dr. Yaacov Okon for suggestions on the manuscript.

References

Adesemoye AO, Torbert HA, Kloepper JW (2008) Enhanced plant nutrient use efficiency with PGPR and AMF in an integrated nutrient management system. Can J Microbiol 54:876–886

Adesemoye AO, Torbert HA, Kloepper JW (2009) Plant growth promoting rhizobacteria allow reduced application rates of chemical fertilizers. Microb Ecol 58(4):921–929

Amir HG, Shamsuddin ZH, Halimi MS, Marziah M, Ramlan MF (2005) Enhancement in nutrient accumulation and growth of oil palm seedlings caused by PGPR under field nursery conditions. Commun Soil Sci Plant Anal 36:2059–2066

Aseri GK, Jain N, Panwar J, Rao AV, Meghwal PR (2008) Biofertilizers improve plant growth, fruit yield, nutrition, metabolism and rhizosphere enzyme activities of pomegranate (*Punica granatum* L.) in Indian Thar Desert. Sci Hortic 117:130–135

Atkinson D, Watson CA (2000) The beneficial rhizosphere: a dynamic entity. Appl Soil Ecol 15:99–104

Bakker PAHM, Pieterse CMJ, van Loon LC (2007) Induced systemic resistance by fluorescent *Pseudomonas* spp. Phytopathology 97:239–243

Barassi CA, Ayrault G, Creus CM, Sueldo RJ, Sobrero MT (2006) Seed inoculation with *Azospirillum* mitigates NaCl effects on lettuce. Sci Hortic 109:8–14

Barea JM, Andrade G, Bianciotto V, Dowling D, Lohrke S, Bonfante P, O'Gara F, Azcon-Anguilar C (1998) Impact on arbuscular mycorrhiza formation of *Pseudomonas* strains used as inoculants for biocontrol of soil-borne fungal plant pathogens. Appl Environ Microbiol 64:2304–2307

Bashan Y, Holguin G, de Bashan LE (2004) *Azospirillum*–plant relationships: physiological, molecular, agricultural, and environmental advances (1997–2003). Can J Microbiol 50:521–577

Berg G (2009) Plant–microbe interactions promoting plant growth and health: perspectives for controlled use of microorganisms in agriculture. Appl Microbiol Biotechnol 84:11–18

Beringer JE (1986) Plant–microbe interactions. In: Silver S (ed) Biotechnology: potentials and limitations. Springer, Berlin, pp 259–273

Bodelier PLE, Wijlhuizen AG, Blom CWPM, Laanbroek HJ (1997) Effects of photoperiod on growth of and denitrification by *Pseudomonas chlororaphis* in the root zone of *Glyceria maxima*, studied in a gnotobiotic microcosm. Plant Soil 190:91–103

Bolton GW, Nester EW, Gordon MP (1986) Plant phenolic compounds induce expression of the *Agrobacterium tumefaciens* loci needed for virulence. Science 232:983–985

Bonfante P (2003) Plants, mycorrhizal fungi, and endobacteria: a dialog among cells and genomes. Biol Bull 204:215–220

Bruehl GW (1987) Soil-borne plant pathogens. Macmillan, New York

Caballero-Mellado J, Onofre-Lemus J, Estrada-de los Santos P, Martinez-Aguilar L (2007) The tomato rhizosphere, an environment rich in nitrogen-fixing *Burkholderia* species with capabilities of interest for agriculture and bioremediation. Appl Environ Microbiol 73:5308–5319

Carletti SM, Llorente BE, Rodríguez Cáceres EA, Tandecarz JS (1998) Jojoba inoculation with *Azospirillum brasilense* stimulates *in vitro* root formation. Plant Tiss Cult Biotech 4:165–174

Carletti SM, Murray A, Llorente BE, Rodríguez Cáceres EA, Puglia ML (2006) Propagación de dos gramíneas ornamentales: *Muhlenbergia dumosa* y *Eustachys distychophylla* inoculadas con *Azospirillum brasilense*. In: UN La Plata, MAA, INTA San Pedro (eds) 3 Congress Argentine of floriculture and 8 National J of flowers. Buenos Aires, pp 380–383

Cassán F, Perrig D, Sgroy V, Masciarelli O, Penna C, Luna V (2009) *Azospirillum brasilense* Az39 and *Bradyrhizobium japonicum* E109, inoculated singly or in combination, promote seed germination and early seedling growth in corn (*Zea mays* L.) and soybean (*Glycine max* L.). Eur J Soil Biol 45:28–35

Coenye T, Vandamme P (2003) Diversity and significance of *Burkholderia* species occupying diverse ecological niches. Environ Microbiol 5:719–729

Compant S, Duffy B, Nowak J, Clément C, Barka EA (2005) Use of plant growth-promoting bacteria for biocontrol of plant diseases: principles, mechanisms of action, and future prospects. Appl Environ Microbiol 71:4951–4959
Correa OS, Montecchia MS, Berti MF, Fernández Ferrari MC, Pucheu NL, Kerber NL, García AF (2009) *Bacillus amyloliquefaciens* BNM122, a potential microbial biocontrol agent applied on soybean seeds, causes a minor impact on rhizosphere and soil microbial communities. Appl Soil Ecol 41:185–194
Dardanelli MS, Angelini J, Fabra A (2003) A calcium dependent rhizobia surface protein is involved in the peanut crack entry infection process. Can J Microbiol 49:399–405
Dardanelli MS, Fernández FJ, Espuny MR, Rodríguez MA, Soria ME, Gil Serrano AM, Okon Y, Megías M (2008a) Effect of *Azospirillum brasilense* coinoculated with *Rhizobium* on *Phaseolus vulgaris* flavonoids and Nod factor production under salt stress. Soil Biol Biochem 40:2713–2721
Dardanelli MS, Rodríguez Navarro DN, Megías M, Okon Y (2008b) Influence of co-inoculation *Azospirillum*-rhizobia to growth and nitrogen fixation of agronomic legume. In: Cassán FD, García de Salamone I (eds) *Azospirillum* sp.: cell physiology, plant interactions and agronomic research in Argentine. Argentine Microbiology Society, Buenos Aires, pp 141–151
Dardanelli MS, Manyani H, González-Barroso S, Rodríguez-Carvajal MA, Gil-Serrano AM, Espuny MR, López-Baena FJ, Bellogín RA, Megías M, Ollero FJ (2009) Effect of the presence of the plant growth promoting rhizobacterium (PGPR) *Chryseobacterium balustinum* Aur9 and salt stress in the pattern of flavonoids exuded by soybean roots. Plant Soil 328:483–493
Di Ciocco C, Rodríguez Cáceres E (1994) Field inoculation of *Setaria italica* with *Azospirillum* spp in Argentine Humid Pampas. Field Crops Res 37:253–257
Diaz-Zorita M, Baliña RM, Fernández-Canigia M, Perticari A (2004) Field inoculation of wheat (*Triticum aestivum* L.) and corn (*Zea mays* L.) with *Azospirillum brasilense* in the Pampas region, Argentina. In: 22nd Latin-American conference on Rhizobiology and 1st Brazilian conference on biological nitrogen fixation (RELAR). Rio de Janeiro, Brazil, p 125
Dobbelaere S, Croonenborghs A, Thys A, Ptacek D, Vanderleyden J, Dutto P, Labandera-Gonzalez C, Caballero-Mellado J, Aguirre JF, Kapulnik Y, Brener S, Burdman S, Kadouri D, Sarig S, Okon Y (2001) Response of agronomically important crops to inoculation with *Azospirillum*. Aust J Plant Physiol 28:871–879
Dommergues YR, Subba Rao NS (2000) Introduction of N_2-fixing trees in non-N_2-fixing tropical plantations. In: Subba Rao NS, Dommergues YR (eds) Microbial interactions in agriculture and forestry. Science Publishers, Enfield, pp 131–154
Doran JW, Sarrantonio M, Liebig MA (1996) Soil health and sustainability. Adv Agron 56:2–54
Figueiredo VB, Buritya HA, Martínez CR, Chanway CP (2008) Alleviation of drought stress in the common bean (*Phaseolus vulgaris* L.) by co-inoculation with *Paenibacillus polymyxa* and *Rhizobium tropici*. Appl Soil Ecol 40:182–188
Fisher RF, Long SR (1992) *Rhizobium*-plant signal exchange. Nature 357:655–660
Flessa H, Ruser R, Dörsch P, Kamp T, Jimenez MA, Munch JC, Beese F (2002) Integrated evaluation of greenhouse gas emissions (CO_2, CH_4, N_2O) from two farming systems in southern Germany. Agric Ecosyst Environ 91:175–189
Fravel DR (2005) Commercialization and implementation of biocontrol. Annu Rev Phytopathol 43:337–359
Frink CR, Waggoner PE, Ausubel JH (1999) Nitrogen fertilizer: retrospect and prospect. Proc Natl Acad Sci USA 96:1175–1180
Gerhardson B (2002) Biological substitutes for pesticides. Trends Biotechnol 20:338–343
Glick BR (1995) The enhancement of plant growth by free-living bacteria. Can J Microbiol 41:109–117
Graham PH, Vance CP (2000) Nitrogen fixation in perspective: on overview of research and extension needs. Field Crops Res 65:93–106
Groppa MD, Zawoznik MS, Tomaro ML (1998) Effect of co-inoculation with *Bradyrhizobium japonicum* and *Azospirillum brasilense* on soybean plants. Eur J Soil Biol 34:75–80

Gruhn P, Goletti F, Yudelman M (2000) Integrated nutrient management, soil fertility, and sustainable agriculture: current issues and future challenges. Food, agriculture, and the environment. Discussion paper 32. International Food Policy Research Institute, Washington, DC, pp 15–16

Gyaneshwar P, Kumar GN, Parekh LJ, Poole PS (2002) Role of soil microorganisms in improving P nutrition of plants. Plant Soil 245:83–93

Han HS, Lee KD (2005) Phosphate and potassium solubilizing bacteria effect on mineral uptake, soil availability, and growth of egg plant. Res J Agric Biol Sci 1:176–180

Hiltner L (1904) Über neuere Erfahrungen und Probleme auf dem Gebiete der Bodenbakteriologie unter besonderer Berücksichtigung der Gründüngung und Brache. Arbeiten der Deutschen Landwirtschaftlichen Gesellschaft 98:59–78

Hodge A, Campbell CD, Fitter AH (2001) An arbuscular mycorrhizal fungus accelerates decomposition and acquires nitrogen directly from organic material. Nature 413:297–299

Hungria M, Vargas MAT (2000) Environmental factors affecting N_2 fixation in grain legumes in the tropics, with an emphasis on Brazil. Field Crops Res 65:151–164

Jarecki MK, Parkin TB, Chan ASK, Hatfield JL, Jones R (2008) Greenhouse gas emissions from two soils receiving nitrogen fertilizer and swine manure slurry. J Environ Qual 37:1432–1438

Jetiyanon K, Fowler WD, Kloepper JW (2003) Broad-spectrum protection against several pathogens by PGPR mixtures under field conditions. Plant Dis 87:1390–1394

Jones DL, Hinsinger P (2008) The rhizosphere: complex by design. Plant Soil 312:1–6

Kennedy AC (1998) The rhizosphere and spermosphere. In: Sylvia DM, Fuhrmann JJ, Hartel PG, Zuberer DA (eds) Principles and applications of soil microbiology. Prentice Hall, Inc., New Jersey, pp 389–407

Kloepper JW, Schroth MN (1978) Plant growth-promoting rhizobacteria on radishes. In: Station de pathologie vegetale et phyto-bacteriologie (ed) Proceedings of the 4th International Conference on Plant Pathogenic Bacteria, vol II. Gilbert-Clarey, Tours, pp 879–882

Kloepper JW, Rodriguez-Kábana R, Zehnder GW, Murphy JF, Sikora E, Fernández C (1999) Plant root–bacterial interactions in biological control of soilborne diseases and potential extension to systemic and foliar diseases. Australas Plant Pathol 28:21–26

Kloepper JW, Ryu CM, Zhang S (2004) Induced systemic resistance and promotion of plant growth by *Bacillus* spp. Phytopathology 94:1259–1266

Lal S, Tabacchioni S (2009) Ecology and biotechnological potential of *Paenibacillus polymyxa*: a minireview. Indian J Microbiol 49:2–10

Lam ST, Gaffney TD (1993) Biological activities of bacteria in plant pathogen control. In: Chet I (ed) Biotechnology in plant disease control. New York, Wiley-Liss Press, pp 291–320

Larraburu EE, Carletti SM, Rodríguez Cáceres EA, Llorente BE (2007) Micropropagation of *Photinia* employing rhizobacteria to promote root development. Plant Cell Rep 26:711–717

Lewis JA, Papavizas GC (1991) Biocontrol of plant diseases: the approach for tomorrow. Crop Prot 10:95–105

Lugtenberg BJJ, Dekkers LC (1999) What makes *Pseudomonas* bacteria rhizosphere competent? Environ Microbiol 1:9–13

Lynch JM, Whipps JM (1990) Substrate flow in the rhizosphere. Plant Soil 129:1–10

Ma J, Li XL, Xu H, Han Y, Cai ZC, Yagi K (2007) Effects of nitrogen fertilizer and wheat straw application on CH_4 and N_2O emissions from a paddy rice field. Australas J Soil Res 45:359–367

Mahaffee WF, Kloepper JW (1994) Applications of plant growth promoting rhizobacteria in sustainable agriculture. In: Pankhurst CE, Doube BM, Gupta VVSR, Grace PR (eds) Soil biota: management in sustainable farming systems. CSIRO, Melbourne, pp 23–31

Mahenthiralingam E, Urban TA, Goldberg JB (2005) The multifarious multireplicon *Burkholderia cepacia* complex. Nat Rev Microbiol 3:144–156

Mannion AM (1998) Future trends in agriculture: the role of agriculture. Outlook Agric 27:219–224

Mantelin S, Touraine B (2004) Plant growth-promoting bacteria and nitrate availability: impacts on root development and nitrate uptake. J Exp Bot 55:27–34
McLaughlin A, Mineau P (1995) The impact of agricultural practices on biodiversity. Agric Ecosyst Environ 55:201–212
Mitchell CC, Tu S (2006) Nutrient accumulation and movement from poultry litter. Soil Sci Soc Am J 70:2146–2153
Mittler R (2006) Abiotic stress, the field environment and stress combination. Trends Plant Sci 11:15–19
Morris PF, Bone E, Tyler BM (1998) Chemotropic and contact responses of *Phytophthora sojae* hyphae to soybean isoflavonoids and artificial substrates. Plant Physiol 117:1171–1178
Morrissey JP, Dow M, Mark GL, O'Gara F (2004) Are microbes at the root of a solution to world food production? Rational exploitation of interactions between microbes and plants can help to transform agriculture. EMBO Rep 5:922–926
Oerke EC (2005) Crop losses to pests. J Agric Sci 144:31–43
Oerke EC, Dehne HW (2004) Safeguarding production: losses in major crops and the role of crop protection. Crop Prot 23:275–285
Okon Y (1994) *Azospirillum*/plant associations. CRC Press, Boca Raton, Florida
Ongena M, Jacques P (2007) *Bacillus* lipopeptides: versatile weapons for plant disease biocontrol. Trends Microbiol 16:115–125
Perin L, Martinez-Aguilar L, Castro-Gonzalez R, Estrada-de los Santos P, Cabellos-Avelar T, Guedes HV, Reis VM, Caballero-Mellado J (2006) Diazotrophic *Burkholderia* species associated with field-grown maize and sugarcane. Appl Environ Microbiol 72:3103–3110
Perret X, Staehelin C, Broughton WJ (2000) Molecular basis of symbiotic promiscuity. Microbiol Mol Biol Rev 64:180–201
Perrott KW, Sarathchandra SU, Dow BW (1992) Seasonal and fertilizer effects on the organic cycle and microbial biomass in a hill country soil under pasture. Australas J Soil Res 30:383–394
Peters NK, Frost J, Long SR (1986) A plant flavone, luteolin, induces expression of *Rhizobium meliloti* nodulation genes. Science 233:977–980
Raaijmakers JM, Weller DM, Thomashow LS (1997) Frequency of antibiotic-producing *Pseudomonas* spp. in natural environments. Appl Environ Microbiol 63:881–887
Raaijmakers JM, Paulitz TC, Steinberg C, Alabouvette C, Moënne-Loccoz Y (2009) The rhizosphere: a playground and battle field for soilborne pathogens and beneficial microorganisms. Plant Soil 321:341–361
Rivas R, Gutiérrez C, Abril A, Mateos PF, Martínez-Molina E, Ventosa A, Velázquez E (2005) *Paenibacillus rhizosphaerae* sp. nov., isolated from the rhizosphere of *Cicer arietinum*. Int J Syst Evol Microbiol 55:1305–1309
Rodríguez Cáceres EA (1982) Improved medium for isolation of *Azospirillum* spp. Appl Environ Microbiol 44:990–991
Rodríguez Cáceres EA, González Anta G, López JR, Di Ciocco C, Pacheco Basurco J, Parada J (1996) Response of field-grown wheat to inoculation with *Azospirillum brasilense* and *Bacillus polymyxa* in semiarid region of Argentina. Arid Soil Res Rehab 10:13–20
Rodríguez Cáceres EA, Di Ciocco C, Carletti S (2008) 25 years of research of *Azospirillum brasilense* Az39 in Argentina. In: Cassán FD, García de Salamone I (eds) *Azospirillum* sp.: cell physiology, plant interactions and agronomic research in Argentine. Argentine Microbiology Society, Buenos Aires, pp 179–188
Rodríguez-Navarro DN, Dardanelli MS, Ruiz-Sainz JE (2007) Attachment of bacteria to the roots of higher plants. FEMS Microbiol Lett 272:127–136
Romero D, de Vicente A, Rakotoaly RV, Dufour SE, Veening JW, Arrebola E, Cazorla FM, Kuipers OP, Paquot M, Pérez-García A (2007) The iturin and fengycin families of lipopeptides are key factors in antagonism of *Bacillus subtilis* toward *Podosphaera fusca*. Mol Plant Microbe Interact 20:430–440

Rosas SB, Andrés JA, Rovera M, Correa NS (2006) Phosphate-solubilizing *Pseudomonas putida* can influence the rhizobia–legume symbiosis. Soil Biol Biochem 38:3502–3505

Ryan AD, Kinkel LL (1997) Inoculum density and population dynamic of suppressive and pathogenic *Streptomyces* strains and their relationships to biological control of potato scab. Biol Control 10:180–186

Sanchis V, Bourguet D (2008) *Bacillus thuringiensis*: Applications in agriculture and insect resistance management: A review. Agron Sustain Dev 28:11–20

Sauer K, Camper AK (2001) Characterization of phenotypic changes in *Pseudomonas putida* in response to surface associated growth. J Bacteriol 183:6579–6589

Saxena A, Shende R, Grover M (2006) Interactions among beneficial microorganisms. In: Mukerji KG, Manoharachary C, Singh J (eds) Microbial activity in the rhizosphere, vol 7, Soil Biology. Springer, Berlin, Heidelberg, pp 121–132

Schippers B, Bakker AW, Bakker PAHM (1987) Interactions of deleterious and beneficial microorganisms and the effect of cropping practices. Annu Rev Phytopathol 25:339–358

Shennan C (2008) Biotic interactions, ecological knowledge and agriculture. Philos Trans R Soc Lond B 363:717–739

Sommers E, Vanderleyden J, Srinivasan M (2004) Rhizosphere bacterial signalling: a love parade beneath our feet. Crit Rev Microbiol 30:205–240

Spaepen S, Vanderleyden J, Okon Y (2009) Plant growth-promoting actions of rhizobacteria. In: van Loon LC (ed) Advances in botanical research, vol 51. Academic, Burlington, pp 283–320

Steenhoudt O, Vanderleyden J (2000) *Azospirillum*, a free-living nitrogen fixing bacterium closely associated with grasses: genetic, biochemical and ecological aspects. FEMS Microbiol Rev 24:487–506

Steinshamn H, Thuen E, Bleken MA, Brenoe UT, Ekerholt G, Yri C (2004) Utilization of nitrogen (N) and phosphorus (P) in an organic dairy farming system in Norway. Agric Ecosyst Environ 104:509–522

Thakore Y (2006) The biopesticide market for global agricultural use. Ind Biotechnol 2:194–208

Tilman D (1998) The greening of the green revolution. Nature 396:211–212

Tisdale SL, Nelson WL (1975) Soil fertility and fertilizers, 3rd edn. Macmillan Publishing, New York

Touré Y, Ongena M, Jacques P, Guiro A, Thonart P (2004) Role of lipopeptides produced by *Bacillus subtilis* GA1 in the reduction of grey mould disease caused by *Botrytis cinerea* on apple. J Appl Microbiol 96:1151–1160

van Elsas JD, Trevors JY (1997) Modern soil microbiology. Marcel Dekker Inc., New York

van Loon LC, Bakker PA, Pieterse CM (1998) Systemic resistance induced by rhizosphere bacteria. Annu Rev Phytopathol 36:453–483

Vance CP (1998) Legume symbiotic nitrogen fixation: agronomic aspects. In: Spaink HP, Kondorosi A, Hooykaas PJJ (eds) The Rhizobiaceae. Kluwer Academic Publishers, Dordrecht, pp 509–530

Vessey JK (2003) Plant growth promoting rhizobacteria as biofertilizers. Plant Soil 255:571–586

Vidhyasekaran P, Rabindran R, Muthamilan M, Nayar K, Rajappan K, Subramanian N, Vasumathi K (1997) Development of a powder formulation of *Pseudomonas fluorescens* for control of rice blast. Plant Soil 46:291–297

Vitousek PM, Aber JD, Howarth RW, Likens GE, Matson PA, Schindler DW, Schlesinger WH, Tilman DG (1997) Technical report: human alteration of the global nitrogen cycle: sources and consequences. Ecol Appl 7:737–750

Wei G, Kloepper JW, Tuzun S (1991) Induction of systemic resistance of cucumber to *Colletotrichum orbiculare* by selected strains of plant growth-promotig rhizobacteria. Phytopathology 81:1508–1512

Weller D (1988) Biological control of soilborne plant pathogens in the rhizosphere with bacteria. Annu Rev Phytopathol 26:379–407

Weller DM (2007) *Pseudomonas* biocontrol agents of soilborne pathogens: looking back over 30 years. Phytopathology 97:250–256

Whipps JM (1990) Carbon economy. In: Lynch JM (ed) The rhizosphere. Wiley, Chichester, pp 59–97
Whipps JM (2001) Microbial interactions and biocontrol in the rhizosphere. J Exp Bot 52:487–511
Wilson M, Lindow SE, Hirano SS (1991) The proportion of different phyllosphere bacteria in sites on or within bean leaves protected from surface sterilization. Phytopathology 81:1222
Yang J, Kloepper JW, Ryu CM (2009) Rhizosphere bacteria help plants tolerate abiotic stress. Trends Plant Sci 14:1–4
Zhu GY, Dobbelaere S, Vanderleyden J (2002) Use of green fluorescent protein to visualized rice root colonization by *Azospirillum irakense* and *A. brasilense*. Funct Plant Biol 29:1279–1285

Plant Growth Promoting Rhizobacteria: Fundamentals and Applications

Márcia do Vale Barreto Figueiredo, Lucy Seldin, Fabio Fernando de Araujo, and Rosa de Lima Ramos Mariano

Contents

1 Introduction 22
2 Coinoculation of PGPR and Rhizobia: Improving Nodulation 23
3 Identification and Characterization of Beneficial Bacterial Strains for Agriculture 24
3.1 Taxonomy of PGPR 24
3.2 Phenotypic Features 25
3.3 Chemotaxonomic Characters 26
3.4 Genetic Approaches 27
4 Prospective Biocontrol Agents of Plant Diseases 28
5 Induced Systemic Resistance as a Mechanism of Disease Suppression by Rhizobacteria 31
6 Bacterial Biofertilizers 33
7 Concluding Remarks 36
References 36

Abstract Plant growth promoting rhizobacteria (PGPR) have gained worldwide importance and acceptance for agricultural benefits. This is due to the emerging demand for dependence diminishing of synthetic chemical products, to the growing

M. do Vale Barreto Figueiredo (✉)
National Research and Technological Development, Brazil Agronomical Institute of Pernambuco, IPA/CARHP, 1371, Gen. San Martin Avenue, Recife, PE 50761-000, Brazil
e-mail: mbarreto@elogica.com.br

L. Seldin
National Research and Technological Development, Brazil Federal University of Rio de Janeiro, Rio de Janeiro, RJ, Brazil

F.F. de Araujo
National Research and Technological Development, Brazil University of West Paulista, UNOESTE, São Paulo, SP, Brazil

R. de Lima Ramos Mariano
National Research and Technological Development, Brazil Federal Rural University of Pernambuco, Recife, PE, Brazil

D.K. Maheshwari (ed.), *Plant Growth and Health Promoting Bacteria*,
Microbiology Monographs 18, DOI 10.1007/978-3-642-13612-2_2,

necessity of sustainable agriculture within a holistic vision of development and to focalize environmental protection. Scientific researches involve multidisciplinary approaches to understand adaptation of PGPR, effects on plant physiology and growth, induced systemic resistance, biocontrol of plant pathogens, biofertilization, and potential green alternative for plant productivity, viability of coinoculating, plant microorganism interactions, and mechanisms of root colonization. By virtue of their rapid rhizosphere colonization and stimulation of plant growth, there is currently considerable interest in exploiting these rhizosphere bacteria to improve crop production. The main groups of PGPR can be found along with the phyla Cyanobacteria, Actinobacteria, Bacteroidetes, Firmicutes, and Proteobacteria. Therefore, the examples coming up next are related to these microorganisms. Although taxonomic affiliation of validated genera containing PGPR strains described in literature is vast, phenotypic and genotypic approaches are now available to characterize these different rhizobacteria. The progress to date in using PGPR in a variety of applications is summarized and discussed here.

1 Introduction

The use of microorganisms with the aim of improving nutrients availability for plants is an important practice and necessary for agriculture (Freitas et al. 2007). During the past couple of decades, the use of plant growth promoting rhizobacteria (PGPR) for sustainable agriculture has increased tremendously in various parts of the world. Significant increases in growth and yield of agronomically important crops in response to inoculation with PGPR have been repeatedly reported (Kloepper et al. 1980; Seldin et al. 1984; Chen et al. 1994; Zhang et al. 1996; Amara and Dahdoh 1997; Chanway 1998; Pan et al. 1999; Bin et al. 2000; Gupta et al. 2000; Biswas et al. 2000; Mariano and Kloepper 2000; Asghar et al. 2002; Vessey 2003; Gray and Smith 2005; Silva et al. 2006; Figueiredo et al. 2008; Araújo 2008). Studies have also shown that the growth-promoting ability of some bacteria may be highly specific to certain plant species, cultivar and genotype (Bashan 1998; Gupta et al. 2000; Lucy et al. 2004).

PGPR can affect plant growth by different direct and indirect mechanisms (Glick 1995; Gupta et al. 2000). Some examples of these mechanisms, which can probably be active simultaneously or sequentially at different stages of plant growth, are (1) increased mineral nutrient solubilization and nitrogen fixation, making nutrients available for the plant; (2) repression of soilborne pathogens (by the production of hydrogen cyanide, siderophores, antibiotics, and/or competition for nutrients); (3) improving plant stress tolerance to drought, salinity, and metal toxicity; and (4) production of phytohormones such as indole-3-acetic acid (IAA) (Gupta et al. 2000). Moreover, some PGPR have the enzyme 1-aminocyclopropane-1-carboxylate (ACC) deaminase, which hydrolyses ACC, the immediate precursor of ethylene in plants (Glick et al. 1995). By lowering ethylene concentration in seedlings and thus its inhibitory effect, these PGPR stimulate seedlings root length (Glick et al. 1999).

The bacteria presenting one or more of these characteristics are known as plant growth promoting rhizobacteria – PGPR (Kloepper and Schroth 1978).

Bashan and Holguin (1998) proposed the division of PGPR into two classes: biocontrol-PGPB (plant growth promoting bacteria) and PGPB. This classification may include beneficial bacteria that are not rhizosphere bacteria but it does not seem to have been widely accepted. According to Vessey (2003), numerous species of soil bacteria which flourish in the rhizosphere of plants, but which may grow in, on, or around plant tissues, and stimulate plant growth by a plethora of mechanisms are collectively known as PGPR. Gray and Smith (2005) have recently shown that the PGPR associations range in the degree of bacterial proximity to the root and intimacy of association. In general, these can be separated into extracellular (ePGPR), existing in the rhizosphere, on the rhizoplane, or in the spaces between cells of the root cortex, and intracellular (iPGPR), which exist inside root cells, generally in specialized nodular structures.

There are several PGPR inoculants currently commercialized that seem to promote growth through at least one mechanism: suppression of plant disease (bioprotectants), improved nutrients acquisition (biofertilizers), or phytohormone production (biostimulants). Bacteria in the genera *Bacillus*, *Streptomyces*, *Pseudomonas*, *Burkholderia*, and *Agrobacterium* are the biological control agents predominantly studied and increasingly marketed. They suppress plant disease through at least one mechanism, production of antibiotics or siderophores and induction of systemic resistance (Tenuta 2003).

Biofertlilizers are also available for increasing crop nutrient uptake of nitrogen from nitrogen-fixing bacteria associated with roots (Bashan and Holguin 1997), iron uptake from siderophore-producing bacteria (Scher and Baker 1982), sulfur uptake from sulfur-oxidizing bacteria (Stamford et al. 2008), and phosphorus uptake from phosphate-mineral solubilizing bacteria (Chabot et al. 1996). Biofertlilizers, that can cater different needs of growing plant, act as a consortium along with other microorganisms in the rhizosphere. Understanding the interaction between consortium of microbial inoculants and plant systems will pave way to harness more benefits from microbial inoculants for improving plant growth and yield (Raja et al. 2006).

2 Coinoculation of PGPR and Rhizobia: Improving Nodulation

Coinoculation studies with PGPR and Rhizobia have shown increased plant nodulation and N fixation (Li and Alexander 1988; Araújo and Hungria 1999; Vessey and Buss 2002; Silva et al. 2006; Figueiredo et al. 2007). Coinoculation of some *Bacillus* strains with effective *Bradyrhizobium* resulted in enhanced nodulation and plant growth of green gram (*Vigna radiata* L.) (Sindhu et al. 2002). A variety of rhizosphere microorganisms, including *Bacillus* and *Pseudomonas* species, are commonly found in the rhizosphere of leguminous and nonleguminous crops (Li and Alexander 1988). By virtue of their rapid colonization of the rhizosphere and stimulation of plant growth, there is currently considerable interest in exploiting these rhizosphere bacteria to improve crop production. Application of *Bacillus* and/or *Paenibacillus* species to seeds or roots has been shown to cause alteration in the composition of rhizosphere leading to increase in growth and yield of different crops (Li and Alexander 1988; Vessey and Buss 2002). Disease suppression of alfalfa by *B. cereus* enhanced nodulation and seedling emergence in common bean (Camacho et al. 2001; Figueiredo et al. 2007), soybean (Araújo and Hungria 1999; Vessey and Buss 2002), cowpea (Silva et al. 2006, 2007), and pea (Cooper and Long 1994) have been demonstrated as beneficial effects on plants. Bacilli are also very attractive as potential inoculants in agriculture, as they produce very hardy spores that can survive for prolonged periods in soil and in storage containers (Nelson 2004).

Araújo and Hungria (1999) demonstrated the viability of coinoculating soybean seeds with crude or formulated metabolites, or with cells of *Bacillus subtilis*, to increase the contribution of the biological nitrogen fixation process.

PGPR, in combination with efficient rhizobia, could improve the growth and nitrogen fixation by inducing the occupancy of introduced rhizobia in the nodules of the legume (Tilak et al. 2006). According to Saravana-Kumar and Samiyappan (2007), *Bradyrhizobium* promoted the nodulation and growth of legumes in combination with active ACC deaminase containing PGPR. It has also been established that certain rhizobacteria possess an enzyme ACC-deaminase that hydrolyses ACC into ammonia and α-ketobutyrate (Mayak et al. 1999). ACC-deaminase activity in PGPR plays an important role in the host nodulation response (Remans et al. 2007). PGPR containing ACC-deaminase could suppress accelerated endogenous ethylene synthesis and thus may facilitate root elongation a nodulation and improve growth and yield of plant (Zafar-ul-Hye 2008).

3 Identification and Characterization of Beneficial Bacterial Strains for Agriculture

Identification and characterization of beneficial bacteria involves morphological, physiological and molecular characteristics based on fatty acid analysis, mol (%), G + C contents, DNA–DNA hybridization, and 16S rRNA sequencing. These characteristics help in defining the taxonomy and nomenclature of PGPR.

3.1 Taxonomy of PGPR

Taxonomy is defined as the science dedicated to the study of relationships among organisms and has to do with their classification, nomenclature, and identification (Mayr and Ashlock 1991; Coenye et al. 2005). The accurate comparison of organisms depends on a reliable taxonomic system. Although many new characterization methods have been developed over the last 30 years, the principle of identification remains the same. Current schemes for identifying different bacterial strains may be roughly divided into four categories effectively based upon (1) traditional biochemical, morphological, and physiological characters, (2) miniaturized versions of traditional biochemical tests (e.g., API kits, VITEK cards, and Biolog plates), (3) chemotaxonomic characters (such as polyacrylamide gel electrophoresis [PAGE], and fatty acid methyl ester [FAME] profiles), and (4) genomic characters (16S rRNA gene sequencing, and DNA–DNA relatedness, and other techniques). Since the fifties, it was becoming clear that no one phenotypic technique would be suitable for identifying all bacterial species. Therefore, the potentials of chemotaxonomic analyses and studies of nucleic acids have been investigated. However, it is impossible to set up standardized conditions to accommodate the growth of all bacterial strains of all species for chemotaxonomic work, and a polyphasic approach is now imperative for a confident classification study. Polyphasic approach refers to the integration of genotypic, chemotypic, and phenotypic information of a microbe in order to perform reliable grouping of the organism (Colwell 1970). Some of the features used for polyphasic characterization of rhizobacteria are presented below. For overviews of modern taxonomy, recent papers can be referred, such as Vandamme et al. (1996), Prakash et al. (2007), Rodríguez-Díaz et al. (2008), and Logan et al. (2009).

3.2 *Phenotypic Features*

Phenotype includes morphological, physiological, and biochemical properties of the microorganism (de Vos et al. 2009). Traditional phenotypic tests used comprise colony morphology (color, dimensions, form) and microscopic appearance of the cells (shape, endospore, flagella, inclusion bodies), characteristics of the organism on different growth substrates, growth range of microorganisms on different conditions of salt, pH, and temperature, and susceptibility toward different kinds of antimicrobial agents, etc. Even if cell wall composition is analyzed, the Gram reaction is still a valuable diagnostic character. Biochemical tests in bacterial identification include the relationship with oxygen, fermentation reactions, and nitrogen metabolism. Other tests may be performed as appropriate, depending on the bacterial strains studied (Heritage et al. 1996; Rodríguez-Díaz et al. 2008). However, reproducibility of results from phenotypic tests between different laboratories is a great problem, and only standardized procedure should be used during execution of experiment. Other major disadvantage with phenotypic methods is the conditional nature of gene expression wherein the same organism might show different phenotypic characters in different environmental conditions. Therefore, phenotypic data must be compared with similar set of data from type strain of closely related organism(s).

Miniaturized versions of traditional biochemical tests are available for taxonomical studies and mostly contain a battery of dehydrated reagents. Addition of a standardized inoculum initiates the reaction (growth, production of enzymatic activity, etc.). The results are interpreted as recommended by the manufacturer and are readily accessible with a minimal input of time. The phenotypic fingerprinting systems API 50CH – composed of 49 different carbohydrates and one negative control – have been used to identify *Bacillus* (Logan and Berkeley 1984) and *Paenibacillus* strains (Seldin and Penido 1986), while the API 20NE system has yielded the highest rate of correct identification of *Pseudomonas* species (Barr et al. 1989). In the same way, Biolog assay is considered a much less laborious system for bacterial identification (Miller and Rhoden 1991). This technique is based on the differential utilization of 95 carbon sources and a redox dye, tetrazolium violet, permits colorimetric determination of the increased respiration that occurs when cells are oxidizing a carbon source. The Biolog system was very useful for the identification of PGPR strains belonging to the species *P. azotofixans* (Pires and Seldin 1997).

3.3 Chemotaxonomic Characters

Some chemotaxonomic fingerprinting techniques applied to PGPR identification include FAME profiling, PAGE analysis of whole-cell proteins, polar lipid analysis, quinone content, cell wall diamino acid content, pyrolysis mass spectrometry, Fourier transform infrared spectroscopy, Raman spectroscopy, and matrix-assisted laser desorption/ionization time-of-flight (MALDI-TOF) mass spectrometry.

Fatty acids are the major constituents of lipids and lipopolysaccharides and have been used extensively for taxonomic purposes. FAME analysis is presently the only chemotaxonomic technique that is linked to a commercial database for identification purposes. Fatty acid profiles showing variability in chain length, double-bond position, and substituent groups are perfectly suitable for taxon description and also for comparative analyses of profiles that have been obtained under identical growth conditions (Suzuki et al. 1993).

Sodium dodecyl sulfate-PAGE of whole-cell proteins requires standardized conditions of growth, combined with a rigorously standardized procedure for analysis, and normalization of the data for computer-assisted comparison of the results. Nevertheless, it has made important contributions to polyphasic taxonomic studies among the aerobic endospore formers (Logan et al. 2009).

Determination of the cell wall composition has traditionally been important in Gram-positive bacteria which contain various peptidoglycan types. The peptidoglycan type of Gram-negative bacteria is rather uniform and provides little information. Preparation of cell wall samples and determination of peptidoglycan structure is usually carried out using the methods described by Schleifer and Kandler (1972).

Isoprenoid quinones occur in the cytoplasmic membranes of most prokaryotes and play important roles in electron transport, oxidative phosphorylation, and, possibly, active transport (Collins and Jones 1981). There are two major structural groups, the naphthoquinones (subdivided into two types: the phylloquinones and the menaquinones) and the benzoquinones. The large variability of the side chains (differences in length, saturation, and hydrogenation) can be used to characterize bacteria at different taxonomic levels (Collins and Jones 1981).

The taxonomic importance of polar lipids has now been demonstrated for some novel genera among the *Bacillaceae*, although many polar lipids detected have not yet been structurally characterized. Likewise, quinones (MK-7, MK-8, and MK-9) have so far been reported for representatives of *Bacillaceae* (Logan et al. 2009).

Finally, pyrolysis mass spectrometry, Fourier transform infrared spectroscopy, and UV resonance Raman spectroscopy are sophisticated analytical techniques which examine the total chemical composition of bacterial cells. These methods have been used for taxonomic studies of particular groups of bacteria, including the members of the family *Bacillaceae* (Vandamme et al. 1996; Logan et al. 2009).

3.4 Genetic Approaches

Genotypic methods are those that are directed toward DNA or RNA molecules. Undoubtedly, these methods have revolutionized the bacterial identification system and taxonomy. Different techniques are now available to subtype bacteria up to strain level, such as restriction fragment length polymorphism (RFLP), plasmid profiling, ribotyping, amplified ribosomal DNA restriction analysis (ARDRA), pulsed field gel electrophoresis (PFGE), and randomly amplified polymorphic DNA (RAPD). Different PGPR have already been characterized by one or more of these methods (Oliveira et al. 2000; von der Weid et al. 2000; Depret and Laguerre 2008; Monteiro et al. 2009; and many others). For a detailed description of these methods, the reviews by Vandamme et al. (1996), Prakash et al. (2007), Rodríguez-Díaz et al. (2008), and Logan et al. (2009) can be referred.

For the description of bacterial taxa, other methods are essentially used. Determination of the moles percent guanosine plus cytosine is one of the classical genotypic methods. Generally, the range observed is not more than 3% within a well-defined species and not more than 10% within a well-defined genus (Stackebrandt and Goebel 1994).

DNA–DNA hybridization or DNA–DNA reassociation technique is based on the fact that at high temperatures DNA can be denatured, but the molecule can be brought back to its native state by lowering down the temperature (reassociation). This technique considers the comparison between whole genome of two bacterial species (Stackebrandt and Liesack 1993). A bacterial species, generally, would include the strain with 70% or greater DNA–DNA hybridization values with 5°C or less ΔTm values, and both the values must be considered. There are many different methods for DNA–DNA hybridization [presented and compared by Mora (2006)], but it is important to state that this technique gives the relative % of similarity but not the actual sequence identity.

DNA microarray is a method which was lined up to overcome the shortcomings of DNA–DNA hybridization. Although DNA microarray also involves hybridization of DNA, it uses fragmented DNA instead of whole genomic DNA. Numerous DNA fragments can be hybridized on a single microarray and gives resolution up to strain level. However, it is still an expensive methodology.

Indeed, taxonomy was revolutionized when the gene sequences of rRNA molecules were introduced to compare evolutionary similarities among strains (phylogenetic comparisons). All the three kinds of rRNA molecules, i.e., 5S, 16S, and 23S and spacers between these can be used for phylogenetic analyses, but 16S rRNA gene (1,650 bp) is the most commonly used marker. It has a universal distribution, highly conserved nature, fundamental role of ribosome in protein synthesis, no horizontal transfer, and its rate of evolution which represents an appropriate level of variation between organisms (Stackebrandt and Goebel 1994). The 16S rRNA molecule comprises of variable and conserved regions, and universal primers for the amplification of full 16S rRNA gene are usually chosen from conserved region while the variable region is used for comparative taxonomy. The 16S rRNA gene sequence is deposited in databases such as Ribosomal Database Project II (http://rdp.cme.msu.edu/) and GenBank (http://www.ncbi.nlm.nih.gov/). Sequences of related species for comparative phylogenetic analysis can also be retrieved from these databases. Thereafter, sequence comparing software packages such as BLAST and CLUSTAL X are used for alignment of 16S rRNA gene sequence. The extent of relatedness between bacterial species can be scrutinized by the construction of phylogenetic trees or dendrograms. The phylogenetic tree ascertains the genus to which the strain belongs and its closest neighbors, i.e., those sharing the clade or showing >97% 16S rRNA gene sequence similarity, are obtained from various culture collections to perform further genotypic, chemotaxonomic, and phenotypic analysis. At present, by correlation with experimental data obtained in the comparison of total genomic DNA (DNA–DNA hybridization), it is proposed that a similarity below 98.7–99% on the 16S rRNA gene sequences of two bacterial strains is sufficient to consider them as belonging to different species. On the other hand, two strains showing similarities above the 98.7% threshold may represent two different species. In these cases, total genome DNA–DNA hybridization must be performed and those strains for which similarities are below 70% are considered to belong to different species (Stackebrandt and Liesack 1993; Stackebrandt and Goebel 1994).

Finally, sequences of other highly conserved housekeeping or other protein-encoding genes, such as *rpoB*, *gyrB*, *recA*, have also great potential for taxonomic analysis at the species level. For example, Mota et al. (2005) obtained clustering patterns for *Paenibacillus* based upon *rpoB* sequence comparisons that were similar to those obtained with 16S rRNA gene sequences. Moreover, Wang et al. (2007) included *gyrB* sequence comparisons in the studies of the *B. subtilis* group and Cerritos et al. (2008) included *recA* sequence comparisons in the work that led to the proposal of a new *Bacillus* species.

4 Prospective Biocontrol Agents of Plant Diseases

Since 1987 in China, PGPR, called yield increasing bacteria (YIB) have been largely applied in 48 different crops over 3.35 millions of hectares (Wenhua and Hetong 1997). In that country, productivity gains as high as 23.1% and 22.5% were obtained, respectively, in sweet potatoes and potatoes. Additionally, remarkable 85.5% and 80.3% reduction levels of diseases caused by *Xanthomonas oryzae* pv. *oryzae* and *Glomerella cingulata*, respectively, were recorded (Zhang et al. 1996).

Rhizobacteria are effective competitors in the rhizosphere which can establish and persist on roots of agronomically grown plants (Kloepper and Mariano 2000). PGPR may promote plant growth directly on healthy plants or indirectly when controlling phytopathogens or pests in different crops (Kloepper 1993; Medeiros et al. 2005; Zhender et al. 1997; Keel and Maurhofer 2009). They can be isolated from any other plant part besides the roots as well as from the plant surface or interior. According to Hallman et al. (1997), the endophytic bacteria involved in biological control showed advantages of having the same ecological niche of the pathogen and could be protected from diverse abiotic influences.

The PGPR mechanisms for plant growth improvement were already discussed in this chapter. PGPR also exhibit several mechanisms of biological disease control, most of which involve competition and production of metabolites which affect the pathogen directly. Examples of such metabolites include antibiotics, cell wall-degrading enzymes, siderophores, and HCN (Enebak et al. 1998; Kloepper 1993; Weller 1988). It is noteworthy to state that different mechanisms may be found in a single strain and act simultaneously. Some PGPR do not produce metabolites against the pathogens and are spatially separated from them. These two traits suggest that alteration of host defense mechanisms account for the observed disease protection. Induced systemic resistance (ISR) or systemic acquired resistance (SAR) is defined as the activation of chemical and physical defenses of the plant host by an inducer which could be a chemical or a microorganism, leading to the control of several pathogens (Kloepper et al. 1992). Several PGPR strains can act as inducers of ISR (Kloepper et al. 1992), and PGPR-mediated ISR may be an alternative to the use of chemical inducers or pathogens for inducing SAR. This mechanism is discussed separately in this chapter.

Two cases of study will be discussed here: black rot of crucifers, a foliar disease, and Fusarium wilt of banana, a vascular disease. Black rot caused by *Xanthomonas campestris* pv. *campestris* (*Xcc*) causes severe economic losses in all developmental crucifer stages (Mariano et al. 2001). *Bacillus* spp. isolated from healthy cabbage, kale, and radish had reduced black rot incidence in kale and cabbage in greenhouse and field experiments (Assis et al. 1996). Monteiro et al. (2005) showed that four of these *Bacillus* strains produced lipopeptides active against *Xcc* during its late growth phase. These peptide antibiotics are amphiphilic compounds with surfactant activity (Zuber et al. 1993). Recently, it was demonstrated that lipopeptides can stimulate ISR in plants, probably by interacting with plant cell membranes and inducing temporary alterations in the plasma membrane which could raise plant defenses (Ongena et al. 2009).

Fusarium wilt of banana caused by *Fusarium oxysporum* f. sp. *cubense* is a very destructive disease in Brazil and other parts of the world. The rhizomes and pseudostems of infected plants used for propagation are the principal sources of inoculum and disease dispersion. Therefore, micropropagated health plantlets are used to prevent or delay the introduction of this pathogen in soils. However, these plantlets are more susceptible to this and other soilborne pathogens and should be protected before transplanting. PGPR are an alternative for improving this system. In greenhouse studies, endophytic and epiphytic bacteria applied, isolated or in mixtures, as root and substrate treatments, significantly increased the growth of micropropagated banana plantlets and controlled fusarium wilt (Mariano et al. 2004) (Fig. 1). According to Nowak and Shulaev (2003), the production of high-quality propagules with disease resistance may be achieved among others methods by their "in vitro" and "ex vitro" biopriming (priming with beneficial microorganisms).

Commonly, control is based on the use of single biocontrol agents. This strategy must be changed because, from the ecological point of view, the disease is part of a complex agroecosystem. As reported by Fravel (2007), a holistic view of this system can help take correct decisions about management. Therefore, a special approach for improving the PGPR efficiency is the use of mixtures containing different genera or species that presents additive or synergistic effects such as nitrogen-fixing bacteria and mycorrhiza helper bacteria (MHB). Another strategy is to use PGPR, mixed or alternated with fungicides, integrating biological and chemical control.

MHB are those which either assist mycorrhiza formation or promote the functioning of their symbiosis. They exist in arbuscular and ectomycorrhizal systems. MHB present three significant functions: nutrient mobilization from soil minerals, fixation of atmospheric nitrogen, and plant protection against root pathogens (Frey-Klett et al. 2007). According to these authors, PGPR induced increases in mycorrhizal root colonization from 1.1 to 17.5 fold in different interactions. Some of the MHB cited were

Fig. 1 Biocontrol of Fusarium wilt in micropropagated banana plantlet cv. Pacovan treated with *Bacillus pumilus* ENF24 (*right*) compared with plantlet not treated (*left*). Plantlets were vertically sliced to show rhizome discoloration, an internal disease symptom

Pseudomonas fluorescens, *P. monteilii*, *Bacillus coagulans*, *B. subtilis*, *Paenibacillus brasilensis*, *Rhizobium leguminosarum*, and *Bradyrrhizobium japonicum*.

Wheat seeds treated with different mixtures of *Paenibacillus macerans* and difenoconazole showed significant reduced incidences of pathogens (Luz 2003a), and in field all treatments promoted germination and grain yield except for difenoconazole alone that increased only yield. Similar results were obtained when corn seeds were bacterized with the same bioprotector + fludioxonil + metalaxyl M (Luz 2003b). Also *Bacillus*-based treatments have been successfully combined with traditional chemical seed treatments (Bugg et al. 2009). Therefore, the use of such mixtures may lead to a substantial reduction of pesticide use in several crops.

It is also important to focus on the critical stages of commercialization of biocontrol agents. Screening for new agents should consider the biology and ecology of the pathosystem, as well as agricultural practices associated with the crop (Fravel 2007). This knowledge will help prevent variation in field performance which is responsible for lack of wider adoption of biocontrol for disease management. The formulation stage aim is to deliver the biocontrol agent in a physiologically active state to provide the needed control. The formulation must be economical and present good shelf-life and a suitable form for shipping, storage, and application. Risk assessment to human health and to the environment are needed before releasing the new product, and early in the screening; even microorganisms with good biocontrol potential but capable of growing at human body temperature should be eliminated (Fravel 2007). In the United States, organisms currently registered for biocontrol and active compounds isolated from plants or other organisms are listed at http://www.epa.gov/oppbppd1/biopesticides/ingredients/index.htm. A few examples of PGPR and biocontrol products are: *Agrobacterium radiobacter* K1026 (Nogall®), *Bacillus pumilus* QST 2808 (Sonata® TM), *B. pumilus* GB34 (YieldShield®), *B. subtilis* GBO3(Kodiak®), *Pantoea agglomerans* C9-1 (BlightBan C9-1®), *P. agglomerans* E325 (Bloomtime®), *Pseudomonas aureofaciens* Tx-1(Spot-Less®T), *P. syringae* ESC-10 and ESC-11 (Bio-save®), *P. fluorescens* A506 (BlightBan®), *P. chlororaphis* MA 342 (Cedomon®), *Streptomyces griseoviridis* K61 (Mycostop®), and *S. lydicus* WYEC 108 (Actinovate®).

5 Induced Systemic Resistance as a Mechanism of Disease Suppression by Rhizobacteria

The increased level of resistance using external agents, without modifying the genome of the plant, is known as induced or acquired resistance. The expression of induced resistance can be local or systemic when it is expressed at sites not directly exposed to the inducers agent (Stadnik 2000). This agent may be a chemical activator, extracts of cells of living organisms or microorganisms (Romeiro 2000). The event of ISR has been demonstrated in various plants inoculated with different species of rhizobacteria (Liu et al. 1995; Raj et al. 2003; Halfeld-Vieira et al. 2006). This type of induced resistance can occur under controlled conditions and in the field, and shows advantages such as: effectiveness against various pathogens; stability due to the action of different mechanisms of resistance, systemicity, energy economy; and metabolic utilization of genetic potential for resistance in all susceptible plants (Bonaldo et al. 2005).

The ISR occurs when plants previously exposed to biotic and abiotic agents are induced to defense against pathogens, which are spatially separated from the inducer agent (Pieterse and Van Loon 1999; Stadnik 2000). PGPR that inhabit the soil and are often isolated from the rhizosphere of several plants have been studied as potential biotic agents of ISR (Mariano and Kloepper 2000). *Bacillus* and *Pseudomonas* are among the most studied genera of PGPR.

It is known that susceptible plants have genetic information for efficient mechanisms of resistance to diseases and that these mechanisms can be systematically expressed for long periods of time by prior inoculation with avirulents pathogens, microbial components, and chemical substances (Kuc 1995). The ISR is persistent and presents complex components in different locations which are responsible for the activity of various defense compounds. Consequently, it is more stable when compared with the few pathways arising from the use of chemical pesticides.

Despite the many studies in this area, only in 1961 the induced resistance was first analyzed, by preinoculation of tobacco plants with tobacco mosaic virus (Ross 1961). This procedure protected the plant against other viruses and resulted in the conception of "Systemic Acquired Resistance" (SAR). The activation of defense mechanisms induced by fungi, bacteria, viruses, and nematodes can be achieved by different routes, which may occur alone or concomitantly (Bonaldo et al. 2005).

Problems of variability in the effectiveness of induced resistance to diseases in plants in different soil and climatic conditions may occur, similar to that found in biological control (Kuc 1995). In agriculture, the use of biological products on the induction of resistance in plants has one more benefit that can be added to the already known to reinforce the plant growth promotion. Induction of resistance by the application of chemical inducers has been used in some crops in the integrated management of diseases and pests. The use of biological inducers may be an option in the management of diseases in plants. The positive effects of PGPR on plants usually are included in two categories: promotion of growth and biological control (Mariano and Kloepper 2000). In practice, these effects are often induced by the same strain of PGPR; therefore, some PGPR selected to promote growth also are able to control diseases and vice versa. The presence of the PGPR in the rhizosphere makes the entire plant, including the shoot, more resistant to pathogens.

Induction of resistance promoted by PGPR is active and signaling in the route of salicylic acid with induction of PR-proteins (proteins related to the pathogenesis) or route of the jasmonic acid and ethylene (Hoffland et al. 1995; Pieterse et al. 1998). When the PGPR colonize the root system, constituents of bacterial cell molecules or synthesized by elicitors act as a biochemical signal. This time, the genes that encode for the synthesis of components of the dynamic resistance are activated and ISR is expressed (Romeiro 2000). Wei et al. (1991) working with cucumber and anthracnose caused by *Colletotrichum orbiculare* showed that this plant could be used as a model for ISR.

In addition to the PR-proteins, the plants produce other enzymes of the defense, including peroxidases, phenylalanine ammonia-lyase (PAL), and polyphenol-oxidase (PPO). Peroxidase and PPO are catalysts in the formation of lignin. PAL and other enzymes are involved in the formation of phytoalexins. Chen et al. (2000) reported that ISR mediated by PGPR against *Pythium aphanidermatum* in cucumber was associated with an increase of peroxidases, PPO and PAL. Metabolic changes involved in the defense mechanism of plants are correlated with changes in activity of key enzymes in primary and secondary metabolism. The production of enzymes related to pathogenesis (PR-proteins) by strains of rhizobacteria is considered the largest property of the antagonistic strains (Saikia et al. 2004). Among these enzymes can be highlighted chitinases, lipoxygenases, peroxidases, and glucanases. Plants express the activity of peroxidase during pathogen–host interaction (Saikia et al. 2006), where this enzyme has been implicated in the oxidation of phenols (Schmid and Feucht 1980), lignification (Saparrat and Guillen 2005), plant protection (Hammerschmidt et al. 1982), and elongation of plant cells (Goldberg et al. 1986). Increased activity of peroxidase has been correlated with resistance in many plant species, including rice and wheat (Young et al. 1995). The action of lipoxygenase products contributes to the defense reactions involving the inhibition of growth of the pathogen and induction of phytoalexins (Li et al. 1991). The phytoalexins are secondary metabolites, antibiotics, low molecular weight produced by plants in response to physical stress, chemical, or biological. They are able to prevent or reduce the activity of pathogens, the rate of production dependent on the genotypes of host and/or pathogen (Daniel and Purkayastha 1995). The phytoalexin compounds are biocides and are directly related to the defense mechanisms of plants.

In several studies, the quantification of activity of enzymes involved in the induction of resistance has been used as a parameter to assess the induction mechanism (biotic or abiotic) involved (Knorzera et al. 1999; Campos et al. 2004; Nakkeeran et al. 2006; Silva et al. 2004; Halfeld-Vieira et al. 2006; Saikia et al. 2006). The increase in activity and accumulation of these enzymes depend mainly on the inducing agent but also the genotype of the plant, physiological conditions, and the pathogen (Tuzun 2001). Depending of pathosystem studied, a variety of substances are produced by rhizobacteria and has been linked to activation of mechanisms of disease suppression in plants which reduce the damage caused by phytopathogens. Thus, the application of PGPR in agriculture via soil or seed inoculation can be characterized as a beneficial component in the integrated management of diseases.

6 Bacterial Biofertilizers

Before initiating a review of PGPR as biofertilizers, it is necessary to define the term biofertilizer. It is proposed frequently here that biofertilizers designate the biological products which contain microorganisms providing direct and indirect gains in yield from crops. Vessey (2003) defines biofertilizers as a substance which contains living microorganisms which, when applied to seed, plant surfaces, or soil colonizes the rhizosphere or the interior of the plant and promotes growth by increasing the supply or availability of primary nutrients the host plant. Rhizobacteria, associated with rhizosphere, can fix nitrogen, and solubilizing phosphorus has been used as inoculum in nonleguminous species such as maize, rice, wheat, and sugar cane (Dobereiner 1997). Biofertilizers have been an alternative to mineral fertilizers to increase the yield and plant growth in sustainable agriculture (Canbolat et al. 2006).

The mechanisms by which PGPR promote plant growth are not fully understood but include among others: ability to produce or change the concentration of plant hormones (Mordukhova et al. 1991); asymbiotic N_2 fixation (Boddey and Dobereiner 1995); and solubilization of mineral phosphate and other nutrients (De Freitas et al. 1997). The production of hormones in PGPR in numerous studies reports the importance of indolacetic acid (IAA) in the roots development (Aloni et al. 2006). The effect of exogenous IAA in the plant can stimulate or inhibit growth and is often a function of hormones concentration available; in addition, the sensitivity of plant tissue changes according to hormones concentration (Persello-Cartieux et al. 2003). It was reported that isolates of *Pseudomonas* (fluorescent) produced exudates in roots of maize in response to IAA (Pan et al. 1999). Gibberellins were detected in several cultures of *B. subtilis*, but were not detected in the presence of auxin (Broadbent et al. 1971). Analyzing the sources of IAA with bacterial origin, Loper and Schroth (1986) found two strains of *Pseudomonas* spp. producing high concentrations of IAA (5–10 mg/ml), which reduced roots elongation and increased shoot/root proportion in sugar beet plants (*Beta vulgaris*) when applied as seed inoculant in this culture. Araújo et al. (2005) detected auxin production in two strains of *B. subtilis* which provided benefits in growth of soybean, in addition to be antagonists of phytopathogenic fungi in culture. Araújo and Hungria (1999) found that *B. subtilis* (AP-3) or its metabolites provided increase in nodulation and yield of soybean in the field.

Gains in nutrition in plants inoculated with rhizobacteria have also been demonstrated as a benefit of the presence of this group of microorganisms in the rhizosphere. In relation to nitrogen for several years has been discovered the potential of bacteria from the genus *Azospirillum*; fixing nitrogen when in free-living (Boddey and Dobereiner 1995), which when associated with the rhizosphere may contribute to nitrogen nutrition of plants. Concerning phosphate nutrition, Rodriguez and Fraga (1999) mention that strain from the genus *Pseudomonas*, *Bacillus* and *Rhizobium* are among the bacteria with the greatest potential of solubilization of phosphorus in the soil.

The solubilization of insoluble phosphates mediated by microorganisms is associated with the detachment of organic acids which are often combined with other metabolites, as found in vitro, that the potential for P solubilization by microorganisms is directly related to production of siderophores, lytic enzymes, and phytohormones (Vassilev et al. 2006). With the increased availability of nutrients in the soil by the action of *B. subtilis*, was shown higher absorption of nutrients such as phosphorus and nitrogen in plants inoculated with rhizobacteria on seeds (Araújo 2008). Richardson (2000) reported that most soils are poor in available phosphorus and phosphate fertilizer represents a high cost to the farmer; therefore, it is interesting to take advantage of soil microorganisms used as inoculum for the mobilization of phosphorus in poor soils. In addition to phosphorus solubilization, other mechanisms are also related to the microbial metabolism in soil, such as enzymes production (nitrogenase, chitinases, and glucanases) (Cattelan et al. 1999).

Some failures derived from the use of biofertilizers containing PGPR may be related to interspecific genetic interaction by the rhizobacteria and the host plant. Previous studies have documented phenotypic variation within cultivars with respect to health and nutrition of plants from microbial inoculation (Remans et al. 2008). Different cultures and species or cultivars may produce different types of root exudates, which may support the activity of the inoculum or serve as substrate for the formation of biologically active substances by the inoculum (Khalid et al. 2004). Dalmastri et al. (1999) reported that different maize cultivars could provide variation in the rhizosphere colonization by *Burkholderia*. Phenotypic variation among cultivars may be partly due to genetic variation and suggested that the breeding of the host was performed in conjunction with PGPR in biofertilizers (Remans et al. 2008). Another strategy to reduce the effects of phenotypic variation can be the use of biofertilizers with more than two isolates in their composition. Studies conducted for 2 years with the application of biofertilizers originating from a mixture of isolates of *Bacillus* showed increase in plant growth and productivity (Adesemoye et al. 2008).

A major problem for massive use of PGPR has been formulated for its commercial use. These include production in the scale of fermentation microorganisms with management of the quality, stability, and effectiveness of the product. *B. subtilis* has been assessed as of great potential for use in agriculture and has been used in the formulation of commercial products for agricultural use in several countries (Lazzareti and Bettiol 1997). Several substances have been used in experimental formulations such as lactose, peptone, gum arabic and xanthan, cellulose, and others (Schisler et al. 2004). This formulation may require a significant value to determine the effectiveness of the final product based on rhizobacteria such as the *B. subtilis*.

Development of formulations with a potential PGP to ensure survival and activity in the field and compatibility with chemical treatment of seeds has been the focus of researches with application of PGPR in agriculture. The research among other things optimizes growth conditions before the formulation, development of vehicles, and appropriate technology for application (Date 2001). In registration and marketing of products with PGPR, a large number of constraints are found (Mathre et al. 1999).

The U.S. market based on the information of the committee of biological products from the American Phytopatology Society (APS) in 2005 has registered the following products: ten products based on the *Bacillus* (*BioYield*, *Companion*, *EcoGuard*, *HiStick N/T*, *Kodiak* , *Mepplus*, *Serenade*, *Sonata*, *Subtilex*, *Yield-Shield*), two products with *Burkholderia cepacia* (*Deny* and *Intercept*), and six products based on *Pseudomonas* (*AtEze*, *Bio-save*, *BlightBan*, *Frostban*, *Spot-Less*). Most of these products has been disposed in powder solubleformulate. Different genera of bacteria have been studied as PGPR; however, investments in research and development of bioproducts have been higher in projects on *Pseudomonas* and *Bacillus*. Works on *Pseudomonas* have been focused on alternatives to improve the survival of this species of bacteria in commercial formulations. Furthermore, bacteria from the genus *Bacillus*, which are tolerant to desiccation and heat, have a longer life in commercial formulations; this explains the greater availability of commercial products based on *Bacillus*.

Currently, biofertilizers with PGPR are still not a reality of extensive commercialization – unlike the agricultural use of legume inoculants using rhizobia already a reality for almost a century – except for *Azospirillum* inoculants that are available for a variety of crops in Europe and Africa (Vessey 2003). There is no doubt that the lack of consistent responses in different host cultivars (Remans et al. 2008) and different field sites (Hilali et al. 2001) are reasons that limit expansion of the marketing of biofertilizers with PGPR. For these, it would be necessary to carry out more studies on ecology and colonization of microorganisms in the rhizosphere at different situations, since the biofertilizers with PGPR are restrictive for certain cultivars, climate, and soil conditions.

7 Concluding Remarks

PGPRs are the potential tools for sustainable agriculture and trend for the future. For this reason, there is an urgent need for research to clear definition of what bacterial traits are useful and necessary for different environmental conditions and plants, so that optimal bacterial strains can either be selected and/or improved. Combinations of beneficial bacterial strains that interact synergistically are currently being devised and numerous recent studies show a promising trend in the field of inoculation technology.

References

Adesemoye AO, Torbert HA, Kloepper JW (2008) Enhanced plant nutrient use efficiency with PGPR and AMF in an integrated nutrient management system. Can J Microbiol 54:876–886

Aloni R, Aloni E, Langhans M, Ulrich CI (2006) Role of cytokinin and auxin in shaping root architecture: regulating vascular differentiation, lateral root initiation, root apical dominance and root gravitropism. Ann Bot 97:883–893

Amara MAT, Dahdoh MSA (1997) Effect of inoculation with plant growth-promoting rhizobacteria (PGPR) on yield and uptake of nutrients by wheat grown on sandy soil. Egypt J Soil 37:467–484

Araújo FF (2008) Inoculação de sementes com *Bacillus subtilis*, formulado com farinha de ostras e desenvolvimento de milho, soja e algodão. Ciênc Agrotec 32:456–462

Araujo FF, Hungria M (1999) Nodulação e rendimento de soja co-infectada com *Bacillus subtilis* e *Bradyrhizobium japonicum / Bradyrhizobium elkanii*. Pesq Agropec Bras 34:1633–1643
Araujo FF, Henning AA, Hungria M (2005) Phytohormones and antibiotics produced by *Bacillus subtilis* and their effects on seed pathogenic fungi and on soybean root development. World J Microbiol Biotechnol 21:1639–1645
Asghar HN, Zahir ZA, Arshad M, Khalig A (2002) Plant growth regulating substances in the rhizosphere: microbial production and functions. Adv Agron 62:146–151
Assis SMP, Mariano RLR, Michereff SJ, Coelho RSB (1996) Biocontrol of *Xanthomonas campestris* pv. *campestris* on kale with *Bacillus* spp. and endophytic bacteria. In: Tang W, Cook RJ, Rovira A (eds) Advances in biological control of plant diseases. China Agricultural University Press, Beijing, China, pp 347–353
Barr JG, Emmerson AM, Hogg GM, Smyth E (1989) API-20NE and sensititre autoidentification systems for identifying *Pseudomonas* spp. J Clin Pathol 42:1113–1114
Bashan Y (1998) Inoculants of plant growth-promoting bacteria for use in agriculture. Biotechnol Adv 16:729–770
Bashan Y, Holguin G (1997) *Azospirillum*-plant relationships: environmental and physiological advances. Can J Microbial 43:103–121
Bashan Y, Holguin G (1998) Proposal for the division of plant growth-promoting rhizobacteria into two classification: biocontrol-PGPB (plant growth-promoting bacteria) and PGPB. Soil Biol Biochem 30:1225–1228
Bin L, Smith DL, Ping-Qui F (2000) Application and mechanism of silicate bacteria in agriculture and industry. Guizhou Sci 18:43–53
Biswas JC, Ladha JK, Dazzo FB (2000) Rhizobial inoculation influences seedling vigor and yield of rice. Agron J 92:880–886
Boddey RM, Dobereiner J (1995) Nitrogen fixation associated with grasses and cereals: recent progress and perspectives for the future. Fert Res 42:241–250
Bonaldo SM, Pascholati SF, Romeiro RS (2005) Indução de resistência: Noções básicas e perspectivas. In: Cavalcanti LS, di Piero RM, Cia P, Pascholati SF, Resende MLV, Romeiro RS (eds) Indução de resistência em plantas a patógenos e insetos. FEALQ, Piracicaba, pp 11–28
Broadbent P, Baker KF, Waterworth Y (1971) Bacteria and actinomycetes antagonistic to root pathogens in Australian soils. Aust J Biol Sci 24:925–944
Bugg K, Hairston W, Riggs J (2009) Succeeding in a traditional Ag-chemical company despite the "snake oil"/"foo-foo dust" concepts of biological-based products. In: Weller D, Thomashow L, Loper J, Paulitz T, Mazzola M, Mavrodi D, Landa BB, Thompson J (eds) 8th International PGPR Workshop. Portland, USA, p 17
Camacho M, Santamaria C, Temprano F, Daza A (2001) Co-inoculation with *Bacillus sp.* CECT 450 improves nodulation in *Phaseolus vulgaris* L. Can J Microbiol 47:1058–1062
Campos AD, Ferreira AG, Hampe MMV, Antunes IF, Brancão N, Silveira EP, Osório VA, Augustin E (2004) Atividade de peroxidase e polifenoloxidase na resistência do feijão a antracnose. Pesq Agrop Bras 39:637–643
Canbolat MY, Barik KK, Cakmarci R, Sabin F (2006) Effects of mineral and biofertilizers on barley growth on compacted soil. Act Agric Scand 56:324–332
Cattelan AJ, Hartel PG, Fuhrmann JJ (1999) Screening for plant growth-promoting rhizobacteria to promote early soybean growth. Soil Sci Soc Am J 63:1670–1680
Cerritos R, Vinuesa P, Eguiarte LE, Herrera-Estrella L, Alcaraz-Peraza LD, Arvizu-Gómez JL, Olmedo G, Ramirez E, Siefert JL, Souza V (2008) *Bacillus coahuilensis* sp. nov., a moderately halophilic species from a desiccation lagoon in the Cuatro Ciénegas Valley in Coahuila, Mexico. Int J Syst Evol Microbiol 58:919–923
Chabot R, Anroun H, Cesces MC (1996) Growth promotion of maize and lettuce by phosphate solubilizing *Rhizobium leguminosarum biovar phaseoli*. Plant Soil 184:31–121
Chanway CP (1998) Bacterial endophytes: ecological and practical implications. Sydowia 50:149–170

Chen C, Bauske EM, Musson G, Rodriguez-Kabaña R, Kloepper JW (1994) Biological control of Fusarium on cotton by use of endophytic bacteria. Biol Control 5:83–91

Chen J, Abawi GS, Zucherman BM (2000) Efficacy of *Bacillus thuringiensis*, *Paecilomyces marquandii* and *Streptomyces costaricanus* with organic amendment against *Meloidogyne hapla* infecting lettuce. J Nematol 32:70–77

Coenye T, Gevers D, Van de Peer Y, Vandamme P, Swings J (2005) Towards a prokaryotic genomic taxonomy. FEMS Microbiol Rev 29:147–167

Collins MD, Jones D (1981) Distribution of isoprenoid quinine structural types in bacteria and their taxonomic implications. Microbiol Rev 45:316–354

Colwell RR (1970) Polyphasic taxonomy of the genus *Vibrio*: numerical taxonomy of *Vibrio cholerae*, *Vibrio parahaemolyticus* and related *Vibrio* species. J Bacteriol 104:410–433

Cooper JB, Long SR (1994) Morphogenetic rescue of *Rhizobium-meliloti* nodulation mutants by trans-zeatin secretion. Plant Cell 6:215–225

Dalmastri C, Chiarini L, Cantale C, Bevinino A, Tabacchioni S (1999) Soil type and maize cultivar affect the genetic diversity of maize root-associated *Burkholderia cepacia* populations. Microb Ecol 38:273–284

Daniel M, Purkayastha RP (1995) Handbook of phytoalexin metabolism and action. Marcel Dekker, New York, p 615p

Date RA (2001) Advances in inoculant technology: a brief review. Austral J Exp Agric 41:321–325

de Freitas JR, Banerjee MR, Germida JJ (1997) Phosphate-solubilizing rhizobacteria enhance the growth and yield but not phosphorus uptake of canola (*Brassica napus*). Biol Fertil Soils 36:842–855

de Vos P, Garrity G, Jones D, Krieg NR, Ludwig W, Rainey FA, Schleifer K-H, Whitman WB (2009) Bergey's manual of systematic bacteriology. Volume 3: The Firmicutes. 2nd edn. XXVI, 1450 p. 393 illus., Hardcover. Originally published by Williams & Wilkins, 1984

Depret G, Laguerre G (2008) Plant phenology and genetic variability in root and nodule development strongly influence genetic structuring of *Rhizobium leguminosarum* biovar viciae populations nodulating pea. New Phytol 179:224–235

Dobereiner J (1997) Biolgical nitrogen fixation in the tropics: social and economic contributions. Soil Biol Biochem 29:771–774

Enebak SA, Wei G, Kloepper JW (1998) Effects of plant growth-promoting rhizobacteria on loblolly and slash pine seedlings. Forest Sci 44:139–144

Figueiredo MVB, Burity HA, Martinez CR, Chanway CP (2007) Plant growth-promoting rhizobacteria for improving nodulation and nitrogen fixation in the common bean (*Phaseolus vulgaris* L.). World J Microbiol Biotechnol 24:1187–1193

Figueiredo MVB, Burity HA, Martinez CR, Chanway CP (2008) Alleviation of water stress effects in common bean (*Phaseolus vulgaris* L.) by co-inoculation *Paenibacillus* x *Rhizobium tropici*. Applied Soil Ecol 40:182–188

Fravel D (2007) Commercialization of biocontrol agents for use against plant pathogens. In: IX Reunião Brasileira sobre Controle Biológico de Doenças de Plantas, Campinas, S. Paulo, Brasil, CD-ROM, pp 1–2

Freitas ADS, Vieira CL, Santos CERS, Stamford NP, Lyra MCCP (2007) Caracterização de rizóbios isolados de Jacatupé cultivado em solo salino no Estado de Pernanbuco, Brasil. Bragantia 66:497–504

Frey-Klett P, Garbaye J, Tarkka M (2007) The mycorrhiza helper bacteria revisited. New Phytol 176:22–36

Glick BR (1995) The enhancement of plant-growth by free-living bacteria. Can J Microbiol 41:109–117

Glick BR, Karaturovic DM, Newell PC (1995) A novel procedure for rapid isolation of plant growth promoting *Pseudomonas*. Can J Microbiol 41:533–536

Glick BR, Patten CL, Holgin G, Penrose DM (1999) Biochemical and genetic mechanisms used by plant growth promoting bacteria. Imperial College Press, London, 267 p

Goldberg R, Imberty A, Liberman M, Prat R (1986) Relationships between peroxidatic activities and cell wall plasticity. In: Greppin H, Peneland C, Gaspar T (eds) Molecular and physiological aspects of plant peroxidases. University of Geneva, Geneva, pp 208–220

Gray EJ, Smith DL (2005) Intracellular and extracellular PGPR: commonalities and distinctions in the plant–bacterium signaling processes. Soil Biol Biochem 37:395–412

Gupta A, Gopal M, Tilak KV (2000) Mechanism of plant growth promotion by rhizobacteria. Indian J Exp Biol 38:856–862

Halfeld-Vieira BA, Vieira JR Jr, Romeiro RS, Silva HSA, Baract-Pereira MC (2006) Induction of systemic resistance in tomato by autochthonus phylloplane residente *Bacillus cereus*. Pesq Agrop Bras 41:1247–1252

Hallman J, Quadt-Hallman A, Mahafee WF, Kloepper JW (1997) Bacterial endophytes in agricultural crops. Can J Microbiol 43:895–914

Hammerschmidt R, Nucckles F, Kuc I (1982) Association of enhance peroxidase activity with induced systemic resistance of cucumber to *Colletotrichum largenarium*. Physiol Plant Pathol 20:73–82

Heritage J, Evans EGV, Killington RA (1996) Introductory microbiology. Cambridge University Press, England, 234 p

Hilali A, Prevost D, Broughton WJ, Antoun H (2001) Effects of inoculation with strains of *Rhizobium leguminosarun* biovar *trifolli* on the growth of wheat in two different Morrocan soils. Can J Microbiol 47:590–593

Hoffland E, Hakulinen J, van Pelt JA (1995) Comparison of systemic resistance induced by avirulent and nonpathogenic *Pseudomonas* species. Phytopathology 86:757–762

Keel C, Maurhofer M (2009) Insecticidal activity in biocontrol pseudomonads. In: Weller D, Thomashow L, Loper J, Paulitz T, Mazzola M, Mavrodi D, Landa BB, Thompson J (eds) 8th International PGPR Workshop. Portland, USA, p 51

Khalid A, Arshad M, Kahir ZA (2004) Screening plant growth-promoting rhizobacteria for improving growth and yield of wheat. Appl Soil Ecol 96:473–480

Kloepper JW (1993) Plant growth-promoting rhizobacteria as biological control agents. In: Metting B (ed) Soil microbial technologies. Marcel Dekker, New York, USA, pp 255–274

Kloepper JW, Mariano RLR (2000) Rhizobacteria to induce plant disease resistance and enhance growth – theory and practice. In: International symposium on biological control for crop protection, Rural Development Administration, Suwon, South Korea, pp 99–116

Kloepper JW, Schroth MN (1978) Plant growth promoting rhizobacteria on radishes, In: Proceedings of the 4th international conference on plant pathogenic bacteria, Angers, France, pp 879–882

Kloepper JW, Schroth MN, Miller TD (1980) Effects of rhizosphere colonization by plant growth-promoting rhizobacteria on potato plant development and yield. Phytopathology 70: 1078–1082

Kloepper JW, Tuzun S, Kuc J (1992) Proposed definitions related to induced disease resistance. Biocontrol Sci Technol 2:349–351

Knorzera OC, Lederera B, Durnerb J, Bogera P (1999) Antioxidative defense activation in soybean cells. Physiol Plant 107:294–302

Kuc J (1995) Induced systemic resistance – an overview. In: Hammerschmidt R, Kuc J (eds) Induced resistance to disease in plants. Kluwer, Dordrecht, pp 169–175

Lazzareti E, Bettiol W (1997) Tratamento de sementes de arroz, trigo, feijão e soja com um produto formulado a base de células e de metabólitos de *Bacillus subtilis*. Sci Agricola 54:89–96

Li DM, Alexander M (1988) Co-inoculation with antibiotic producing bacteria to increase colonization and nodulation by rhizobia. Plant Soil 108:211–219

Li WX, Kodama O, Akatsuka T (1991) Role of oxygenated fatty acids in rice phytoalexin production. Agric Biol Chem 55:1041–1147

Liu L, Kloepper JW, Tuzun S (1995) Induction of systemic resistance in cucumber against *Fusarium* wilt by plant growth promoting rhizobacteria. Phytopathology 85:695–698

Logan NA, Berkeley RCW (1984) Identification of *Bacillus* strains using the API system. J Gen Microbiol 130:1871–1882

Logan NA, Berge O, Bishop AH, Busse HJ, De Vos P, Fritze D, Heyndrickx M, Kampfer P, Rabinovitch L, Salkinoja-Salonen MS, Seldin L, Ventosa A (2009) Proposed minimal standards for describing new taxa of aerobic, endospore-forming bacteria. Int J Syst Evol Microbiol 59(8):2114–2121

Loper JE, Schroth MN (1986) Influence of bacterial sources of indole-3-acetic acid on root elongation of sugar beet. Phytopathology 76:386–389

Lucy M, Reed E, Glick BR (2004) Applications of free living plant growth-promoting rhizobacteria. Review Antonie Van Leeuwenhoek 86:1–25

Luz WC (2003a) Avaliação dos tratamentos biológico e químico na redução de patógenos em semente de trigo. Fitopatol Bras 28:093–095

Luz WC (2003b) Combinação dos tratamentos biológico e químico de semente de milho. Fitopatol Bras 28:37–40

Mariano RLR, Kloepper JW (2000) Método alternativo de biocontrole: Resistência sistêmica induzida por rizobactérias. Revisão Anual de Patologia de Plantas 8:121–137

Mariano RLR, Silveira EB, Assis SMP, Gomes AMA, Oliveira IS, Nascimento ARP (2001) Diagnose e manejo de fitobacterioses de importância no Nordeste Brasileiro. In: Michereff SJ, Barros R (eds) Proteção de Plantas na Agricultura Sustentável. UFRPE, Recife, Brasil, pp 141–169

Mariano RLR, Medeiros FHV, Albuquerque VV, Assis SMP, Mello MRF (2004) Growth-promotion and biocontrol of diseases in fruits and ornamentals in the states of Pernambuco and Rio Grande do Norte, Northeastern Brazil. In: Kobayashi K, Gasoni L, Terashima H (eds) Biological control of soilborne plant diseases. JICA, Buenos Aires, Argentina, pp 70–80

Mathre DE, Cook RJ, Callan NW (1999) From discovery to use. Traversing the world of commercializing biocontrol agents for plant disease control. Plant Dis 83:972–983

Mayak S, Tirosh T, Glick BR (1999) Effect of wild type and mutant plant growth promoting rhizobacteria on the rooting of mung bean cuttings. J Plant Growth Regul 18:49–53

Mayr E, Ashlock PD (1991) Principles of systematic zoology, 2nd edn. McGraw-Hill, New York, pp 1–12

Medeiros FHV, Silva G, Mariano RLR, Barros R (2005) Effect of bacteria on the biology of diamondback moth (*Plutella xylostella*) on cabbage (*Brassica oleraceae* var. *capitata*) cv. Midori. An Acad Pernamb Ciênc Agronôm 2:204–212

Miller JM, Rhoden DL (1991) Preliminary evaluation of Biolog, a carbon source utilization method for bacterial identification. J Clin Microbiol 29:1143–1147

Monteiro L, Mariano RLR, Souto-Maior AM (2005) Antagonism of *Bacillus* spp. against *Xanthomonas campestris* pv. *campestris*. Braz Arch Biol Technol 48:23–29

Monteiro JM, Vollú RE, Coelho MRR, Alviano CS, Blank AF, Seldin L (2009) Culture-dependent and -independent approaches to analyze the bacterial community of different genotypes of *Chrysopogon zizanioides* (L.) Roberty (vetiver) rhizospheres. J Microbiol 47:363–370

Mora RR (2006) DNA-DNA reassociation methods applied to microbial taxonomy and their critical evaluation. In: Stackebrandt E (ed) Molecular identification, systematics, and population structure of prokaryotes. Springer, Berlin, Heidelberg, pp 23–49

Mordukhova EA, Skvortsova VV, Kochetkov AN, Dubeikovskii AN, Boronin AM (1991) Synthesis of the phytohormone índole-3-acetic acid by rhizosphere bactéria of the genus *Pseudomonas*. Mikrobiologiya 60:494–500

Mota FF, Gomes EA, Paiva E, Seldin L (2005) Assessment of the diversity of *Paenibacillus* species in environmental samples by a novel *rpoB*-based PCR-DGGE method. FEMS Microbiol Ecol 53:317–328

Nakkeeran S, Kavitha K, Chandrasekar G, Renukadevi P, Fernando WGD (2006) Induction of plant defence compounds by *Pseudomonas chloraphis* PA23 and *Bacillus subtilis* BSCBE4 in controling damping-off of hot pepper caused by *Pythium aphanidermatum*. Biocontrol Sci Technol 16:403–416

Nelson LM (2004) Plant growth promoting rhizobacteria (PGPR): prospects for new inoculants. 2004 Plant Management Network, online doi:10.1094/CM-2004-0301-05-RV

Nowak J, Shulaev V (2003) Priming for transplant stress resistance in *in vitro* propagation. In Vitro Cell Dev Biol Plant 39:122–130

Oliveira IA, Vasconcellos MJ, Seldin L, Paiva E, Vargas MAT, Sá NMH (2000) Random amplified polymorphic DNA analysis of effective *Rhizobium* sp. associated with beans cultivated in Brazilian cerrado soils. Braz J Microbiol 31:39–44

Ongena M, Henry G, Adam A, Jourdan E, Thonart P (2009) Plant defense reactions stimulated following perception of *Bacillus* lipopeptides. In: Weller D, Thomashow L, Loper J, Paulitz T, Mazzola M, Mavrodi D, Landa BB, Thompson J (eds) 8th International PGPR Workshop. Portland, USA, p 43

Pan B, Bai YM, Leibovitch S, Smith DL (1999) Plant growth promoting rhizobacteria and kinetic as ways to promote corn growth and yield in short season areas. Eur J Agron 11:179–186

Persello-Cartieux F, Nussaume L, Robaglia C (2003) Tales from the underground: molecular plant-rhizobacteria interactions. Plant Cell Environ 26:189–199

Pieterse CMJ, Van Loon LC (1999) Salicylic acid-independent plant defense pathways. Trends Plant Sci 4:52–58

Pieterse CMJ, van Pelt JA, Knoester M, Laan R, Gerrits H, Weisbeek PJ, van Loon LCA (1998) Novel signaling pathway controlling induced systemic resistance in *Arabidopsis*. Plant Cell 10:1571–1580

Pires MN, Seldin L (1997) Evaluation of Biolog system for identification of strains of *Paenibacillus azotofixans*. Antonie Leeuwenhoek 71:195–200

Prakash O, Verma M, Sharma P, Kumar M, Kumari K, Singh A, Kumari H, Jit S, Gupta SK, Khanna M, Lal R (2007) Polyphasic approach of bacterial classification – an overview of recent advances. Indian J Microbiol 47:98–108

Raj SN, Chaluvaraju G, Amruthesh KN, Shetty HS (2003) Induction of growth promotion and resistance against downy mildew on pearl millet (*Penninsetum glaucum*) by rhizobacteria. Plant Dis 87:380–384

Raja P, Una S, Gopal H, Govindarajan K (2006) Impact of BioInoculants consortium on rice root exudates, biological nitrogen fixation and plant growth. J Biol Sci 6:815–823

Remans R, Croonenborghs A, Gutierrez RT, Michiels J, Vanderleyden J (2007) Effects of plant growth-promoting rhizobacteria on nodulation of *Phaseolus vulgaris* [L.] are dependent on plant P nutrition. Europ J Plant Pathol 119:341–351

Remans S, Blair MW, Manrique G, Tovar LE, Rao IM, Croomenborghs A, Torres GR, El-Howeity M, Michiels J, Vanderleyden J (2008) Physiological and genetic analysis of root responsiveness to auxin-producing plant growth-promoting bacteria in common bean (*Phaseolus vulgaris* L.). Plant Soil 302:149–161

Richardson AE (2000) Prospects for using soil microorganisms to improve the acquisition of phosphorus by plants. Aust J Plant Physiol 28:897–906

Rodriguez H, Fraga R (1999) Phosphate solubilizing bacteria and their role in plant growth promotion. Biotechnol Adv 17:319–339

Rodríguez-Díaz M, Rodelas B, Pozo C, Martínez-Toledo MV, González-López J (2008) A review on the taxonomy and possible screening traits of plant growth promoting rhizobacteria. In: Ahmad I, Pichtel J, Hayat S (eds) Plant-bacteria interactions: strategies and techniques to promote plant growth. Wiley, UK

Romeiro RS (2000) PGPR e indução de resistência sistêmica em plantas a patógenos. Summa Phytopathol 26:177–184

Ross AF (1961) Localizated acquired resitance to plant virus infection in hypersensitive hosts. Virology 14:340–358

Saikia R, Kumar R, Singh T, Srivastava AK, Arora DK, Gogoi DK, Lee MW (2004) Induction of defense related enzymes and pathogenesis related proteins in *Pseudomonas fluorescens*-treated chickpea in response to infection by *Fusarium oxysporum* F. sp. *Ciceri*. Mycobiology 32:47–52

Saikia R, Kumar R, Arora DK, Gogoi DK, Azad P (2006) *Pseudomonas aeruginosa* inducing rice resistance against *Rhizoctonia solani*: production of salicylic acid and peroxidases. Folia Microbiol 51:375–380

Saparrat MCN, Guillen F (2005) Lignolitic ability and potential biotechnology applications of the South American fungus *Pleurotus lacioniatocrenatus*. Folia Microbiol 50:155–160

Saravana-Kumar D, Samiyappan R (2007) ACC deaminase from *Pseudomonas fluorescens* mediated saline resistance in groundnut (*Arachis hypogea*) plants. J Appl Microbial 102:1283–1292

Scher FM, Baker R (1982) Effect of *Pseudomonas putida* and synthetic iron chelator on induction of soil suppressiveness to *Fusarium* wilt pathogens. Phytopathology 72:1567–1573

Schisler DA, Slininger PJ, Behle RW, Jackson MA (2004) Formulation of *Bacillus* spp. for biological control of plant diseases. Phytopathology 94:1267–1271

Schleifer KH, Kandler O (1972) Peptidoglycan types of bacterial cell walls and their taxonomic implications. Bacteriol Rev 36:407–477

Schmid PS, Feucht W (1980) Tissues-specific oxidation browning of polyphenols by peroxiase in cherry shoots. Gartenbauwisenschaft 45:68–73

Seldin L, Penido EGC (1986) Identification of *Bacillus azotofixans* using API tests. Antonie Leeuwenhoek 52:403–409

Seldin L et al (1984) *Bacillus azotofixans* sp. nov. a nitrogen fixing species from Brazilian soils and grass roots. Int J Syst Bacteriol 34:451–456

Silva HSA, Romeiro RS, Macagnan D, Halfeld-vieira BA, Pereira MCB, Mounteer A (2004) Rhizobacterial induction of systemic resitance in tomato plants: non-specific protection and increase in enzyme activities. Biol Control 29:288–295

Silva VN, Silva LESF, Figueiredo MVB (2006) Atuação de rizóbios com rizobactérias promotoras de crescimento em plantas na cultura do caupi (*Vigna unguiculata* L. Walp). Acta Sci Agron 28:407–412

Silva VN, Silva LESF, Martínez CR, Seldin L, Burity HA, Figueiredo MVB (2007) Estirpes de *Paenibacillus* promovem a nodulação específica na simbiose *Bradyrhizobium*-caupi. Acta Sci Agron 29:331–338

Sindhu SS, Gupta SK, Suneja S, Dadarwal KR (2002) Enhancement of green gram nodulation and growth by *Bacillus* species. Biol Plant 45:117–120

Stackebrandt E, Goebel BM (1994) Taxonomic note: a place for DNA-DNA reassociation and 16S rRNA sequence analysis in the present species definition in bacteriology. Int J Syst Bacteriol 44:846–849

Stackebrandt E, Liesack W (1993) Nucleic acids and classification. In: Goodfellow M, O'Donnell AG (eds) Handbook of new bacterial systematics. Academic, London, pp 151–194

Stadnik MJ (2000) Indução de resistência a Oídios. Summa Phytopathol 26:175–177

Stamford NP, Santos CERS, Lira Junior MA, Figueiredo MVB (2008) Effect of rhizobia and rock biofertilizers with *Acidithiobacillus* on cowpea nodulation and nutrients uptake in a tabeland soil. World J Microbiol Biotechnol 24:1857–1865

Suzuki K, Goodfellow M, O'Donnell AG (1993) Cell envelopes and classification. In: Goodfellow M, O'Donnell AG (eds) Handbook of new bacterial systematics. Academic, London, pp 195–250

Tenuta M (2003) http://www.umanitoba.ca/afs/agronomists_conf/2003/pdf/tenuta_rhizobac-teria.pdf

Tilak KVBR, Rauganayaki N, Manoharachari C (2006) Synergistic effects of plant-growth promoting rhizobacteria and *Rhizobium* on nodulation and nitrogen fixation by pigeonpea (*Cajanus cajan*). Europ J Soil Sci 57:67–71

Tuzun S (2001) The relationship between pathogen-induced systemic resistance (ISR) and multigenic (horizontal) resistance in plants. Eur J Plant Pathol 107:85–93

Vandamme P, Pot B, Gillis M, de Vos P, Kersters K, Swings J (1996) Polyphasic taxonomy, a consensus approach to bacterial systematics. Microbiol Rev 60:407–438

Vassilev N, Vassileva M, Nikolaeva I (2006) Simultaneous p-solubilizing and biocontrol activity of microorganisms: potential and future trends. Appl Microbiol Biotechnol 71:137–144

Vessey JK (2003) Plant growth-promoting rhizobacteria as biofertilizers. Plant Soil 255:571–586

Vessey JK, Buss TJ (2002) *Bacillus cereus* UW85 inoculation effects on growth, nodulation, and N accumulation in grain legumes. Controlled environment studies. Can J Plant Sci 82:282–290

von der Weid I, Paiva E, Nobrega A, van Elsas JD, Seldin L (2000) Diversity of *Paenibacillus polymyxa* strains isolated from the rhizosphere of maize planted in Cerrado soil. Res Microbiol 151:369–381

Wang L-T, Lee F-L, Tai C-J, Kasai H (2007) Comparison of *gyr*B gene sequences, 16S rRNA gene sequences and DNA-DNA hybridization in the *Bacillus subtilis* group. Int J Syst Evol Microbiol 57:1846–1850

Wei G, Kloepper JW, Tuzun S (1991) Induction of systemic resistance of cucumber to *Colletotrichum orbiculare* by seven strains of plant growth-promoting rhizobacteria. Phytopathology 81:1508–1512

Weller DM (1988) Biological control of soilborne plant pathogens in the rhizosphere with bacteria. Ann Rev Phytopathol 26:379–407

Wenhua T, Hetong Y (1997) Research and application of biocontrol of plant diseases and PGPR in China. In: Ogoshi A, Kobayashi K, Homma Y, Kodama F, Kondo N, Akino S (eds) Plant growth-promoting rhizobacteria – present status and future prospects. OECD-OCDE, Sapporo, Japan, pp 2–9

Young SA, Guo A, Guikema JA, White F, Leach IE (1995) Rice cationic peroxidase accumulation in xylem vessels during incompatible interaction with *Xanthomonas oryzae*. Plant Physiol 107:1333–1341

Zafar-ul-Hye M. (2008). Improving nodulation in lentil through co-inoculation with rhizobia and ACC-deaminase containing plant growth promoting Rhizobacteria. PhD Thesis. University of Agriculture, Faisalabad, Pakistan, p 198

Zhang F, Dashti N, Hynes RK, Smith DL (1996) Plant growth-promoting rhizobacteria and soybean [*Glycine max* (L.) Merr.]. Nodulation and fixation at suboptimal root zone temperatures. Ann Bot 7:453–459

Zhender G, Kloepper J, Changbin Y, Wei G (1997) Induction of systemic resistance in cucumber against cucumber beetles (Coleoptera: Crysomelidae) by Plant-Growth-Promoting-Rhizobacteria. J Econ Entomol 90:391–396

Zuber P, Nakano MM, Marahiel MA (1993) Peptide antibiotics. In: Sonenshein AL, Hoch JA, Losick R (eds) *Bacillus subtilis* and other Gram-positive Bacteria. ASM, Washington, USA, pp 897–916

Potential of PGPR in Agricultural Innovations

Haluk Caglar Kaymak

Contents

1 Introduction 46
2 Direct Plant Growth Promotion 47
  2.1 Biological Nitrogen Fixation 47
  2.2 Solubilization of Phosphates 48
  2.3 Plant Growth Regulators 50
  2.4 Effects on Plant Growth 53
3 Indirect Plant Growth Promotion 57
  3.1 Induced Systemic Resistance 57
  3.2 Suppression of Plant Diseases, Insects, and Nematodes by PGPR 58
4 Conclusions and Future Prospects 67
References 67

Abstract Plant growth promoting rhizobacteria (PGPR) are a group of free-living bacteria that colonize the rhizosphere and benefit the root growth in plants. Bacteria of diverse genera such as *Azospirillium*, *Bacillus*, *Burkholderia*, *Klebsiella*, *Pseudomonas*, etc., were identified as PGPR. These PGPR exert a direct effect on plant growth by inducing the production of phytohormones, supplying biologically fixed nitrogen, and increasing the phosphorous uptake by the solubilization of inorganic phosphates. These bacteria affect plant growth by indirect mechanisms that involve suppression of bacterial, fungal, viral, and nematode pathogens. A lot of study showed that inoculation with PGPR resulted in significant yield increases in different crops, rooting of hardwood and semi-hardwood cuttings, increased germination and enhanced emergence of seeds under different conditions, promoted nutrient uptake of roots, total biomass of the plants, increased seed weight, induced early flowering, etc. In this review, the importance of PGPR is discussed for

H.C. Kaymak
Faculty of Agriculture, Department of Horticulture, Atatürk University, 25240 Erzurum, Turkey
e-mail: hckaymak@atauni.edu.tr; hckaymak@yahoo.com

D.K. Maheshwari (ed.), *Plant Growth and Health Promoting Bacteria*,
Microbiology Monographs 18, DOI 10.1007/978-3-642-13612-2_3,

agricultural innovations with special references that utilises direct and indirect plant growth promotion.

1 Introduction

The rhizosphere, volume of soil surrounding roots influenced chemically, physically, and biologically by the plant root, is a highly favorable habitat for the reproduction of microorganisms, which exerts a potential impact on plant health and soil fertility (Sorensen 1997; Antoun and Prevost 2005; Podile and Kishore 2006). This environment is relatively rich in nutrients released by the plant roots, and its microbial communities are different from those that are not influenced by the roots (Alexander 1977; Burdman et al. 2000).

In the rhizosphere, very important and intensive interactions occur among the plant, soil, microorganisms, and soil microfauna (Antoun and Prevost 2005). These interactions can significantly influence plant growth and crop yields. In the rhizosphere, bacteria are the most abundant microorganisms. Rhizobacteria are rhizosphere-competent bacteria that aggressively colonize plant roots, could be free-living, parasitic, or saprophytic, and their diversity remains dynamic with a frequent shift in community structure and species abundance (Kunc and Macura 1988). These microbial communities are beneficial for plant growth, yield, and crop quality, and they have been called "plant growth promoting rhizobacteria (PGPR)" (Kloepper and Schroth 1978) including numerous strains of the genera *Acinetobacter*, *Aeromonas*, *Alcaligenes*, *Arthrobacter*, *Azospirillium*, *Azotobacter*, *Azoarcus*, *Bacillus*, *Beijerinckia*, *Burkholderia*, *Clostridium*, *Enterobacter*, *Erwinia*, *Flavobacterium*, *Gluconacetobacter*, *Klebsiella*, *Pseudomonas*, *Serratia*, *Rhizobium*, etc. (Burdman et al. 2000; Sudhakar et al. 2000; Hamaoui et al. 2001; Bertrand et al. 2001; Mirza et al. 2001; Bonaterra et al. 2003; Esitken et al. 2003a; Murphy et al. 2003; Raj et al. 2004; Joo et al. 2004; Esitken et al. 2006; Podile and Kishore 2006; Saleem et al. 2007).

PGPR can be divided into two groups according to their relationship with the plants: symbiotic bacteria and free-living rhizobacteria (Khan 2005). A lot of work have been done to study about the mechanisms and principles of the PGPR–plant relationship, which was widely accepted as the rhizosphere effect (Zhuang et al. 2007). Glick (1995) reported that PGPR function in three different ways: synthesizing particular compounds for the plants, facilitating the uptake of certain nutrients from the environment (Cakmakci et al. 2006; Garcia et al. 2004a, b; Siddiqui and Mahmood 2001), and preventing the plants from diseases (Guo et al. 2004; Jetiyanon and Kloepper 2002; Raj et al. 2003a, b).

In other words, these mentioned bacteria can directly cause plant growth, seed emergence, or improvement in crop yields by producing and secreting plant growth regulators such as auxins, gibberellins (GAs), and cytokinins. They elicit the root metabolic activities, supply biologically fixed nitrogen, and increase the phosphorous uptake by solubilization of inorganic phosphates (Burdman et al. 2000;

Podile and Kishore 2006). The near direct effect of PGPR is that these bacteria affect plant growth by indirect mechanisms that involve suppression of bacterial, fungal, viral, and nematode pathogens (Burdman et al. 2000; Kirankumar et al. 2008).

In this review, the importance of PGPR is discussed for agriculture innovations with special reference to their utilization in direct plant growth promotion such as seed emergence, secretion of plant growth regulators, and indirect plant growth promotion such as suppression of pest and disease.

2 Direct Plant Growth Promotion

PGPR influence direct growth promotion of plants by fixing atmospheric nitrogen, solubilizing insoluble phosphates, secreting hormones such as IAA, GAs, and Kinetins besides ACC deaminase production, which helps in regulation of ethylene.

2.1 Biological Nitrogen Fixation

Nitrogen is a well-known and essential key nutrient for plant growth and development. However, the global nitrogen cycle pollutes groundwater and increases the risk of chemical spills. The production of chemical fertilizers is a highly energy-intensive process using large amounts of fossil energy. High-input farming practices achieving high yields have created environmental problems and degradation in natural resources (Şahin et al. 2004). Thus, Figueiredo et al. (2008) reported that during the past couple of decades, the use of PGPR for sustainable and environment friendly agriculture has been increased tremendously in various parts of the world. Increasing and extending the role of bio-fertilizing with PGPR would reduce the need for chemical fertilizers and decrease their adverse environmental effects. Microorganisms are gaining importance in agriculture to promote the circulation of plant nutrients and reduce the need for chemical fertilizers (Şahin et al. 2004; Orhan et al. 2006).

Rhizosphere associated N-fixing bacteria have increasingly been used in nonlegume crop species such as sugar beet, sugar cane, rice, maize, and wheat (Döbereiner 1997; Hecht-Buchholz 1998; Şahin et al. 2004). For example, experiments with *Bacillus* species indicated yield increases in cereals (Belimov et al. 1995; Cakmakci et al. 2001; Öztürk et al. 2003) and maize (Pal 1998).

N-fixation is the first mechanism suggested to promote the growth of plants by *Azospirillum*. The majority of evidence collected during the last 3 decades concerning this mechanism has generated controversy (Bashan et al. 2004). At the same time, *Azospirillum* lead the list of PGPR assessed in worldwide experiments (Burdman et al. 2000; Dobbelaere et al. 2003; Vessey 2003; Lucy et al. 2004; Ramirez and Mellado 2005). *Pseudomonas* and *Bacillus* species (Alam et al. 2001; Cakmakci et al. 2001; Glick et al. 1994; Kokalis-Burelle et al. 2002), and the other PGPR and endophytic bacteria, such as *Enterobacter*, *Klebsiella*, *Burkholderia*, and *Stenotrophomonas*, have been gaining attention in the recent years, because of their association with important crops and potential to enhance the plant growth (Chelius and Triplett 2000; Sturz et al. 2001; Verma et al. 2001; Dong et al. 2003; Ramirez and Mellado 2005).

Some greenhouse and field experiments have shown repeatedly that the transfer of nitrogen fixed by *Azospirillum* spp. to the plant is not enough (Bashan and Holguin 1997; Kennedy et al. 1997; Kennedy and Chellapillai 1998; Bashan et al. 2004). Yet other studies showed that the bacteria cannot fulfil all of the nitrogen requirements of the plants; nevertheless, it can contribute only significant amounts of nitrogen. For example, seed inoculation of chickpea with *Rhizobium*, N-fixing *Bacillus subtilis* (OSU-142) significantly increased N percentage compared with the control treatment and may substitute costly N fertilizers in chickpea production even in cold highland areas (Elkoca et al. 2008).

Similarly, N-fixing bacterial strains *Pseudomonas putida* RC06, *Paenibacillus polymyxa* RC05 and RC14, and *Bacillus* OSU-142 have great potential, and as formulations, they are used as biofertilizers for better yield and the quality of wheat, sugar beet, and spinach growth (Cakmakci et al. 2007; Cakmakci et al. 2006). The N-fixing *Bacillus* strains and *A. brasilense* sp246 have a potential on plant growth activity of spring wheat and barley cultivation in organic and low-N input agriculture (Öztürk et al. 2003; Canbolat et al. 2006). Inoculation with the *Rhizobium leguminosarum* E11 and *Azotobacter* sp. S8, strain E11 increased root dry weight, root length, and growth in cotton (Hafeez et al. 2004). Significant positive effects on growth, nodule number, and yield of soybean were obtained after inoculation with *Bradyrhizobium* spp strains S62 and S63 (Egamberdiyeva et al. 2004).

Furthermore, inoculation commonly and significantly reduced the required doses of nitrogen fertilization in numerous greenhouse and field experiments in a lot of plant species (Bashan and Levanony 1990; Bashan and Holguin 1997; Bashan et al. 2004).

The strain(s), soil types, climate, and the development of appropriate formulations as well as strategies of field experimentations should be considered for a successful application of PGPR when using as fertilizers.

2.2 *Solubilization of Phosphates*

Phosphorous (P), next to nitrogen, is one of the major and key nutrients limiting plant growth (Kumar and Narula 1999; Sundara et al. 2002; Podile and Kishore 2006). Even in phosphorous rich soil, most of the P is unavailable for the plants, as large amount of soil P is found in its insoluble form. Phosphate solubilizing bacteria (PSB) are common in the rhizosphere and can be used to overcome this problem (Vessey 2003).

PSB secretes organic acids and phosphatases that converts the insoluble phosphates into soluble monobasic and dibasic ions and may also solubilize inorganic phosphate and makes soil phosphorus, which otherwise remain fixed, available to the plants (Kumar and Narula 1999; Whitelaw 2000; Gyaneshwar et al. 2002). In other words, phosphate solubilizing microorganisms convert insoluble phosphates into soluble forms through the process of acidification, chelation, exchange reactions, and production of gluconic acid (Rodriguez et al. 2004; Chung et al. 2005; Hameeda et al. 2008).

PSB are ubiquitous (Gyaneshwar et al. 2002), and *Bacillus*, *Enterobacter*, *Erwinia*, and *Pseudomonas* spp. are among the most potent strains (Podile and Kishore 2006). PSB is common in rhizospheres of crop plants, and few examples of beneficial association with phosphate solubilizing PGPR and plants include *B. megaterium* (M-3) and chickpea (Elkoca et al. 2008), *B. licheniformis* RC08 and *B. megaterium* RC07, and wheat and spinach (Cakmakci et al. 2007), *Enterobacter agglomerans* and tomato (Kim et al. 1998), *P. chlororaphis*, *P. putida*, and soybean (Cattelan et al. 1999), *Avena sativa* and PGPR strains isolated from the rhizosphere of forage (WenXing et al. 2008), *Serratia marcescens* EB 67, *Pseudomonas* sp. CDB 35, and maize (Hameeda et al. 2008).

In the controlled environment and in the field trials, single and dual N-fixing *B. subtilis* (OSU-142) and P-solubilizing *B. megaterium* (M-3) inoculations significantly increased all the parameters investigated in chickpea (plant height, shoot, root and nodule dry weight, N%, chlorophyll content, pod number, seed yield, total biomass yield, and seed protein content) compared with the control treatment, equal to or higher than N, P, and NP treatments (Elkoca et al. 2008).

In another research, Orhan et al. (2006) reported that plant growth promoting effects of two *Bacillus* strains OSU-142 (N-fixing) and M3 (N-fixing and phosphate solubilizing) were tested alone or in combinations of organically grown primocane fruiting raspberry (cv. Heritage) plants and a significant increase in yield (33.9 and 74.9%), cane length (13.6 and 15.0%), number of cluster per cane (25.4 and 28.7%), and number of berries per cane (25.1 and 36.0%) were observed when compared with that of the control.

Hameeda et al. (2008) reported that plant biomass increased with *Serratia marcescens* EB 67 and *Pseudomonas* sp CDB 35 under both glasshouse and field conditions. And also, seed treatment with EB 67 and CDB 35 increased the grain yield of field-grown maize by 85 and 64% compared with the uninoculated control.

Furthermore, four strains namely, *Arthrobacter aureofaciens*, *Phyllobacterium myrsinacearum*, *Rhodococcus erythropolis*, and *Delftia* sp. are being reported for the first time as PSB after confirming their capacity to solubilize considerable amount of tricalcium phosphate in the medium by secreting organic acids (Chen et al. 2006). Peix et al. (2001) notified that *Mesorhizobium mediterraneum* strain PECA21 was able to mobilize phosphorous efficiently in barley and chickpea when tricalcium phosphate was added to the soil. Also, treating with insoluble phosphates and inoculating with strain PECA21, the phosphorous content, dry matter, nitrogen, potassium, calcium, and magnesium content in both plants were significantly increased.

It was known that natural phosphate rocks have been identified as an alternative for P fertilizers. For example, there are almost 40 million tons of phosphatic rock deposits in India (Rodríguez and Fraga 1999), and this material should provide a cheap source of phosphate fertilizer for crop production (Halder et al. 1990); especially, should be considered in organic production of horticulture and the other crops.

2.3 Plant Growth Regulators

Several stages of plant growth and development such as cell elongation, cell division, tissue differentiation, and apical dominance are controlled by the plant hormones, especially auxins and cytokinins. The biosynthesis and the underlying mechanism of auxins and cytokinins action are subjects of intense investigation. Auxins and cytokinins can be synthesized by both the plants and the microorganisms. Although the role of phytohormone biosynthesis by microorganisms is not fully explained, it is stated that direct mechanisms of plant growth by PGPR include production of plant hormones such as auxins, cytokinins, GAs, and lowering of plant ethylene levels (Glick 1995; Costacurta and Vanderleyden 1995; Lucy et al. 2004). A list of examples of plant growth stimulating phytohormones produced by PGPR is given in Table 1.

Auxin, indole-3-acetic acid (IAA), is a quantitatively important phytohormone produced by a member of PGPR, and treatment with auxin-producing rhizobacteria increased the plant growth (Vessey 2003; Erturk et al. 2008). On the one hand, most beneficial bacteria such as *Rhizobium*, *Bradyrhizobium*, and *Azospirillum* synthesize IAA via the Indole-3-pyruvic acid (IPyA) pathway (Manulis et al. 1991; Costacurta and Vanderleyden 1995; Patten and Glick 1996; Burdman et al. 2000). On the other hand, some pathogenic bacteria such as *Pseudomonas syringae*, *Agrobacterium tumefaciens*, and *Erwinia herbicola* synthesize IAA predominantly via the indole-3-acetamide (IAM) pathway (Dobbelaere et al. 2003).

The role of IAA in the observed plant growth promotion was obtained by attempting to mimic the effect of the bacterium for the root growth by the direct application of IAA on the roots. Inoculation with *Bacillus* RC23, *Paenibacillus polymyxa* RC05, *B. subtilis* OSU142, *Bacillus* RC03, *Comamonas acidovorans* RC41, *B. megaterium* RC01, and *B. simplex* RC19 with tea (*Camellia sinensis)* cuttings enhanced rooting percentages when compared with control because of IAA production of bacteria. Similarly, treatments of hardwood stem cuttings of kiwifruit cv. Hayward, stem cuttings of two rose selections (ERS 14, *Rosa canina*, and ERS 15, *Rosa dumalis*), sour cherry (*Prunus cerasus*) softwood and semi-hardwood cuttings and *Pistacia vera* cuttings with *Agrobacterium rubi* (A1, A16, and A18) and *Bacillus subtilis* OSU142 promoted rooting ratio and increased the numbers of lateral roots (Ercisli et al. 2000; Ercisli et al. 2003; Esitken et al. 2003b; Ercisli et al. 2004; Orhan et al. 2007).

In addition, *Azospirillum* is not only capable of nitrogen fixation but also codes for plant growth hormone auxins (Elmerich 1984). Strains of *Azospirillum* showed that production depended on the type of culture media and availability of tryptophan as a precursor. *A. brasilense* Cd produced the highest level of IAA among the

Table 1 Examples of plant growth stimulating phytohormones produced by PGPR

Phytohormones	PGPR	References
Gibberellin	*Acetobacter diazotropicus*	
	Herbospirillum seropedicae	Bastian et al. (1998)
	Bacillus lichenifirmis	
	Bacillus pumilus	Gutierrez-Manero et al. (2001)
	Bacillus cereus MJ-1	
	Bacillus macroides CJ-29	Joo et al. (2004)
	Bacillus pumilus CJ-69	
IAA	*Agrobacterium* sp.	
	Alcaligenes piechaudii	Barazani and Friedman (1999)
	Comamonas acidovorans	
	Azospirillum brasilense	Kaushik et al. (2000)
	Aeromonas veronii	
	Enterobacter cloacae	Mehnaz et al. (2001)
	Enterobacter sp.	Mirza et al. (2001)
	Comamonas acidovorans RC41	
	Paenibacillus polymyxa RC05	
	Bacillus RC23	Erturk et al. (2008)
	Bacillus simplex RC19	
	Bacillus RC03	
	Bacillus megaterium RC01	
Cytokinin	*Paenibacillus polymyxa*	Timmusk et al. (1999)
	Pseudomonas fluorescens	de Salamone et al. (2001)
		Bent et al. (2001)
ACC deaminase	*Pseudomonas putida*	Mayak et al. (1999)
	Pseudomonas cepacia	Cattelan et al. (1999)
	Enterobacter cloacae	Saleh and Glick (2001)
	Pseudomonas brassicacearum Am3	Belimov et al. (2007)
	Variovorax paradoxus 5C-2	Belimov et al. (2009)
	Pseudomonas putida Biovar B	Rodriguez et al. (2008)
	Pseudomonas putida N21	
	Pseudomonas aeruginosa N39	Zahir et al. (2009)
	Serratia proteamaculans M35	

Azospirillum strains tested (El-Khawas and Adachi 1999; Radwan 1998; Bashan et al. 2004).

The isolation and quantification of cytokinins in nonpathogenic soil bacteria in general and diazotrophic bacteria in particular has received a little attention. Cytokinins are a diverse group of labile compounds that are usually presented in small amounts in biological samples and are often difficult to identify and quantify (Dobbelaere et al. 2003).

Cytokinins are produced by bacteria such as *Azospirillum* and *Pseudomonas* spp. (Gaudin et al. 1994). Moreover, a few PGPR strains were reported to produce cytokinins, such as *Rhizobium leguminosarum*, *Paenibacillus polymyxa*, and *Pseudomonas fluorescens* (Noel et al. 1996; Timmusk et al. 1999; de Salamone et al. 2001; Bent et al. 2001; Vessey 2003). These studies sufficiently cloud the production of cytokinins, compared with IAA or GAs, in PGPR. Also, it appears that more work is necessary before proving for the role of PGPR-produced cytokinins in plant growth promotion.

Also in the case of GAs, the bacterial genetic determinants have not been identified so far. Therefore, no mutants are available to demonstrate the role of this phytohormone in plant growth promotion (Dobbelaere et al. 2003). Also the evidence of GA production by PGPR is rare (Vessey 2003). On the other hand, PGPR such as *R. phaseoli*, *A. lipoferum*, *Azotospirillum brasilense*, *Acetobacter diazotropicus*, *Herbospirillum seropedicae*, *Bacillus licheniformis*, *B. pumilus*, *Bacillus cereus* MJ-1, *Bacillus macroides* CJ-29 were reported to produce GAs (Atzhorn et al. 1988; Bottini et al. 1989; Janzen et al. 1992; Bastian et al. 1998; Gutierrez-Manero et al. 2001; Joo et al. 2004 and Table 1). However, this is not a strong evidence of GA production in a common method of growth promotion by PGPR.

Nevertheless, in recent studies, Gutierrez-Manero et al. (2001) provide an evidence that four different forms of GAs are produced by *B. pumilus* and *Bacillus licheniformis*. Inoculation of alder (*Alnus glutinosa*) with these PGPR could effectively reverse a chemically induced inhibition of stem growth. In addition to this research, Joo et al. (2004) reported that the growth of red pepper plug seedlings was increased by *Bacillus cereus* MJ-1, *B. macroides* CJ-29, and *B. pumilus* CJ-69, though the number of leaves and stem diameter were not significantly changed. The greatest increase is in the height and the root fresh weight of the seedlings was by *B. pumilus*, which could increase the height by 12% and the root fresh weight by 20%.

In the last few years, a new mechanism of plant growth promotion involving ethylene has been proposed (Burdman et al. 2000). Showing that some soil bacteria contain 1-aminocyclopropane-1-carboxylate (ACC) deaminase (Klee et al. 1991) and Glick et al. (1998) put forward the theory that the mode of action of some PGPR was the production of 1-aminocyclopropane-1-carboxylate (ACC) deaminase, an enzyme that could cleave ACC, the immediate precursor to ethylene in the biosynthetic pathway for ethylene in plants. They submitted that ACC deaminase activity would decrease ethylene production in the roots of host plants and results in root lengthening. In some cases, the growth promotion effects of ACC deaminase-producing PGPR is the best expressed in stress conditions including drought (Zahir et al. 2008) and salt (Nadeem et al. 2007; Zahir et al. 2009) stress.

PGPR (containing ACC deaminase) boost plant growth particularly under stressed conditions by the regulation of accelerated ethylene production in response to a multitude of abiotic and biotic stresses such as salinity, drought, waterlogging, temperature, pathogenicity, and contaminants (Saleem et al. 2007). For example, under salinity stress, 1-aminocyclopropane-1-carboxylic acid-deaminase activity of *P. putida* (N21), *P. aeruginosa* (N39) and *Serratia proteamaculans* (M35) might have caused reduction in the synthesis of stress (salt)-induced inhibitory levels of ethylene (Zahir et al. 2009). Similarly, inoculation with *Variovorax paradoxus* 5C-2 improved growth, yield, and water-use efficiency of droughty peas (Belimov et al. 2009). It is reported that inoculation with *P. fluorescens* was found to be more effective in promoting root growth than that with *P. putida* as it caused up to 46% increase in root elongation and up to 94% increase in root weight of pea over the respective uninoculated drought stressed control (Arshad et al. 2008).

In addition to stress factors, recent studies indicated that canola plants inoculated with the *P. putida* strain HS-2 produced an increase in plant biomass (Rodriguez et al. 2008). The ACC-utilizing PGPR *Pseudomonas brassicacearum* strain Am3 increased in-vitro root elongation and root biomass of soil-grown tomato cv. Ailsa Craig at low bacterial concentrations but had negative effects on in-vitro root elongation at higher bacterial concentrations (Belimov et al. 2007).

2.4 Effects on Plant Growth

Since the last few decades, the response of agriculturally important crops to inoculation with PGPR was investigated in numerous field and greenhouse experiments carried out in various countries. On the basis of the given data, it was concluded that inoculation with PGPR resulted in significant yield increases in different crops, enhanced rooting of hardwood and semi-hardwood cuttings, seed germination and emergence under different conditions. In other words, they can affect plant growth and yield in a number of ways and enhancement of vegetative and reproductive growth is documented in a range of crops such as cereals or vegetables. Treatments with PGPR increase germination percentage, seedling vigor, emergence, plant stand, root and shoot growth, total biomass of the plants, seed weight, early flowering, grains, fodder and fruit yields, etc., (van Loon et al. 1998; Ramamoorthy et al. 2001). Applications of PGPR in relation to the plant growth in different subjects are described later with recent studies.

2.4.1 Yield and Yield Components

In crop production, there is a continuous demand of increasing crop productivity and quality. There are lot of agricultural practices applied for increasing the yield and the yield components. Recently, one of them is applications of PGPR for increasing yield and environment friendly crop production.

Floral and foliar applications of PGPR strains *Pseudomonas* BA-8 and *Bacillus* OSU-142 on apple trees significantly increased yield per trunk cross-section area (13.3–118.5%), fruit weight (4.2–7.5%), shoot length (20.8–30.1%), and shoot diameter (9.0–19.8%) in "Starkrimson" and yield per trunk cross-sectional area (TCSA; 14.9%) and fruit weight (6.5–8.7%) in "Granny Smith" compared with the control (Pırlak et al. 2007). Karlıdağ et al. (2007) reported similar results in apple. Thus, *Bacillus* M3 and/or OSU-142 and/or *Microbacterium* FS01 in combination have the potential to increase the yield and growth of apple trees.

In addition, Esitken et al. (2003a, 2005, 2006) and Orhan et al. (2006) reported that *Pseudomonas* BA-8, *Bacillus* OSU-142 and M3 increased the shoot length, crop yield and improved fruit quality of apricot, sweet cherry, and raspberry.

In another research, Cakmakci et al. (2006) suggested that in the greenhouse, inoculations with PGPR increased sugar beet root weight by 2.8–46.7% depending on the species. Leaf, root, and sugar yield were increased by the bacterial inoculation by 15.5–20.8%, 12.3–16.1%, and 9.8–14.7%, respectively. Effective *Bacillus* species, such as OSU-142, RC07 and M-13, *Paenibacillus polymyxa* RC05, *P. putida* RC06, and *Rhodobacter capsulatus* RC04 may be used in organic and sustainable sugar beet agriculture.

The average weight of tomato fruit per plant treated with *Rhodopseudomonas* sp KL9 strain (82.7 g) was higher than those of others including the uninoculated control. The content of lycopene in the ripe tomato fruit increased by 48.3% with the application of *Rhodopseudomonas* sp. KL9, but *Rhodopseudomonas* sp BL6 did not show any effect on lycopene content although the lycopene content in the cells of *Rhodopseudomonas* sp BL6 were 1.12 mg/g (Lee et al. 2008a).

Dursun et al. (2008) reported that the highest rocket yield, average leaf weight, leaf length, leaf stem diameter, leaf area and root weight were obtained from *Pseudomonas* BA-7 applications when compared with *P. putidae* BA-8, *B. subtilis* OSU-142 and MFD-5, *B. megatorium* M3, *A. rubi* A-1, A-16, and A-18. The highest leaf number (8.23), leaf dry matter (6.70%), and root dry matter (11.85%) were determined in A-18, OSU-142 and MFD-5 applications, respectively, and especially *Burkholderia gladii* BA-7, *Pseudomonas* BA-8, and *Bacillus* OSU-142 have a great potential to increase the parameters of plant growth of rocket.

Although the examples of relations between the yield and PGPR applications can be increased, other recent studies such as de Freitas (2000), Herman et al. (2008), and Yıldırım et al. (2008) clearly demonstrated the potential of PGPR in increasing the plant growth and yield.

2.4.2 Seed Germination and Emergence

Sivritepe and Dourado (1995) reported that priming (osmoconditioning) is one of the physiological methods, which improves seed performance and provides faster and synchronized germination in vegetables. However, bio-priming with different genera, especially PGPR, have a great potential over other priming methods.

Nelson (2004) noted that PGPR were able to exert a beneficial effect upon plant growth such as increase in seed germination rate and percentage. Rodriguez et al. (2001) reported that using *Azospirillum* spp. gave better germination in both tomato and pepper seeds. Also, Vargas et al. (2001) mentioned that *Hafnia alvei* strain P3 increased germination by 36.5% when compared with the control in lettuce and inoculation of the soybean plants either with *Pseudomonas* strain PMZ2 or with *B. japonicum* increased seed emergence (Zaidi 2003). Similarly, Basavaraju et al. (2002) reported that inoculation of *Azotobacter chroococcum* strain C2 significantly increased the germination percentage in radish. The greenhouse inoculation experiment with pepper and maize pointed out that *Azotobacter* sp. strains 17 and 20 promoted pepper germination, while the *Azospirillum* strains 1 and 23 promoted maize germination (Reyes et al. 2008). Although studies were mentioned about the effect of bacterial strains on germination of different vegetable species that were conducted out under optimum conditions, Kaymak et al. (2009) suggested that bio-priming with *A. rubi* strain A16, *Burkholderia gladii* strain BA7, *P. putida* strain BA8,

B. subtilis strain BA142, *B. megaterium* strain M3 under saline stress could be useful to obtain higher seed germination percentage in radish.

Also, PGPR can be used under pathogenic factor. Thus, different isolates of plant growth-promoting rhizobacteria (i.e., *B. pumilus* (INR-7), *B. subtilis* (GBO-3), *B. subtilis* (IN937b), *B. pumilus* (SE-34), *Brevibacillus brevis* (IPC-11), *B. pumilus* (T-4), and *B. amyloliquefaciens* (IN937a)) were used for seed treatment to suppress the seedling diseases caused by fungi. Among them, isolates GBO3, IPC-11, and INR-7 increased seed germination and seedling vigour to the greatest extent (Lokesh et al. 2007). Alike, Begum et al. (2003) reported that PGPR, *B. pumilus* (SE-34), *B. pasteurii* (T4), *B. subtilis* (IN937-b), and *B. subtilis* (GBO3) strains reduced the incidence of seed mycoflora, which indirectly enhanced the seed germination percentage and vigour index of the seedlings in okra. In another recent study, de Araujo (2008) reported that the inoculation of seeds with *B. subtilis* is a promising technological alternative for seed treatment owing to the fact that inoculation with *B. subtilis*, formulated with oyster meal, increased emergence in cotton and soybean.

2.4.3 Rooting of Cuttings

There are many physiological and environmental factors that influence root formation, with exogenous treatments on cuttings being particularly important (Couvillon 1998). Growers have attempted to stimulate rooting by applying growth regulators, various chemical substances, etc. However, the use of chemicals can produce environmental problems and increase proportion costs. Ecological problems have raised interest in environmental friendly sustainable agricultural practices (Salantur et al. 2005). Therefore, use of PGPR can overcome such problems associated with environment (Kaymak et al. 2008).

Recent studies showed that bacteria in several genera (*Agrobacterium*, *Bacillus*, *Streptomyces*, *Pseudomonas*, and *Alcaligenes*) induce root formation and growth in stem cuttings (Bassil et al. 1991; Hatta et al. 1996; Rinallo et al. 1999). More recently, PGPR such as *A. rubi* (A1, A16 and A18), *B. subtilis* (OSU142), *Bacillus* (BA16, RC03, RC23), *B. gladii* (BA7), *P. putida* (BA8), *B. megatorium* (M3 and RC01), *Paenibacillus polymyxa* (RC05), *Comamonas acidovorans* RC41, and *B. simplex* RC19 were effectively used for both hardwood and semi-hardwood cuttings to obtain higher rooting percentages in sour cherry (Ercisli et al. 2000; Esitken et al. 2003b), kiwifruit (Ercisli et al. 2003), grapevine (Köse et al. 2003), rose (Ercisli et al. 2004), pistachio (Orhan et al. 2006), tea (*Camellia sinensis* var. *Sinensis*) (Erturk et al. 2008), and mint (*Mentha piperita* L.) (Fig. 1) (Kaymak et al. 2008).

2.4.4 Nutrient Uptake

Living plants require 16 essential elements to survive. Three of 16 elements (carbon, hydrogen, and oxygen) are supplied primarily from air and water. The remaining 13 are normally absorbed by plant roots. Each of these essential elements has at least one specially defined role in plant growth (Swaider et al. 1992; Decateau 2000).

Fig. 1 Effect of inoculation with PGPR (*Agrobacterium rubi* A16, *Burkholderia gladii* BA7, *Peseudomonas putida* BA8, *Bacillus subtilis* OSU142, and *Bacillus megatorium* M3) on root formation of mint cuttings

PGPR have been promised as a component in approaching for maintaining adequate plant nutrition and reducing the negative environmental effects of fertilizers. PGPR might increase nutrient uptake from soils, thus reducing the need for fertilizers and preventing the accumulation of nitrates and phosphates in agricultural soils (Yang et al. 2009). It is known that phosphorous and nitrogen is the major and key nutrients limiting plant growth and important macronutrient required for plant growth (Kumar and Narula 1999; Sundara et al. 2002; Podile and Kishore 2006).

Additionally, some PGPR promote root development (Mantelin and Touraine 2004) by the production of phytohormones such as indole acetic acid (Kloepper et al. 2007). Given that root tips and root surfaces are sites of nutrient uptake, it is likely that one mechanism by which PGPR lead to increased nutrient uptake is via stimulation of root development (Yang et al. 2009). It has also been suggested that PGPR increase uptake of mineral ions via stimulation of the proton pump ATPase (Mantelin and Touraine 2004), although experimental evidence for this is lacking (Yang et al. 2009).

Several studies can be given about the relations with PGPR and enhancement of nutrient uptake. For example, Naveed et al. (2008) notified that PGPR application significantly enhanced N, P, and K uptakes. The *Pseudomonas fluorescens* biotype G (N-3) was found to be the best in increasing the grain yield of maize and nutrient uptake. In addition, the inoculation process with *Azospirillum* and *Bacillus* spp. showed positive response in enhancing higher accumulation of nitrogen, phosphorus, and potassium in the plant tissues, enhanced root dry weight and top growth of the oil palm seedlings under field nursery conditions (Amir et al. 2005).

In other recent study, Dursun et al. (2008) reported that *Burkholderia gladii* BA-7, *P putidita* BA-8, *B. subtilis* OSU-142 and MFD-5, *B. megaterium* M3, *A. rubi* A-1, A-16, and A-18 applications increased mineral contents particularly N, K, P, Zn, Fe, Mn, Na, Ca, and Mg in rocket leaves when compared with the control.

In a study aimed at assessment of effects of foliar application of bacteria *Bacillus* OSU-142, *Burkholderia* OSU-7, and *Pseudomonas* BA-8 on yield and growth of apricot, it was stated that application of bacteria resulted in an increase of N, P, K, Ca, and Mg contents of leaves (Esitken et al. 2005). In a similar study, Esitken et al. (2003a) suggested that N, P, K, Ca, and Mg contents of leaves were higher on OSU 142-treated trees than on the untreated control and OSU 142 has the potential to increase the yield of apricot trees.

Therefore, PGPR contributed significantly to the reducing nutrient build up in the soil. Several studies are underway that will further define the utility of PGPR in nutrient management strategies aimed at reducing fertilizer application rates and nutrient runoff from agricultural sources (Yang et al. 2009; Kumar et al. 2009).

3 Indirect Plant Growth Promotion

Induced systemic resistance (ISR), antibiosis, competition for nutrients, parasitism, production of metabolites suppressive to deleterious rhizobacteria are some of the mechanism that indirectly benefit plant growth.

3.1 Induced Systemic Resistance

More recently, biological control has been considered as an alternative strategy to manage soil-borne plant diseases. Available literature revealed positive effects of specific strains of rhizobacteria on growth of many plant species in soils in which more or less defined pathogens cause substantial losses. For this reason, several rhizobacteria have extensively been used as biological agents to control many soil-borne plant pathogens (Jeun et al. 2004; Dell'Amico et al. 2005; Rajkumar et al. 2005).

A strain, *P. fluorescens* WCS417, active against *Fusarium oxysporum* f. sp. *dianthi* was tested on carnation and results showed that bacteria, while remaining confined to the plant root system, were still protective when the pathogen was slash-inoculated into the stem (Van Peer et al. 1991). This protective effect had to be plant-mediated because in this case the rhizobacteria and the pathogenic fungus were never found to contact each other on the plant (Van Loon and Bakker 2006). Several strains of PGPR, which applied to roots of cucumber, and the leaves were subsequently challenged inoculation with the anthracnose fungus *Colletotrichum orbiculare* (Gang et al. 1991). The phenomenon was called ISR. (Van Loon et al. 1998; Vallad and Goodman 2004; Van Loon and Bakker 2006; Choudhary et al. 2007)

It is thought that the inducing rhizobacteria in the plant roots produce signal, which spreads systemically within the plant and increases the defensive capacity of the distant tissues from the subsequent infection by the pathogens. ISR thus extended the protective action of PGPR from their antagonistic activity against soil-borne pathogens in the rhizosphere to a defense-stimulating effect above the surface of the ground tissues against foliar pathogens (Van Loon and Bakker 2006).

ISR appears phenotypically similar to SAR, which is the phenomenon that once a plant has been infected by a pathogen and been able to effectively resist it, it has become more resistant to subsequent challenge inoculation by the same and other pathogens and, in some instances, even insects (Sticher et al. 1997; Van Loon et al. 1998; Van Loon and Bakker 2006). SAR occurs in distal plant parts following localized infection by a necrotizing pathogen. It is controlled by a signaling pathway that depends upon the accumulation of salicylic acid and the regulatory protein NPR1. In contrast, ISR is induced by selected strains of nonpathogenic PGPR. ISR functions independent from SA, but requires NPR1 and is regulated by jasmonic acid and ethylene (Walters and Heil 2007).

To reduce crop loss, pesticides are generally used. They are cost-effective and thus have become an integral part of modern agriculture. Environmental and human health-related concerns associated with use of hazardous chemicals have necessitated the search for eco-friendly alternatives. Such approaches must enhance and sustain agricultural productivity and at the same time be safe from environmental and health perspectives (Raj et al. 2003a).

Therefore, for economic reasons biological crop protectants can only seldom compete with highly effective chemicals. However, ISR is only one of the mechanisms that may be mobilized to counteract plant pathogens in an environmentally friendly and durable way. Integrating ISR-triggering PGPR into disease management programs in conjunction with other strategies will be a worthwhile approach to explore (Van Loon and Bakker 2006).

3.2 Suppression of Plant Diseases, Insects, and Nematodes by PGPR

Biocontrol is the process by which a pathogenic organism is maintained at low inoculum density or controlled or eradicated by beneficial organisms. Several microorganisms such as PGPR and insects present in the natural environment serve as potential biocontrol agents.

3.2.1 Bacterial Plant Diseases

The bacteria associated with plants exist as epiphytes, endophytes, and pathogens. Phytopathogens are comparatively few in both type and number, and all bacterial phytopathogens described to date fall within the domain Bacteria, formerly known as the *Eubacteria*. Bacterial phytopathogens that possesses a cell wall can be subdivided into Gram-positive (*Clavibacter*, *Curtobacterium*, *Rathayibacter*, and *Streptomyces*) and Gram-negative (*Acidovorax*, *Agrobacterium*, *Burkholderia*, *Enterobacter*, *Erwinia*, *Pantoea*, *Pseudomonas*, *Ralstonia*, and *Xanthomonas*) (Saddler 2002).

Bacterial soft rot of vegetables; blackleg of potato; fire blight of pome fruits; angular leaf spot or black arm, of cotton; bacterial blights of bean, lack rot of crucifers, southern bacterial wilt, bacterial wilt of cucurbits, ring rot of potato, bacterial canker of tomato, crown gall, hairy root, and cane gall, and common scab of potato are the more common bacterial diseases (Walker 1957; Waller et al. 2002).

Several cultural practises such as crop rotation, mixed cropping and intercropping, selection of cultivar, tillage, planting time, fertilization and irrigation, or highly effective chemical substances affect some diseases in different ways depending on the form of their application (Termorshuizen 2002). Recently, many microorganisms are increasingly used as inoculants for biocontrol (Romero et al. 2003; Chinnasamy 2005; Aliye et al. 2008; Xue et al. 2009). PGPR are nonpathogenic, environmental-friendly, cheaper to produce and easy to handle, and may create long-lasting effects (Chinnasamy 2005).

For instance, tomato is prone to a number of bacterial diseases, among which bacterial canker disease caused by *Clavibacter michiganensis* ssp. *michiganensis* is one of the most important diseases and nearly 100% crop loss can occur (Boudyach et al. 2001; Umesha 2006). Utkhede and Koch (2004) reported that treatments with *B. subtilis* (Quadra 136 and 137) and *Trichoderma harzianum* (R), *Rhodosporidium diobovatum* (S33), applied as a spray at 0.3, 0.6, 10 g^{-1}, have the ability to prevent the incidence of bacterial canker of tomato plants caused by *C. michiganensis* subsp. *michiganensis* under greenhouse conditions. Similarly, tomato seeds were treated with PGPR strains *B. subtilis* GBO3, *B. amyloliquefaciens* IN937a and *Brevibacillus brevis* IPC11 were recorded for maximum disease protection for bacterial canker under greenhouse conditions (Girish and Umesha 2005). Recent studies about the relations with bacterial diseases and PGPR are given in Table 2.

3.2.2 Fungal Plant Diseases

Fungal pathogens found on plants can be classified in different taxonomic groups. A few fungal pathogens such as rusts, powdery and downy mildews are obligate parasites. However, most of the plant pathogens are necrotrophs, killing plant tissues for their nutrition (Waller and Cannon 2002).

Exclusion or eradication of a disease from production areas, highly effective chemical substances or biological control of plant diseases have been suggested to protect the plants from fungal pathogens. Recently, PGPRs are increasingly and extensively used in biological control of fungal plant diseases (Altindag et al. 2006; Lourenco et al. 2006; Saravanakumar et al. 2007; Akgul and Mirik 2008; Sang et al. 2008; Dutta et al. 2008).

For example, apricot is the most important fruit crop grown in Anatolia, with approximately 600,000 tons of fruit produced annually, and Turkey dominates

Table 2 Examples of suppression of bacterial diseases by PGPR in different plant species

Phytopathogens	Species	PGPR	References
Pseudomonas syringae pv. *lachrymans*	Cucumber	*Pseudomonas putida* 89B-27 *Serratia marcescens* 90–166	Liu et al. (1995)
Pseudomonas syringae pv. *glycinea*	Soy bean	*Pseudomonas* sp. *Erwinia herbicola*	May et al. (1996)
Xanthomonas albilineans	Sugar cane	*Pentoena dispersa*	Zhang and Birch (1997)
Erwinia amylovora	Apple	*Erwinia herbicola* C9-1 *Pseudomonas fluorescens* A506	Pusey (1997)
		Single-strain treatments and three-way mixture of *Bacillus pumilus* INR7, *Curtobacterium flaccumfaciens* ME1 and *Bacillus subtilis* GB03	Raupach and Kloepper (1998, 2000)
Ralstonia solanacearum	Tomato	*Bacillus subtilis* B2G	Lemessa and Zeller (2007)
		Pseudomonas sp. (APF1) *Acinetobacter* sp. (Xa6) *Enterobacter* sp. (Xy3)	Xue et al. (2009)
Clavibacter michiganensis subsp. *michiganensis*		*Azospirillum brasilense* Sp7	Romero et al. (2003)
Xanthomonas campestris pv. *vesicatoria*		*Azospirillum* sp. (BNM-65)	
Ralstonia solanacearum	Eucalyptus	*Pseudomonas fluorescens* WCS417r *Pseudomonas putida* WCS358r	Ran et al. (2005)
	Potato	*Bacillus subtilis* PFMRI *Paenibacillus macerans* BS-DFS and PF9	Aliye et al. (2008)
Xanthomonas axonopodis pv. *malvacearum*	Cotton	*Bacillus cereus* MT5-5, MT5-6, L2-1 *Achromobacter xylosoxidans* L2-2, *Brevibacterium* sp. MT5-11	Ishida et al. (2008)

apricot production in the world (Ercisli 2009). Therewithal, brown rot caused by *Moniliana laxa* Ehr. is one of the most destructive diseases of apricot in Turkey. This pathogen is able to destroy the whole annual crop in the phase of blossom, although it can kill shoots up to 30 cm beyond the initial blossom infection, and management of brown rot in Turkey is in general carried out by fungicide application (Gulcan et al. 1999). Altindag et al. (2006) suggested that *Burkholdria gladii* OSU 7 has the potential to be used as biopesticide for effective management of brown rot disease on apricot.

Similarly, pepper (*Capsicum annum* L.) is one of the most important market vegetables grown worldwide, but the yield and quality of marketable peppers are frequently limited by Phytophthora blight. The incidence of this disease has continued to increase production areas since the pathogen can infect roots, crowns, and even foliar parts of pepper plants through splashing rains or overhead irrigation waters (Ristaino and Johnston 1999; Hausbeck and Lamour 2004). Control of this disease has usually depended on chemical and cultural measures such as use of phenylamide fungicides or metalaxyl as well as crop rotation, soil amendments, use of protective mulches and water management (Matheron and Porchas 2000; Hausbeck and Lamour 2004). In a recent study, Sang et al. (2008) reported that *Pseudomonas corrugata* (CCR04 and CCR80), *Chryseobacterium indologenes* (ISE14), and *Flavobacterium* sp. (GSE09) showed consistently good control efficacy against *Phytophthora capsici*. Also, these strains could be applied by either drench or root-dip treatment as alternatives to agricultural chemicals to control Phytophthora blight of pepper. In another recent study, Akgul and Mirik (2008) also reported that *Bacillus megaterium* strains could be used for biocontrol of *Phytophthora capsici*.

The combination of *Pseudomonas* strains Pf1, TDK1, and PY15 was more effective in reducing sheath rot (*Sarocladium oryzae*) disease in rice plants compared with individual strains under glasshouse and field conditions (Saravanakumar et al. 2009).

Hernandez-Rodriguez et al. (2008) obtained that *Burkholderia* sp. MBf21, MBp1, MBf15, and *P. fluorescens* MPp4 stood out for their plant growth stimulation in maize and for the biological control exerted on *Fusarium verticillioides* M1. The strains *Burkholderia* sp. MBf21 and MBf15 showed the best results in disease suppression, which was achieved up to 80%.

The combined use of PGPR (*Bacillus cereus* strain BS 03 and a *Pseudomonas aeruginosa* strain RRLJ 04) and rhizobia (strain RH 2) were recommended for induction of systemic resistance against fusarial wilt (*Fusarium udum*) in pigeon pea (Dutta et al. 2008). Recent studies and more examples about the suppression of fungal diseases by PGPR are given in Table 3.

3.2.3 Viral Plant Diseases

Viruses are obligate parasites of submicroscopic size, with one dimension smaller than 200 nm. Virus particles, or virions, consist of segments of double or single-stranded RNA or DNA encased in protein structures, in some cases with lipid and additional substances (Waller 2002). So far at least 700 plant viruses has been discovered, many of which cause catastrophic diseases and have wide host ranges. They have been classified into three families and 32 groups (Martelli 1992; Waller 2002).

Some chemicals are used to produce virus-free plant material because they inhibit virus replication in agricultural crops. However, there are no therapeutic agents or viricides that can be applied to plants to control virus diseases. Consequently, control measures are based mainly on avoiding infection by using host plant resistance or disrupting the epidemic cycle of the disease. The use of

Table 3 Examples of suppression of fungal diseases by PGPR in different plant species

Phytopathogens	Species	PGPR	References
Rhizoctonia solani sclerotia *Fusarium oxysporum* f. sp. *cyclaminis*	Cyclamen	*Serratia marcescens* B2	Someya et al. (2000)
Fusarium oxysporium	Soybean	*Pseudomonas* PMZ2 *Bradyrhizobium japonicum*	Zaidi (2003)
Sclerospora graminicola	Pearl millet	*Bacillus pumilus* INR7 and SE34 *Bacillus subtilis* GB03	Raj et al. (2003b)
		Pseudomonas fluorescens UOM SAR 14	Raj et al. (2004)
Cronartium quercuum f.sp. *fusiforme*	Loblolly pine	*Bacillus pumilus* SE34 and T4	Enebak and Carey (2004)
Puccinia psidii	Eucalyptus	*Pseudomonas aeruginosa* FL2 *Pseudomonas* sp. MF4	Teixeira et al. (2005)
Didymella bryoniae	Muskmelon	Pseudomonas fluorescens	Sudisha et al. (2006)
Pythium ultimum, *Pythium debaryanum*, *Rhizoctonia solani*, *Fusarium oxysporum*, *Phytophthora capsici*, *Botrytis cinerea*, *Botrytis allii*, *Cladosporium fulvum*, *Aspergillus* sp.	Sesame (in vitro)	*Paenibacillus polymyxa* E681	Ryu et al. (2006)
Exobasidium vexans	Tea	*Pseudomonas fluorescens* Pf1	Saravanakumar et al. (2007)
Fusarium spp. *Didymella bryoniae* *Myrothecium* spp.	Watermelon	*Bacillus subtilis* GBO-3 and *Brevibacillus brevis* IPC-11 *Bacillus pumilus* SE34 and T4	Lokesh et al. (2007)
Fusarium oxysporum f. sp. *lycopersici*	Tomato	*Paenibacillus lentimorbus* GBR158	Son et al. (2008)
Phytopthora capsici	Red pepper	*Bacillus subtilis* R33 and R13	Lee et al. (2008b)
Phytopthora capsici	Chili pepper	*Paenibacillus polymyxa* GBR-462	Kim et al. (2009)
Fusarium oxysporum L sp. *lycopersici*	Tomato	*Azospirillum brasilense* *Bacillus subtilis*	Abo-Elyousr and Mohamed (2009)
Rhizoctonia solani	Wheat	*Azotobacter* sp. WPR-51	Fatima et al. (2009)

genetically resistant cultivars provides effective control of many viral diseases. Mechanisms of resistance vary, some are explained to effects on vectors, whereas others may inhibit viral replication (Waller 2002).

Kirankumar et al. (2008) reported that *Pseudomonas* B-25 was highly efficient in promoting growth, fruit yield, and nutrient uptake of tomato in the presence of tobacco mosaic virus (TMV) pathogen, and the incidence of pathogenesis was markedly less after PGPR treatment. Similarly, biological control using PGPR protected tomato plants against cucumber mosaic virus (CMV) under greenhouse and to a limited extent in the field conditions (Sikora and Murphy 2005). In another research, *P. fluorescens* strains were investigated for biocontrol efficacy against tomato spotted wilt virus (TSWV) in tomato. Virus concentration value clearly showed a reduction in viral antigen concentration in *P. fluorescens*-treated tomato plants corresponding to reduced disease ratings. All the *P. fluorescens*-treated tomato plants also showed enhanced growth and yield compared with control plants. Hence, it was suggested that PGPR could play a major role in reducing TSWV and increasing yield in tomato plants (Kandan et al. 2005). Banana bunchy top disease (BBTD) caused by Banana bunchy top virus (BBTV) is the most serious virus disease of banana plantations world wide. *P. fluorescens* Pf1 and CHA0 were significantly effective in reducing BBTV under field conditions, recording 33.33% infection with 60% reduction over control (Harish et al. 2008).

In a greenhouse experiment, *P. fluorescens* FB11 and *Rhizobium leguminosarum* FBG05 were tested alone and in combination as seed inoculants to induce systemic resistance in faba bean against bean yellow mosaic potyvirus (BYMV). The results demonstrated that BYMV challenged plants emerged from *Pseudomonas* inoculated seeds not only showed a pronounced and significant reduction in percent disease incidence but also a significant reduction in virus concentration in the challenged plants, compared with the nonbacterized seeds. *Rhizobium* alone also showed a significant reduction in both in percent disease incidence and in viral concentration value, but the reduction was less pronounced than that resulting from *Pseudomonas* inoculation (Elbadry et al. 2006).

In a recent study, the PGPR combinations (combinations included *B. subtilis* GB03 and IN937b, *B. pumilus* SE34, INR7 and T4, *B. amyloliquefaciens* IN937a) formulated with chitosan were referred to as biopreparations. The result indicated that treatment of tomato plants with biopreparations resulted in significant enhancement of plant growth and protection against infection by CMV (Murphy et al. 2003). Zehnder et al. (2000) reported that CMV symptom development was significantly reduced on PGPR-treated (*B. pumilus* SE34, *Kluyvera cryocrescens* IN114, *B. amyloliquefaciens* IN937a, and *B. subtilus* IN937b) plants compared with control, but the percentage of infected plants and tomato yields were not significantly different among treatments, suggested that PGPR-mediated induced resistance against CMV infection following mechanical inoculation into tomato can be maintained under field conditions.

Tomato plants treated with PGPR (*B. amyloliquefaciens* 937a, *B. subtills* 937b, and *B. pumilus* SE34), applied as an industrially formulated seed treatment, a spore preparation mixed with potting medium (referred to as powder), or a combined seed-powder treatment, were evaluated under field conditions for induced resistance to tomato mottle virus (ToMoV), resulted in reduced ToMoV incidence and disease severity. In some cases, a corresponding increase in fruit yield was observed. The use of PGPR could become a component of an integrated program for management of this virus in tomato (Murphy et al. 2000)

It was known that there are no highly effective chemical substances that can be applied to plants to control viral disease of agricultural or horticultural crops. For exclusion or eradication of a viral disease from production areas, highly effective chemical substances cannot be suggested; however, biological control with PGPR may be suggested to protect these areas or plants from viral pathogens. Nevertheless, it is recommended that more work must be conducted because of the complexity and variability of virus diseases.

3.2.4 Nematodes

Plant–parasitic nematodes cause serious crop losses in production areas,e.g., yield loss of tomato due to root-knot nematodes (*Meloidogyne* spp.) ranges from 39.7 to 46.0% in India (Reddy 1985), and are among the most important agricultural pests (Koenning et al. 1999; Siddiqui and Akhtar 2008). The control of nematodes is difficult because nematodes mostly inhabit the soil and generally attack and settle around or inside the roots of the plants. During the last few decades, plant disease control has been based largely on the use of chemicals (Siddiqui et al. 2001). Although chemical nematicides are effective, easy to apply, and show rapid effects, they have begun to be withdrawn from the market in some developed countries owing to concerns about public health and environmental safety (Schneider et al. 2003; Nico et al. 2004). The search for novel, environmentally friendly alternatives with which to manage plant–parasitic nematode populations has, therefore, become increasingly important (Tian et al. 2007).

Biological control using microbial antagonists is one potential alternative to chemical nematicides (Burkett-Cadena et al. 2008). PGPR can also be used for the biological control of plant parasitic nematodes. Among the biological control agents that have been assessed are *B.* spp. and *Pseudomonas* spp. dominant populations in the rhizosphere that are able to antogonize nematodes (Tian et al. 2007).

Recently, rhizobacteria-mediated ISR in plants has been shown to be active against nematode pests. Plant growth-promoting rhizobacteria can bring about ISR by strengthening the physical and mechanical resistance of the cell wall of plants. They also change the physiological and biochemical ability of the host to promote the synthesis of defence chemicals against the challenge pathogen (van Loon et al. 1998; Ramamoorthy et al. 2001; Tian et al. 2007). Siddiqui and Shaukat (2004) concluded that fluorescent Pseudomonads ISR against root-knot nematode via a signal transduction pathway, which is independent of SA accumulation in roots.

In other words, PGPR may suppress pests and pathogens of plants and promote plant growth. For example, *P. aeruginosa* and *B. subtilis* exhibited nematicidal activity by killing the second stage larvae of *Meloidogyne javanica* to a varying degree. Especially, *B. subtilis* significantly suppressed root-knot infection and nematode population densities under greenhouse and field conditions and thereby enhanced plant growth and yield in mungbean (Siddiqui et al. 2001).

In a different example, *P. putida* promoted tomato growth in nematode-infected and nematode-free plants but growth promotion was higher in the infected ones. *P. putida* was better in reducing galling and nematode multiplication than arbuscular mycorrhizal fungus (Siddiqui and Akhtar 2008).

In another recent study, Li et al. (2005) reported that *Brevibacillus brevis* and *B. subtilis* exhibited strong nematicidal activity by killing the second stage larvae of *Meloidogyne javanica* to varying degrees in the greenhouse. The toxic principles of bacterium *B. subtilis* B7 that showed the highest juvenile mortality were partially characterized.

The influence of *P. fluorescens* as the treatment on the seed germination, migration, and penetration of *Meloidogyne incognita* in aubergine was evaluated under laboratory conditions. The results revealed that *P. fluorescens* promoted germination (87.5%) and was effective in reducing root penetration by *M. incognita* and the number of gall formation was also controlled by 70.3% (Inam-ul-Haq et al. 2003).

Rhizobacteria reported to show antagonistic effects against nematodes include members of different genera are given in Table 4.

3.2.5 Insects

Next to phytochemical insecticides, biocontrol agents of microbial origin play a role in pest management (Gandhi et al. 2006). Among the biocontrol agents, the strains of PGPR, *P. fluorescens* is the promising one (Commarea et al. 2002). They activate systemic resistance (Raupach and Kloepper 1998) by inducing plants' latent defense mechanisms and to control insect pests (Zehnder et al. 1997; Commarea et al. 2002) in addition to exerting beneficial effect on plant growth promotion (Gandhi et al. 2006).

Herman et al. (2008) notified that there are several examples of plants treated with PGPR, which showed a decrease in insect herbivory. Zehnder et al. (1997) used PGPR to reduce feeding by the spotted cucumber beetle, *Diabrotica undecimpunctata howardi* Barber. Boughton et al. (2006) reported that plants treated with defence elicitors caused the green peach aphid, *Myzus persicae*, to significantly decrease in their population growth when compared with that of the control plants. Similarly, Herman et al. (2008) notified that *B. subtilis* and *B. amyloliquefaciens* could be useful in *Myzus persicae* management program, for pepper plants grown in locations with consistently high aphid pressure. Additionally, white clover and *Medicago* plants grown in the presence of a *Pseudomonas*-like PGPR were better able to resist the effects of blue-green aphids (Kempster et al. 2002).

The talc-based formulation of two *P. fluorescens* PF1, FP7 and its mixture were tested against leaffolder in rice. The application of talc-formulation through seed, root, soil, and foliar spray significantly reduced leaffolder incidence both under greenhouse and field conditions. The mixture of two strains performed better than the individual strains. Additionally, *Pseudomonas* treated leaves altered the feeding behavior of leaffolder larvae and reduced larval and pupal weight, increased larval mortality and incidence of malformed adults under in vitro conditions. An increased population of natural enemies of leaffolder and predatory spider was noticed in *Pseudomonas* treated plots under field conditions, which yielded 12–21% more rice (Commarea et al. 2002). PGPR belonging to *Pseudomonas* spp. are being exploited

Table 4 Reported PGPR show antagonistic effects against nematodes

Nematodes	Species	PGPR	References
Meloidogyne incognita	Lettuce and tomato	*Pseudomonas* sp. W34 *Bacillus cereus* S18	Hoffmann-Hergarten et al. (1998)
Globodera pallida	Potato	*Agrobacterium radiobacter* G12A *Rhizobium etli* G12	Hackenberg et al. (1999) Reitz et al. (2000)
Meloidogyne incognita	Tomato and banana	*Bacillus cereus*, *B. subtilis*, *Pseudomonas fluorescens*, *Pseudomonas chlororaphis* *Burkholderia cepacia*	Jonathan et al. (2000)
	Bell pepper	*Burkholderia cepacia* Bc-2 *Burkholderia cepacia* Bc-F	Meyer et al. (2001)
Meloidogyne javanica	Tomato	*Pseudomonas aeruginosa* IE-6S(+) *Pseudomonas fluorescens* CHA0	Siddiqui and Shaukat (2002) Siddiqui and Shaukat (2003)
		Pseudomonas aeruginosa strain 7NSK2 *Pseudomonas fluorescens* CHA0	Siddiqui and Shaukat (2004)
Globodera rostochiensis	Potato	*Pseudomonas oryzihabitans*	Andreoglou et al. (2003)
Meloidogyne javanica	Lentil	*Pseudomonas putida*, *P. alcaligenes*, *Paenibacillus polymyxa*, *Bacillus pumilus*	Siddiqui et al. (2007)
Meloidogyne incognita	Tomato and soybean	*Pseudomonas fluorescens* CHA0	Siddiqui et al. (2005)
	Tomato	*Rhizobium etli* G12 *Bacillus amyloliquefaciens* FZB42	Reimann et al. (2008) Burkett-Cadena et al. (2008)
	Chickpea	*Pseudomonas alcaligenes* *Bacillus pumilus*	Akhtar and Siddiqui (2008)
Meloidogyne javanica	Chickpea	*Pseudomonas putida* 3604 *Pseudomonas alcaligenes* 493	Siddiqui and Akhtar (2009a)
Meloidogyne incognita	Tomato	*Bacillus subtilis*, *Paenibacillus polymyxa* *Burkholderia cepacia*	Siddiqui and Akhtar (2009b)
	Chickpea	*Pseudomonas putida* *Pseudomonas alcaligenes*	Akhtar and Siddiqui (2009)

commercially for plant protection to induce ISR against various pests and diseases. The performance of PGPR has been successful against certain pathogens, insect, and nematode pests under field conditions (Ramamoorthy et al. 2001). Murphy et al. (2000) studied the effects of PGPR treatment on whitefly nymphs number in field trials in Florida. They recorded significantly lower numbers of whitefly nymphs on PGPR-treated plants compared with the untreated tomato.

The metabolic pathways associated with insect-active secondary plant metabolites may be affected by induction of SAR or ISR, which could in turn effect changes in plant concentrations of insect feeding stimulants. Because induction of SAR and ISR involves different metabolic pathways, it is not unexpected that plants treated with PGPR or other elicitors will vary in their suitability as insect host plants (Stout et al. 2002).

Consequently, it can be said that PGPR would be of great potential, especially to conserve natural enemies and to avoid potential problems encountered when some insecticides fail to control populations that have developed resistance (Wang et al. 2002).

4 Conclusions and Future Prospects

Since Kloepper and Schroth (1978) reported that microbial communities that exert benefit for plant growth have been called PGPR, there has been an increasing effort in advancing bacterial inoculants such as *Azotobacter*, *Azoarcus*, *Bacillus*, *Burkholderia*, *Enterobacter*, *Erwinia*, *Gluconacetobacter*, *Klebsiella*, *Pseudomonas*, *Serratia*, *Rhizobium*, etc., for plant growth promotion in agriculture. Significant advances in the explanation of the mechanisms involved in plant growth promotion have been made, especially when using molecular biology approaches (Dobbelaere and Okon 2003). Mechanisms involved in plant growth promotion include biological nitrogen fixation, solubilization of insoluble phosphates, production of phytohormones, suppression of diseases, rooting of cuttings, increase germination and emergence of seeds under different conditions, promoted nutrient uptake of roots, total biomass of the plants, induce early flowering, increase in yield, etc.

Different PGPR have been examined under controlled and field conditions, and generally plant growth promotion such as yield increases in different crops, reduction of fertilizer and pesticides have been clearly demonstrated. The scientific basis of PGPR should continue to be investigated, tested, and explored for better and effective use of strains in the future, and free exchange of PGPR strains between researchers and countries (Podile and Kishore 2006) may help this. There is good possibility that the commercial mix of PGPR for various aims such as improved crop yield or suppression of pests and disease developed will be used extensively in the production of different crops in sustainable and environment friendly agriculture.

References

Abo-Elyousr KAM, Mohamed HM (2009) Biological control of *Fusarium* wilt in tomato by plant growth-promoting yeasts and rhizobacteria. Plant Pathol J 25:199–204

Akgul DS, Mirik M (2008) Biocontrol of *Phytophthora capsici* on pepper plants by *Bacillus megaterium* strains. J Plant Pathol 90:29–34

Akhtar MS, Siddiqui ZA (2008) *Glomus intraradices*, *Pseudomonas alcaligenes*, and *Bacillus pumilus*: effective agents for the control of root-rot disease complex of chickpea (*Cicer arietinum* L.). J Gen Plant Pathol 74:53–60. doi:10.1007/s10327-007-0062-4

Akhtar MS, Siddiqui ZA (2009) Use of plant growth-promoting rhizobacteria for the biocontrol of root-rot disease complex of chickpea. Australas Plant Path 38:44–50. doi:10.1071/AP08075

Alam MS, Cui ZJ, Yamagishi T, Ishii R (2001) Grain yield and related physiological characteristics of rice plants *Oryza sativa* L. inoculated with free-living rhizobacteria. Plant Prod Sci 4:126–130

Alexander M (1977) Introduction to soil microbiology. Wiley, New York

Aliye N, Fininsa C, Hiskias Y (2008) Evaluation of rhizosphere bacterial antagonists for their potential to bioprotect potato (*Solanum tuberosum*) against bacterial wilt (*Ralstonia solanacearum*). Biol Control 47:282–288. doi:10.1016/j.biocontrol.2008.09.003

Altindag M, Sahin M, Esitken A, Ercisli S, Guleryuz M, Donmez MF, Sahin F (2006) Biological control of brown rot (*Moniliana laxa* Ehr.) on apricot (*Prunus armeniaca* L. cv. Hacıhaliloğlu) by *Bacillus*, *Burkholdria*, and *Pseudomonas* application under in vitro and in vivo conditions. Biol Control 38:369–372. doi:10.1016/j.biocontrol.2006.04.015

Amir HG, Shamsuddin ZH, Halimi MS, Marziah M, Ramlan MF (2005) Enhancement in nutrient accumulation and growth of oil palm seedlings caused by PGPR under field nursery conditions. Commun Soil Sci Plan 36(15–16):2059–2066. doi:10.1080/00103620500194270

Andreoglou FI, Vagelas IK, Wood M, Samaliev HY, Gowen SR (2003) Influence of temperature on the motility of *Pseudomonas oryzihabitans* and control of *Globodera rostochiensis*. Soil Biol Biochem 35:1095–1101. doi:10.1016/S0038-0717(03)00157-3

Antoun H, Prevost D (2005) Ecology of plant growth promoting rhizobacteria. In: Siddiqui ZA (ed) PGPR: biocontrol and biofertilization. Springer, Dordrecht, Netherlands, pp 1–38

Arshad M, Shaharoona B, Mahmood T (2008) Inoculation with *Pseudomonas* spp. containing ACC-Deaminase partially eliminates the effects of drought stress on growth, yield, and ripening of pea (*Pisum sativum* L.). Pedosphere 18:611–620

Atzhorn R, Crozier A, Wheeler CT, Sandberg G (1988) Production of gibberellins and indole-3-acetic acid by *Rhizobium phaseoli* in relation to nodulation of *Phaseolus vulgaris* roots. Planta 175:532–538

Barazani O, Friedman J (1999) Is IAA the major root growth factor secreted from plant-growth-mediating bacteria? J Chem Ecol 25:2397–2406

Basavaraju O, Rao ARM, Shankarappa TH (2002) Effect of *Azotobacter* inoculation and nitrogen levels on growth and yield of radish (*Raphanus sativus* L.). In: Rajak R (ed) Proceedings of microbial technology for sustainable development and productivity. Biotechnology of Microbes and Sustainable Utilization, Jabalpur, pp 155–160

Bashan Y, Holguin G (1997) *Azospirillum*-plant relationships: environmental and physiological advances (1990–1996). Can J Microbiol 43:103–121

Bashan Y, Holguin G, de-Bashan LE (2004) *Azospirillum*-plant relationships: physiological, molecular, agricultural, and environmental advances (1997–2003). Can J Microbiol 50:521–577. doi:10.1139/W04-035

Bashan Y, Levanony H (1990) Current status of *Azospirillum* inoculation technology: *Azospirillum* as a challenge for agriculture. Can J Microbiol 36:591–608

Bassil NV, Proebsting WM, Moore LW, Lightfoot DA (1991) Propagation of hazelnut stem cuttings using *Agrobacterium rhizogenes*. HortScience 26:1058–1060

Bastian F, Cohen A, Piccoli P, Luna V, Baraldi R, Bottini R (1998) Production of indole-3-acetic acid and gibberellins A1 and A3 by *Acetobacter diazotrophicus* and *Herbaspirillum seropedicae* in chemically defined media. Plant Growth Regul 24:7–11

Begum M, Rai VR, Lokesh S (2003) Effect of plant growth promoting rhizobacteria on seedborne fungal pathogens in okra. Indian Phytopathol 56:156–158

Belimov AA, Dodd IC, Hontzeas N, Theobald JC, Safronova VI, Davies WJ (2009) Rhizosphere bacteria containing 1-aminocyclopropane-1-carboxylate deaminase increase yield of plants grown in drying soil via both local and systemic hormone signalling. New Phytol 181:413–423. doi:10.1111/j.1469-8137.2008.02657.x

Belimov AA, Dodd IC, Safronova VI, Hontzeas N, Davies WJ (2007) *Pseudomonas brassicacearum* strain Am3 containing 1-aminocyclopropane-1-carboxylate deaminase can show both pathogenic and growth-promoting properties in its interaction with tomato. J Exp Bot 58:1485–1495. doi:10.1093/jxb/erm010

Belimov AA, Kojemiakov PA, Chuvarliyeva CV (1995) Interaction between barley and mixed cultures of nitrogen fixing and phosphate-solubilizing bacteria. Plant Soil 17:29–37
Bent E, Tuzun S, Chanway CP, Enebak S (2001) Alterations in plant growth and in root hormone levels of lodgepole pines inoculated with rhizobacteria. Can J Microbiol 47:793–800
Bertrand H, Nalin R, Bally R, Cleyet-Marel JC (2001) Isolation and identification of the most efficient plant growth-promoting bacteria associated with canola (*Brassica napa*). Biol Fertil Soils 33:152–156
Bonaterra A, Ruz L, Badosa E, Pinochet J, Montesinos E (2003) Growth promotion of *Prunus* rootstocks by root treatment with specific bacterial strains. Plant Soil 255:555–569
Bottini R, Fulchieri M, Pearce D, Pharis RP (1989) Identification of gibberellins A1, A3, and iso-A3 in cultures of *Azospirillum lipoferum*. Plant Physiol 90:45–47
Boudyach EH, Fatmi M, Akhayat O, Benizri E, Aoumar AAB (2001) Selection of antagonistic bacteria of *Clavibacter michiganensis* ssp *michiganensis* and evaluation of their efficiency against bacterial canker of tomato. Biocontrol Sci Technol 11:141–149. doi:10.1080/09583150020029817
Boughton AJ, Hoover K, Felton GW (2006) Impact of chemical elicitor applications on greenhouse tomato plants and population growth of the green peach aphid, *Myzus persicae*. Entomol Exp Appl 120:175–188
Burdman S, Jurkevitch E, Okon Y (2000) Recent advences in the use of plant growth promoting rhizobacteria (PGPR) in agriculture. In: Subba Rao NS, Dommergues YR (eds) Microbial interactions in agriculture and forestry, vol 2. Science publishers Inc., Enfield, New Hampshire, pp 229–250
Burkett-Cadena M, Kokalis-Burelle N, Lawrence KS, van Santen E, Kloepper JW (2008) Suppressiveness of root-knot nematodes mediated by rhizobacteria. Biol Control 47:55–59. doi:10.1016/j.biocontrol.2008.07.008
Cakmakci R, Dönmez F, Aydın A, Şahin F (2006) Growth promotion of plants by plant growth-promoting rhizobacteria under greenhouse and two different field soil conditions. Soil Biol Biochem 38(6):1482–1487. doi:10.1016/j.soilbio.2005.09.019
Cakmakci R, Erat M, Erdogan U, Donmez MF (2007) The influence of plant growth-promoting rhizobacteria on growth and enzyme activities in wheat and spinach plants. J Plant Nutr Soil Sci 170:288–295. doi:10.1002/jpln.200625105
Cakmakci R, Kantar F, Sahin F (2001) Effect of N_2-fixing bacterial inoculations on yield of sugar beet and barley. J Plant Nutr Soil Sci 164:527–531
Canbolat MY, Bilen S, Cakmakci R, Sahin F, Aydin A (2006) Effect of plant growth-promoting bacteria and soil compaction on barley seedling growth, nutrient uptake, soil properties and rhizosphere microflora. Biol Fertil Soils 42:350–357. doi:10.1007/s00374-005-0034-9
Cattelan AJ, Hartel PG, Fuhrmann JJ (1999) Screening for plant growth-promoting rhizobacteria to promote early soybean growth. Soil Sci Soc Am J 63:1670–1680
Chelius MK, Triplett EW (2000) Immunolocalization of dinitrogenase reductase produced by Klebsiella pneumoniae in association with *Zea mays* L. Appl Environ Microb 66:783–787
Chen YP, Rekha PD, Arun AB, Shen FT, Lai WA, Young CC (2006) Phosphate solubilizing bacteria from subtropical soil and their tricalcium phosphate solubilizing abilities. Appl Soil Ecol 34:33–41. doi:10.1016/j.apsoil.2005.12.002
Chinnasamy G (2005) A proteomics perspective on biocontrol and plant defense mechanism. In: Siddiqui ZA (ed) PGPR: biocontrol and biofertilization. Springer, Dordrecht, Netherlands, pp 233–256
Choudhary DK, Prakash A, Johri BN (2007) Induced systemic resistance (ISR) in plants: mechanism of action. Indian J Microbiol 47:289–297. doi:10.1007/s12088-007-0054-2
Chung H, Park M, Madhaiyan M, Seshadri S, Song J, Cho H, Sa T (2005) Isolation and characterization of phosphate solubilizing bacteria from the rhizosphere of crop plants of Korea. Soil Biol Biochem 37:1970–1974. doi:10.1016/j.soilbio.2005.02.025

Commarea RR, Nandakumara R, Kandana A, Sureshb S, Bharathib M, Raguchandera T, Samiyappana R (2002) *Pseudomonas fluorescens* based bio-formulation for the management of sheath blight disease and leaffolder insect in rice. Crop Prot 21:671–677

Costacurta A, Vanderleyden J (1995) Synthesis of phytohormones by plant-associated bacteria. Crit Rev Microbiol 21:1–18

Couvillon GA (1998) Rooting responses to different treatments. Acta Hort 227:187–196

de Araujo FF (2008) Seed inoculation with *Bacillus subtilis*, formulated with oyster meal and growth of corn, soybean and cotton. Ciênc Agrotec 32:456–462

de Freitas JR (2000) Yield and N assimilation of winter wheat (*Triticum aestivum* L., var. Norstar) inoculated with rhizobacteria. Pedobiologia 44:97–104

de Salamone IEG, Hynes RK, Nelson LM (2001) Cytokinin production by plant growth promoting rhizobacteria and selected mutants. Can J Microbiol 47:404–411

Decateau RD (2000) Vegetable crops. Prentice-Hall Inc, Upper Saddle River, New Jersey

Dell'Amico E, Cavalca L, Andreoni V (2005) Analysis of rhizobacterial communities in perennial *Graminaceae* from polluted water meadow soil, and screening of metal resistant, potentially plant growth-promoting bacteria. FEMS Microbiol Ecol 52:153–162. doi:10.1016/j.femsec.2004.11.005

Dobbelaere S, Okon Y (2003) The plant growth promoting effect and plant responses. In: Elmerich C, Newton WE (eds) Associative and endophytic nitrogen-fixing bacteria and cyanobacterial associations. Kluwer Academic, Netherlands, pp 1–26

Dobbelaere S, Vanderleyden J, Okon Y (2003) Plant growth-promoting effects of diazotrophs in the rhizosphere. Crit Rev Plant Sci 22:107–149

Döbereiner J (1997) Biological nitrogen fixation in the tropics: social and economic contributions. Soil Biol Biochem 29:771–774

Dong Y, Iniguez AL, Triplett EW (2003) Quantitative assessments of the host range and strain specificity of endophytic colonization by *Klebsiella pneumoniae* 342. Plant Soil 257:49–59

Dursun A, Ekinci M, Donmez MF (2008) Effects of inoculation bacteria on chemical content, yield and growth in rocket (*Eruca vesicaria* subsp *sativa*). Asian J Chem 20:3197–3202

Dutta S, Mishra AK, Kumar BSD (2008) Induction of systemic resistance against fusarial wilt in pigeon pea through interaction of plant growth promoting rhizobacteria and rhizobia. Soil Biol Biochem 40:452–461. doi:10.1016/j.soilbio.2007.09.009

Egamberdiyeva D, Qarshieva D, Davranov K (2004) Growth and yield of soybean varieties inoculated with *Bradyrhizobium* spp in N-deficient calcareous soils. Biol Fert Soils 40:144–146. doi:10.1007/s00374-004-0755-1

Elbadry M, Taha RM, Eldougdoug KA, Gamal-Eldin H (2006) Induction of systemic resistance in faba bean (*Vicia faba* L.) to bean yellow mosaic potyvirus (BYMV) via seed bacterization with plant growth promoting rhizobacteria. J Plant Dis Protect 113:247–251

El-Khawas H, Adachi K (1999) Identification and quantification of auxins in culture media of *Azospirillum* and *Klebsiella* and their effect on rice roots. Biol Fert Soils 28:377–381

Elkoca E, Kantar F, Sahin F (2008) Influence of nitrogen fixing and phosphorus solubilizing bacteria on the nodulation, plant growth, and yield of chickpea. J Plant Nutr 31:157–171. doi:10.1080/01904160701742097

Elmerich C (1984) Molecular biology and ecology of diazotrophs associated with non-leguminous plants. Biotechnol 11:967–978. doi:10.1038/nbt1184-967

Enebak SA, Carey WA (2004) Plant growth-promoting rhizobacteria may reduce fusiform rust infection in nursery-grown loblolly pine seedlings. South J Appl Forest 28:185–188

Ercisli S (2009) Apricot culture in Turkey. Sci Res Essays 4:715–719

Ercisli S, Esitken A, Cangi R, Sahin F (2003) Adventitious root formation of kiwifruit in relation to sampling date, IBA and *Agrobacterium rubi* inoculation. Plant Growth Regul 41:133–137

Ercisli S, Esitken A, Sahin F (2000) Effect of IBA and bacteria (*Agrobacterium rubi*) on the rooting of cuttings of sour cherry cv. Kutahya. Bahce 29:75–80

Ercisli S, Esitken A, Sahin F (2004) Exogenous IBA and inoculation with *Agrobacterium rubi* stimulate adventitious root formation on hardwood stem cuttings of two rose genotypes. HortScience 39:533–534

Erturk Y, Ercisli S, Sekban R, Haznedar A, Donmez MF (2008) The effect of plant growth promoting rhizobacteria (PGPR) on rooting and root growth of tea (*Camellia sinensis* var. *Sinensis*) cuttings. Roum Biotechnol Lett 13:3747–3756

Esitken A, Ercisli S, Karlidag H, Sahin F (2005) Potential use of plant growth promoting rhizobacteria (PGPR) in organic apricot production. In: Libek A, Kaufmane E, Sasnauskas A (eds) International conference on environmentally friendly fruit growing. Tartu, Estonia, pp 90–97

Esitken A, Ercisli S, Sevik I, Sahin F (2003a) Effect of Indole-3-butyric acid and different strains of *Agrobacterium rubi* on adventive root formation from softwood and semi-hardwood wild sour cherry cuttings. Turk J Agric Forest 27:37–42

Esitken A, Karlidag H, Ercisli S, Turan M, Sahin F (2003b) The effect of spraying a growth promoting bacterium on the yield, growth and nutrient element composition of leaves of apricot (*Prunus armeniaca* L. cv. Hacihaliloglu). Aust J Agr Res 54:377–380. doi:10.1071/AR02098

Esitken A, Pirlak L, Turan M, Sahin F (2006) Effects of floral and foliar application of plant growth promoting rhizobacteria (PGPR) on yield, growth and nutrition of sweet cherry. Sci Hortic 110:324–327. doi:10.1016/j.scienta.2006.07.023

Fatima Z, Saleemi M, Zia M, Sultan T, Aslam M, Rehman RU, Chaudhary MF (2009) Antifungal activity of plant growth-promoting rhizobacteria isolates against *Rhizoctonia solani* in wheat. Afr J Biotechnol 8:219–225

Figueiredo MV, Martinez CR B, Burity HA, Chanway CP (2008) Plant growth-promoting rhizobacteria for improving nodulation and nitrogen fixation in the common bean (*Phaseolus vulgaris* L.). World J Microb Biot 24:1187–1193. doi:10.1007/s11274-007-9591-4

Gandhi PI, Gunasekaran K, Tongmin S (2006) Neem oil as a potential seed dresser for managing Homopterous sucking pests of Okra (*Abelmoschus esculentus* (L.) Moench). J Pest Sci 79:103–111. doi:10.1007/s10340-006-0122-0

Gang W, Kloepper JW, Tuzun S (1991) Induction of systemic resistance of cucumber to *Colletotrichum orbiculare* by selected strains of plant growth-promoting rhizobacteria. Phytopathology 81:1508–1512

Garcia JAL, Domenech J, Santamaría C, Camacho M, Daza A, Gutierrez Mañero FJ (2004a) Growth of forest plants (pine and holm-oak) inoculated with rhizobacteria: relationship with microbial community structure and biological activity of its rhizosphere. Environ Exp Bot 52:239–251. doi:10.1016/j.envexpbot.2004.02.003

Garcia JAL, Probanza A, Ramos B, Barriuso J, Gutierrez Mañero FJ (2004b) Effects of inoculation with plant growth promoting rhizobacteria (PGPRs) and *Sinorhizobium fredii* on biological nitrogen fixation, nodulation and growth of *Glycine max* cv. Osumi. Plant Soil 267:143–153

Gaudin V, Vrain D, Jouanin L (1994) Bacterial genes modifying hormonal balance in plant. Plant Physiol Biochem 32:11–29

Girish N, Umesha S (2005) Effect of plant growth promoting rhizobacteria on bacterial canker of tomato. Arch Phytopathol Plant Protect 38:235–243

Glick BR, Jacobson CB, Schwarze MMK, Pasternak JJ (1994) 1-Aminocyclopropane-1-carboxylic acid deaminase mutants of the plant growth promoting rhizobacterium *Pseudomonas putida* GR12-2 do not stimulate canola root elongation. Can J Microbiol 40:911–915

Glick BR (1995) The enhancement of plant growth by free-living bacteria. Can J Microbiol 41:109–17

Glick BR, Penrose DM, Li JP (1998) A model for the lowering of plant ethylene concentrations by plant growth-promoting bacteria. J Theor Biol 190:63–68. doi:10.1006/jtbi.1997.0532

Gulcan R, Mısırlı A, Demir T (1999) A research on resistance of Hacıhaliloğlu apricot variety against Monilinia (*Sclerotinia laxa*, Aderh et Ruhl) through cross pollination. Acta Hort 488:673–676

Guo JH, Qi HY, Guo YH, Ge HL, Gong LY, Zhang LX (2004) Biocontrol of tomato wilt by plant growth-promoting rhizobacteria. Biol Control 29:66–72. doi:10.1016/S1049-9644(03)00124-5

Gutierrez-Manero FJ, Ramos-Solano B, Probanza A, Mehouachi J, Tadeo FR, Talon M (2001) The plant-growth promoting rhizobacteria *Bacillus pumilus* and *Bacillus licheniformis* produce high amounts of physiologically active gibberellins. Physiol Plant 111:206–211

Gyaneshwar P, Naresh Kumar G, Parekh LJ, Poole PS (2002) Role soil microorganisms in improving P nutrition of plant. Plant Soil 245:83–93

Hackenberg C, Vrain TC, Sikora RA (1999) Rhizosphere colonization pattern of *Agrobacterium radiobacter* strain G12A, an antagonistic rhizobacterium to the potato cyst nematode *Globodera pallid*. Microbiol Res 154:57–61

Hafeez FY, Safdar ME, Chaudhry AU, Malik KA (2004) Rhizobial inoculation improves seedling emergence, nutrient uptake and growth of cotton. Aust J Exp Agric 44:617–622. doi:10.1071/EA03074

Halder AK, Mishra AK, Bhattacharyya P, Chakrabartty PK (1990) Solubilization of rock phosphate by *Rhizobium* and *Bradyrhizobium*. J Gen Appl Microbiol 36:81–92

Hamaoui B, Abbadi JM, Burdman S, Rashid A, Sarig S, Okon Y (2001) Effects of inoculation with *Azospirillum brasilense* on chickpeas (*Cicer arietinum*) and faba beans (*Vicia faba*) under different growth conditions. Agronomie 21:553–560

Hameeda B, Harini G, Rupela OP, Wani SP, Reddy G (2008) Growth promotion of maize by phosphate solubilizing bacteria isolated from composts and macrofauna. Microbiol Res 163:234–242. doi:10.1016/j.micres.2006.05.009

Harish S, Kavino M, Kumar N, Saravanakumar D, Soorianathasundaram K, Samiyappan R (2008) Biohardening with plant growth promoting rhizosphere and endophytic bacteria induces systemic resistance against banana bunchy top virus. Appl Soil Ecol 39:187–200. doi:10.1016/j.apsoil.2007.12.006

Hatta M, Beyl CA, Garton S, Diner AM (1996) Induction of roots on jujube softwood cuttings using *Agrobacterium rhizogenes*. J Hort Sci 71:881–886

Hausbeck MK, Lamour KH (2004) *Phytophthora capsici* on vegetable crops: research progress and management challenges. Plant Dis 88:1292–1303

Hecht-Buchholz C (1998) The apoplast-habitat of endophytic dinitrogen-fixing bacteria and their significance for the nitrogen nutrition of nonlegumious plants. J Plant Nutr Soil Sci 161:509–520

Herman MAB, Nault BA, Smart CD (2008) Effects of plant growth-promoting rhizobacteria on bell pepper production and green peach aphid infestations in New York. Crop Prot 27:996–1002. doi:10.1016/j.cropro.2007.12.004

Hernandez-Rodriguez A, Heydrich-Perez M, Acebo-Guerrero Y, Velazquez-del Valle MG, Hernandez-Lauzardo AN (2008) Antagonistic activity of Cuban native rhizobacteria against *Fusarium verticillioides* (Sacc.) Nirenb. in maize (*Zea mays* L.). Appl Soil Ecol 39:180–186. doi:10.1016/j.apsoil.2007.12.008

Hoffmann-Hergarten S, Gulati MK, Sikora RA (1998) Yield response and biological control of Meloidogyne incognita on lettuce and tomato with rhizobacteria. Z Pflanzenk Pflan 105:349–358

Inam-ul-Haq M, Tariq JA, Javed N, Khan NA, Imran-ul-Haq Khan HU (2003) In-vitro interrelationship between plant growth promoting rhizobacteria and root knot nematode (*Meloidogyne incognita*) and their effect on growth parameters of brinjal. Mycopath 1:191–193

Ishida AKN, de Souza RM, de Resende MLV, Zacaroni AB, Boas CHV, de Souza JT (2008) Rhizobacteria to control cotton bacterial blight. Ciênc Agrotec 32:149–156

Janzen RA, Rood SB, Dormaar JF, McGill WB (1992) *Azospirillum brasilense* produces gibberellin in pure culture on chemically defined medium and in co-culture on straw. Soil Biol Biochem 24:1061–1064

Jetiyanon K, Kloepper JW (2002) Mixtures of plant growth-promoting rhizobacteria for induction of systemic resistance against multiple plant diseases. Biol Control 24:285–291

Jeun YC, Park KS, Kim CH, Fowler WD, Kloepper JW (2004) Cytological observations of cucumber plants during induced resistance elicited by rhizobacteria. Biol Control 29:34–42. doi:10.1016/S1049-9644(03)00082-3

Jonathan EI, Barker KR, Abdel-Alim FF, Vrain TC, Dickson DW (2000) Biological control of *Meloidogyne incognita* on tomato and banana with rhizobacteria actinomycetes, and *Pasteuria penetrans*. Nematropica 30:231–240

Joo GJ, Kim YM, Lee IJ, Song KS, Rhee IK (2004) Growth promotion of red pepper plug seedlings and the production of gibberellins by *Bacillus cereus*, *Bacillus macroides* and *Bacillus pumilus*. Biotechnol Lett 26:487–491

Kandan A, Ramiah M, Vasanthi VJ, Radjacommare R, Nandakumar R, Ramanathan A, Samiyappan R (2005) Use of *Pseudomonas fluorescens*-based formulations for management of tomato spotted wilt virus (TSWV) and enhanced yield in tomato. Biocontrol Sci Technol 15 (6):553–569. doi:10.1080/09583150500088546

Karlıdağ H, Esitken A, Turan M, Sahin F (2007) Effects of root inoculation of plant growth promoting rhizobacteria (PGPR) on yield, growth and nutrient element contents of leaves of apple. Sci Hortic 114:16–20. doi:10.1016/j.scienta.2007.04.013

Kaushik R, Saxena AK, Tilak KVBR (2000) Selection of Tn5::lacZ mutants isogenic to wild type *Azospirillum brasilense* strains capable of growing at sub-optimal temperature. World J Microb Biot 16:567–570

Kaymak HC, Guvenc I, Yarali F, Donmez MF (2009) The effects of bio-priming with PGPR on germination of radish (*Raphanus sativus* L.) seeds under saline conditions. Turk J Agric Forest 33:173–179

Kaymak HC, Yarali F, Guvenc I, Donmez MF (2008) The effect of inoculation with plant growth rhizobacteria (PGPR) on root formation of mint (*Mentha piperita* L.) cuttings. Afr J Biotechnol 7:4479–4483

Kempster VN, Scott ES, Davies KA (2002) Evidence for systemic, cross-resistance in white clover (*Trifolium repens*) and annual medic (*Medicago truncatula* var *truncatula*) induced by biological and chemical agents. Biocontrol Sci Technol 12:615–623. doi:10.1080/09583150210000016270

Kennedy IR, Pereg-Gerk LL, Wood C, Deaker R, Gilchrist K, Katupitiya S (1997) Biological nitrogen fixation in nonleguminous field crops: facilitating the evolution between *Azospirillum* and wheat. Plant Soil 194:65–79

Kennedy RW, Chellapillai KL (1998) Synergistic effect of VAM, *Azospirillum*, and phosphobacteria on growth response and nutrient uptake of shola tree species. Indian J Forest 21:308–312

Khan AG (2005) Role of soil microbes in the rhizospheres of plants growing on trace metal contaminated soils in phytoremediation. J Trace Elem Med Biol 18:355–364. doi:10.1016/j.jtemb.2005.02.006

Kim KY, Jordan D, McDonald GA (1998) *Enterobacter agglomerans*, phosphate solubilising bacteria, and microbial activity in soil: effect of carbon sources. Soil Biol Biochem 30:995–1003

Kim SG, Khan Z, Jeon YH, Kim YH (2009) Inhibitory effect of *Paenibacillus polymyxa* GBR-462 on *Phytophthora capsici* causing *Phytophthora* Blight in Chili Pepper. J Phytopathol 157:329–337. doi:10.1111/j.1439-0434.2008.01490.x

Kirankumar R, Jagadeesh KS, Krishnaraj PU, Patil MS (2008) Enhanced growth promotion of tomato and nutrient uptake by plant growth promoting rhizobacterial isolates in presence of tobacco mosaic virus pathogen. Karnataka J Agric Sci 21:309–311

Klee HJ, Hayford MB, Kretzmer KA, Barry GF, Kiskore GM (1991) Control of ethylene synthesis by expression of a bacterial enzyme in transgenic tomato plants. Plant Cell 3:1187–1193

Kloepper JW, Gutierrez-Estrada A, McInroy JA (2007) Photoperiod regulates elicitation of growth promotion but not induced resistance by plant growth-promoting rhizobacteria. Can J Microbiol 53:159–167. doi:10.1139/W06-114

Kloepper JW, Schroth MN (1978) Plant growth-promoting rhizobacteria on radishes. In: Proceedings of the fourth international conference on plant pathogen bacteria. vol 2, INRA, p 879–882

Koenning SR, Overstreet C, Noling JW, Donald PA, Becker JO, Fortnum BA (1999) Survey of crop losses in response to phytoparasitic nematodes in the United States for 1994. J Nematol 31:587–618

Kokalis-Burelle N, Vavrina EN, Rosskopf EN, Shelby RA (2002) Field evaluation of plant growth-promoting rhizobacteria amended transplant mixes and soil solarization for tomato and pepper production in Florida. Plant Soil 238:257–266
Köse C, Guleryuz M, Sahn F, Demirtas I (2003) Effects of some plant growth promoting rhizobacteria (PGPR) on rooting of grapevine rootstocks. Acta Agrobot 56:47–52
Kumar V, Narula N (1999) Solubilization of inorganic phosphates and growth emergence of wheat as affected by *Azotobacter chrococcum*. Biol Fert Soils 28:301–305
Kumar S, Pandey P, Maheshwari DK (2009) Reduction in dose of chemical fertilizers and growth enhancement of sesame (*Sesamum indicum* L.) with application of rhizospheric competent *Pseudomonas aeruginosa* LES4. Eur J Soil Biol 45:334–340
Kunc F, Macura J (1988) Mechanisms of adaptation and selection of microorganisms in the soil. In: Vancura V, Kunc F (eds) Soil microbial associations. Elsevier, Amsterdam, pp 281–299
Lee KH, Koh RH, Soh HG (2008a) Enhancement of growth and yield of tomato *Rhodopseudomonas* sp. under greenhouse conditions. J Microbiol 46:641–646. doi:10.1007/s12275-008-0159-2
Lee KJ, Kamala-Kannan S, Sub HS, Seong CK, Lee GW (2008b) Biological control of *Phytophthora* blight in red pepper (*Capsicum annuum* L) using *Bacillus subtilis*. World J Microb Biot 24:1139–1145. doi:10.1007/s11274-007-9585-2
Lemessa F, Zeller W (2007) Screening rhizobacteria for biological control of *Ralstonia solanacearum* in Ethiopia. Biol Control 42:336–344. doi:10.1016/j.biocontrol.2007.05.014
Li B, Xie GL, Soad A, Coosemans J (2005) Suppression of *Meloidogyne javanica* by antagonistic and plant growth-promoting rhizobacteria. J Zhejiang Univ Sci B 6:496–501. doi:10.1631/jzus.2005.B0496
Liu L, Kloepper JW, Tuzun S (1995) Induction of systemic resistance in cucumber against bacterial angular leaf-spot by plant growth-promoting rhizobacteria. Phytopathology 85:843–847
Lokesh S, Bharath BG, Raghavendra VB, Govindappa M (2007) Importance of plant growth-promoting rhizobacteria in enhancing the seed germination and growth of watermelon attacked by fungal pathogens. Acta Agron Hung 55:243–249
Lourenco V, Maffia LA, Romeiro RD, Mizubuti ESG (2006) Biocontrol of tomato late blight with the combination of epiphytic antagonists and rhizobacteria. Biol Control 38:331–340. doi:10.1016/j.biocontrol.2006.04.005
Lucy M, Reed E, Glick BR (2004) Applications of free living plant growth-promoting rhizobacteria. Anton Leeuw Int J G 86:1–25
Mantelin S, Touraine B (2004) Plant growth-promoting bacteria and nitrate availability: impacts on root development and nitrate uptake. J Exp Bot 55:27–34. doi:10.1093/jxb/erh010
Manulis S, Valinski L, Gafni Y, Hershenhorn J (1991) Indole-3-acetic acid biosynthetic patways in erwinia herbicola in relation to pathogenity in *Gypsophia paniculata*. Physiol Mol Plant P 39:161–171
Martelli GP (1992) Classification and nomenclature of plant viruses state of the art. Plant Dis 76:436–442
Matheron ME, Porchas M (2000) Impact of azoxystrobin, dimethomorph, fluazinam, fosetyl-Al, and metalaxyl on growth sporulation and zoospore cyst germination of three *Phytophthora* spp. Plant Dis 84:454–458
May R, Völksck B, Kampmann G (1996) Antagonistic activities of epiphytic bacteria from soy bean leaves against *Pseudomonas syringae* pv. *syringae* in vitro and in planta. Microb Ecol 34:118–124
Mayak S, Tirosh T, Glick BR (1999) Effect of wild-type and mutant plant growth- promoting rhizobacteria on the rooting of mung bean cuttings. J Plant Growth Regul 18:49–53
Mehnaz S, MirzaMS HJ, Bally R, Normand P, Bano A, Malik KA (2001) Isolation and 16S rRNA sequence analysis of the beneficial bacteria from the rhizosphere of rice. Can J Microbiol 472:110–117

Meyer SLF, Roberts DP, Chitwood DJ, Carta LK, Lumsden RD, Mao WL (2001) Application of *Burkholderia cepacia* and *Trichoderma virens*, alone and in combinations, against *Meloidogyne incognita* on bell pepper. Nematropica 31:75–86

Mirza MS, Ahmad W, Latif F, Haurat J, Bally R, Normajd P, Malik KA (2001) Isolation, partial characterization, and the effect of plant growth-promoting bacteria (PGPB) on micropropagated sugarcane in vitro. Plant Soil 237:47–54

Murphy JF, Reddy MS, Ryu CM, Kloepper JW, Li RH (2003) Rhizobacteria-mediated growth promotion of tomato leads to protection against Cucumber mosaic virus. Phytopathology 93:1301–1307

Murphy JF, Zehnder GW, Schuster DJ, Sikora EJ, Polston JE, Kloepper JW (2000) Plant growth-promoting rhizobacterial mediated protection in tomato against tomato mottle virus. Plant Dis 84:779–784

Nadeem SM, Zahir ZA, Naveed M, Arshad M (2007) Preliminary investigations on inducing salt tolerance in maize through inoculation with rhizobacteria containing ACC deaminase activity. Can J Microbiol 53:141–1149. doi:10.1139/W07-081

Naveed M, Khalid M, Jones DL, Ahmad R, Zahir ZA (2008) Relative efficacy of *Pseudomonas* spp., containing ACC-Deaminase for improving growth and yield of maize (*Zea mays* L.) in the presence of organic fertilizer. Pak J Bot 40:1243–1251

Nelson LM (2004) Plant growth promoting rhizobacteria (PGPR): prospects for new inoculants. Crop Management. doi:10.1094/CM-2004-0301-05-RV

Nico AI, Rafael RM, Jiménez-Díaza M, Castillo P (2004) Control of root-knot nematodes by composted agro-industrial wastes in potting mixtures. Crop Prot 23:581–587. doi:10.1016/j.cropro.2003.11.005

Noel TC, Sheng C, Yost CK, Pharis RP, Hynes MF (1996) *Rhizobium leguminosarum* as a plant growth-promoting rhizobacterium: direct growth promotion of canola and lettuce. Can J Microbiol 42:279–283

Orhan E, Esitken A, Ercisli S, Sahin F (2007) Effects of indole-3-butyric acid (IBA), bacteria and, radicle tip-cutting on lateral root induction in *Pistacia vera*. J Hortic Sci Biotech 82:2–4

Orhan E, Esitken A, Ercisli S, Turan M, Sahin F (2006) Effects of plant growth promoting rhizobacteria (PGPR) on yield, growth and nutrient contents in organically growing raspberry. Sci Hortic 111:38–43. doi:10.1016/j.scienta.2006.09.002

Öztürk A, Caglar O, Sahin F (2003) Yield response of wheat and barley to inoculation of plant growth promoting rhizobacteria at various levels of nitrogen fertilization. J Plant Nutr Soil Sci 166:262–266

Pal SS (1998) Interaction of an acid tolerant strain of phosphate solubilizing bacteria with a few acid tolerant crops. Plant Soil 198:169–177. doi:10.1023/A:1004318814385

Patten CL, Glick BR (1996) Bacterial biosynthesis of indole-3-acetic acid. Can J Microbiol 42:207–220

Peix A, Rivas-Boyerob AA, Mateos PF, Rodriguez-Barrueco C, Martínez-Molina E, Velazquez E (2001) Growth promotion of chickpea and barley by a phosphate solubilizing strain of *Mesorhizobium mediterraneum* under growth chamber conditions. Soil Biol Biochem 33:103–110

Pırlak L, Turan M, Sahin F, Esitken A (2007) Floral and foliar application of plant growth promoting rhizobacteria (PGPR) to apples increases yield, growth, and nutrient element contents of leaves. J Sustain Agric 30:145–155. doi:10.1300/J064v30n04-11

Podile AR, Kishore GK (2006) Plant growth-promoting rhizobacteria. In: Gnanamanickam SS (ed) Plant-associated bacteria: rhizosphere bacteria. Springer, Netherlands, pp 195–230

Pusey PL (1997) Crap apple Blossoms as a model for research on biological control of fire blight. Phytopathology 87:1096–1102

Radwan FI (1998) Response of some maize cultivars to VAmycorrhizal inoculation, biofertilization and soil nitrogen application. Alex J Agric Res 43:43–56

Raj NS, Shetty NP, Shetty HS (2004) Seed bio-priming with *Pseudomonas fluorescens* isolates enhances growth of pearl millet plants and induces resistance against downy mildew. Int J Pest Manag 50:41–48. doi:10.1080/09670870310001626365

Raj SN, Chaluvaraju G, Amruthesh KN, Shetty HS, Reddy MS, Kloepper JW (2003a) Induction of growth promotion and resistance against downy mildew on pearl millet (*Pennisetum glaucum*) by rhizobacteria. Plant Dis 87:380–384

Raj SN, Deepak SA, Basavaraju P, Shetty HS, Reddy MS, Kloepper JW (2003b) Comparative performance of formulations of plant growth promoting rhizobacteria in growth promotion and suppression of downy mildew in pearl millet. Crop Prot 22:579–588

Rajkumar M, Lee WH, Lee KJ (2005) Screening of bacterial antagonists for biological control of *Phytophthora* blight of pepper. J Basic Microb 45:55–63. doi:10.1002/jobm.200410445

Ramamoorthy V, Viswanathan R, Raghuchander T, Prakasam V, Samiyappan R (2001) Induction of systemic resistance by plant growth promoting rhizobacteria in crop plants against pests and diseases. Crop Prot 20:1–11

Ramirez LEF, Mellado JC (2005) Bacterial biofertilizers. In: Siddiqui ZA (ed) PGPR: biocontrol and biofertilization. Springer, Dordrecht, Netherlands, pp 143–172

Ran LX, Liu CY, Wu GJ, van Loon LC, Bakker PAHM (2005) Suppression of bacterial wilt in *Eucalyptus urophylla* by fluorescent *Pseudomonas* spp. in China. Biol Control 32:111–120. doi:10.1016/j.biocontrol.2004.08.007

Raupach GS, Kloepper JW (1998) Mixtures of plant growth promoting rhizobacteria enhance biological control of multiple cucumber pathogens. Phytopathology 88:1158–1164

Raupach GS, Kloepper JW (2000) Biocontrol of cucumber diseases in the field by plant growth-promoting rhizobacteria with and without methyl bromide fumigation. Plant Dis 84:1073–1075

Reddy DDR (1985) Analysis of crop losses in tomato due to *Meloidogyne incognita*. Indian J Nematol 15:55–59

Reimann S, Hauschild R, Hildebrandt U, Sikora RA (2008) Interrelationships between *Rhizobium etli* G12 and *Glomus intraradices* and multitrophic effects in the biological control of the root-knot nematode *Meloidogyne incognita* on tomato. J Plant Dis Protect 115:108–113

Reitz M, Rudolph K, Schroder I, Hoffmann-Hergarten S, Hallmann J, Sikora RA (2000) Lipopolysaccharides of *Rhizobium etli* strain G12 act in potato roots as an inducing agent of systemic resistance to infection by the cyst nematode *Globodera pallid*. Appl Environ Microbiol 66:3515–3518

Reyes I, Alvarez L, El-Ayoubi H, Valery A (2008) Selection and evaluation of growth promoting rhizobacteria on pepper and maize. Bioagro 20:37–48

Rinallo C, Mittempergher L, Frugis G, Mariotti D (1999) Clonal propagation in the genus *Ulmus*: improvement of rooting ability by *Agrobacterium rhizogenes* T-DNA genes. J Hortic Sci Biotech 74:502–506

Ristaino JB, Johnston SA (1999) Ecologically based approaches to management of *Phytophthora* blight on bell pepper. Plant Dis 83:1080–1089

Rodríguez H, Fraga R (1999) Phosphate solubilizing bacteria and their role in plant growth promotion. Biotechnol Adv 17:319–339

Rodriguez H, Gonzalez T, Goire I, Bashan Y (2004) Gluconic acid production and phosphate solubilization by the plant growth-promoting bacterium *Azospirillum* spp. Naturewissenschaften 91:552–555. doi:10.1007/s00114-004-0566-0

Rodriguez H, Vessely S, Shah S, Glick BR (2008) Effect of a Nickel-tolerant ACC deaminase-producing *Pseudomonas* strain on growth of nontransformed and transgenic Canola plants. Curr Microbiol 57:170–174. doi:10.1007/s00284-008-9181-1

Rodriguez MN, Villalonga RD, Castillo RAJ, Marques AJL, Gonzalez LR, Llanes SP, Peguero FM (2001) Influence of application of a biofertilizer based on *Azospirillum* on germination of seed and production of vegetable crops. Centro Agricola 28:38–41

Romero AM, Correa OS, Moccia S, Rivas JG (2003) Effect of *Azospirillum*-mediated plant growth promotion on the development of bacterial diseases on fresh-market and cherry tomato. J Appl Microbiol 95(4):832–838. doi:10.1046/j.1365-2672.2003.02053.x

Ryu CM, Kima J, Choi O, Kima SH, Park CS (2006) Improvement of biological control capacity of *Paenibacillus polymyxa* E681 by seed pelleting on sesame. Biol Control 39:282–289. doi:10.1016/j.biocontrol.2006.04.014

Saddler G (2002) Bacteria and plant disease. In: Waller JM, Lenné JM, Waller SJ (eds) Plant pathologist's pocketbook, 3rd edn. CABI Publishing, Wallingford, Oxon, UK, pp 94–106

Şahin F, Çakmakçi R, Kantar F (2004) Sugar beet and barley yields in relation to inoculation with N2-fixing and phosphate solubilizing bacteria. Plant Soil 265:123–129

Salantur A, Ozturk A, Akten S, Sahin F, Donmez F (2005) Effect of inoculation with non-indigenous and indigenous Rhizobacteria of Erzurum (Turkey) origin on growth and yield of spring barley. Plant Soil 275:147–156. doi:10.1007/s11104-005-8094-z

Saleem M, Arshad M, Hussain S, Bhatti AS (2007) Perspective of plant growth promoting rhizobacteria (PGPR) containing ACC deaminase in stress agriculture. J Ind Microbiol Biotechnol 34:635–648. doi:10.1007/s10295-007-0240-6

Saleh SS, Glick BR (2001) Involvement of gacS and rpoS in enhancement of the plant growth-promoting capabilities of *Enterobacter cloacae* CAL2 and UW4. Can J Microbiol 47:698–705

Sang MK, Chun SC, Kim KD (2008) Biological control of Phytophthora blight of pepper by antagonistic rhizobacteria selected from a sequential screening procedure. Biol Control 46:424–433. doi:10.1016/j.biocontrol.2008.03.017

Saravanakumar D, Lavanya N, Muthumeena K, Raguchander T, Samiyappan R (2009) Fluorescent pseudomonad mixtures mediate disease resistance in rice plants against sheath rot (*Sarocladium oryzae*) disease. Biocontrol 54:273–286. doi:10.1007/s10526-008-9166-9

Saravanakumar D, Vijayakumar C, Kumar N, Samiyappan R (2007) PGPR-induced defense responses in the tea plant against blister blight disease. Crop Prot 26:556–565. doi:10.1016/j.cropro.2006.05.007

Schneider SM, Rosskopf EN, Leesch JG, Chellemi DO, Bull CT, Mazzola M (2003) United States department of agriculture - agricultural research service research on alternatives to methyl bromide: pre-plant and post-harvest. Pest Manag Sci 59:814–826. doi:10.1002/ps.728

Siddiqui IA, Ehetshamul-Haque S, Shaukat SS (2001) Use of rhizobacteria in the control of root rot-root knot disease complex of mungbean. J Phytopathol 149:337–346

Siddiqui IA, Haas D, Heeb S (2005) Extracellular protease of *Pseudomonas fluorescens* CHA0, a biocontrol factor with activity against the root-knot nematode *Meloidogyne incognita*. Appl Environ Microbiol 71:5646–5649. doi:10.1128/AEM.71.9.5646-5649.2005

Siddiqui IA, Shaukat SS (2002) Rhizobacteria-mediated induction of systemic resistance (ISR) in tomato against *Meloidogyne javanica*. J Phytopathol 150:469–473

Siddiqui IA, Shaukat SS (2003) Suppression of root-knot disease by *Pseudomonas* fluorescens CHA0 in tomato: importance of bacterial secondary metabolite 2, 4-diacetylphloroglucinol. Soil Biol Biochem 35:1615–1623. doi:10.1016/j.soilbio.2003.08.006

Siddiqui IA, Shaukat SS (2004) Systemic resistance in tomato induced by biocontrol bacteria against the root-knot nematode, *Meloidogyne javanica* is independent of salicylic acid production. J Phytopathol 152:48–54

Siddiqui ZA, Akhtar MS (2008) Effects of organic wastes, *Glomus intraradices* and *Pseudomonas putida* on the growth of tomato and on the reproduction of the root-knot nematode *Meloidogyne incognita*. Phytoparasitica 36:460–471

Siddiqui ZA, Akhtar MS (2009a) Effect of plant growth promoting rhizobacteria, nematode parasitic fungi and root-nodule bacterium on root-knot nematodes *Meloidogyne javanica* and growth of chickpea. Biocontrol Sci Technol 19:511–521. doi:10.1080/09583150902887792

Siddiqui ZA, Akhtar MS (2009b) Effects of antagonistic fungi and plant growth-promoting rhizobacteria on growth of tomato and reproduction of the root-knot nematode, *Meloidogyne incognita*. Australas Plant Pathol 38:22–28. doi:10.1071/AP08072

Siddiqui ZA, Baghel G, Akhtar MS (2007) Biocontrol of *Meloidogyne javanica* by *Rhizobium* and plant growth-promoting rhizobacteria on lentil. World J Microb Biot 23:435–441. doi:10.1007/s11274-006-9244-z

Siddiqui ZA, Mahmood I (2001) Effects of rhizobacteria and root symbionts on the reproduction of *Meloidogyne javanica* and growth of chickpea. Bioresour Technol 79:41–45

Sikora EJ, Murphy JF (2005) Identification and management of cucumber mosaic virus in Alabama. Acta Hort 695:191–194

Sivritepe HO, Dourado AM (1995) The effect of priming treatments on the viability and accumulation of chromosomal damage in aged pea seeds. Ann Bot 75:165–171

Someya N, Kataoka N, Komagata T, Hirayae K, Hibi T, Akutsu K (2000) Biological control of cyclamen soilborne diseases by *Serratia marcescens* strain B2. Plant Dis 84(3):334–340

Son SH, Khan Z, Vim SG, Kim YH (2008) Effects of seed treatment with rhizobacterium, *Paenibacillus* species on management of root-knot nematode-*Fusarium* wilt fungus disease complex in tomato plants. Russ J Nematol 16:97–105

Sorensen J (1997) The rhizosphere as a habitat for soil microorganisms. In: van Elsas JD, Trevors JT, Welington EMH (eds) Modern soil ecology. Marcel Dekker Inc., New York, pp 21–46

Sticher L, Mauch-Mani B, Métraux JP (1997) Systemic acquired resistance. Annu Rev Phytopathol 35:235–270

Stout MJ, Zehnder GW, Baur ME (2002) Potential for the use of elicitors of plant resistance in arthropod management programs. Arch Insect Biochem 51:222–235. doi:10.1002/arch.10066

Sturz AV, Mathenson BG, Arsenault W, Kimpinski J, Christie BR (2001) Weeds as a source of plant growth promoting rhizobacteria in agricultural soils. Can J Microbiol 47:1013–1024

Sudhakar P, Chattopadhyay GN, Gangwar SK, Ghosh JK (2000) Effect of foliar application of *Azotobacter*, *Azospirillum* and *Beijerinckia* on leaf yield and quality of mulberry (*Morus alba*). J Agr Sci 134:227–234

Sudisha J, Niranjana SR, Umesha S, Prakash HS, Shetty HS (2006) Transmission of seed-borne infection of muskmelon by *Didymella bryoniae* and effect of seed treatments on disease incidence and fruit yield. Biol Control 37:196–205. doi:10.1016/j.biocontrol.2005.11.018

Sundara B, Natarajan V, Hari K (2002) Influence of phosphorus solubilizing bacteria on the changes in soil available phosphorus and sugarcane and sugar yields. Field Crop Res 77:43–49

Swaider JM, Ware GW, McCollum JP (1992) Producing vegetable crops. Interstate Publishers Inc, Danville, IL

Teixeira DA, Alfenas AC, Mafia RG, Maffia LA, Ferreira EM (2005) Evidence of induction of systemic resistance to eucalyptus rust by plant growth promoting rhizobacteria. Fitopatol Bras 30:350–356. doi:10.1590/S0100-41582005000400003

Termorshuizen AJ (2002) Cultural control. In: Waller JM, Lenné JM, Waller SJ (eds) Plant pathologist's pocketbook, 3rd edn. CABI Publishing, CAB International, Wallingford, Oxon, UK, pp 318–327

Tian B, Yang J, Ke-Qin Z (2007) Bacteria used in the biological control of plant-parasitic nematodes: populations, mechanisms ofaction, and future prospects. FEMS Microbiol Ecol 61:197–213. doi:10.1111/j.1574-6941.2007.00349.x

Timmusk Someya NS, Nicander B, Granhall U, Tillberg E (1999) Cytokinin production by *Paenibacillus polymyxa*. Soil Biol Biochem 31:1847–1852

Umesha S (2006) Occurrence of bacterial canker in tomato fields of Karnataka and effect of biological seed treatment on disease incidence. Crop Prot 25:375–381. doi:10.1016/j.cropro.2005.06.005

Utkhede R, Koch C (2004) Biological treatments to control bacterial canker of greenhouse tomatoes. Biocontrol 49:305–313

Vallad GE, Goodman RM (2004) Systemic acquired resistance and induced systemic resistance in conventional agriculture. Crop Sci 44:1920–1934

van Loon LC, Bakker PAHM (2006) Induced systemic resistance as a mechanism of disease suppression by rhizobacteria. In: Siddiqui ZA (ed) PGPR: biocontrol and biofertilization. Springer, Dordrecht, Netherlands, pp 39–66

van Loon LC, Bakker PAHM, Pieterse CMJ (1998) Systemic resistance induced by rhizosphere bacteria. Annu Rev Phytopathol 36:453–483 van Peer R, Niemann GJ, Schippers B (1991) Induced resistance and phytoalexin accumulation in biological control of fusarium wilt of carnation by *Pseudomonas* sp. strain WCS417r. Phytopathology 81:728–734

Vargas DP, Ferrera-Cerrato R, Almaraz-Suarez JJ, Gonzalez AG (2001) Inoculation of plant growth-promoting bacteria in lettuce. Terra 19:327–335

Verma SC, Ladha JK, Tripathi AK (2001) Evaluation of plant growth promoting and colonization ability of endophytic diazotrophs from deep water rice. J Biotechnol 91:127–141

Vessey JK (2003) Plant growth promoting rhizobacteria as biofertilizers. Plant Soil 255:571–586

Walker JC (1957) Plant pathology, 2nd edn. McGraw-Hill Book Company Inc., New York

Waller JM (2002) Virus diseases. In: Waller JM, Lenné JM, Waller SJ (eds) Plant pathologist's pocketbook, 3rd edn. CABI Publishing, CAB International, Wallingford, Oxon, UK, pp 108–125

Waller JM, Cannon PF (2002) Fungi as plant pathogens. In: Waller JM, Lenné JM, Waller SJ (eds) Plant pathologist's pocketbook, 3rd edn. CABI Publishing, Wallingford, Oxon, UK, pp 318–327

Waller JM, Lenné JM, Waller SJ (2002) Plant pathologist's pocketbook, 3rd edn. CABI Publishing, Wallingford, Oxon

Walters D, Heil M (2007) Costs and trade-offs associated with induced resistance. Physiol Mol Plant P 71:3–17. doi:10.1016/j.pmpp. 2007.09.008

Wang KY, Liu TX, Yu CH, Jiang XY, Yi MQ (2002) Resistance of *Aphis gossypii* (Homoptera: Aphididae) to fenvalerate and imidacloprid and activities of detoxification enzymes on cotton and cucumber. J Econ Entomol 95:407–413

WenXing H, Tuo Y, HongYang S, LiNa S (2008) PGPR bio-fertilizers producing and its effect on *Avena sativa* growth and quality development. Acta Pratac Sin 17:75–84

Whitelaw MA (2000) Growth promotion of plants inoculated with phosphate-solubilizing fungi. Adv Agron 69:99–151

Xue QY, Chen Y, Li SM, Chen LF, Ding GC, Guo DW, Guo JH (2009) Evaluation of the strains of *Acinetobacter* and *Enterobacter* as potential biocontrol agents against *Ralstonia* wilt of tomato. Biol Control 48:252–258. doi:10.1016/j.biocontrol.2008.11.004

Yang JW, Kloepper JW, Ryu CM (2009) Rhizosphere bacteria help plants tolerate abiotic stress. Trends Plant Sci 14:1–4. doi:10.1016/j.tplants.2008.10.004

Yıldırım E, Turan M, Donmez MF (2008) Mitigation of salt stress in radish (*Raphanus sativus* l.) by plant growth promoting rhizobacteria. Roum Biotechnol Lett 13:3933–3943

Zahir ZA, Ghani U, Naveed M, Nadeem SM, Asghar HN (2009) Comparative effectiveness of *Pseudomonas* and *Serratia* sp. containing ACC-deaminase for improving growth and yield of wheat (*Triticum aestivum* L.) under salt-stressed conditions. Arch Microbiol 191:415–424. doi:10.1007/s00203-009-0466-y

Zahir ZA, Munir A, Asghar HN, Shaharoona B, Arshad M (2008) Effectiveness of rhizobacteria containing ACC deaminase for growth promotion of peas (*Pisum sativum*) under drought conditions. J Microb Biotechnol 18:958–963

Zaidi SFA (2003) Biocontrol of *Fusarium oxysporium* by plant growth promoting rhizobacteria (PGPRs) in soybean. Ann Agr Res 24:676–678

Zehnder G, Kloepper J, Tuzun S, Yao C, Wei G, Chambliss O, Shelby R (1997) Insect feeding on cucumber mediated by rhizobacteria-induced plant resistance. Ent Exp Appl 83:81–85

Zehnder GW, Yao CB, Murphy JF, Sikora ER, Kloepper JW (2000) Induction of resistance in tomato against cucumber mosaic cucumovirus by plant growth-promoting rhizobacteria. Biocontrol 45:127–137

Zhang L, Birch RG (1997) Mechanisms of biocontrol by *Pentoena dispersa* of sugar cane leaf scald disease caused by *Xanthomonas albilineans*. J Appl Microbiol 82:448–454

Zhuang X, Chen J, Shim H, Bai Z (2007) New advances in plant growth-promoting rhizobacteria for bioremediation. Environ Int 33:406–413. doi:10.1016/j.envint.2006.12.005

Importance of Biofilm Formation in Plant Growth Promoting Rhizobacterial Action

Gamini Seneviratne, M.L.M.A.W. Weerasekara, K.A.C.N. Seneviratne, J.S. Zavahir, M.L. Kecskés, and I.R. Kennedy

Contents

1 Introduction 82
2 Occurrence of PGPR Biofilms in Plant–Microbe Interaction 83
3 PGPR Biofilms in Futuristic Agriculture 85
3.1 PGPR Biofilms as Biofertilizers 86
3.2 PGPR Biofilms as Plant Growth Promoting Agents 88
3.3 PGPR Biofilms as Biocontrolling Agents 89
4 Conclusions 91
References 91

Abstract Among the diverse soil microflora, plant growth promoting rhizobacteria (PGPR) mark an important role in enhancing plant growth through a range of beneficial effects. This is often achieved by forming biofilms in the rhizosphere, which has advantages over planktonic mode of bacterial existence. However, the biofilm formation of PGPR has been overlooked in past research. This chapter

G. Seneviratne (✉)
Institute of Fundamental Studies, Hantana Road, Kandy, Sri Lanka
e-mail: gaminis@ifs.ac.lk

M.L.M.A.W. Weerasekara
Laboratory of Soil Microbiology, Faculty of Agriculture, Saga University, Saga, Japan
e-mail: anjani6466@gmail.com

K.A.C.N. Seneviratne
Royal Botanic Gardens, Peradeniya, Sri Lanka
e-mail: champase@yahoo.com

J.S. Zavahir
Geocycle (SBF), Dandenong South, VIC 3174, Australia
e-mail: bungali@rediffmail.com

M.L. Kecskés and I.R. Kennedy
SUNFix Centre for Nitrogen Fixation, Faculty of Agriculture, Food and Natural Resources, University of Sydney, Sydney, NSW 2006, Australia
e-mail: m.kecskes@usyd.edu.au; i.kennedy@usyd.edu.au

D.K. Maheshwari (ed.), *Plant Growth and Health Promoting Bacteria*, Microbiology Monographs 18, DOI 10.1007/978-3-642-13612-2_4,

focuses on new insights and concepts with reference to improved PGPR effects caused by the biofilm formation by PGPR and its impact on overall plant growth promotion, compared with the planktonic lifestyle of PGPR. Beneficial PGPR play a key role in agricultural approaches through quorum sensing in their biofilm mode. The in vitro production of biofilmed PGPR can be used to give increased crop yields through a range of plant growth mechanisms. They can be used as biofertilizers through improved N_2 fixation and micro- and macronutrient uptake. Further, higher levels of plant growth with PGPR have been observed due to their production of plant growth regulators and their abilities to act as biocontrol agents, which are carried out by the production of antibiotics and other antimicrobial compounds. The microbial inoculant industry would also benefit greatly by developing biofilmed PGPR with N_2 fixing microbes. Biofilmed PGPR can be manipulated to achieve results in novel agricultural endeavors and hence is as an area which needs a deeper probing into its potential.

1 Introduction

The soil represents a favorable habitat for diverse populations of microbes which have made inquisitive minds probe into their function and activities since time immemorial. The intrinsic roles they play in terrestrial ecosystems have a direct effect on plant growth and soil quality. This feature has led to considerable attention being paid to improve plant growth promotion using effective microorganisms in sustainable agriculture. By and large, this is attributed to the ability of microbes to "turnover" nutrients and to bind particles in soil which is essential for plant growth.

Among the plant associated soil microbial communities, root colonizing beneficial bacteria (rhizobacteria), known as plant growth promoting rhizobacteria (PGPR) (Lugtenberg and Kamilova 2009), are recognized as one of the predominant groups that wield a range of beneficial effects in enhancing plant growth. This is achieved by an array of activities including N_2 fixation, increasing the availability of phosphate and other nutrients in the soil, phytostimulation, suppression of plant diseases, synthesis of antibiotics and the production of phytohormones (Sivan and Chet 1992; Zehnder et al. 2001). Excellent reviews on the PGPR action on roots and mycorrhizosphere are found in Bending et al. (2006) and Spaepen et al. (2009). The success of PGPR in agriculture is attributed to their effective colonization of plant roots (Raaijmakers et al. 1995; Bolwerk et al. 2003) and subsequent growth to form microcolonies or biofilms, which is their common occurrence in a successful plant–microbe interaction (Saleh-Lakha and Glick 2006).

Biofilms are mass colonies of single or multispecies of microbial cells adherent to biotic or abiotic surfaces and/or in intimate contact with each other, encased in a self produced matrix of extracellular polymeric substances (EPS). Less complex biofilms with lower numbers of cells are variably described as microcolonies, aggregates, or cell clusters (Morris and Monier 2003; Ramey et al. 2004). The microcolony is the basic growth unit of a biofilm, and this mode of biofilms is predominant in almost all natural environments (Lappin-Scott and Costerton 1995). The colonization of plant surfaces by plant-associated microbial populations shows similarities to the formation of biofilms on abiotic surfaces with certain genetic determinants common to both processes (Molina et al. 2003).

As outlined by Saleh-Lakha and Glick (2006), these bacterial assemblages have the capability to communicate chemically with one another through quorum sensing, functioning as a single unit. Thus, PGPR when they are in biofilm mode should perform well in inhibiting competing organisms, nutrient uptake, quick responses, and adaptation to changing environmental conditions. However, the natural existence of PGPR in the soil has not been adequately investigated, and the knowledge of biofilmed mode of PGPR and their actions is vastly unexplored. Some reports have highlighted that the plant-associated biofilms have a higher ability to protect themselves from external stress and microbial competition that are characteristic of the rhizosphere, and also to produce beneficial effects in plant growth promotion (Ramey et al. 2004; Seneviratne et al. 2008a, b, 2009). Additionally, it has been shown that naturally occurring or in vitro produced effective PGPR inocula have many potential uses evidently in agricultural and biotechnological settings (Seneviratne et al. 2008b).

Most bacteria appear to form biofilms and this multicellular mode of growth likely predominates in nature as a protective mechanism against hostile environmental conditions (e.g., *Pseudomonas aeruginosa*, Costerton et al. 1995; Costerton and Stewart 2000; Walker et al. 2004). Biofilms, in general, have unique developmental characteristics that are different to freely swimming planktonic cells or nonbiofilm-forming cells. Molecular and genetic studies have identified that biofilms differ considerably from individual microbes in planktonic mode of growth in vital characteristics such as gene expression (Davies et al. 1993; Vilain and Brözel 2006) and physiological functions (Dow et al. 2007). Further, Stoodley et al. (2002) reported that as a result of biofilm structure, physiological adaptation, and the adherent nature of microbial cells in biofilms, they show an elevated antimicrobial tolerance.

Thus, the role of biofilm architecture in plant–microbe interactions cannot be negligible and identification of plant growth improvements through developed biofilmed inocula would have a great scope in plant growth promotion. The impact of microbial biofilms in plant growth promotion has not received adequate attention and studies of beneficial biofilm communities are thus of special interest. This chapter focuses on new insights and concepts with reference to improved PGPR effects caused by the biofilm formation by PGPR and its impact on overall plant growth promotion, compared with the planktonic lifestyle of PGPR. In addition, their potentials in agricultural innovations are also discussed.

2 Occurrence of PGPR Biofilms in Plant–Microbe Interaction

It is well known that most microorganisms in the rhizosphere exist as biofilms rather than their planktonic mode (Watnick and Kolter 1999; Davey and O'Toole 2000). Biofilms associated with the plant roots have been found to be beneficial for plant growth, yield, and crop quality. PGPR biofilm formation and plant growth promotion are governed by effective root colonization of the host plant (Saleh-Lakha and Glick 2006). However, to date biofilm-mediated PGPR actions have not been described adequately. Therefore, evidences found in literature for occurrence of PGPR biofilms in plant–microbe interactions and their possible mechanisms are discussed in this section.

Common plant-associated bacteria found on leaves, roots, and the soil such as *P. putida*, *P. fluorescens*, and related pseudomonads, together with a majority of other natural isolates, have been reported to form effective biofilms (Ude et al. 2006). Bloemberg et al. (2000) noted that the plant growth promoting *P. fluorescens* discontinuously colonized the root surface, developing as small biofilms along epidermal tissues. In contrast, dense and confluent biofilms on root surfaces were observed in studies analyzing pathogenic *Pseudomonas* spp. (Bais et al. 2004; Walker et al. 2004). Although the fundamental cause of these different observations is uncertain, it is evident that the root biofilms of *Pseudomonas* spp. can range from relatively small multicellular clusters to extensive biofilm networks.

Microbes in root-associated biofilms depend basically on root exudates for food and nutrition (Bais et al. 2006). Although the quantities of organic compounds exuding from plant roots are not large, seldom exceeding 0.4% of the C photosynthesized, they exert a very strong influence on the soil microorganisms and may be significant in affecting plant nutrient availability (Rovira 1969). By providing organic compounds as a nutrient source, these root exudates take a central role in being a major plant-derived factor and in triggering of root colonization (Lugtenberg et al. 1999) and biofilm associations (Walker et al. 2004). Some studies have also suggested that the biofilm formation at root sites is triggered by a plant-derived component similar to that seen in *Rhizobium*-legume and other bacterial interactions (de Ruijter et al. 1999), which has happened to be organic compounds of root exudates in this case. The role played by root exudates is further confirmed by Espinosa-Urgel et al. (2002) by observing that *P. putida* can respond rapidly to the presence of root exudates in soils, converging at root colonization sites and establishing stable biofilms.

Most species of bacteria use the quorum sensing to coordinate their gene expression according to the local density of their population. This signaling mechanism modulates and coordinates bacterial interactions with plants, including the control of tissue maceration, antibiotic production, toxin release, and horizontal gene transfer (HGT) (von Bodman et al. 2003). It is one of the main regulatory mechanisms in the formation of biofilms and it is seen that most beneficial phenotypes of PGPR are under its control (Loh et al. 2002). Quorum sensing of PGPR is mediated by an array of signal molecules which include (a) acylated homoserine lactones (AHLs) among proteobacteria; (b) gamma-butyrolactones in Streptomyces species; (c) *cis*-11-methyl-2-dodecanoic acid (also called DSF) in species of *Xanthomonas*, *Xylella*, and their relatives; and (d) oligopeptides among gram-positive microbes (Danhorn and Fuqua 2007). The AHLs-mediated cell-to-cell communication is mostly common among rhizospheric bacteria. The AHLs act as population density sensors and facilitate the communication between different cells (Pierson et al. 1998). Although the AHLs-based quorum sensing is characterized by the proteins LuxI-type protein, AHL synthase, and LuxR-type protein, exceptions have been reported for *Vibrio harveyi* and *P. fluorescens* F113, as they replace the LuxI-type with LuxM AHL and HdtS AHL synthase, respectively (Case et al. 2008). The AHLs-mediated quorum sensing is widely detected in *Pseudomonas* spp. than any other root colonizing bacteria (Juhas et al. 2005). The root-associated biocontrol agent *P. fluorescens* 2P24 requires AHLs for biofilm formation and therefore control of take-all disease on wheat (Wei and Zhang 2006).

It is evident from above information that biofilm formation by PGPR is common in the rhizosphere and that quorum-sensing-based cell-to-cell communication could play a key role in the action of PGPR in green agricultural approaches. The importance of discovering effective forms of PGPR biofilms leads us to the next section, where we focus on their potential applications in futuristic agricultural systems.

3 PGPR Biofilms in Futuristic Agriculture

The current public concerns on the detrimental side effects in the use of agrochemicals have lead to search other avenues of gaining better crop productivity. Of these, an increasing interest has been shown in the use of biofertilizers comprising of beneficial microorganisms, which improves plant growth through the supply of plant nutrients in a manner sustaining environmental health and soil productivity (O'Connell 1992). However, an inconsistency in the field application of such microbial inocula has limited its widespread commercial application, most probably due to the incapability of such inocula to successfully compete with indigenous microflora in establishing themselves in the rhizosphere (Van Elsas et al. 1986; Bent and Chanway 1998).

This failure can be overcome by the introduction of bacterial inoculants in the form of biofilms, thus protecting the inoculants against adverse environmental conditions such as high salinity, tannin concentrations, low pH, heavy metals, predation by earthworms, the competition by native soil populations (Seneviratne et al. 2008b), and the resistance to protozoan grazing (Matz et al. 2004). In this respect, the use of well-characterized PGPR biofilms is remarkable than solitary PGPR since the biofilm formation is an added advantage for PGPR to colonize effectively on or in the plant root, where they can compete well with indigenous microflora along with improved plant growth promotion. This has been made evident by Timmusk et al. (2005) who reported that *Paenibacillus polymyxa* forms biofilms around the root tip and behaves as a root-invading bacterium attributing a possible mechanism in biocontrol and drought tolerance-enhancing activities. Apart from the root colonization, recent observations have been made that the bacterial colonization of biotic fungal surfaces leading to the formation of fungal–bacterial biofilms (FBB) renders the biofilms enhanced metabolic activities in comparison to monocultures, leading to improved biofertilizing and biocontroling effects (Seneviratne et al. 2008a, b). Further, as speculated by Seneviratne and Jayasinghearachchi (2003), the establishment of such biofilms of rhizobia with common soil fungi provides a plausible strategy for the rhizobial survival.

This leads us to confirm that the in vitro production of such biofilmed inocula with PGPR can be utilized to satisfy the future demand of augmented crop production attributed to increased N_2 fixation, nutrient uptake, plant growth promoting agents, and biocontrol of diseases, through a range of mechanisms described below.

3.1 PGPR Biofilms as Biofertilizers

The plant growth promoting rhizobacterial species which flourish in the rhizosphere of plants have been seen to stimulate plant growth, yield, and crop quality by a plethora of mechanisms. This has led to a considerable number of PGPR being tested as biofertilizers, mainly because they provide inorganic nutrients to plants by mineralizing organic and insoluble inorganic forms of nitrogen, phosphorous, and sulfur that plants cannot utilize directly (Mendez-Castro and Alexander 1983) as well as providing essential micro and macro nutrients. This has been made evident by the possession of N_2-fixing properties in many PGPR species including *Bacillus* spp., *Azotobacter* spp., *Azospirillum* spp., *Beijerinckia* spp., *Pseudomonas* spp. (Dobereiner 1997; Reis et al. 1994; Vance 1997), and *Rhizobium* and *Bradyrhizobium* spp. (Dobereiner 1997; Vance 1997).

Such PGPR have been seen to valuably carry out their N_2-fixing ability in the biofilm mode as well, as shown by many studies. Jayasinghearachchi and Seneviratne (2004a) demonstrated that a fungal-rhizobial biofilm (FRB) (*Bradyrhizobium elkanii* SEMIA 5019 and *Penicillium* spp.) biologically fixed N_2 more effectively, as revealed by nitrogenase activity and N accumulation, in comparison to the rhizobial strain grown as a monoculture. A developed biofilmed inoculant of this FRB was also seen to significantly increase N_2 fixation (by ca. 30%) compared with a rhizobium-only (conventional) inoculant when applied to soybean (Jayasinghearachchi and Seneviratne 2004b). The ability to increase N availability by ca. twofold and a high nitrogenase activity, even under a very high soil NO_3^- concentration, were observed in the direct application of FRBs to soil, compared with the monocultures (Seneviratne and Jayasinghearachchi 2005). Yet another PGPR *Azospirillum brasilense*, a free-living N_2 fixer, was found to interact with roots of wheat and maize, forming dense biofilms and thereby promoting their host plant's growth (Assmus et al. 1995; Burdman et al. 2000).

Of the PGPR used to date, two genus most widely known are *Rhizobium* and *Bradyrhizobium* and their symbiotic N_2 fixation through inoculation to legume crops is well known (Dobereiner 1997; Vance 1997). Recent reports have indicated that these symbiotic bacteria may have the potential to be used as PGPR with nonlegumes. Seneviratne et al. (2009) have recently observed the heavy colonization of FBBs/FRBs on root hairs of rice (*Oryza sativa*), tea (*Camellia sinensis*),

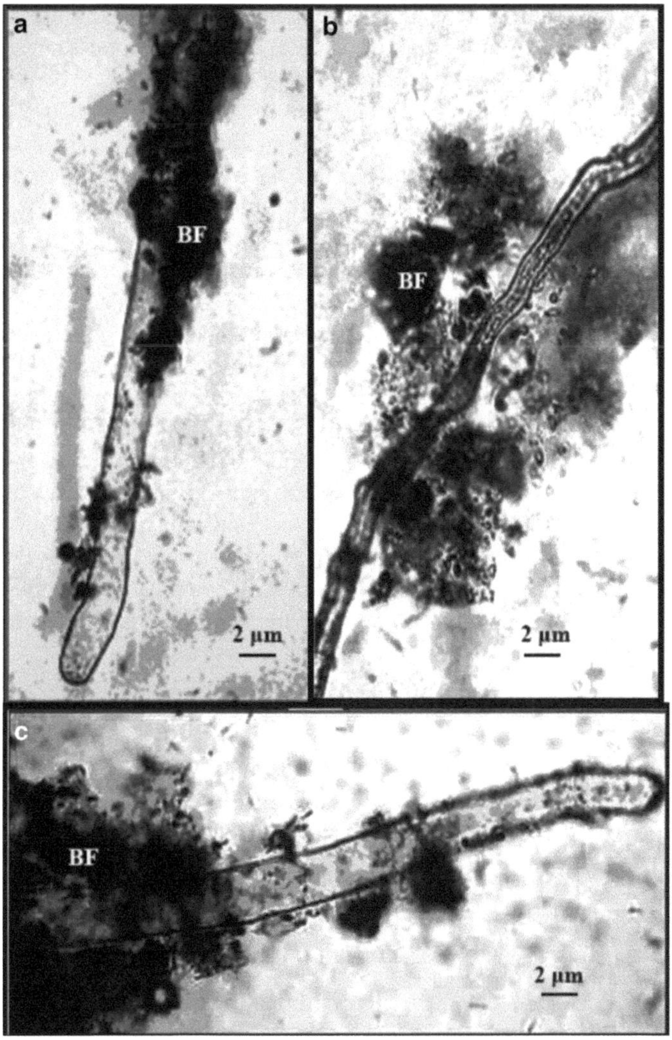

Fig. 1 Root hairs of rice (**a**), tea (**b**), and anthurium (**c**) colonized by microbial biofilms (BF), when fungal–bacterial biofilms (FBB) or fungal–rhizobial biofilms (FRB) were inoculated under axenic conditions. Darkness is due to cotton blue stain absorbed by the extra cellular polymeric substances (EPS) produced by the BF. Reprinted from Seneviratne et al. (2009)

Anthurium (*Anthurium andraeanum*), and wheat (*Triticum aestivum*) (Fig. 1). It has been suggested that such FRBs may act as "pseudonodules," fixing N_2 biologically on the roots of nonlegumes. Further, it was found that recommended chemical fertilizers may be reduced by about 50% by applying such in vitro produced biofilmed biofertilizers (BBs). When BBs were applied with chemical fertilizers to micropropagated Anthurium plantlets, leaf number and chlorophyll content increased by ca. 60% and 100%, respectively, compared with chemical fertilizers alone application (KACN Seneviratne, unpublished). The BBs alone application increased root length of Anthurium by ca. 65%, compared with chemical fertilizers alone application.

Phosphorus (P) is a highly limited nutrient in some soils and hence phosphate-solubilizing bacteria play an important role in the P nutrition in plant growth. Seneviratne and Jayasinghearachchi (2005) have shown that the application of FRBs directly into soil increased P availabilities by 15-fold and that the biofilmed inocula can be effectively used in biosolubilisation of rock phosphate. This was amply demonstrated by an increased P solubilisation (up to ca. 230%) when biofilms developed from *Penicillium* spp., *Pleurotus ostreatus*, and *Xanthoparmelia mexicana*, a lichen fungus, were used compared with the fungus-only cultures (Jayasinghearachchi and Seneviratne 2006a; Seneviratne and Indrasena 2006).

Apart from the major nutrients required for plant growth, some studies have also shown that coinoculation of PGPR inocula enhanced the uptake of micronutrient such as Zn, Cu, and Fe (Bashan 1998). Coinoculation of *Pseudomonas* BA-8 + *Bacillus* OSU-142 increased Fe and Zn contents of leaves up to 50.5 and 35.5%, respectively, compared with the control (Esitken et al. 2005). Investigations of the modes of action by PGPR are increasing at a rapid pace to exploit them commercially as biofertilizers. The benefits of such combinations of mixed cultures or biofilms can be manipulated to overcome the challenges facing for more widespread utilization of PGPR as biofertilizers.

3.2 PGPR Biofilms as Plant Growth Promoting Agents

Numerous studies have demonstrated an improvement in plant growth and development in response to seed or root inoculation with various microbial inoculants capable of producing plant growth regulators (Zahir et al. 2004). Important plant growth promoting substances commonly produced by rhizosphere bacteria include auxins (indolyl-3-acetic acid), gibberellins, and cytokinins (Brown 1974).

Studies by Bandara et al. (2006) revealed that higher acidity and PGP hormone levels were produced by FBBs of beneficial rice endophytes than their mono- or mixed cultured forms with no biofilm formation. Their studies on a large collection of microbes also revealed the existence of a significant negative relationship between the production of indoleacetic-acid-like substances (IAAS) and pH in liquid culture media of FBBs, but not in mixed cultures with no biofilm formation. This high acidity reflects high IAAS production when biofilms are formed. Thus, the use of biofilmed inocula, rather than the conventional practice of plant inoculation with monocultures or mixed cultures of effective microbes, may help achieve the highest microbial effect. Another recent study on early growth of rice showed that the contribution of developed biofilmed inocula in enhanced release of organic acids and PGP substances led to ca. 25% increase in plant dry weight compared with conventional monocultured inocula (Seneviratne et al. 2009). In further studies, biofilmed inocula showed lower pH, higher IAAS, and rice plant dry weights than the monocultured inocula (MLMAW Weerasekara, unpublished). The biofilmed inocula showed a fourfold increase in H^+ secretion to the culture medium, compared with the monocultured inocula. Negative relationships were observed between pH of both types of the inocula and plant dry weight (Fig. 2a) or soil NH_4^+ (Fig. 2b). This implies that the inoculated biofilmed inocula colonize the rhizosphere, producing high acidity and IAAS (Seneviratne et al. 2008a), and the high acidity in microsites causes to an increase of plant available NH_4^+ (Xu et al. 1997) in the soil solution near root hairs, which helps increase the plant growth. Therefore, in vitro production and application of more effective combinations of such beneficial biofilmed inocula would play an important role in the inoculant industry. However, this needs further research to fully understand the effects and potentials of the biofilmed inocula in the plant growth promotion. It is clear from the above studies that one of the most plausible mechanisms of plant growth promotion by PGPR is the production of plant growth regulators. Further, the effectiveness of using such PGPR in their biofilmed mode in the production of higher levels of plant growth promoting substances is also noticeable.

3.3 PGPR Biofilms as Biocontrolling Agents

Biocontrolling has been seen as a well-suited alternative or supplement in contrast to conventional methods of disease control of which microbial biocontrolling

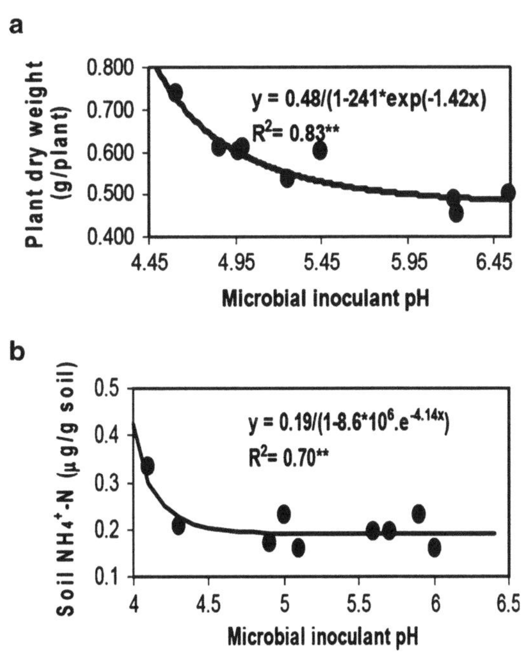

Fig. 2 Relationships between (**a**) microbial inoculant pH of both biofilmed and conventional inocula and rice plant dry weight, and (**b**) the microbial inoculant pH and soil NH_4^+, when inoculated in a soil pot experiment. The biofilmed inocula represent relatively low pHs

agents have emerged as favored options due to their complex mode of action and success in bringing out a reduced risk of resistance. For example, the extensive studies of root-associated pseudomonads have revealed that many of these promote the growth of host plants or are used as biocontrol agents (Lugtenberg et al. 2001). *P. fluorescens* has been reported to coat plant roots by forming a biofilm, which may protect roots against soil bacterial and fungal pathogens (O'Toole and Kolter 1998; Walker et al. 2004). The promising nature of PGPR strains as means of plant protection via disease suppression was amply demonstrated by Raupach and Kloepper (1998) in finding the occurrence of a consistent protection against pathogens when mixtures of PGPR were present, possibly in the biofilm mode.

An array of studies has confirmed that bacteria when they are in the biofilm mode perform well as biocontrol agents, mainly because the plant is made less sensitive to infection by the formation of biofilms by bacteria on the plant root (Bais et al. 2004; Rudrappa et al. 2008). Owing to the heterogeneous nature of biofilms, it is likely that the biofilm formation on the plant roots protects the plants against soil borne diseases through resistance mechanisms such as cell–cell communication via quorum-sensing (Danhorn and Fuqua 2007) and production of antibiotics against pathogens (Russo et al. 2006).

Biofilms bring about disease suppression through a variety of roles played by antibiotics. Such microbial communities have a significant resistance to antibiotics compared with planktonic bacteria of the same species (Stewart and Costerton 2001), while some biofilms have the ability to produce different antibiotics (Leifert et al. 1995; Raaijmakers et al. 2002; Yu et al. 2002; Risøen et al. 2004; Roberts and Stewart 2005).

In addition, biocontrolling agents of PGPR have been shown to successfully establish in plants, when they were applied as biofilmed inocula. Jayasinghearachchi and Seneviratne (2006b) confirmed this in vitro by using a *Pleurotus ostreatus* – *Pseudomonas fluorescens* biofilm which was seen to increase endophytic colonization of tomato by *P. fluorescens*, a biocontrolling agent, by over tenfold compared with inoculation of *P. fluorescens* alone. The PGPR *Paenibacillus polymyxa* provides protection from pathogens through the synthesis of several antibiotics, when it forms biofilms by predominantly colonizing the root tips of *Arabidopsis thaliana*, as revealed by fluorescence microscopy and electron scanning microscopy (Timmusk et al. 2005). *Bacillus subtilis*, another biocontrolling PGPR, protects plant roots from pathogenic bacteria by mechanisms which include biofilm formation and antibiotic and surfactin production (Bais et al. 2004; Cavaglieri et al. 2005). Surfactin possesses antimicrobial activity, and pathogens those reach inside the biofilms are killed by high surfactin concentrations (Bais et al. 2004).

Bacteria used to accomplish biocontrolling exert their action also through producing antimicrobial secondary metabolites, which target the competing microorganisms (Mazzola et al. 1992; Raaijmakers et al. 2002; Haas and Keel 2003). Some *Pseudomonas* strains secrete antimicrobial compounds such as exoproteases, antibiotics, HCN, or metabolites with antifungal activity known as phenazines (Molina et al. 2003). These compounds have the capacity to eliminate competititors from the rhizosphere with a plethora of studies demonstrating their prospect as biocontrol agents (Thomashow 1996; Chin-A-Woeng et al. 2000; Kremer and Souissi 2001). Studies outlined above highlight the potential of using biofilmed PGPR with increased microbial action to carry out biocontrol feats in conventional agriculture and organic farming systems.

4 Conclusions

Although developing biofilms has been the axis around which many recent studies have evolved in diverse areas of biotechnology, the investigation of the involvement of PGPR in such biofilms is yet in its infancy. The capability of PGPR to colonize plant roots proficiently and carry out a range of benefits to the plant has made it one of the predominant soil microbial groups. Regulatory mechanisms, such as quorum sensing, exhibited by PGPR have made them stable partners in biofilms, placing them on a higher pedestal compared with their existence alone. PGPR biofilms have been shown to play a fundamental role in futuristic agricultural approaches such as biofertilizers, plant growth promoters, and biocontrolling agents. A heightened interest in recent times in inoculant technology has thrown much importance on the designing and developing of PGPR biofilmed inocula. The beneficial results they yield encourage the deeper delving into its applications and the innovative future perspectives. The importance of biofilm formation in PGPR action is thus an area which needs much more in depth exploration to bring out its true potential.

References

Assmus B, Hutzler P, Kirchhof G, Amann R, Lawrence JR, Hartmann A (1995) In situ localization of *Azospirillum brasilense* in the rhizosphere of wheat with fluorescently labeled rRNAtargeted oligonucleotide probes and scanning confocal laser microscopy. Appl Environ Microbiol 61:1013–1019

Bais HP, Fall R, Vivanco JM (2004) Biocontrol of *Bacillus subtilis* against infection of Arabidopsis roots by *Pseudomonas syringae* is facilitated by biofilm formation and surfactin production. Plant Physiol 134:307–319

Bais HP, Weir TL, Perry LG, Gilroy S, Vivanco JM (2006) The role of root exudates in rhizosphere interactions with plants and other organisms. Annu Rev Plant Biol 57:233–266

Bandara WMMS, Seneviratne G, Kulasooriya SA (2006) Interactions among endophytic bacteria and fungi: effects and potentials. J Biosci 31:645–650

Bashan Y (1998) Inoculants of plant growth-promoting bacteria for use in agriculture. Biotechnol Adv 16:729–770

Bending GD, Aspray TJ, Whipps JM (2006) Significance of microbial interactions in the mycorrhizosphere. Adv Appl Microbiol 60:97–132

Bent E, Chanway CP (1998) The growth-promoting effects of a bacterial endophyte on lodgepole pine are partially inhibited by the presence of other rhizobacteria. Can J Microbiol 44:980–988

Bloemberg GV, Wijfjes AHM, Lamers GEM, Stuurman N, Lugtenberg BJJ (2000) Simultaneous imaging of *Pseudomonas fluorescens* WCS365 populations expressing three different autofluorescent proteins in the rhizosphere: new perspectives for studying microbial communities. Mol Plant Microbe Interact 13:1170–1176

Bolwerk A, Lagopodi AL, Wijfjes AH, Lamers GE, Chin AWTF, Lugtenberg BJ, Bloemberg GV (2003) Interactions in the tomato rhizosphere of two Pseudomonas biocontrol strains with the phytopathogenic fungus *Fusarium oxysporum* f. sp. radicis-lycopersici. Mol Plant Microbe Interact 16:983–993

Brown ME (1974) Seed and root bacterization. Annu Rev Phytopathol 12:181–197

Burdman S, Okon Y, Jurkevitch E (2000) Surface characteristics of *Azospirillum brasilense* in relation to cell aggregation and attachment to plant roots. Crit Rev Microbiol 26:91–110

Case RJ, Labbate M, Kjelleberg S (2008) AHL-driven quorum-sensing circuits: their frequency and function among the Proteobacteria. ISME J 2:345–349

Cavaglieri L, Orlando J, Rodriguez MI, Chulze S, Etcheverry M (2005) Biocontrol of *Bacillus subtilis* against *Fusarium verticillioides* in vitro and at the maize root level. Res J Microbiol 156:748–754

Chin-A-Woeng TFC, Bloemberg GV, Mulders IHM, Dekkers LC, Lugtenberg BJJ (2000) Root colonisation is essential for biocontrol of tomato foot and root rot by the phenazine-1-carboxamide-producing bacterium *Pseudomonas chlororaphis* PCL1391. Mol Plant Microbe Interact 13:1340–1345

Costerton JW, Stewart PS (2000) Bacterial biofilms. In: Nataro JP, Blaser MJ, Cunningham-Rundles S (eds) Persistent bacterial infections. American Society of Microbiologists, Washington, pp 423–439

Costerton JW, Lewandowski Z, Caldwell DE, Korber DR, Lappin-Scott HM (1995) Microbial biofilms. Annu Rev Microbiol 49:711–745

Danhorn T, Fuqua C (2007) Biofilm formation by plant-associated bacteria. Annu Rev Microbiol 61:401–422

Davey ME, O'Toole AG (2000) Microbial biofilms: from ecology to molecular genetics. Microbiol Mol Biol Rev 64:847–867

Davies DG, Chakrabarty AM, Geesey GG (1993) Exopolysaccharide production in biofilms: substratum activation of alginate gene expression by *Pseudomonas aeruginosa*. Appl Environ Microbiol 59:1181–1186

de Ruijter NCA, Bisseling T, Emons AMC (1999) Rhizobium nod factors induce an increase in sub-apical fine bundles of actin filaments in *Vicia sativa* root hairs within minutes. Mol Plant Microbe Interact 12:829–832

Dobereiner J (1997) Biological nitrogen fixation in the tropics: social and economic contributions. Soil Biol Biochem 29:771–774

Dow JM, Fouhy Y, Lucey J et al (2007) Cyclic di-GMP as an intracellular signal regulating bacterial biofilm formation. In: Kjelleberg S, Givskov M (eds) The biofilm mode of life: mechanisms and adaptations. Horizon Bioscience, Norwich, pp 71–94

Esitken A, Ercisli S, Karlidag H, Sahin F (2005) Potential use of plant growth promoting rhizobacteria (PGPR) in organic apricot production. In: Proceedings of the international scientific conference of environmentally friendly fruit growing, Tartu-Estonia, pp 90–97

Espinosa-Urgel M, Kolter R, Ramos JL (2002) Root colonization by *Pseudomonas putida*: love at first sight. Microbiology 148:341–343

Haas D, Keel C (2003) Regulation of antibiotic production in root-colonized *Pseudomonas* spp. and relevance for biological control of plant disease. Annu Rev Phytopathol 41:117–153

Jayasinghearachchi HS, Seneviratne G (2004a) Can mushrooms fix atmospheric nitrogen? J Biosci 23:293–296

Jayasinghearachchi HS, Seneviratne G (2004b) A *Bradyrhizobial- Penicillium* spp. biofilm with nitrogenase activity improves N_2 fixing symbiosis of soybean. Biol Fertil Soils 40:432–434

Jayasinghearachchi HS, Seneviratne G (2006a) Fungal solubilization of rock phosphate is enhanced by forming fungal–rhizobial biofilms. Soil Biol Biochem 38:405–408

Jayasinghearachchi HS, Seneviratne G (2006b) A mushroom-fungus helps improve endophytic colonization of tomato by *Pseudomonas fluorescenc* through biofilm formation. Res J Microbiol 1:83–89

Juhas M, Eberl L, Tümmler B (2005) Quorum sensing: the power of cooperation in the world of *Pseudomonas*. Environ Microbiol 7:459–471

Kremer RJ, Souissi T (2001) Cyanide production by rhizobacteria and potential for suppression of weed seedling growth. Curr Microbiol 43:182–186

Lappin-Scott HM, Costerton JW (1995) Microbial biofilms. Cambridge University Press, Cambridge, p 324

Leifert C, Li H, Chidburee S, Hampson S, Workman S, Sigee D, Epton HAS, Harbour A (1995) Antibiotic production and biocontrol activity by *Bacillus subtilis* CL27 and *Bacillus pumilus* CL45. J Appl Bacteriol 78:97–108

Loh J, Pierson EA, Pierson LS III, Stacey G, Chatterjee A (2002) Quorum sensing in plant-associated bacteria. Curr Opin Plant Biol 5:285–290

Lugtenberg B, Kamilova F (2009) Plant-growth-promoting rhizobacteria. Annu Rev Microbiol 63:541–556

Lugtenberg BJJ, Kravchenko LV, Simons M (1999) Tomato seed and root exudate sugars: composition, utilization by Pseudomonas biocontrol strains and role in rhizosphere colonization. Environ Microbiol 1:439–446

Lugtenberg BJJ, Dekkers L, Bloemberg GV (2001) Molecular determinants of rhizosphere colonization by *Pseudomonas*. Annu Rev Phytopathol 39:461–490

Matz C, Bergfeld T, Rice SA et al (2004) Microcolonies, quorum sensing and cytotoxicity determine the survival of *Pseudomonas aeruginosa* biofilms exposed to protozoan grazing. Environ Microbiol 6:218–226

Mazzola M, Cook RJ, Thomashow LS, Weller DM, Pierson LS (1992) Contribution of phenazine antibiotic biosynthesis to the ecological competence of fluorescent pseudomonads in soil habitats. Appl Environ Microbiol 58:2616–2624

Mendez-Castro FA, Alexander M (1983) Method for establishing a bacterial inoculum on corn roots. Appl Environ Microbiol 45:248–254

Molina MA, Ramos JL, Espinosa-Urgel M (2003) Plant-associated biofilms. Rev Environ Sci Biotechnol 2:99–108

Morris CE, Monier JM (2003) The ecological significance of biofilm formation by plant-associated bacteria. Annu Rev Phytopathol 41:455–482

O'Connell PF (1992) Sustainable agriculture – a valid alternative. Outlook Agric 21:5–12

O'Toole GA, Kolter R (1998) Flagellar and twitching motility are necessary for *Pseudomonas aeruginosa* biofilm development. Mol Microbiol 30:295–304

Pierson EA, Wood DW, Cannon JA, Blachere FM, Pierson LS III (1998) Interpopulation signaling via N-Acyl-Homoserine lactones among bacteria in the wheat rhizosphere. Mol Plant Microbe Interact 11:1078–1084

Raaijmakers JM, Leeman M, van Oorschot MMP, van der Siuls I, Schippers B, Bakker PAHM (1995) Dose-response relationships of biological control of Fusarium wilt of radish by *Pseudomonas* spp. Phytopathol 85:1075–1081

Raaijmakers JM, Vlami M, de Souza JT (2002) Antibiotic production by bacterial biocontrol agents. Antonie Leeuwenhoek 81:537–547

Ramey BE, Matthysse AG, Fuqua C (2004) The FNR-type transcriptional regulator SinR controls maturation of *Agrobacterium tumefaciens* biofilms. Mol Microbiol 52:1495–1511

Raupach GS, Kloepper JW (1998) Mixtures of plant growth-promoting rhizobacteria enhance biological control of multiple cucumber pathogens. Phytopathol 88:1158–1164

Reis MY, Olivares FL, Dobereiner J (1994) Improved methodology for isolation of *Acetobacter diazatrophicus* and confirmation of its endophytic habitat. World J Microbiol Biotechnol 10:101–105

Risøen PA, Rønning P, Hegna IK, Kolstø AB (2004) Characterization of a broad range antimicrobial substance from *Bacillus cereus*. J Appl Microbiol 96:648–655

Roberts ME, Stewart PS (2005) Modelling protection from antimicrobial agents in biofilms through the formation of persister cells. Microbiology 51:75–80

Rovira AD (1969) Plant root exudates. Bot Rev 35:35–57

Rudrappa T, Biedrzycki ML, Bais HP (2008) Causes and consequences of plant-associated biofilms. FEMS Microbiol Ecol 641:53–166

Russo DM, Williams A, Edwards A, Posadas DM, Finnie C, Dankert M, Downie JA, Zorreguieta A (2006) Proteins exported via the PrsD-PrsE type I secretion system and the acidic exopolysaccharide are involved in biofilm formation by *Rhizobium leguminosarum*. J Bacteriol 188:4474–4486

Saleh-Lakha S, Glick BR (2006) Plant growth-promoting bacteria. In: van Elsas JD, Jansson JK, Trevors JT (eds) Modern soil microbiology. CRC/Thomson Publishing, Boca Raton, FL/UK, pp 503–520

Seneviratne G, Indrasena IK (2006) Nitrogen fixation in lichens is important for improved rock weathering. J Biosci 31:639–643

Seneviratne G, Jayasinghearachchi HS (2003) Mycelial colonization by bradyrhizobia and azorhizobia. J Biosci 28:243–247

Seneviratne G, Jayasinghearachchi HS (2005) A rhizobial biofilm with nitrogenase activity alters nutrient availability in a soil. Soil Biol Biochem 37:1975–1978

Seneviratne G, Kecskés ML, Kennedy IR (2008a) Biofilmed biofertilisers: novel inoculants for efficient nutrient use in plants. In: Kennedy IR, Choudhury ATMA, Kecskés ML, Rose MT (eds) Efficient nutrient use in rice production in Vietnam achieved using inoculants biofertilisers. Proceedings of a project (SMCN/2002/073) workshop held in Hanoi, Vietnam, 12–13 October 2007. ACIAR Proceeding No. 130, ACIAR, Canberra, pp 126–130

Seneviratne G, Zavahir JS, Bandara WMMS, Weerasekara MLMAW (2008b) Fungal–bacterial biofilms: their development for novel biotechnological applications. World J Microbiol Biotechnol 24:739–743

Seneviratne G, Thilakaratne RMMS, Jayasekara APDA, Seneviratne KACN, Padmathilake KRE, De Silva MSDL (2009) Developing beneficial microbial biofilms on roots of non-legumes: a novel biofertilizing technique. In: Khan MS, Zaidi A, Musarrat J (eds) Microbial strategy for crop improvement. Springer, Berlin, Heidelberg, pp 51–61

Sivan A, Chet I (1992) Microbial control of plant diseases. In: Mitchell R (ed) Environmental microbiology. Wiley-Liss, New York, pp 335–354

Spaepen S, Vanderleyden J, Okon Y (2009) Plant growth-promoting actions of rhizobacteria. Adv Bot Res 51:283–320

Stewart PS, Costerton JW (2001) Antibiotic resistance of bacteria in biofilms. Lancet 358: 135–138

Stoodley P, Sauer K, Davies DG, Costerton JW (2002) Biofilms as complex differentiated communities. Annu Rev Microbiol 56:187–209

Thomashow LS (1996) Biological control of plant root pathogens. Curr Opin Biotechnol 7:343–347

Timmusk S, Grantcharova N, Gerhart E, Wagner H (2005) *Paenibacillus polymyxa* invades plant roots and forms biofilms. Appl Environ Microbiol 71:7292–7300

Ude S, Arnold DL, Moon CD, Timms-Wilson T, Spiers AJ (2006) Biofilm formation and cellulose expression among diverse environmental *Pseudomonas* isolates. Environ Microbiol 8:1997–2011

van Elsas JD, Dijkstra AF, Govarert JM, van Veen JA (1986) Survival of *Pseudomonas fluorescens* and *Bacillus subtilis* introduced into soils of different texture in field microplots. FEMS Microbiol Ecol 38:150–160

Vance CP (1997) Enhanced agricultural sustainability through biological nitrogen fixation. In: biological fixation of nitrogen for economic and sustainable agriculture. Proceedings of a NATO Advanced Research Workshop, Poznan, Poland, pp 179–185

Vilain S, Brözel VS (2006) Multivariate approach to comparing whole-cell proteomes of *Bacillus cereus* indicates a biofilm specific proteome. J Proteome Res 5:1924–1930 von Bodman SB, Bauer WD, Coplin DL (2003) Quorum sensing in plant-pathogenic bacteria. Annu Rev Phytopathol 41:455–482

Walker TS, Bais HP, Déziel E, Schweizer HP, Rahme LG, Fall R, Vivanco JM (2004) *Pseudomonas aeruginosa*-plant root interactions. Pathogenicity, biofilm formation, and root exudation. Plant Physiol 134:320–331

Watnick PI, Kolter R (1999) Steps in the development of a *Vibrio cholerae* El Tor biofilm. Mol Microbiol 34:586–595

Wei HL, Zhang LQ (2006) Quorum-sensing system influences root colonization and biological control ability in *Pseudomonas fluorescens* 2P24. Antonie Leeuwenhoek 89:267–280

Xu JM, Cheng HH, Koskinen WC, Molina JAE (1997) Characterization of potentially bioreactive soil organic carbon and nitrogen by acid hydrolysis. Nutr Cycl Agroecosyst 49:267–271

Yu GY, Sinclair JB, Hartman GL, Bertagnolli BL (2002) Production of iturin A by *Bacillus amyloliquefaciens* suppressing *Rhizoctonia solani*. Soil Biol Biochem 34:955–963

Zahir AZ, Arshad M, Frankenberger WT Jr (2004) Plant growth promoting rhizobacteria: applications and perspectives in agriculture. Adv Agron 81:97–168

Zehnder GW, Murphy IF, Sikora EJ, Kloepper JW (2001) Application to rhizobacteria for induced resistance. Eur J Plant Pathol 107:39–50

Plant Growth Promoting Rhizobacteria: Constraints in Bioformulation, Commercialization, and Future Strategies

Naveen K. Arora, Ekta Khare, and Dinesh K. Maheshwari

Contents

1 Introduction 98
2 Plant Growth Promotory Bioformulations 99
3 Production and Marketing Constraints 99
3.1 High Cost of Production 100
3.2 Shelf Life 100
3.3 Inconsistent Performance: Fate of Inoculant Introduced in Soil 101
4 Research Areas for Development and Optimization of Bioformulations 102
4.1 Microbial Diversity 102
4.2 Metagenomics 104
4.3 Plant–Microbe–Microbe Interactions 105
4.4 Formulation Design 107
5 Integrated Management 108
6 Conclusion 109
References 112

Abstract Bioformulations for plant growth promotion continue to inspire research and development in many fields. Increase in soil fertility, plant growth promotion, and suppression of phytopathogens are the targets of the bioformulation industry that leads to the development of ecofriendly environment. The synthetic chemicals used in the agriculture to increase yields, kill pathogens, pests, and weeds, have a big harmful impact on the ecosystem. But still the chemicals rule the agroindustry. The aim of the review is to assess the constraints associated with the effective

N.K. Arora (✉) and E. Khare
Department of Microbiology, Institute of Biosciences and Biotechnology, CSJM University, Kanpur 208024, Uttar Pradesh, India
e-mail: nkarora_net@rediffmail.com

D.K. Maheshwari
Dapartment of Botany and Microbiology, Gurukula Kangri University, Haridwar 249404, Uttarakhand India
e-mail: maheshwaridk@gmail.com

D.K. Maheshwari (ed.), *Plant Growth and Health Promoting Bacteria*, Microbiology Monographs 18, DOI 10.1007/978-3-642-13612-2_5,

development of bioinoculant industry particularly in developing countries. Another objective of the review is to evaluate what should be explored in the future to uplift the stature of the bioinoculants. Bioformulations offer an environmentally sustainable approach to increase crop production and health, contributing substantially in making the twenty-first century the age of biotechnology.

1 Introduction

The new challenge in the new millennium is to produce more and more food from shrinking per capita arable land, keeping the environment safe. As agricultural production intensified over the past few decades, producers became more and more dependent on agrochemicals. Chemical fertilizers and pesticides are presently accumulating in the environment harming the ecosystem, causing pollution, and spreading disease (Gerhardson 2002). Therefore, the urgent need of biological agents is accepted worldwide. Interest in biological control of plant pathogens has increased considerably over the past years, partly as a response to public concern about the use of hazardous chemical pesticides, but also because it may provide control of diseases that cannot or only partially be managed by other control strategies (Arora et al. 2008b, c).

For many decades, bacteria have been introduced into soil or on seeds, roots, bulbs, or other planting material to improve plant growth and health. The major objectives of bacterization include enhancement of symbiotic or associative nitrogen fixation, degradation of xenobiotic compounds, plant growth promotion, and biological control of plant pathogenic microorganisms (van Elsas and Heijnen 1990; Whipps 2001). To date, many bacterial genera are being used and tested in bacterization, including *Acinetobacter*, *Agrobacterium*, *Alcaligens*, *Arthrobacter*, *Azospirillum*, *Azotobacter*, *Bacillus*, *Bradyrhizobium*, *Frankia*, *Pantoea*, *Pseudomonas*, *Rhizobium*, *Serratia*, *Stenotrophomonas*, *Streptomyces*, and *Thiobacillus* (Whipps 2001; Lutenberg and Kamilova 2009). Some of the fungal taxa that have been successfully commercialized and are currently marketed as Environmental Protection Agency (EPA) registered biopesticides in United States belong to the genera, *Ampelomyces*, *Candida*, *Coniothyrium*, and *Trichoderma* (McSpadden Gardner 2002).

Although plant growth promoting rhizobacteria (PGPR) occur in soil, usually their number is not high enough to compete with other bacteria commonly established in the rhizosphere. Therefore for agronomic utility, inoculation of plants with target microorganisms at a much higher concentration than those normally found in soil is necessary to take advantage of their beneficial properties for plant yield enhancement (Subba Rao 1993). The erratic performances of bioinoculants under field conditions have raised concerns about the practical potential offered by microbial releases into soil (Arora et al. 2007b). Soil is an unpredictable environment and an intended result is sometimes difficult to obtain. The immediate response to soil inoculation with PGP bioformulations varies considerably depending on the bacteria (PGP agent), plant species, soil type, inoculum density, and environment conditions. The inoculated bacteria sometimes do not survive in the soil when competing with the better adapted indigenous microflora (Bashan 1998). An effective PGP strain isolated from one region may not perform in the same way in other soil and climatic conditions (Johnsson et al. 1998). The aim of this review is to point out the status of bioformulations and constrains related with their application and to draw the focus on future research strategies for the development of better inoculants.

2 Plant Growth Promotory Bioformulations

Bioformulations are best defined as biologically active products containing one or more beneficial microbial strains in an easy to use and economical carrier materials. Usually, the term bioformulation refers to preparations of microorganism(s) that may be partial or complete substitute for chemical fertilization/ pesticides. Biological control of plant pests and pathogens continues to inspire research and development of formulations targeted at plant pathogens.

The first objective when considering inoculation with beneficial strain is to find the best bacteria available (Validov et al. 2007). Many potentially useful bacteria reported in the scientific literature never appear on the commercial market, perhaps because of inappropriate formulation. The carrier is the delivery vehicle of live microorganisms from the factory to the field; however, no universal carrier or formulation is presently available for the release of microorganisms into soil (Smith 1992). A good carrier should have the capacity to deliver the right number of viable cells in good physiological condition at the right time (Smith 1992; Arora et al. 2008a). Carriers can be divided into three basic categories: (1) soils (peat, coal, clays, and inorganic soil), (2) plant waste materials, (3) inert materials viz. vermiculite, ground rock phosphate, polyacrylamide gels, and alginate beads (Bashan 1998). Inoculants come in four basic dispersal forms (powders, slurries, granules, and liquids). The use of each type of inoculant depends upon market availability, choice of farmers, cost, and the need of a particular crop under specific environmental conditions.

3 Production and Marketing Constraints

Since bioformulation is a living product, utmost care is needed at all the steps beginning from the production till the end use to maintain the microbial load and vigor. Production technology of bioformulations requires proper care and aid of sophisticated equipments to ensure availability of quality products in the market. Kabi (1997) gave stress on the production of quality inoculants since these are very important not only in providing nutrient supply to the plants but also in rendering sustainability to farming systems. In developing countries, the insufficient knowledge, lack of adequate machinery, and improper distribution and importation laws for live inoculants can lead to loss of their viability and effectiveness. The major constraints associated with effective bioformulation development are as follows:

3.1 High Cost of Production

Because of low price structure, hi-tech instrumentation required for producing bioformulations under completely sterile conditions is not getting acceptance. The potential hazards associated with bacterial contaminants should not be ignored as long as nonsterile carrier inoculants are widely used. However, one should note that the use of nonsterile carrier inoculants has caused no reported health hazards in decades of usage (Bashan 1998). The development of bacterial inoculants is claimed to be cheaper than that of agrochemicals, although the large-scale screening of strains with biological activity is still required (comparable to more than 1:20,000 screened molecules for a new chemical product) (Bashan 1998).

Deficiencies in handling procedure are a major cause of under performance in real life application. The high sensitivity to temperature and other external conditions of these "living" inputs calls for enormous caution at the stage of manufacture/culture, transportation/distribution, and application. This involves investment in packaging, storage, and use of suitable carrier materials (Arora et al. 2001).

Spurring the development of agricultural markets is the key factor for achieving targeted growth in bioformulation usage. In general, firms with larger production facilities are expected to invest more on networks to understand and access the market. A big obstacle is the registration procedure, which is often expensive and time consuming; especially, the cost of registration is the principal obstacle in the development of new products (Ehlers 2006).

3.2 Shelf Life

One of the main barriers faced by the producers of bioformulations and investors is inadequate demand and the inconsistent and seasonal nature of the existing demand, necessitating efficient storage. The storage of bioformulations requires special facilities and skills, which most producers, shopkeepers, and farmers do not possess. Shelf life is a culmination of several factors like production technology, carrier and packaging material used, mode and distance of transport, all these levels are desired to sustain the shelf life.

The most common solutions to this problem of survival time have been air-dried and lyophilized preparations (Kosanke et al. 1992). The lowered water content in the final product is responsible for long-term survival during storage. In this way, the bacteria in the formulation remain inactive, resistant to environmental stresses, insensitive to contamination, and are more compatible with fertilizer application (Bashan 1998). The dehydration phase is perhaps the most critical of the entire formulation process especially for nonspore-forming bacteria (Shah-Smith and Burns 1997). Bacterial survival is affected by several variables: the culture medium used for bacterial cultivation, the physiological state of the bacteria when harvested from the medium, the use of protective materials, the type of drying technology used, and the rate of dehydration (Paul et al. 1993).

3.3 Inconsistent Performance: Fate of Inoculant Introduced in Soil

Inconsistent field performance is the major constrain associated with their marketing. These failures have raised concerns about the perspective of the great practical potential offered by microbial releases into soils. A key factor involved in the lack of success has been the rapid decline of the size of populations of active cells, to levels ineffective to achieve the objective, following introduction into soil. As soil is a heterogeneous system with a mixed biota under fluctuating local conditions, temporal and spatial aspects pertaining to the introduction should be critically evaluated for each release. This growth/survival-inhibitory effect of soil has been called soil microbiostasis (Ho and Ko 1985). It has been attributed to the scarcity of available nutrient sources to microbes in soil and the hostility of the soil environment to incoming microbes due to a myriad of adverse abiotic and biotic factors. The physiological characteristics of the inoculant organism determine to a great extent its fate and activity in soil. Hence, different species will show varying responses, in terms of survival and activity. The physiological traits that play a role in the capacity of inoculant bacteria to colonize soil and survive are often not well known. Therefore, a thorough selection procedure is required when searching for effective inoculants. Besides the intrinsic physiological characteristics of the organisms, abiotic and biotic soil factors play an important role. Abiotic soil factors (e.g., textural type, pH, temperature, and moisture) exert their (direct) effect on inoculant population dynamics by imposing stresses of various natures on the cells (Evans et al. 1993). They can also act indirectly by affecting the activity of the indigenous soil microflora. A correlation was found between the decline in populations of individual bacterial strains and the activity and increase in the abundance of protozoa in soil (Wright et al. 1995; Lutenberg and Kamilova 2009). Hence, protozoa play an important role as regulators of microbial inoculant population sizes in soil. Another biological factor, in line with the predation process, is the competition between inoculant and indigenous populations for available substrate and biological space.

Moreover, Elliott et al. (1980) showed that trophic interactions in soil, including nematode–protozoan–bacterium interactions, are influenced by the soil type as reflected in the pore space distribution. Colonization of soil particles and aggregates is assumed to be vital to ultimate inoculant survival in soil (Hattori and Hattori 1976). Under similar prevailing climatic conditions, the inoculant revealed higher survival levels in fine-textured (clayey) than in coarser (sandy) soils. Vargas and Hattori (1986) clearly showed that in the presence of a cointroduced grazing protozoan species, the survival of inoculant bacteria localized in the interior parts of 1–2-mm soil aggregates was far better than that of cells present at external aggregate sites. This suggested that cells localized in the interior parts were physically protected from grazing by protozoa, presumably due to their localization in soil pores with small necks. The maintenance of sufficient activity of an inoculants population over a prolonged period after release often represents the main hurdle in the successful use of microbes as PGP agents. Furthermore, efficient introduction into soil during the growing season is a major technical constraint.

4 Research Areas for Development and Optimization of Bioformulations

Although the vast body of research on microbial inoculants deals with their ability to promote plant growth, there has been limited success in developing commercially viable products. For the development of successful bioformulation technology, progress must be made to meet numerous scientific challenges: (1) selection of improved strains having greater crop diversification, (2) survival during seed coating/pelleting, soil application and during storage at ambient temperatures is critical for the development of microbial inoculant products; therefore, it seems logical that these traits should form an integral part of any screening process for the development of new effective bioformulations, (3) more efficient plant growth promotory bacteria compete poorly with the rhizobacteria already in the soil. Ways to improve the competitive ability of inoculant should be explored, (4) study of environmental stresses that negatively affect nodulation, nitrogen fixation, and biocontrol ability such as soil pH, nutritional deficiencies, salinity, high temperature, and presence of toxic elements, (5) efficacy of microbial inoculants varies somewhat from site to site and year to year and this has to be considered and studied elaborately and, (6) understanding of interactions between the plant, beneficial rhizobacteria, and plant pathogens in the highly complex and dynamic rhizosphere environments is the ultimate need to overcome practical problems such as the inconsistency in field performance.

4.1 Microbial Diversity

Over the past 100 years, research has repeatedly demonstrated that phylogenetically diverse microorganisms can act as natural antagonists of various plant pathogens and promote plant growth (Cook 2000). The intensive screening of plant growth promoting microorganisms will allow the development of commercial bioformulation(s). The rhizosphere is known to provide ecologically favorable niche for most of the beneficial soil organisms. The abundance of nitrogen fixing, phosphate solubilizing, and plant disease suppressing bacteria in the rhizosphere of crop plants assumes a natural significance from the agronomic point of view (Subba Rao 1999). A successful PGP agent must be an aggressive colonizer with better competence and storage conditions in its formulation and use. As plant pathogens survive and cause diseases at dry, nutrient-poor, and high soil-temperature conditions (e.g., *Rhizoctonia bataticola*), the biocontrol agent (BCA) must also be able to withstand more competitively in the same adverse environment. Growth at high temperature (45°C) and endospore-producing trait of *Paenibacillus* sp. (endophytic bacteria) makes it a more suitable bioinoculant and ensures its survival in soil when a host is not available (Senthilkumar et al. 2007). Endophytic bacteria probably have evolved an intimate relation with their host plants through coevolutionary process and may influence physiological process of plants. Moreover, their unique ability to survive in plants with no or little microbial competition makes them potential candidates for biological control (Bhowmik et al. 2002).

One important factor to be considered when screening new isolates is their activity in the range of environments in which they would be expected to be used; in particular different soil types (Ross et al. 2000). Saline conditions are known to suppress the growth of plants, causing a diminished yield. *Ochrobactrum* sp., the free-living α-proteobacteria, was reported to have the potential of plant growth promotion in saline soil conditions (Príncipe et al. 2007). Recent reports have described the isolation of *Ochrobactrum* from plant tissue of deep water rice (*Oryza sativa*) (Tripathi et al. 2006) as well as from soils and sediments.

PGPB that are effective in degradation of soil pollutants in laboratory conditions have not done well in presence of soil pollutants is another constraint for field application. Selection of pollutant-degrading rhizobacteria that live on, or are close to the root so that they can use root exudate as their major nutrient source is a promising strategy to solve this problem (Böeltner et al. 2008; Lutenberg and Kamilova 2009). Similar approach resulted in the isolation of novel *Sphingomonas* strains that are relatively efficient in the in situ removal of lindane (Böeltner et al. 2008). *Pseudomonas putida* PCL1444, effectively utilizes root exudate, degrades naphthalene around the root, protects seeds from being killed by naphthalene, and allows the plant to grow normally. Mutants unable to degrade naphthalene did not protect the plant (Kuiper et al. 2001). Validov et al. (2007) isolated two new BCAs, *Pseudomonas rhodesiae* and *Delftia tsuruhatensis*. *P. rhodesiae* was first isolated from natural mineral water and is taxonomically affiliated to the *Pseudomonas fluorescens* group (Anzai et al. 2000). The representatives of this species were known as degraders of aromatic compounds (Kahng et al. 2002) or as isonovalol producers (Fontanille and Larroche 2003), but had not been reported yet as control agents for plant disease. *Delftia* is a newly classified genus closely related to *Comamonas*. These bacteria were isolated for the first time from active sludge as degraders of terephthalate (Shigematsu et al. 2003). *Delftia terephthalate*, also been reported as a diazotrophic PGPR, is able to control blast and blight of rice caused by *Xanthomonas oryzae*, *Rhizoctonia solani*, and *Pyricularia oryzae* (Han et al. 2005).

During the past two decades research on marine bacteria has highlighted the tremendous potential of these microorganisms as a source of new bioactive secondary metabolites (Ahmed et al. 2000) and there is a growing awareness of the need for development of new antimicrobial agents for the treatment of human, animal, and plant diseases. Marine bacteria could represent a new scope of antibiotics, which are currently needed to combat emergent antibiotic-resistant pathogen. The strains of species isolated from different ecological niches also generally showed wide genetic diversity despite some strains having similarity in their biochemical characteristic. It has become essential to understand the bacterial community structure in relation to environmental factors and ecosystem functions to screen, select, and utilize the microbial diversity for development of bioformulations leading to environment safe for life.

4.2 Metagenomics

The majority of microorganisms on earth resist life in captivity, i.e., they cannot be grown in broth or on plates in the laboratory. An often-cited estimate is that as much as 99% or more of microbial life remains unculturable, and therefore, cannot be studied and understood in a way that microbial ecologists have become accustomed to over the past century. The metagenomic toolbox allows accessing, storing, and analyzing the DNA of nonculturable life-forms and thus can provide an otherwise hard-to-attain insight into the biology and evolution of environmental microorganisms, independent of their culturable status (Fig. 1).

Due to a historical bias to study those microorganisms that can be grown in the laboratory, there is limited knowledge on the abundance and activity of not-yet

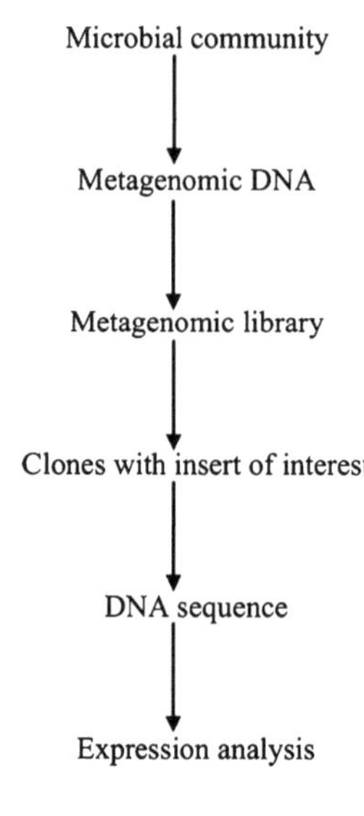

Fig. 1 Schematic representation of the classical metagenomic protocol

culturable PGPR. However, there are several examples of their existence and contribution to plant health, e.g., *Pasteuria penetrans*, a not-yet-culturable bacterium parasitic to plant-pathogenic nematodes (Fould et al. 2001), the nitrogen fixing activity by viable-but-not-culturable *Azoarcus* grass endophytes (Hurek et al. 2002), and the obligate biotrophism of arbuscular mycorrhizal (AM) fungi (Millner and Wright 2002). Bacteria belonging to the *Acidobacteria* and *Verrucomicrobia* are in many rhizospheres among the most abundant, difficult-to-culture representatives (Buckley and Schmidt 2003). However, it is not clear if and how their abundance is correlated to their contribution toward plant health. Several protocols have also been developed for the isolation of metagenomic bacterial DNA from inside plant material. Jiao et al. (2006) described an indirect method based on enzymatic hydrolysis of plant tissues to release associated microorganisms for subsequent DNA isolation and cloning. While optimized for leaves and seeds, this method seems readily adaptable for use with root material, and thus of great use to the metagenomic exploration of microorganisms in the rhizosphere.

There is a clear potential for metagenomics to contribute to the study of microbial communities of the rhizosphere, in particular PGPR. The other possible contributions of metagenomics in the study of PGPR include the discovery of novel PGP genes and gene products, and the characterization of not-yet culturable PGPRs. The tools of metagenomics offer many openings into a broadened view of the rhizosphere in general and of PGPR and their activities in particular. An analysis of the rhizosphere by comparative metagenomics holds the promise to reveal several important questions regarding the unculturable fraction of the rhizosphere community. For one, it could expose what actually constitutes this fraction from a comparison of metagenomic DNA isolated directly from rhizosphere to DNA isolated from all the colonies forming on solid media after plating from that same rhizosphere (i.e., the culturable fraction). The discovery of novel PGPR activities based on DNA sequence information from unculturables will add enormously to our understanding of the mechanistic variation that exists in PGPR phenotypes. It will also benefit our ability to improve existing PGPR, by adding to the pool of exploitable PGPR genes and utilization of this pool to develop PGPRs with enhanced performance (Timms-Wilson et al. 2004).

4.3 Plant–Microbe–Microbe Interactions

Rhizosphere is rich in microbial activity which takes part in biological and ecological processes important for plant health. To develop efficient and reliable bioformulation, understanding of the role of microbes in the panoply of processes and interactions which take place in the rhizosphere is essential. When analyzed within the context of biocontrol, the translocation processes of PGPR bacteria seem to warrant more attention. Motility on surfaces is an important mechanism for bacterial colonization of new environments. Furthermore, the ability to move in a directional manner may confer distinct advantage upon host-adapted prokaryotes.

There are few investigations reporting that motility is essential for the initial steps of development of microbial biofilms, which are often basic condition of beneficial effects of PGPR (Kinsinger et al. 2003). Avoidance of antimicrobial compounds produced either by the host or by competitors inhabiting the same niche also seems to be important for maintaining this contact.

Once a beneficial microbial strain has been able to colonize a host plant, it might be able to display a wide array of activities contributing to plant fitness. Expression of bacterial traits involved in biological control of plant pathogens is tightly regulated and N-acyl-homoserine lactones (AHL) signal molecules play an intriguing role in this respect. These AHL molecules have recently also been implicated in the sensing of bacteria by animals, more specifically *Caenorhabditis elegans* (Beale et al. 2006). Thus, these molecules play a role in communication within and between bacterial populations, in communication between bacteria and plants and vice versa (Teplitski et al. 2000; Schuhegger et al. 2006), and between bacteria and nematodes. In an environment containing all these organisms, like the rhizosphere, studying these interactions and predicting their outcome undoubtedly constitutes an exciting challenge.

The ever-increasing availability of plant and bacterial genome sequences and the development of "omic" technologies permit genome-wide approaches to unveil either microbial or plant functioning in the rhizosphere. Indeed, much has been done to investigate the global gene expression or transcriptomes of various plants when confronted with pathogens, symbiotic nitrogen-fixing bacteria, PGPR, or environmental conditions. However, the gene expression of microbes in the rhizosphere is much less studied largely due to the difficulty to obtain sufficient material under controlled conditions in this otherwise highly variable and irregular niche. The report by Matilla et al. (2007) constitutes the first on bacterial genomics in the rhizosphere. Secondary metabolites are often synthesized by multimodular, multidomain proteins called nonribosomal peptide synthetases (NRPS), and polyketide synthases (PKS). Both NRPS and PKS systems are molecular assembly lines for successive linking of multipleamino/hydroxyl acids or acyl-CoA precursors, respectively, into complex polymers which are often further modified into unique structures. A novel "genomisotopic" approach uses a combination of genomic sequence analysis and isotope-guided fractionation to identify unknown compounds synthesized by NRPS gene clusters (Gross et al. 2007). A phage display method was developed for high-throughput mining of gene clusters encoding PKS and NRPS systems, which can be applied to genomes of unknown sequence and metagenomes (Yin et al. 2007), providing opportunities for exploiting the potentially rich source of natural products from unculturable microbes. The ever-increasing pace of microbial genome sequencing is revealing a plethora of new NRPS/PKS gene clusters, mostly of unknown function. A major challenge for the next decade is to back this up with characterization of the chemical structures and biological activities of these secondary metabolites, so that we can chart Nature's unique repertoire of natural products and exploit them for the directed synthesis of novel molecules of agricultural utility (Arora et al. 2007a). Future developments in functional genomics (including proteomics and metabolomics) will be useful to identify the genes expressed in the rhizosphere, while the use of promoters to drive gene expression specifically at the root–soil interfaces will allow the engineering of microorganisms for beneficial purposes.

4.4 Formulation Design

Formulation is the crucial issue for inoculants containing an effective bacterial strain and can determine the success or failure of a biological agent. Since natural soil commonly represents a hostile environment to inoculant cells (Ho and Ko 1985), the use of inoculant formulations involving carrier materials for the delivery of microbial cells to soil or the rhizosphere is an attractive option. Carrier materials are generally intended to provide a (temporarily) protective niche to microbial inoculants in soil, either physically, via the provision of a protective surface or pore space (creating protective microhabitats), or nutritionally, via the provision of a specific substrate (Trevors et al. 1992). Peat and soil rich in organic matter are generally used in the preparation of legume inoculants and constitute a suitable carrier for the purpose. Peat and lignite, though good carriers, are not easily available and are expensive. The low cost and easily availability of carrier material are the major requirements for bioformulations in developing countries (Saha et al. 2001).

The microbial inoculant is not merely a suitable carrier containing the bacteria. Other materials might be involved in the final formulations. Evidence suggests that the addition of nutrients to seed pellets may be a useful strategy for improving inoculant survival (Moënne-Loccoz et al. 1999). Furthermore, carbon sources and minerals have been shown to have an important role in antifungal metabolite production by *Pseudomonas* BCAs, suggesting that nutrient amendments to formulations may also be a useful strategy for improving biocontrol efficacy (Duffy and Défago 1999). Soil amendment with chitin showed increase of the chitinolytic microbial populations and significantly reduced the incidence of fungal diseases in celery (Bell et al. 1998). Chitin supplementation supports the survival of *Bacillus cereus* and *B. circulans* in the groundnut phylloplane and resulted in better control of early and late leaf spot disease (Kishore et al. 2005). These improved disease control results are associated with an increase in the population of the introduced BCAs in presence of chitin.

Drying is a part of many procedures for development of formulation of microbial inoculants. The drying procedures are not very suitable for incorporation in a formulation protocol. However, Amiet-Charpentier et al. (1998) reported that it is possible to formulate nonsporulating bacteria using both freeze- and spray-drying. It was demonstrated that a methacrylic copolymer carrier, an ethyl-cellulose, and a modified starch product all increase survival of rhizosphere bacteria during spray-drying (Palmfeldt et al. 2003). Remarkably low percentage of endospore formers was observed that survived after drying (Validov et al. 2007). Designing of formulation that allow inoculant survival during drying procedure and support high colony forming units of PGP agents on short storage in the grower's warehouse (which in developing countries usually lack refrigeration) was an important necessity for commercialization of the technology.

One factor which can have a detrimental effect on dried microorganisms over the long term is humidity in the environment; increasing moisture content of the dried sample compromises viability. Storage under vacuum or in an inert atmosphere can prevent this (Johnsson et al. 1998), but is costly and unwieldy. Manzanera et al. (2004) have shown how osmotic preconditioning of bacteria, followed by drying in the presence of glass-forming protectant molecules, such as trehalose or hydroxyectoine, results in a high level of desiccation tolerance, where viability is maintained throughout extended storage periods at above-ambient temperatures and its potential application as a seed coating. This has been termed anhydrobiotic engineering (Fages 1992), in reference to anhydrobiotic organisms which naturally exhibit extreme desiccation tolerance (Validov et al. 2009). Similar observations of García de Castro et al. (2000) demonstrate the potential of this novel biotechnology for stabilizing nonsporulating organisms. Storage of culture collections and libraries could be simplified using a plastic encapsulation procedure, for example, since there is no requirement for freezing or storage under vacuum or in an inert atmosphere.

5 Integrated Management

In the era of integrated use and management of various agro-inputs for maximization of crop yields, a comprehensive knowledge about the compatibility of various components to each other is very much required. Recommendations on combined use of such inputs, like treatment of seeds both with fungicides and biofertilizers, must accompany appropriate information on their compatibility to each other. Inhibitory effects have been observed on some nitrogen fixing microorganisms by insecticides (Sarkar and Balasubramanayam 1978) and seed dressing chemicals (Chitriv 1986). Knowledge of multiple microbial interactions is also of extreme value for development of bioformulations. The majority of interactions considered so far concern a single pathogen and a single BCA in the rhizosphere. However, one way of improving biocontrol in the rhizosphere may be to add combinations of BCAs, particularly those exhibit different or complementary mode of action or abilities to colonize root microsides. Application of a combination of *Paenibacillus* sp. and a *Streptomyces* sp. suppressed *Fusarium* wilt of cucumber than when either was used alone (Singh et al. 1999). The combination of *Pseudomonas aeruginosa* and *Pochonia chlamydospora* caused greater suppression of fungal phytopathogens and promoted plant growth compared with their individual application (Siddiqui and Shaukat 2002). Combinations of fungi and bacteria have also been shown to provide enhanced biocontrol (Duffy et al. 1996).

However, it is important when considering the use of combinations of strains that no member of the mixture is inhibitory to another or interferes excessively with the existing, normal, and nonpathogenic microbiota associated with the roots. Various reports indicate that coinoculation of beneficial organisms generally increased plant growth and/or decreased plant disease relative to single inoculation with a sole beneficial organism (Whipps 2004; Raimam et al. 2007). Most of the effects of the individual microorganisms in coinoculation are additive, although a synergistic effect has been reported in some cases (Ravnskov et al. 2006; Kohler et al. 2007). However, neutral or negative effects have been reported (Akköprü and Demir 2005) indicating that the outcome of coinoculation of these microorganisms on plant health and productivity should be determined on a case-by-case basis. There is evidence to suggest that *Pseudomonas* BCAs can affect the growth and subsequent nodule occupancy of certain *Sinorhizobium meliloti* strains in gnotobiotic systems (Neimann et al. 1997). Within commercial scale field trials, however, a *Pseudomonas* BCA did not affect nodulation in the foliage of a red clover rotation crop (Moënne-Loccoz et al. 1998), again demonstrating the necessity of conducting impact analysis experiments within agronomically relevant parameters. Only when the symbiosis is well understood are we likely to be able to exploit it to provide optimum growth enhancement of the host and control of phytopathogens (Arora et al. 2008a).

6 Conclusion

Because of current public concerns about the side effects of agrochemicals, there is an increasing interest in improving the understanding of cooperative activities among rhizosphere microbial populations and how these might be applied to agriculture. Certain cooperative microbial activities can be exploited as a low-input biotechnology and form basis for a strategy to help sustainable, environmentally friendly practice fundamental to the stability and productivity of both agricultural systems and natural ecosystems (Kennedy and Smith 1995). Recent survey of both conventional and organic growers indicates an interest in using biological products (Rzewnicki 2000), suggesting that the market potential of bioformulations will increase in coming years. It is estimated that the total global market for synthetic pesticides which was valued at US$ 26.7 billion in 2005 will decline to US$ 25.3 billion in 2010. On the other hand, the global market for biopesticides will increase from US$ 672 million in 2005 to over US$ 1 billion in 2010 (Fig. 2). While Europe, at an average annual growth rate (AAGR) of 15%, is projected to lead the growth in biopesticide use, Asia will be no far behind with an average AAGR of 12%. The global market is divided into 43.5% of sales in North American Free trade Agreement countries (including Mexico), 20.7% in Europe, 12.2% in Asia, 11.2% in Oceania (including Australia), 8.3% in Latin America (excluding Mexico), and 3.9% in Africa (Bolckmans 2008). Furthermore, a detailed report about nitrogen-fixing bacteria as biofertilizers, for which the market is also growing, was published by Bhattacharjee et al. (2008).

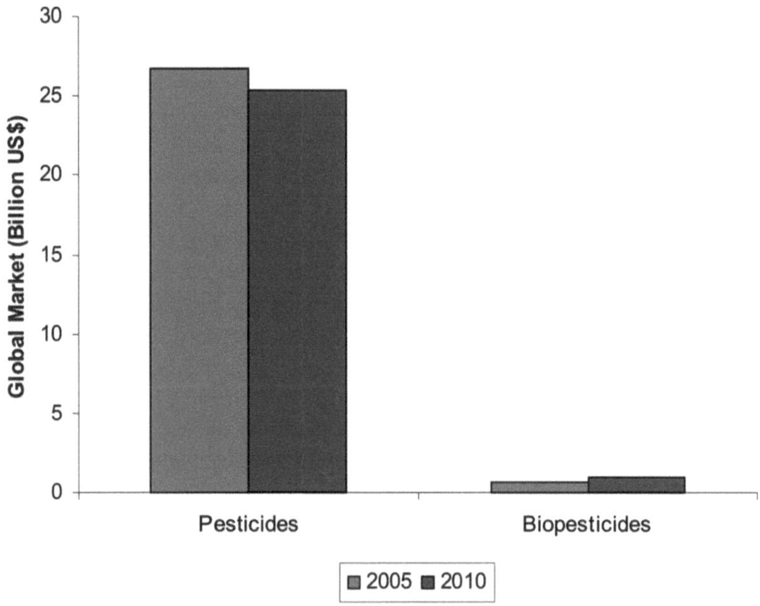

Fig. 2 Change in global market of synthetic pesticides and biopesticides in 5 years

The effects of soil on the physiology and ecology of introduced microorganisms are still poorly understood at the microscale (pore) level. Future research in this area should aim for a better understanding of the in situ physiology of inoculant cells, as well as for possible ways to manipulate it. Molecular techniques are being used in microbial ecology to understand the soil ecosystem, for the production of microbial inoculates, and for monitoring these inoculates after field release. These inoculants may or may not be genetically modified strains. Thus, future research in rhizosphere biology will rely on the development of molecular and biotechnological approaches to increase our knowledge of rhizosphere biology and to achieve an integrated management of soil microbial populations. Future investigation in the field of development of microbial formulation for plant growth promotion will include: (1) advances in visualization technology; (2) analysis of the molecular basis of root colonization; (3) signaling in the rhizosphere; (4) functional genomics; (5) mechanisms involved in beneficial cooperative microbial activities; (6) engineering of microorganisms for beneficial purposes; and (7) biotechnological developments for integrated management. A variety of research questions remain to be fully answered about the nature of bioformulations and the means to most effectively manage it under production conditions. As our understanding of the complex environment of the rhizosphere, of the mechanisms of action of PGPR, and of the practical aspects of inoculant formulation and delivery increases, we can expect to see new PGPR products becoming available.

On the applied side, and given the history of failures or variabilities of previous microbial releases, it is interesting to test the concept of application of mixtures of ecologically diverse strains with similar functions instead of limited function of single strains. Such consortia might consist either of mixtures of completely natural strains or of different strains into which similar functions had been engineered. By this way, beneficial functions might be expressed more continually in a soil or rhizosphere system, even under ecologically different and/or variable conditions.

One of the major challenges for the inoculant industry is to develop an improved formulation that provides high shelf-life, high number of viable cells, protection against soil environment, convenience to use, and cost effective (Smith 1992). More studies on the practical aspects of mass-production and formulation need to be undertaken to make new bioformulations that are stable, effective, safer, and more cost-effective. There is an urgent need to develop a definite correlation between agriculturists, microbiologists, biotechnologists, industrialists, and farmers (Fig. 3).

Generally, inoculants are being used for legume crops and to a certain extent for cereal crops. Fresh alternatives should be explored for the use of bioinoculants for other high value crops such as vegetables, fruits, and flowers. This will not only

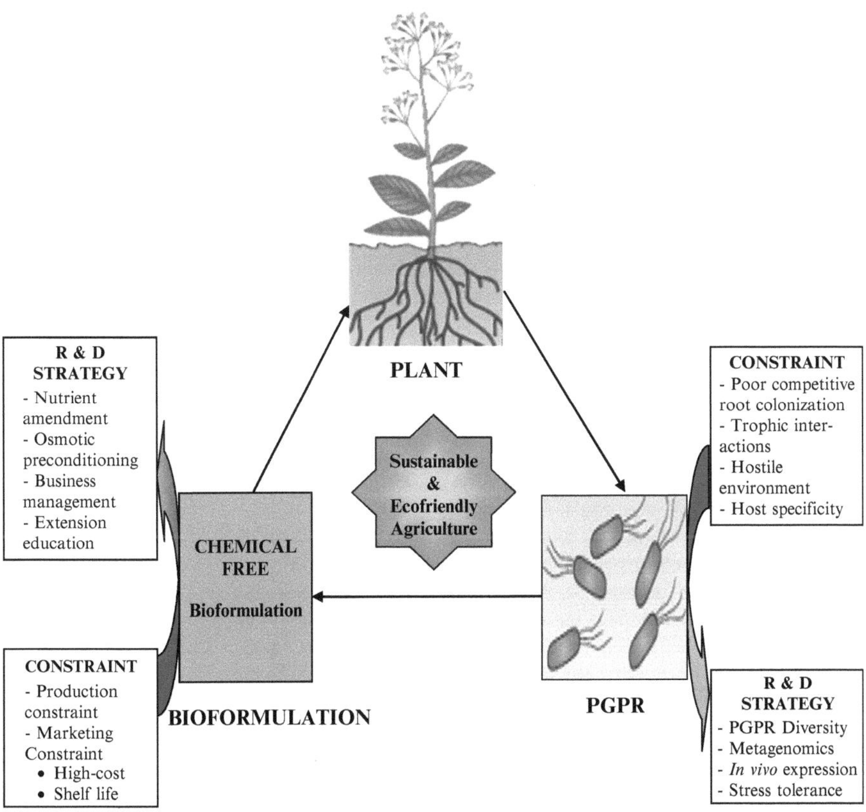

Fig. 3 Research and development strategies for commercialization of bioformulation technology

increase the field of the inoculants but also create confidence among the farmers for their use. As a recent approach, new environment friendly, genetically modified, microbial inoculates are being produced commercially and used to protect plants from disease and to promote plant growth. Numerous studies on technological evolution emphasized the developmental role of a firm and the strength of its sales network, creating market and drawing market feedback, for its success. In addition, future marketing of bioinoculant products and release of these transgenics into the environment as eco-friendly alternations to agrochemicals will depend on the generation of biosafety data required for the registration of PGP agents. Clearly, the future success of the bioformulation industry will depend on innovative business management, product marketing, quality product, extension education, and research.

Acknowledgment Thanks are due to Council of Scientific and Industrial Research, New Delhi, Council of Science & Technology, Lucknow and UCOST, Dehradun, India. Authors are grateful to Vice Chancellor, CSJM University, Kanpur, India for their support.

References

Ahmed N, Jamil N, Khan OY, Yasmen S, Haq Z-Ul, AhmedmVU, Rahman AT (2000) Commercially important products from marine bacteria. Marine Biotechnology. In: Proceedings of National ONR Symposium on Arabian Sea a resource of biological diversity, pp 104–114

Akköprü A, Demir S (2005) Biological control of Fusarium wilt in tomato caused by *Fusarium oxysporum* f.sp. *lycopersici* by AMF *Glomus intraradices* and some rhizobacteria. J Phytopathol 153:544–550

Amiet-charpentier C, Gadille P, Digat B, Benoit JP (1998) Microencapsulation of rhizobacteria by spray-drying: formulation and survival studies. J Microencapsul 15:639–659

Anzai Y, Kim H, Park JY, Wakabayashi H, Oyaizu H (2000) Phylogenetic affiliation of the pseudomonads based on 16s rRNA sequence. Int J Syst Evol Microbiol 50:1563–1589

Arora NK, Kumar V, Maheshwari DK (2001) Constraints, development and future of the inoculants with special reference to rhizobial inoculants. In: Maheshwari DK, Dubey RC (eds) Innovative approaches in microbiology. Singh and Singh, Dehradun, India, pp 241–245

Arora NK, Kim MJ, Kang SC, Maheshwari DK (2007a) Role of chitinase and β-1,3-glucanase activities produced by a fluorescent pseudomonad and in vitro inhibition of Phytophthora capsici and Rhizoctinia solani. Can J Microbiol 53:207–212

Arora NK, Khare E, Verma A (2007b) Biofertilizer technology for economical and environmentally viable agriculture production. Kurukshetra 55(4):20–24

Arora NK, Khare E, Naraian R, Maheshwari DK (2008a) Sawdust as a superior carrier for production of multipurpose bioinoculant using plant growth promoting rhizobial and pseudomonad strains and their impact on productivity of *Trifolium repense*. Curr Sci 95(1):90–94

Arora NK, Khare E, Oh JH, Kang SC, Maheshwari DK (2008b) Diverse mechanisms adopted by fluorescent *Pseudomonas* PGC2 during the inhibition of *Rhizoctonia solani* and *Phytophthora capsici*. World J Microbiol Biotechnol 24:581–585

Arora NK, Khare E, Verma A, Sahu RK (2008c) *In vivo* control of *Macrophomina phaseolina* by a chitinase and β-1,3-glucanase-producing pseudomonad NDN1. Symbiosis 46:129–135

Bashan Y (1998) Inoculants of plant growth promoting bacteria use in agriculture. Biotech Adv 6:729–770

Beale E, Li G, Tan MW, Rumbaugh KP (2006) *Caenorhabditis elegans* senses bacterial autoinducers. Appl Environ Microbiol 72:5135–5137

Bell A, Hubbard JC, Liu L, Davis RM, Subbarao KV (1998) Effects of chitin and chitosan on the incidence and severity of *Fusarium* yellows of celery. Plant Dis 82:322–328

Bhattacharjee RB, Sing A, Mukhopadyay SN (2008) Use of nitrogen-fixing bacteria as biofertilizer for non-legumes: prospects and challenges. Appl Microbiol Biotechnol 80:199–209

Bhowmik B, Singh RP, Jayaraman J, Verma JP (2002) Population dynamics of cotton endophytic *Pseudomonas*, their antagonism and protective action against the major pathogens of cotton. Indian Phytopathol 55(2):24–132

Böeltner D, Godoy P, Muñoz-Rojas J, Duque E, Moreno-Morillas S, Sánchez L, Ramos JL (2008) Rhizoremediation of lindane by root-colonizing *Sphingomonas*. Microb Biotechnol 1:87–93

Bolckmans K (2008) Biocontrol files. Can Bull Ecol Pest Manag 13:1–10

Buckley DH, Schmidt TM (2003) Diversity and dynamics of microbial communities in soils from agroecosystems. Environ Microbiol 5:441–452

Chitriv AJ (1986) Sensitivity of mungbean rhizobia to seed dressing chemicals. In: Singh R, Nanawati HS, Sawhrey SK (eds) Proceedings of national symposium on current status of biological nitrogen fixation research, Hissar, pp 116

Cook RJ (2000) Advances in plant health management in the 20th century. Ann Rev Phytopathol 38:95–116

Duffy BK, Défago G (1999) Environmental factors modulating antibiotic and siderophore biosynthesis by *Pseudomonas fluorescens* biocontrol strains. Appl Environ Microbiol 65:2429–2438

Duffy BK, Simon A, Weller DM (1996) Combination of *Trichoderma koningii* with fluorescent pseudomonads for control off take-all on wheat. Phytopathology 86:88–194

Ehlers RU (2006) Einsatz der Biotechnologie im biologischen Pflanzenschuz. Schnreihe dtsch Phytomed Ges 8:17–31

Elliott ET, Anderson RV, Coleman DC, Cole CV (1980) Habitable pore space and microbial trophic interaction. Oikos 35:327–335

Evans J, Wallace C, Dobrowolski N (1993) Interaction of soil type and temperature on the survival of *Rhizobium leguminosarum* bv. *viciae*. Soil Biol Biochem 25:1153–1160

Fages J (1992) An industrial view of *Azospirillum* inoculants formulation and application technology. Symbiosis 13:15–26

Fontanille P, Larroche C (2003) Optimization of isonovalal production from alpha-pinene oxide using permeabilized cells of *Pseudomonas rhodesiae* CiolIP 107491. Appl Microbiol Biotechnol 60:534–540

Fould S, Dieng AL, Davies KG, Normand P, Mateille T (2001) Immunological quantification of the nematode parasitic bacterium *Pasteuria penetrans* in soil. FEMS Microbiol Ecol 37:187–195

García de Castro A, Lapinski J, Tunnacliffe A (2000) Anhydrobiotic engineering. Nat Biotechnol 18:473

Gerhardson B (2002) Biological substitutes for pesticides. Trends Biotechnol 20:338–343

Gross H, Stockwell VO, Henkels MD, Nowak-Thompson B, Loper JE, Gerwick WH (2007) The genomisotopic approach: a systematic method to isolate products of orphan biosynthetic gene clusters. Chem Biol 14:53–63

Han J, Sun L, Dong X, Cai Z, Sun X, Yang H, Wang Y, Song W (2005) Characterization of a novel plant growth promoting bacteria strain *Delftia tsuruhatensis* HR4 both as a diazotroph and a potential biocontrol agent against various plant pathogens. Syst Appl Microbiol 28:66–76

Hattori T, Hattori R (1976) The physical environment in soil microbiology: an attempt to extend principles of microbiology to soil microorganisms. CRC Crit Rev Microbiol 4:423–461

Ho WC, Ko WH (1985) Soil microbiostasis: effects of environmental and edaphic factors. Soil Biol Biochem 17:167–170

Hurek T, Handley LL, Reinhold-Hurek B, Piche Y (2002) Azoarcus grass endophytes contribute fixed nitrogen to the plant in an unculturable state. Mol Plant Microbe Interact 15:233–242

Jiao JY, Wang HX, Zeng Y, Shen YM (2006) Enrichment for microbes living in association with plant tissues. J Appl Microbiol 100:830–837

Johnsson L, Hokeberg M, Gerhardson B (1998) Performance of the *Pseudomonas chlororaphis* biocontrol agent MA 342 against cereal seed borne disease in field experiments. Eur J Plant Pathol 104:701–711

Kabi MC (1997) Impact of biofertilizer on rural development. In: Proceedings of National Conference on impact of biotechnology and modern horticulture in rural development. Jadavpur University, Calcutta

Kahng HY, Nam K, Kukor JJ, Yoon BJ, Lee DH, Oh DC, Kam SK, Oh KH (2002) PAH utilization by *Pseudomonas rhodesiae* KK1 isolated from a former manufactured-gas plant site. Appl Microbiol Biotechnol 60:475–480

Kennedy AC, Smith KL (1995) Soil microbial diversity and the sustainability of agricultural soils. Plant Soil 170:75–86

Kinsinger RF, Shirk MC, Fall R (2003) Rapid surface motility in *Bacillus subtilis* is dependent on extracellular surfactin and potassium ion. J Bacteriol 185:5627–5631

Kishore GK, Pande S, Podile AR (2005) Biological control of late leaf spot of peanut (*Arachis hypogea* L.) with chitinolytic bacteria. Phytopathology 95:1157–1165

Kohler J, Caravaca F, Carrasco L, Roldán A (2007) Interactions between a plant growth-promoting rhizobacterium, an AM fungus and a phosphate-solubilising fungus in the rhizosphere of *Lactuca sativa*. Appl Soil Ecol 35:480–487

Kosanke JW, Osburn RM, Shuppe GI, Smith RS (1992) Slow rehydration improves the recovery of dried bacterial populations. Can J Microbiol 38:520–525

Kuiper I, Bloemberg GV, Lugtenberg BJJ (2001) Selection of a plant-bacterium pair as a novel tool for rhizostimulation of polycyclic aromatic hydrocarbon-degrading bacteria. Mol Plant Microbe Interact 14:1197–1205

Lutenberg B, Kamilova F (2009) Plant-growth-promoting rhizobacteria. Annu Rev Microbiol 63:541–556

Manzanera M, Vilchez S, Tunnacliffe A (2004) Plastic encapsulation of stabilized *Escherichia coli* and *Pseudomonas putida*. Appl Environ Microbiol 70(5):3143–3145

Matilla MA, Espinosa-Urgel M, Rodríguez-Herva JJ, Ramos JL, Ramos-González MI (2007) Genomic analysis reveals the major driving forces of bacterial life in the rhizosphere. Genome Biol 8:R179.1–R179.13

McSpadden Gardner BB (2002) Biological control of plant pathogens: research, commercialization, and application in the USA. Plant Health Progress. doi:10.1094/PHP-2002-0510-01-R

Millner PD, Wright SF (2002) Tools for support of ecological research on arbuscular mycorrhizal fungi. Symbiosis 33:101–123

Moënne-Loccoz Y, Powell J, Higgins P, Britton J, O' Gara F (1998) Effect of the biocontrol agent Pseudomonas fluorescens F113 released as sugarbeet inoculant on the nutrient contents of soil and foliage of a red clover rotation crop. Biol Fertil Soils 27:380–385

Moënne-Loccoz Y, Naughton M, Higgins P, Powell J, O' Gara F (1999) Effect of inoculum preparation and formulation on survival and biocontrol efficacy of *Pseudomonas fluorescens* F113. J Appl Microbiol 86:108–116

Neimann S, Keel C, Puhler A, Seibitschka W (1997) Biocontrol strain *Pseudomonas fluorescens* CHA0 and its genetically modified derivative with enhanced biocontrol capability exert comparable effects on the structure of a *Sinorhizobium meliloti* population in gnotobiotic systems. Biol Fertil Soils 25:240–244

Palmfeldt J, Radstrom P, Hahn-Hagerdal B (2003) Optimization of initial cell concentration enhances freeze-drying tolerance of *Pseudomonas chlororaphis*. Cryobiology 47:21–29

Paul E, Fages J, Blanc P, Goma G, Pareilleux A (1993) Survival of alginate-entrapped cells of *Azospirillum lipoferum* during dehydration and storage in relation to water properties. Appl Microbiol Biotechnol 40:34–39

Príncipe A, Alvarez F, Castro MG, Zachi L, Fisher SE, Mori GB, Jofré E (2007) Biocontrol and PGPR features in native strains isolated from saline soils of Argentina. Curr Microbiol 55:314–322

Raimam MP, Albino U, Cruz MF, Lovato GM, Spago F, Ferracin TP, Lima DS, Goulart T, Bernardi CM, Miyauchi M, Nogueira MA, Andrade GA (2007) Interaction among free-living N-fixing bacteria isolated from *Drosera villosa* var. *villosa* and AM fungi (*Glomus clarum*) in rice (*Oryza sativa*). Appl Soil Ecol 35:25–34

Ravnskov S, Jensen B, Knudsen IMB, Bødker L, Jensen DF, Karliński L, Larsen J (2006) Soil inoculation with the biocontrol agent *Clonostachys rosea* and the mycorrhizal fungus *Glomus intraradices* results in mutual inhibition, plant growth promotion and alteration of soil microbial communities. Soil Biol Biochem 38:3453–3462

Ross IL, Alami Y, Harvey Achouak PRW, Ryder MH (2000) Genetic diversity and biological control activity of novel species of closely related pseudomonads isolated from wheat field soils in South Australia. Appl Environ Microbiol 66:1609–1616

Rzewnicki P (2000) Ohio organic producers: final survey results. Online. Ohio State University Extention, College of Food Agricultural and Environmental Sciences, Bulletin, Special Circular, pp 174

Saha AK, Deshpande MV, Kapadnis BP (2001) Studies on survival of *Rhizobium* in the carriers at different temperatures using green fluorescent protein marker. Curr Sci 80(5):669–671

Sarkar T, Balasubramanayam A (1978) Interaction of two organophosphorus pesticides with *Rhizobium* species and their degradation *in vitro*. Madras Agric 65:325–328

Schuhegger R, Ihring A, Gantner S, Bahnweg G, Knappe C, Hartmann A, Langebartels C (2006) Induction of systemic resistance in tomato by N-acyl-L-homoserine lactoneproducing rhizosphere bacteria. Plant Cell Environ 29:909–918

Senthilkumar M, Govindasamy V, Annapurna K (2007) Role of antibiosis in suppression of charcoal rot disease by soybean endophyte *Paenibacillus* sp. HKA-15. Curr Microbiol 55:25–29

Shah-Smith DA, Burns RG (1997) Shelf-life of a biocontrol *Pseudomonas putida* applied to sugar beet seeds using commercial coating. Biocontrol Sci Technol 7:65–74

Shigematsu T, Tumihara K, Ueda Y, Numaguchi M, Morimura S, Kida K (2003) *Delftia tsuruhatensis* sp. nov., a terephthalate-assimilating bacterium isolated from activated sludge. Int J Syst Evol Microbiol 53:1479–1483

Siddiqui IA, Shaukat SS (2002) Mixtures of plant disease suppressive bacteria enhances biological control of multiple tomato pathogens. Biol Fertil Soils 36:260–268

Singh PPS, Shin YC, Park CS, Chung YR (1999) Biological control of *Fusarium* wilt of cucumber by chitinolytic bacteria. Phytopathology 89:92–99

Smith RS (1992) Legume inoculant formulation and application. Can J Microbiol 38:485–492

Subba Rao NS (1993) Biofertilizers in agriculture and forestry. Oxford and IBH Publishing, New Delhi, p 242

Subba Rao NS (1999) Soil Microbiology In: Soil microorganisms and plant growth, 4th ed. Oxford & IBH Publishing, New Delhi, pp 78–98

Teplitski M, Robinson JB, Bauer WD (2000) Plants secrete substances that mimic bacterial N-acyl homoserine lactone signal activities and affect population densitydependent behaviours in associated bacteria. Mol Plant Microbe Interact 13:637–648

Timms-Wilson TM, Kilshaw K, Bailey MJ (2004) Risk assessment for engineered bacteria used in biocontrol of fungal disease in agricultural crops. Plant Soil 266:57–67

Trevors JT, van Elsas JD, Lee H, van Overbeck LS (1992) Use of alginate and other carriers for encapsulation of microbial cells for use in soil. Microb Releases 1:61–69

Tripathi A, Verma SC, Chowdhury SP, Lebuhn M, Gattinger A, Schloter M (2006) *Ochrobactrum oryzae* sp. Nov., an endophytic bacterial species isolated from deep-water rice in India. Int J Syst Evol Microbiol 56:1677–1680

Validov S, Kamilova F, Qi S, Stephen D, Wang JJ, Makarova N, Lugtenberg B (2007) Selection of bacteria able to control *Fusarium oxysporum* f. sp. *Radicus-lycopersici* in stone substrate. J Appl Microbiol 102:461–471

Validov SZ, Kamilova F, Lugtenberg BJJ (2009) *Pseudomonas putida* strain PCL1760 controls tomato foot and root rot in stonewool under industrial conditions in a certified greenhouse. Biol Control 48:6–11

van Elsas JD, Heijnen CE (1990) Methods for the introduction of bacteria into soil – a review. Biol Fertil Soils 10:127–133

Vargas R, Hattori T (1986) Protozoan predation of bacterial cells in soil aggregates. FEMS Microbiol Ecol 38:233–242

Whipps JM (2001) Microbial interactions and biocontrol in the rhizosphere. J Exp Bot 52:487–511

Whipps JM (2004) Prospects and limitations for mycorrhizas in biocontrol of root pathogens. Can J Bot 82:1198–1227

Wright DA, Killham K, Glover LA, Prosser JI (1995) Role of pore size location in determining bacterial activity during predation by protozoa in soil. Appl Environ Microbiol 61:3537–3543

Yin J, Straight PD, Hrvatin S, Dorrestein PC, Bumpus SB, Jao C et al (2007) Genome-wide high-throughput mining of natural-product biosynthetic gene clusters by phage display. Chem Biol 14:303–312

Antifungal Compounds of Plant Growth Promoting Rhizobacteria and Its Action Mode

C.S. Quan, X. Wang, and S.D. Fan

Contents

1 Introduction 118
2 Antagonistic PGPR and Its Antifungal Metabolites 119
  2.1 Bacillus and Its Antifungal Metabolites 119
  2.2 Pseudomonas and Its Antifungal Metabolites 120
  2.3 Burkholderia and Its Antifungal Metabolites 130
3 Antifungal Mechanisms 138
4 Understanding Biosynthesis of Antifungal Metabolites at the Molecular Level 145
5 Concluding Remarks and Further Perspectives 146
References 150

Abstract Plant growth promoting rhizobacteria (PGPR) are bacteria that colonize plant roots and then promote plant growth and/or reduce disease or insect damage via exudation of some active metabolites. Antagonistic PGPR have attracted much attention in their role in reducing plant diseases, especially strains of the genus *Bacillus*, *Pseudomonas*, and *Burkholdeira*, and there is now an increasing number of PGPR being commercialized for crops. In this chapter, we present three major antagonistic PGPR (*Bacillus* spp., *Pseudomonas* spp., and *Burkholdeira* spp.) and their antifungal metabolites including the chemical structure first, and then introducing the mode of action and biosynthesis pathway of these antifungals.

C.S. Quan (✉) and S.D. Fan
College of Life Science, Dalian Nationalities University, Dalian 116600, China
e-mail: mikyeken@dlnu.edu.cn

X. Wang
Dalian Institute of Chemical Physics, Chinese Academy of Science, Dalian 116023, China
Graduate School of Chinese Academy of Science, Beijing 100049, China

D.K. Maheshwari (ed.), *Plant Growth and Health Promoting Bacteria*,
Microbiology Monographs 18, DOI 10.1007/978-3-642-13612-2_6,

1 Introduction

Plant fungal diseases reduce yield and productivity of several economical crops all over the world. Resistant plant cultivars, cultural practices, and chemical applications are often used to control plant disease. However, resistant cultivars for every disease are not available and cultural practices are not always economically or technologically feasible. Moreover, available chemical fungicides are often expensive and also have bad effects on human beings. Environmentally friendly control of plant disease is a pressing need for agriculture in new century (Emmert and Handelsman 1999). Biological control using antibiotics and antifungal rhizobacteria to suppress plant diseases offers a powerful alternative to the use of synthetic chemicals.

There have been many studies regarding the use of antagonistic microbes as an alternative to synthetic chemical pesticides in biocontrol systems, because the latter has given rise to human and ecological risk. Many bacteria and fungi have been reported as antagonistic microbes against phytopathogenic fungi (Bonsall et al. 1999; Lee et al. 2001). Most of the interactions between antagonistic and phytopathogenic microbes have been summarized as deriving from the inhibition of the pathogen by antimicrobial materials (Raaijmakers et al. 2002), competition for nutrients (Mondal and Hyakumachi 2000), the inaction of pathogen germinating factors (Whipps 1997), and degradation of the pathogenicity factor (Steijl et al. 1999). The usage of antagonistic microorganisms with antifungal effects as biocontrol agents to inhibit or reduce the rate of propagation of deleterious fungi during storage is considered a safer and more environmentally friendly alternative. Biological control of plant pathogens is strongly based on the production of antifungal factors such as antibiotics, hydrolytic enzymes, and siderophores by the bacterial antagonists (Becker and Cook 1988; Howell and Stipanovic 1980; Keel et al. 1990; Thomashow and Weller 1988; Vincent et al. 1991).

Plant growth promoting rhizobacteria (PGPR) are free-living soil bacteria that can either directly or indirectly facilitate rooting and growth of plants. PGPR indirectly enhance plant growth via suppression of phytopathogens by a variety of mechanisms. These include the ability to produce siderophores that chelate iron, making it unavailable to pathogens; the ability to synthesize antifungal metabolites such as antibiotics, fungal cell wall-lysing enzymes, or production of volatiles such as hydrogen cyanide, which suppress the growth of fungal pathogens; the ability to successfully compete with pathogens for nutrients or specific niches on the root; and the ability to induce systemic resistance (ISR). Taxonomically, PGPR represent a variety of bacterial species from different genera such as *Pseudomonas*, *Bacillus*, *Burkholderia*, *Enterobacter*, and *Azospirillum* (Lodewyckx et al. 2002). Among PGPR bacteria, *Bacillus*, *Pseudomonas*, and *Burkholderia* have been intensively investigated as biological control agents with regard to the production of antimicrobial metabolites.

The purpose of this chapter is to provide an up-to-date overview of the current knowledge of the structural diversity and activity of antifungal compounds produced by plant-associated PGPR, in particular the antagonistic *Pseudomonas* spp., *Bacillus* spp., and *Burkholderia* spp. A detailed description of structures, mechanism of action mode, and genes involved in the biosynthesis of antifungals is presented.

2 Antagonistic PGPR and Its Antifungal Metabolites

PGPR suppress various group of plant pathogens, thus protect the plants against different diseases. This protective effect is mainly due to production of antifungal metabolites produced by various species of *Bacilli, Pseudomonads*, and *Burkholderia* in particular.

2.1 Bacillus and Its Antifungal Metabolites

Spore-forming bacteria, typically *Bacillus* species, are one of the major types of soil bacteria. *Bacillus* species offer several advantages for protection against root pathogens because of their ability to form endospores and the broad-spectrum activity of their antibiotics. *Bacillus* species produce 167 biological compounds active against bacteria, fungi, protozoa, and viruses (Bottone and Peluso 2003).

The first successful application and commercial production of PGPR is a *B. subtilis* strain A13. *B. subtilis* A13 was isolated more than 25 years ago in Australia based on in vitro inhibitory activity to all of nine pathogens tested and was subsequently shown to promote plant growth. Since 1990, *Bacillus* spp. have been developed as fungal disease control agents. Strains of *B. megaterium*, *B. cereus*, and *B. subtilis* have been used for the biocontrol purpose (Idris et al. 2008; Kildea et al. 2008), in the form of the commercial product namely, Serenade, EcoGuard, Kodiak, Yield Shield, and BioYield.

Bacteria of the genus *Bacillus* are capable of producing antibiotics, as well as a variety of fungal cell-wall-degrading enzymes, such as chitinase, cellulases, amylases, glucanses, etc. Most of the antibiotics are peptides effective against Gram-positive, Gram-negative bacterial species, and filamentous fungi, and also with a high stability attributable to their structure. Several antifungal peptides synthesized by *Bacillus* species are active against filamentous fungi and yeasts. According to structural features of peptides, it can be divided into cyclic lipopeptide (CLP), phosphono-oligopeptide, and dipeptide. Many *Bacillus* strains produce small circular peptides (such as Iturin, Fengycin, Bacillopeptins, and Surfactin) with a long fatty acids moiety. They are composed of seven (surfactins and iturins) or ten α-amino acids (fengycins) linked to one unique β-amino (iturins) or β-hydroxy (surfactins and fengycins) fatty acid. The length of this fatty acid chain may vary from C-13 to C-16 for surfactins, from C-14 to C-18 in the case of fengycins (Ongena and Jacques 2007). Iturin and fengycins display a strong antifungal activity and are inhibitory for the growth of a wide range of phytopathogens (Hsieh et al. 2008; Vanittanakoma and Loeffler 1986). Bacilysocin produced by *B. subtilis* has a phospholipid structure, and it may be derived from phosphatidyl glycerol through acyl ester hydrolysis. Phosphatidyl glycerol is the major component of phospholipids in *B. subtilis*, which constitutes about 75% of the total phospholipids (Tamehiro et al. 2002). Bacilysin is a nonribosomally synthesized dipeptide antibiotic that is active against a wide range of bacteria and some fungi (Rajavel et al. 2009).

Production of polyketide-like compounds with antimicrobial activity by wild-type isolates of *Bacillus* spp. has been described previously (Hofemeister et al. 2004). The polyene antibiotics, difficidin and oxydifficidin, are highly unsaturated 22-member macrolides with a rare phosphate group (Wilson et al. 1987). Another antibiotic, bacillaene, is a linear molecule with two amide bonds: the first links an α-hydroxy carboxylic acid to a β-amino carboxylic acid containing a conjugated hexaene, and the second links the hexaene-containing carboxylic acid to an (ω-1) amino carboxylic acid containing a conjugated triene.

Numerous cell-wall-degrading enzymes, especially chitinase, have been isolated from *Bacillus* species. Many strains of *Bacillus* can produce a high level of chitinolytic enzymes (Xiao et al. 2009; Huang et al. 2005). Moreover, many researches have shown that chitinase is involved in antifungal activity and can enhance the insecticidal activity of *Bacillius* sp. (Table 1).

2.2 *Pseudomonas* and Its Antifungal Metabolites

Pseudomonas species are ubiquitous inhabitants of soil, water, and plant surfaces that belong to the Gamma-subclass of Proteobacteria. Many pseudomonades live in a commensal relationship with plants, utilizing nutrients exuded from plant surfaces and surviving environmental stress by occupying protected sites provided by the plant's architecture. Bacteria of *Pseudomonas* genus are the most popular PGPR and some species were also used in practice for biocontrol of *Gaeumannomyces graminis* var *tritici*, *Rhizoctonia solani*, *Erwinia carotovora* var. *carotovora*, *Pythium ultimum*, and *Fusarium oxysporum*. The mechanism suggested for achieving such inhibition includes: production of antibiotics, iron chelating compounds, hydrolytic enzymes, and biosurfactants; competition for favorable nutritional sites; and ISR and even due to their action as mycorrhization-helper bacteria (MHB).

Pseudomonads have an exceptional capacity to produce a wide variety of metabolites, including antibiotics (Pyrrolnitrin, Pyoluteorin (Plt), Phenazine compounds) and chitinase that are toxic to plant pathogens. Antibiotic production by *Pseudomonas* spp. enhances the fitness of the producing strain and suppresses pathogens that would otherwise disserve plant health. Certain antibiotic-producing *Pseudomonas* spp. have received great attention in the world as biological control agents to enhance agricultural productivity.

Table 1 Antifungal metabolites produced by *Bacillus* spp.

Antifungals	Type	Source	Structure	Pathogen	Reference
Bacillomycin D, F, L, Lc	Cyclic lipopetide	*B. subtilis* *B. amyloliquefaciens*	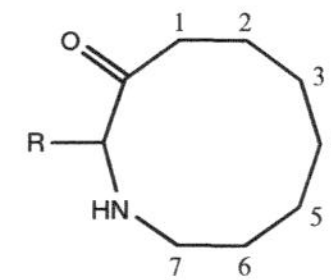 1: L-Asn; 2: D-Tyr; 3: D-Asn; 4: L-Pro; 5: L-Glu; 6: D-Ser; 7: L-Thr R: n-C_{14}, i-C_{15}, ai-C_{15} bacillomycin D 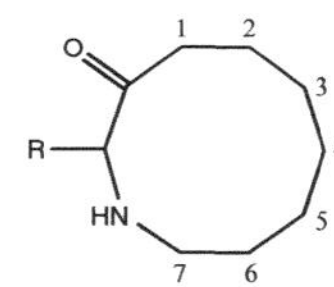 1: L-Asn; 2: D-Tyr; 3: D-Asn; 4: L-Gln; 5: L-Pro; 6: D-Asn; 7: L-Thr R: i-C_{16}, i-C_{17}, ai-C_{17} bacillomycin F 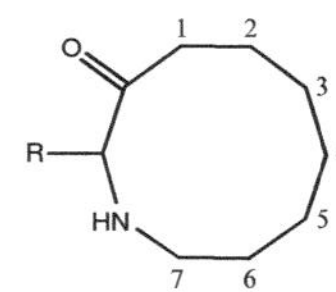 1: L-Asp; 2: D-Tyr; 3: D-Asn; 4: L-Ser; 5: L-Gln; 6: D-Ser; 7: L-Thr R: n-C_{14}, i-C_{15}, ai-C_{15} bacillomycin L	*Fusarium graminearum, Alternaria alternata, Rhizoctonia solani, Cryphonectria parasitica, Phytophthora capsici*	Eshita and Roberto (1995) Zhao et al. (2010) Ongena and Jacques (2007)

(*continued*)

Table 1 (continued)

Antifungals	Type	Source	Structure	Pathogen	Reference
			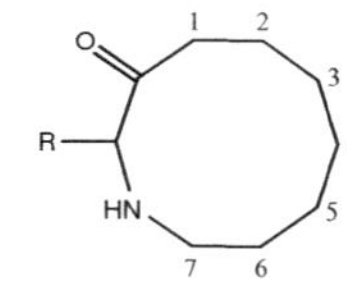 1: L-Asn; 2: D-Tyr; 3: D-Asn; 4: L-Ser; 5: L-Glu; 6: D-Ser; 7: L-Thr R: n-C_{14}, i-C_{15}, ai-C_{15}, i-C_{16} bacillomycin C		
Bacillopeptins A, B, C	Cyclic lipopetide	*B. subtilis*	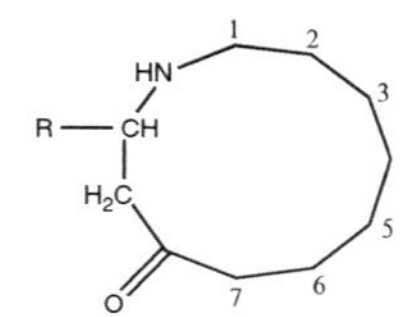 1: L-Thr; 2: D-Ser; 3: L-Glu; 4: L-Ser; 5: D-Asn2; 6: D-Tyr; 7: L-Asn1 R A: $CH_3(CH_2)_9CH_2$ B: $CH_3CH(CH_3)(CH_2)_8CH_2$ C: $CH_3CH(CH_3)(CH_2)_9CH_2$	With broad spectrum antibacterial activity	Kajimura et al. (1995)
Bacitracin	Cyclic lipopetide	*B. subtilis* *B. licheniformis*	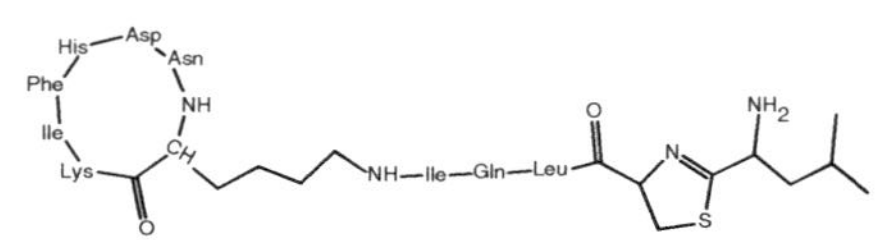	*Streptococcus aureus*, *S. faecalis*	Azevedo et al. (1993) Haavik and Froyshov (1975)
Fengycin	Cyclic lipopettide	*B. thuringiensis* *B. subtilis* *B. amyloliquefaciens*	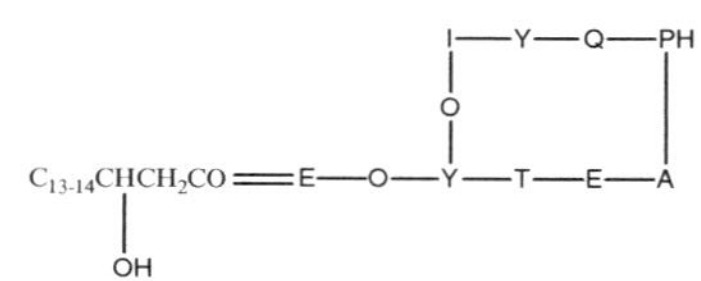	*C. gloeosporioides*, *F. oxysporum*, *F. solani*, *R. solani*, *B. cinerea*, *P. capsici*, *Sclerotium cepivorum*, *Colletotrichum*	Kim et al. (2004) Wang et al. (2007) Vanittanakoma and Loeffler (1986)

				coccodes, *Trichoderma harzianum*	
Iturin(A–E)	Cyclic lipopetide	*B. subtilis* *B. amyloliquefaciens* *B. circulan*	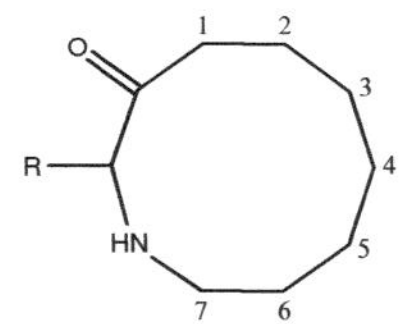 1: L-Asn; 2: D-Tyr; 3: D-Asn; 4: L-Gln; 5: L-Pro; 6: D-Asn; 7: L-Ser R: n-C_{14}, i-C_{15}, ai-C_{15} Iturin A	*Mycobacterium smegmatis*, *Agrobacterium tumefaciens*, *Penicillium chrysogenum*, *P. notatum*, *R. solani*, *B. cinerea*	Peypoux et al. (1978) Besson et al. (1984a, b) Hsieh et al. (2008)
Mycosubtilin	Cyclic lipopetide	*B. subtilis*	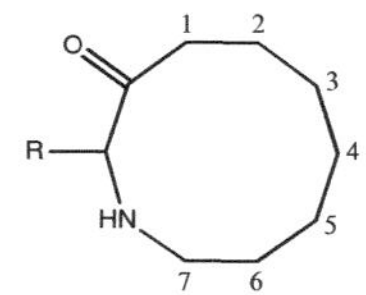 1: L-Asn; 2: D-Tyr; 3: D-Asn; 4: L-Gln; 5: L-Pro; 6: D-Ser; 7: L-Asn R: n-C_{16}, i-C_{15}, ai-C_{17}	*Candida* spp., *Saccharomyces cerevisiae*, *Pythium aphanidermatum*	Duitman et al. (1999)
Mycobacillin	Cyclic peptide	*B. subtilis*	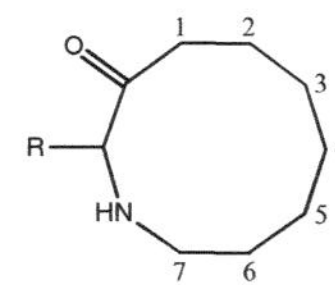 1: L-Asn; 2: D-Tyr; 3: D-Asn; 4: L-Gln; 5: L-Pro; 6: L-Ser; 7: D-Asn; R: n-C_{14}, i-C_{15}, ai-C_{15}	*C. albicans*, *Aspergillus niger*	Majumder et al. (1988)

(*continued*)

Table 1 (continued)

Antifungals	Type	Source	Structure	Pathogen	Reference
Rhizocticins	Phosphono-oligopeptides	*B. subtilis*	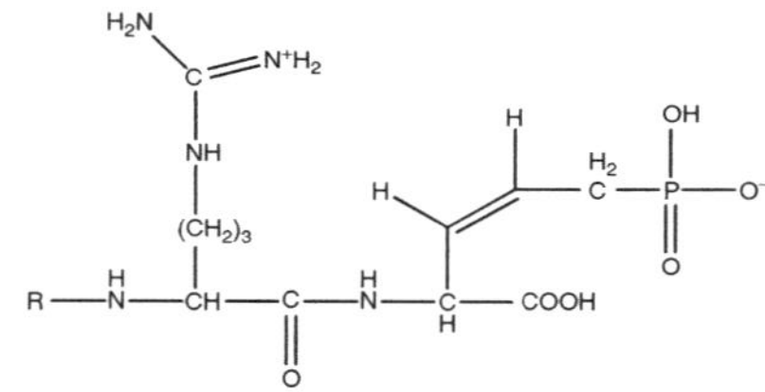	With broad spectrum antifungal activity	Rapp et al. (1988)
Alboleutin	Peptide	*B. subtilis* AF-8	Unknown	Unknown	Omura et al. (1980)
Bacillibactin	Iron-siderophore	*B. amyloliquefaciens*	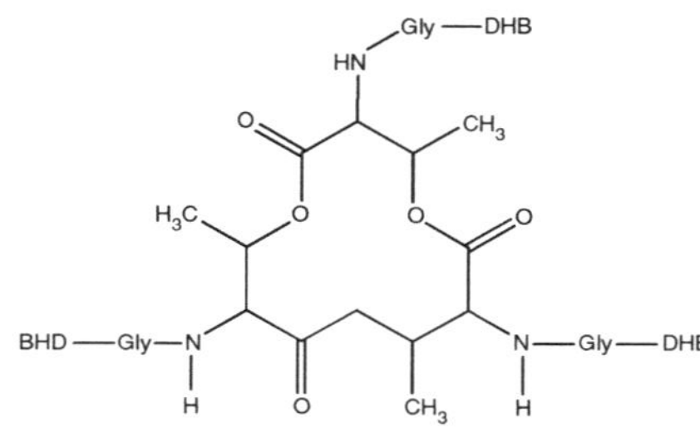 DHB 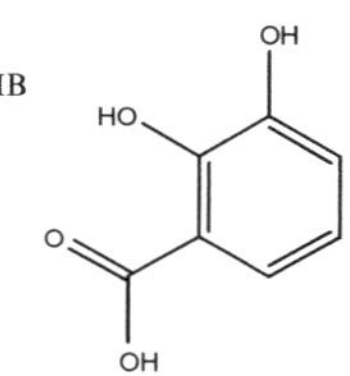	With broad spectrum antibacterial activity	Chen et al. (2009)

Bacillaene	Polyketides	*B. subtilis* *B. amyloliquefaciens*	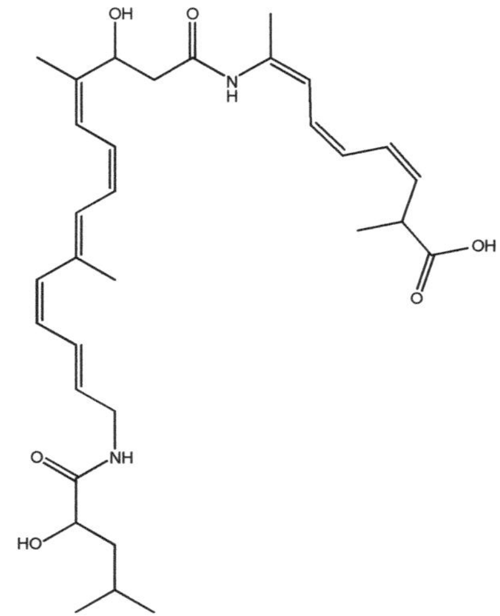	*Escherichia coli*, *Klebsiella pneumoniae*, *Proteus vulgaris*, *Serratia marcescens*, *B. thuringiensis*, *S. aureus*	Chen et al. (2009)
Macrolactin	Polyketides	Marine *Bacillus* strains	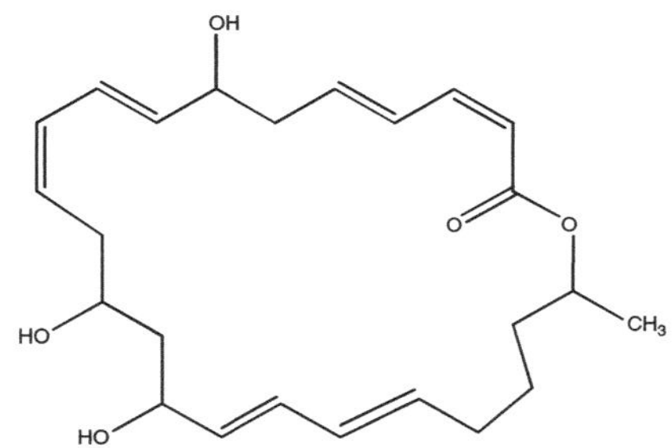	*S. aureus*, murine melanoma cancer cells, *Herpes simplex* viruses	Schneider et al. (2007)

(*continued*)

Table 1 (continued)

Antifungals	Type	Source	Structure	Pathogen	Reference
Difficidin (oxydifficidin)	Polyketides	*B. subtilis* *B. amyloliquefaciens*	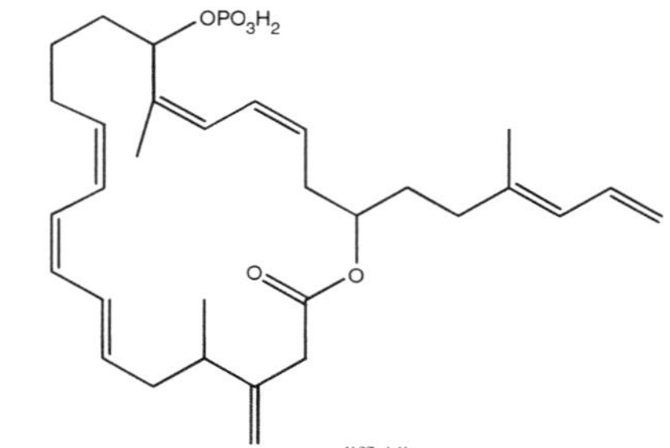 difficidin	With broad spectrum antibacterial activity	Wilson et al. (1987)
Zwittermicin A	Aminopolyol	*B. thuringiensis* *B. cereus*	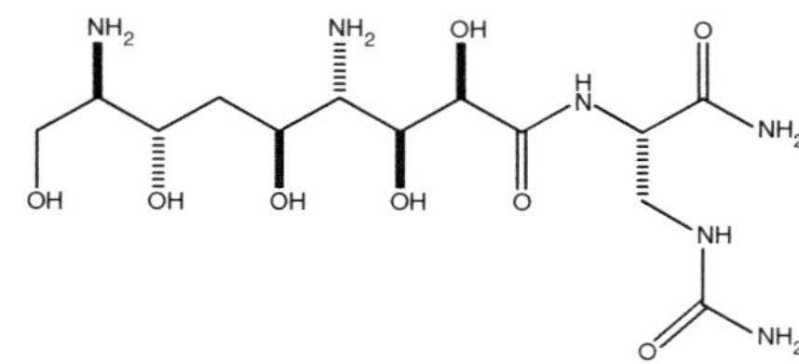	*Phytophthora medicaginis*, *P. aphanidermatum*	Nair et al. (2004) Silo-Suh et al. (1998)
Bacilysin	Dipeptide	*B. subtilis* *B. amyloliquefaciens*	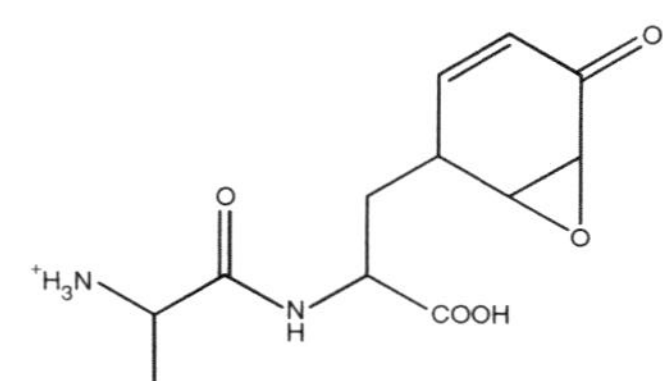	*E. coli* *Absidia* sp.	Phister et al. (2004) Ozcengiz and Alaeddinoglu (1991)

Subtilin	Peptide	*B. subtilis*	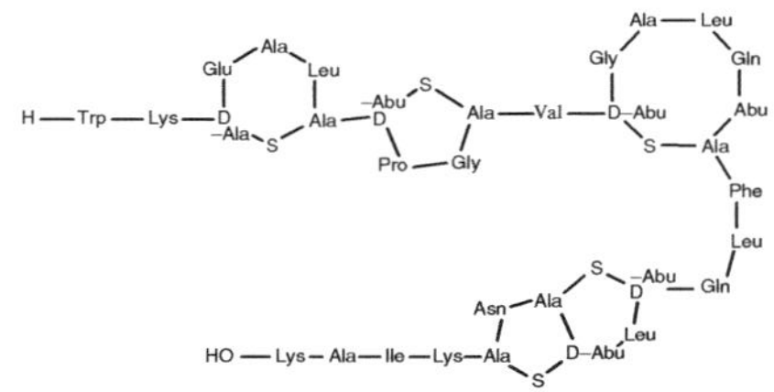	*Micrococcus luteus*, *Lactococcus lactis*	Chan et al. (1993)
Surfactin	Biosurfactant	*B. subtilis* *B. polyfermenticus* *B. amyloliquefaciens*		With broad spectrum antibacterial activity	Nakano et al. (1988) Kim et al. (2009)
Plipastatin	Lipopeptide	*B. subtilis* *B. cereus*	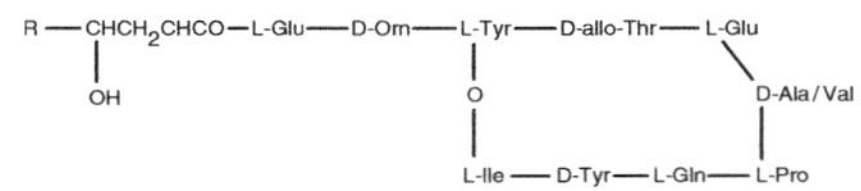	*A. mali*, *B. cinerea*, *Pyricularia oryzae*	Tsuge et al. (2007)
Kurstakins	Lipopeptide	*B. thuringiensis*	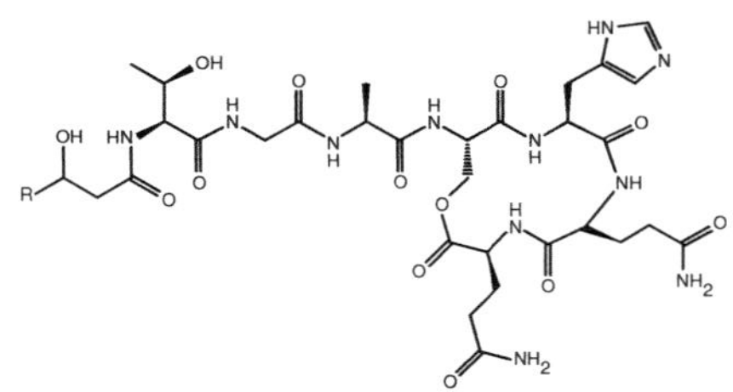	*Stachybotrys charatum*	Hathout et al. (2000)
Fungicin M-4	Peptide	*B. licheniformis*	Unknown	*Microsporum canis*, *Mucor mucedo*, *M. plumbeus*, *Sporothrix schenckii*	Lebbadi et al. (1994)

(*continued*)

Table 1 (continued)

Antifungals	Type	Source	Structure	Pathogen	Reference
Bacisubin	Protein	*B. subtilis*		*Magnaporthe grisease, Sclerotinia sclerotiorum, R. solani, A. oleracea, A. brassicae, B. cinerea*	Liu et al. (2007)
YM-47522		*Bacillus* sp. YL-03709B	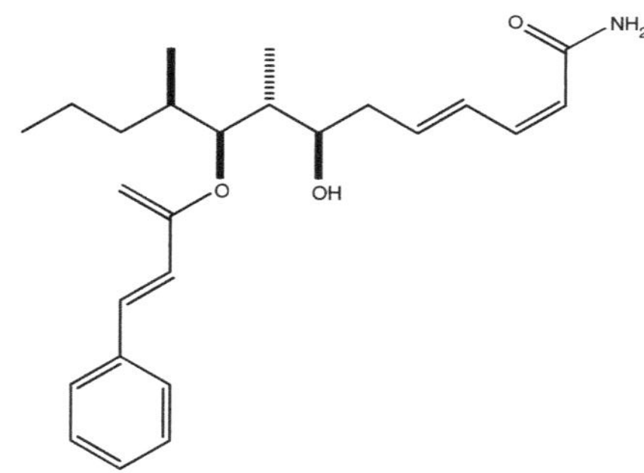	*Rhodotorula species, Pichia angusta, C. albicans, C. tropicalis*	Shibazaki et al. (1996), Sugawara et al. (1996)
Chitinase		*B. licheniformis* *B. subtilis* *B. cereus*		*Gibberella saubinetii, A. niger*	Xiao et al. (2009) Yang et al. (2009) Chang et al. (2007)
Cellulases, Amylases, Glucanases	Fungal cell-wall-degrading enzymes	*B. subtilis* *Bacillus* sp.		With broad spectrum antifungal activity	Fukumori et al. (1986) Roncero 1983
Others	Protein or peptide	*Bacillus* sp.		*Geotrichum candidum, Bipolaris maydis, A. brassicae, A. niger, Cercospora personata*	Maldonado et al. (2009) Zhang et al. (2008)

Pyrrolnitrin (Prn, 3-chloro-4-(29-nitro-39-chlorophenyl)-pyrrole) is a secondary metabolite derived from tryptophan and has strong antifungal activity. Most of *Pseudomonas* and *Burkholderia* strains produce this antibiotic. Production of Prn has been correlated with the ability of some bacteria to control fungal plant pathogens and diseases, including the damping-off pathogen, *R. solani*. Prn and its production by *Pseudomonas* species was first described by Arima et al. (1964). This highly active metabolite has been used as a clinical antifungal agent for the treatment of skin mycoses, and a phenylpyrrole derivative of Prn has been developed as an agricultural fungicide (Tawara et al. 1989).

Pyoluteorin (Plt) is an antibiotic that inhibits *P. ultimum* and suppresses plant diseases. Plt is composed of a resorcinol ring, derived through polyketide biosynthesis, which is linked to a bichlorinated pyrrole moiety whose biosynthesis remains uncharacterized. The production of the antifungal metabolite 2,4-diacetylphloroglucinol (2,4-DAPG) by many fluorescent *Pseudomonas* spp. has been seen to play a major role in the biocontrol of a range of plant pathogens, including *P. ultimum*, *G. graminis* var. *tritici*, and *Thielaviopsis bassicola* (Keel et al. 1990; Vincent et al. 1991; Fenton et al. 1992; Levy et al. 1992).

Kim (2003) reported that *P. aeruginosa* excrete two different types of siderophores (pyoverdine and pyochelin) at low iron concentration. When free iron concentration in environment reached to 10^{-17} mol/L, binding ability of pyochelin with iron ions is 1.5×10^{-7} mol/L, whereas binding ability of pyoverdine is higher than 10^{-20} mol/L (Chen et al. 2008).

Phenazines (Phz) are N-containing heterocyclic pigments. Currently, over 50 naturally occurring Phz compounds have been described and mixtures of as many as ten different Phz derivatives can occur simultaneously in one organism. Growth conditions determine the number and type of Phz synthesized by an individual bacterial strain. For example, *P. fluorescens* 2–79 produces mainly phenazine 1-carboxylic acid (PCA), whereas *P. aureofaciens* 30–38 not only produces PCA but also lesser amounts of 2-OH-phenazines (Dwivedi and Johri 2003).

Antifungal proteins, such as chitinase, are key components of defense and offense mechanisms of many groups of fungi and bacteria. These microbial lytic enzymes are being exploited widely for crop disease management. These enzymes hydrolyze the chitin present in the fungal cell wall, leading to lysis. Production of these lytic enzymes is considered to be the major antagonistic activity of fluorescent pseudomonads belonging to PGPR.

Plant-associated *Pseudomonas* spp. also produces diversified CLPs with potential antimicrobial, cytotoxic, and surfactant properties. Based on the length and composition of the fatty acid tail as well as the number, type, and configuration of the amino acids in the peptide moiety, CLPs of *Pseudomonas* spp. were classified into four major groups (i.e., the viscosin, amphisin, tolaasin, and syringomycin groups). In brief, CLPs of the viscosin group are composed with nine amino acids linked at the N terminus to, in most cases, 3-hydroxy decanoic acid (3-HDA). CLPs in the amphisin group consist of 11 amino acids in the peptide part coupled to 3-HDA. For several members of this group, including amphisin and tensin, the crystal structure has been resolved (Henriksen et al. 2000; Sorensen et al. 2001). For both tensin and amphisin, the structures are mainly helical, with the cyclic peptide wrapping around a water molecule. Compared with the viscosin and amphisin groups, CLPs in the tolaasin group are much more diverse due to multiple variations in both the composition and length of the peptide chain (19–25 amino acids) and the lipid tail (3-HDA or 3-hydroxyoctanoic acid). In terms of shear numbers of amino acids in the peptide moiety, CLPs in the syringomycin group are structurally similar to the CLPs in the viscosin group. However, macrolides, polyenes, quinone-type antibiotics, and hydroxyphenol have not been found so far among the secondary metabolites produced by the *Pseudomonas* (Table 2).

2.3 Burkholderia and Its Antifungal Metabolites

The genus *Burkholderia* comprises more than 40 different species, which inhabit a wide array of ecological niches. Among others, they occur in soil and water, in the plant rhizosphere and endophytically in roots and shoots, but also in and on fungal mycelia. *Burkholderia* species are also well known for their biological and metabolic properties, which can be exploited for biological control of fungal diseases but also for bioremediation and plant-growth promotion. *Burkholderia* spp. can antagonize and repress many soil-borne plant pathogens. Particularly, *B. cepacia* complex (Bcc) is known to be a ubiquitous inhabitant in soil, which has been used as an effective biocontrol agent for *Pythium*-induced damping-off, *Aphanomyces*-induced root rot of pea, and *R. solani*-induced root rot of poinsettia. Bcc is a group of genetically distinct but phenotypically similar bacteria that are divided into at least nine species. The effectiveness of Bcc isolates as biocontrol and PGP agents is based on a wide array of beneficial properties including the production of indole-acetic acid (IAA), the ability to fix atmospheric nitrogen, and the production of a wide array of compounds with antimicrobial activity, including cepacin, cepaciamide, cepacidines, altericidins, pyrrolnitrin, quinolones, phenazine, siderophores, and a lipopeptide. In the early 1990s, these useful properties led to the registration of four Bcc strains for use as biopesticides by the EPA (the U.S. Environmental Protection Agency), three of which were later classified as *B. ambifaria* and one as *B. cenocepacia.*

Cepaciamides A and B are fungitoxic 3-amino-2-piperidinone-containing lipids effective against *Botrytis cinerea* and *Penicillium expansum*, which cause the storage rot of beet roots, and are considered to be a promising biocontrol agent (Toshima et al. 1999). A peptide antibiotic complex, altericidins (altericidin a, B, and C), was isolated from the culture broth of *P. cepacia* KB-1 (Kirinuki et al. 1984). The alteridins inhibit the growth of a wide range of fungi and yeasts, but show no effect on the growth of bacteria species.

Cepacidine A is a novel glycopeptides with a potent antifungal antibiotic produced by *P. cepacia* AF 2001 (Lee et al. 1994). Cepacidine A exhibited a broad antifungal spectrum against various animal and plant pathogenic fungi. In particular, cepacidine A was highly active against dermatophytes, namely

Table 2 Antifungal metabolites produced by *Pseudomonas* spp.

Antifungals	Type	Source	Structure	Pathogen	Reference
Pyrrolnitrin	N-containing heterocycles	*P. fluorescens* *P. cepacia*	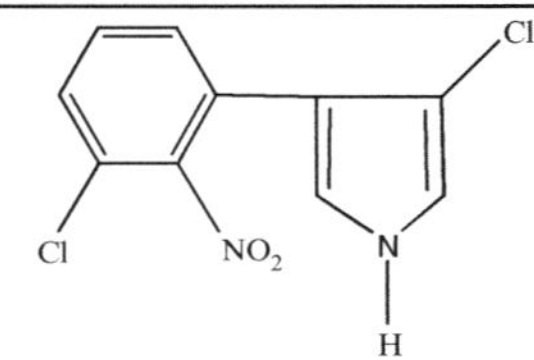	*Fusarium* sp., *Pythium ultimum*, *R. solani*, *Alternaria* spp., *Pyricularia oryzae*, *Botrytis cinerea*, *Penicillium expansum*, *Verticillium dahiae*	Loper and Gross (2007) Hammer et al. (1997) Ligon et al. (2000)
Pyoluteorin	N-containing heterocycles	*P. fluorescens*	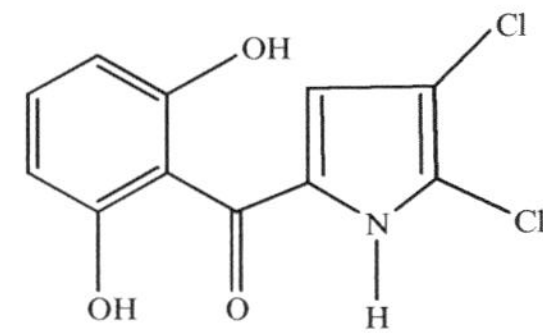	*P. ultimum*, *R. solani*	Howell and Stipanovic (1980) Maurhofer et al. (1995)
2,4-diacetylphloro-glucinol	Polyketide	*P. fluorescens*	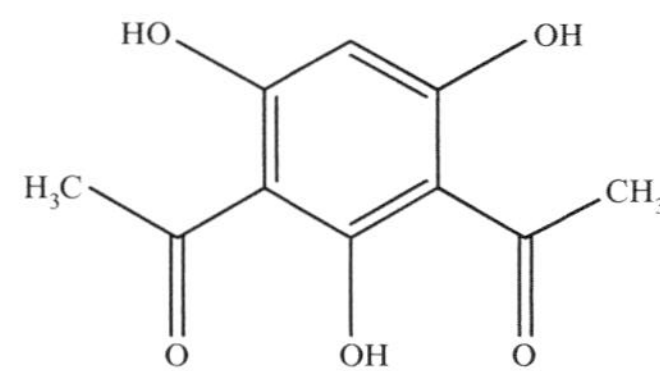	*Gaeumannomyces graminis*, *P. ultimum*	Bangera and Thomashow (1999) Keel et al. (1996)

(*continued*)

Table 2 (continued)

Antifungals	Type	Source	Structure	Pathogen	Reference
Pyoverdine	Siderophore	*P. fluorescens* *P. aeruginosa* *P. putida*	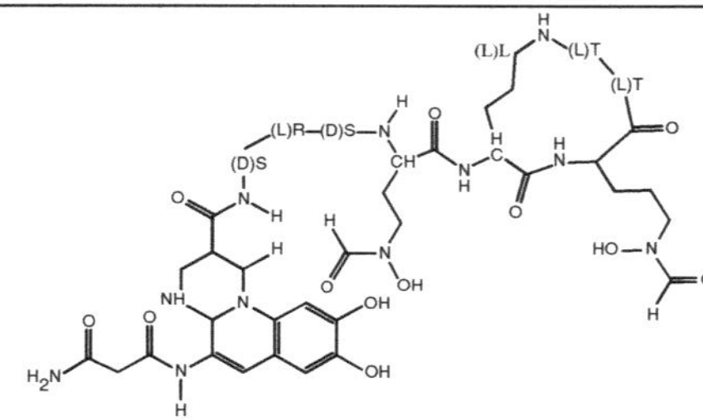	*F. graminearum*	Lamont and Martin (2003)
Pyochelin	Siderophore	*P. fluorescens*	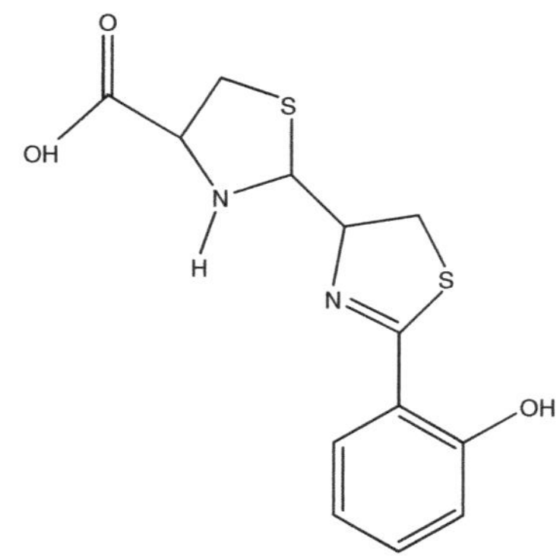	*Candida* sp., *Aspergillus* sp.	Cox et al. (1981)
Pyocyanin	Phenazine compounds	*P. aeruginosa*	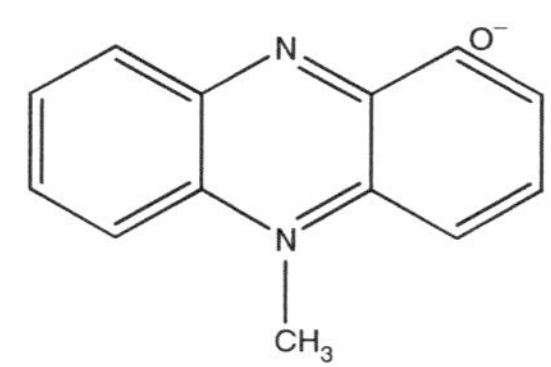	*C. albicans*, *A. fumigatus*	Budzikiewicz (1993)

Phenazine-1-carboxylic acid	Phenazine compounds	*P. aeruginosa* *P. chlororaphis* *P. fluorescens*	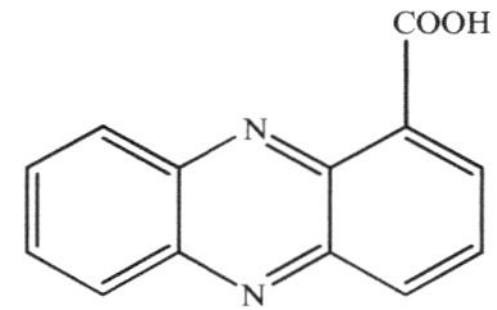	*Caenorhabditis elegans*, *Gaeumannomyces graminis*	Denning et al. (2003) Liu et al. (2008)
1-hydroxyphenazine	Phenazine compounds	*P. aeruginosa*	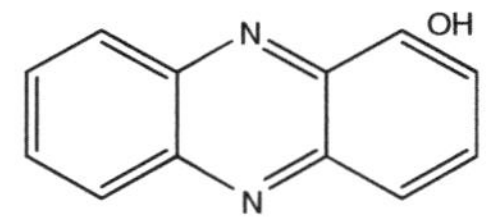	*C. albicans*, *A. fumigatus*	Loper and Gross (2007)
Phenazine-1-carboxamide	Phenazine compounds	*P. aeruginosa* *P. Chlororaphis*	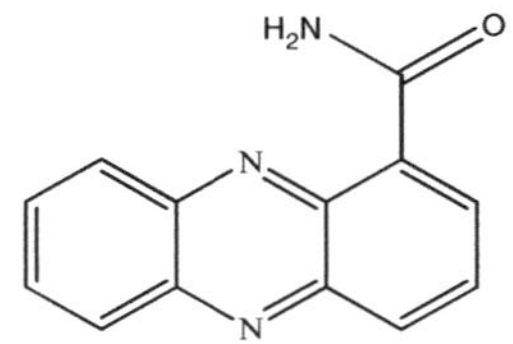	*F. oxysporum*	Loper and Gross (2007) Woeng et al. (1998)
Hydrogen cyanide		*P. fluorescens*		*P. ultimum*	Ramette et al. (2003)
Chitinase		*P. pseudomonuds*		*F. oxysporum*	Ajit et al. (2006)
Viscosin	Cyclic lipopeptide	*P. fluorescens*	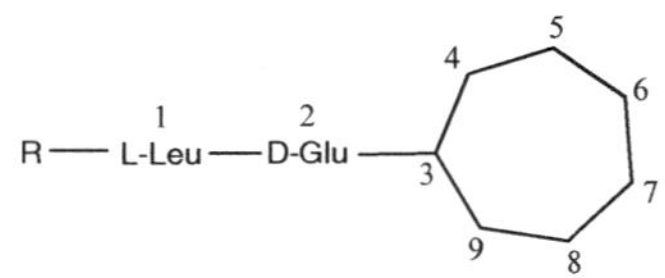 R: C_{10}HO acid; 3: D-aThr; 4: D-Val; 5: L-Leu; 6: D-Ser; 7: L-Leu; 8: D-Ser; 9: L-Ile	*F. culmorum*	Braun et al. (2001)

(*continued*)

Table 2 (continued)

Antifungals	Type	Source	Structure	Pathogen	Reference
Massetolide (A, D)	Cyclic lipopeptide	*P. fluorescens*	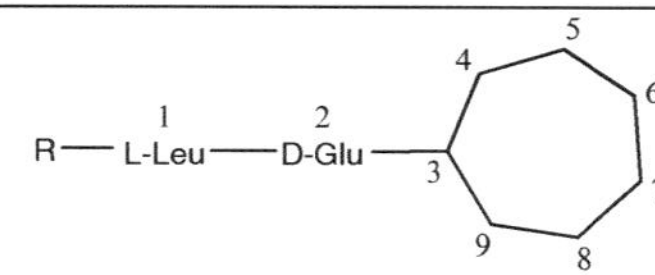 R: $C_{10}HO$ acid; 3: D-aThr; 4: D-aIle; 5: L-Leu; 6: D-Ser; 7: L-Leu; 8: D-Ser; 9: L-Ile(A)/Leu(D)	*Mycobacterium tuberculosis*, *M. aviumintracellulare*	De Souza et al. (2003)
Amphisin	Cyclic lipopeptide	*Pseudomonas* sp.	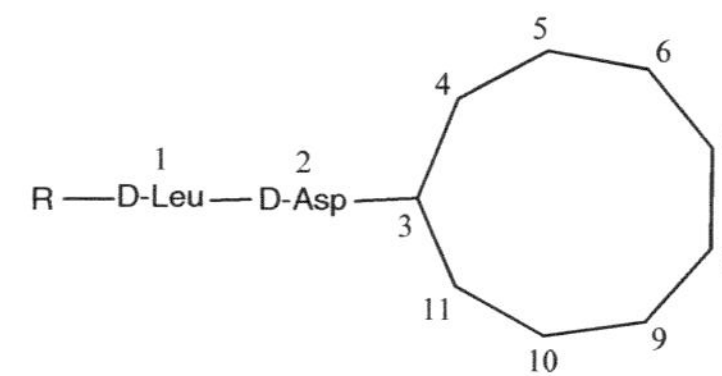 R: $C_{10}HO$ acid; 3: D-aThr; 4: D-Leu; 5: D-Leu: 6: D-Ser; 7: L-Leu; 8: D-Gln; 9: L-Leu; 10: L-Ile; 11: L-Asp	*P. ultimum*, *R. solani*	Koch et al. (2002)
Arthrofactin	Cyclic lipopeptide	*Pseudomonas* sp.	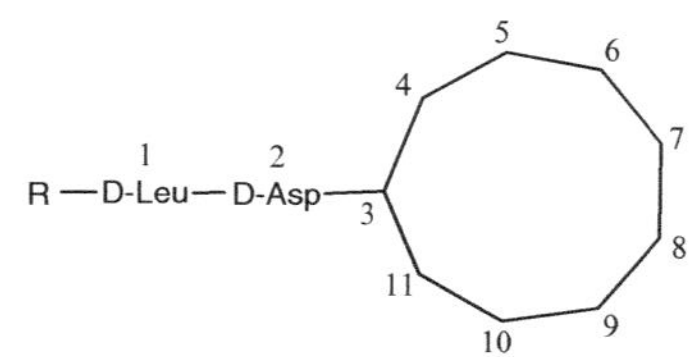 R: $C_{10}HO$ acid; 3: D-Thr; 4: D-Leu; 5: D-Leu; 6: D-Ser; 7 : L-Leu; 8: D-Ser; 9: L-Ile; 10: L-Ile; 11: L-Asp	With broad spectrum antifungal activity	Roongsawang et al. (2003)

Tolaasin	Lipopeptide	*Pseudomonas* sp. *P. tolaasii*	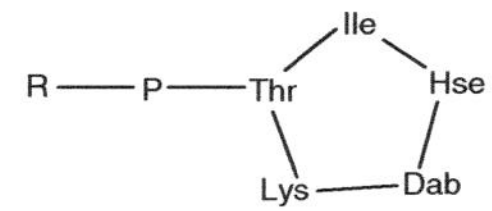 R: C_8HO acid P: Dhb-Pro-Ser-Leu-Val-Ser-Leu-Val-Val-Gln-$(Leu)_5$-Val-Dhb	With broad spectrum antifungal activity	Rainey et al. (1993) Godfrey et al. (2001)
Syringopeptin	Lipopeptide	*P. syringae* pv. *syringae*	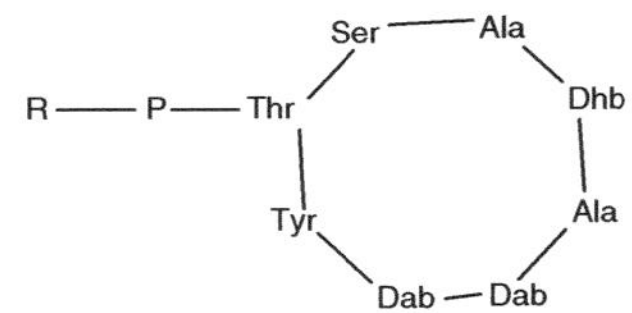 R: $CH_3(CH_2)_8CHOHCH_2CO$ P: Dhb-Pro-Val-Val-Ala-Ala-Val-Val-Dhb-Ala-Val-Ala-Ala-Dhb	With broad spectrum antifungal activity	Feil et al. (2005) Scholz-Schroeder et al. (2003) Kang and Gross (2005)
Syringomycin	Cyclic Lipopeptide	*P. syringae* pv. *syringae*	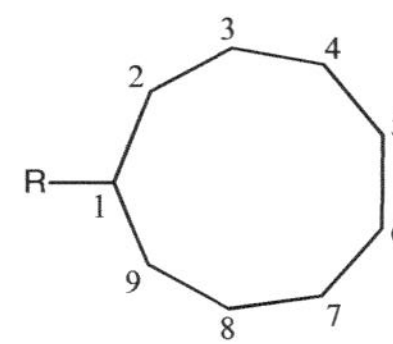 R: 3-hydroxy fatty acyl; 1: L-Ser; 2: D-Ser; 3: D-Dab; 4: L-Dab; 5: L-Arg; 6: L-Phe; 7: Z-Dhb; 8: L-Asp(3-OH); 9: L-Thr(4-Cl)	With broad spectrum antifungal activity	Kang and Gross (2005) Vaillancourt et al. (2005)

(*continued*)

Table 2 (continued)

Antifungals	Type	Source	Structure	Pathogen	Reference
Putisolvin	Cyclic Lipopeptide	*P. putida*	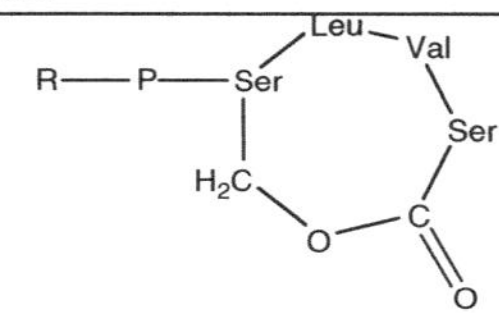 R: $CH_3(CH_2)_4CO$ P: Leu-Glu-Leu-Ile-Gln-Ser-Val-Ile-Ser	*Candida* sp.	Kuiper et al. (2004)
Orfamide A	Cyclic Lipopeptide	*P. fluorescens*	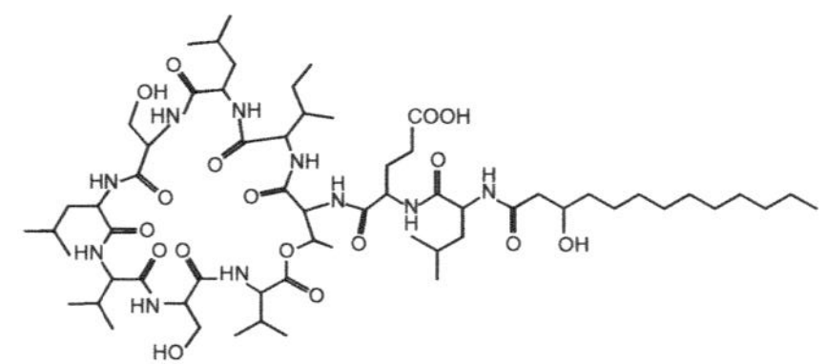	Unknown	Loper and Gross (2007)

Microsporum canis, *Trichophyton* spp., and *Epidermophyton* spp., and true yeast at concentrations lower than 0.049 μg/ml. The activities of cepacidine A were greater than those of amphotericin B in almost all strains. However, no antibacterial activity against *B. subtilis*, *E. coli*, *Staphylococcus aureus*, and *P. aeruguinosa* was detected (MIC > 100 μg/ml).

Two acetylenic antibiotics, cepacins A and B, have been isolated from the fermentation broth of *P. cepacia* SC 11,783. Cepacin A has good activity against *Staphylococci* (MIC 0.2 mg/ml) but weak activity against *Streptococci* (MIC 50 mg/ml) and the majority of Gram-negative organisms. Cepacin B has excellent activity against *Staphylococci* (MIC less than 0.05 mg/ml) and some Gram-negative organisms (Parker et al. 1984). Similarly, Glidobactins are acylated tripeptide derivatives that contain a 12-membered ring structure consisting of the two unique nonproteinogenic amino acids, *erythro*-4-hydroxy-l-lysine and 4(*S*)-amino-2(*E*)-pentenoic acid (Schellenberg et al. 2007). The antibiotic 2-Hydroxymethyl-chroman-4-one isolated from *Burkholderia* sp. MSSP exhibited good activities against phytopathogens such as *P. ultimum*, *Phytophthora capsici*, and *S. sclerotiorum*. 2-Hydroxymethyl-chroman-4-one was used to mediate for synthesis of benzopyranones (Kang et al. 2004).

Most antibiotics isolated from *Burkholderia* culture filtrates, namely, Prn, Phz, Plt, and indole derivatives, belong to the class of N-containing heterocycles and have been shown to originate from intermediates or end products of the aromatic amino acid biosynthetic pathways. Prn is a chlorinated phenylpyrrole antibiotic that was first isolated from *B. pyrrocinia* (Arima et al. 1964) and later from other microorganisms, including *P. fluorescens*, *P. chlororaphis*, *P.aureofaciens*, *B.cepacia*, *Enterobacter agglomerans*, *Myxococcus fulvus*, and *Serratia species*. Prn has activity against several bacteria and soil-borne fungi, in particular *R. solani*. Prn is also effective against postharvest diseases caused by *B. cinerea* on apple, pear, and on cut flowers and has been used to treat humans infected by opportunistic fungi. Plt is a phenolic polyketide that was first isolated from *P. aeruginosa* and later from *P. aeruginosa* strain S10B2 and *P. fluorescens* strains Pf-5 and CHA0 (Takeda 1958). Plt has bactericidal, herbicidal, and fungicidal activities, in particular against *Pythium* spp. Application of pure Plt to cotton seeds resulted in significant suppression of *P. ultimum*-induced damping-off (Howell and Stipanovic 1980).

B. contaminans strain MS14 isolated from disease-suppressive soil produced a cyclic glycopeptide antibiotic, Occidiofungin. This compound inhibited the growth of a broad range of fungal pathogens, and the high-resolution mass spectrometry data revealed the existence of two structural variants of this antifungal peptide (Lu et al. 2009).

Many 4-quinolone compounds have antifungal activity. 4-Hydroxy-2-alkylquinolines (HAQs) have been long known as a class of antimicrobials produced by the opportunistic bacterial pathogen such as *P. aeruginosa* and *B. cepacia*. Many HAQs also act as iron chelators and even immune modulants. HAQs especially 3,4-dihydroxy-2-heptylquinoline (*Pseudomonas* quinolone signal) and its precursor, 4-hydroxy-2-heptylquinoline, have recently attracted much attention because of their role as intercellular signaling molecules in bacteria (Vial et al. 2008).

Members of the Bcc produce up to four different siderophores (ornibactin, pyochelin, cepabactin, and cepaciachelin). The siderophores produced by the Bcc contain most of the bidentate ferric iron-chelating groups commonly present in bacterial siderophores and includes catechols (present in cepaciachelin), linear hydroxamate and ahydroxycarboxylate groups (both present in ornibactin), a cyclic hydroxamate (hydroxypyridonate) moiety (as in cepabactin), and 2-hydroxy-phenyl-thiazoline/-oxazoline and thiazolidine-carboxylate moieties. Pyochelin, 2-(2-o-hydroxyphenyl-2-thiazolin-4-yl)-3-methylthiazolidine-4-carboxylic acid, is derived from the condensation of salicylic acid with two molecules of cysteine, each of which undergoes cyclisation to thiazoline and thiazolidine ring derivatives following their incorporation into the molecule. Natural pyochelin is present as two spontaneously interconvertible stereoisomers, pyochelins I and II, due to isomerisation at the C-2″ position of the thiazolidine ring. Ornibactin, L-Orn1(N5-OH, N5-acyl)-D-threo-Asp(b-OH)-L-Ser-L-Orn4 (N5-OH, N5-formyl)-1,4-diaminobutane, is a linear tetrapeptide derivative that chelates iron by providing three bidentate metal chelation groups. These groups (two hydroxamates and an ahydroxy-carboxylate) are generated by modification of the sidechains of three of the amino acids in the peptide (the N- and C-terminal ornithines, and the D-aspartate), with the serine residue serving only as a spacer. Cepabactin, 1-hydroxy-5-methoxy-6-methyl-2(1H)-pyridinone, is a cyclic hydroxamate (i.e., a hydroxypyridonate) and for that reason can also be considered as a heterocyclic catecholate. It was first identified as a metal-binding antibiotic that is secreted into the culture medium by *P. alcaligenes strain* NCIB 11492 and was termed G1549. Cepaciachelin, 1-N-[2-N′,6-N″-di(2,3-dihydroxybenzoyl)-L-lysyl]-1,4-diaminobutane, is a catecholate siderophore first isolated from the culture supernatant of *B. ambifaria* strain PHP7 (LMG 11351), a rhizosphere isolate, grown under iron-limiting conditions. It is comprised of a single molecule of lysine derivatised with 2,3-dihydroxybenzoic acid (DHBA) on the a and e amino groups, and with diaminobutane (putrescine) on the carboxyl group (Table 3).

3 Antifungal Mechanisms

Antagonistic PGPR, including *Bacillus* and *Pseudomonas*, are often considered microbial factories for the production of a vast array of biologically active CLPs (*Bacillus*: surfactin, iturin, and fengycin families; *Pseudomonas*: viscosin, amphisin, tollasin, and syringomycin group) potentially inhibitory for phytopathogen growth. One of the main modes of action of CLPs produced by antagonistic PGPR involves the formation of ion channels in the host plasma membrane leading to cytolysis. Pore formation results in the alkalization of the intercellular fluid and in the release of multiple cellular compounds. These antibiotics gave important modifications in the membrane permeability which permitted nucleotides proteins, polysaccharides, and lipids to escape from cells. At high concentrations (well above the critical micelle concentration), CLPs can directly solubilize plasma membranes

Table 3 Antifungal metabolites produced by *Burkholderia* spp.

Antifungals	Source	Structure	Pathogen	Reference
Altericidins (peptide)	*B. cepacia* KB-1	Unclear	A wide range of fungi and yeasts	Kirinuki et al. (1984)
Cepaciamides A, B	*B. cepacia* D-202	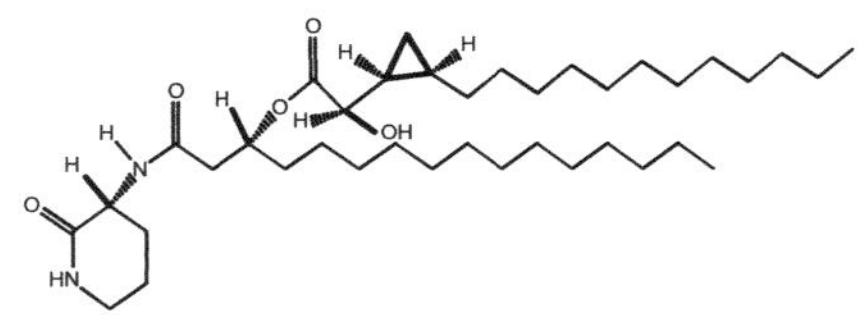	*B. cinerea*, *P. expansum*	Toshima et al. (1999)
Cepacidines/xylocandins (glycopeptide)	*B. cepacia* AF 2001	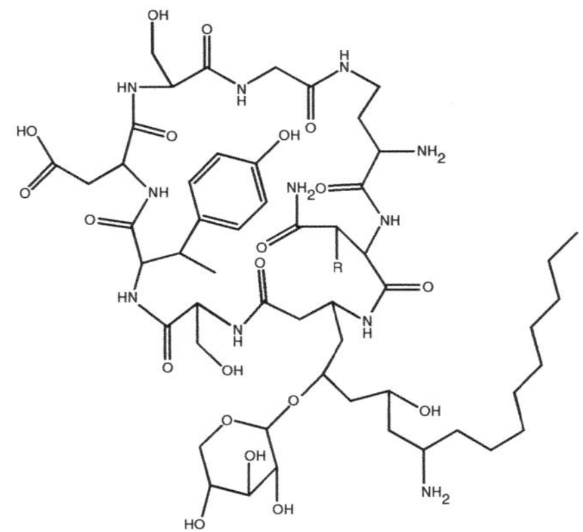	*Microsporum canis*, *Trichophyton* spp., *Epidermophyton* spp.	Lee et al. (1994)
Cepacin	*B. cepacia*	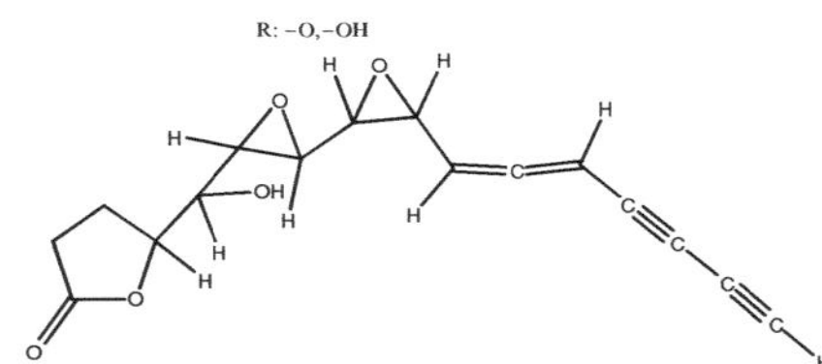	With broad spectrum antibacterial activity	Parker et al. (1984)

(*continued*)

Table 3 (continued)

Antifungals	Source	Structure	Pathogen	Reference
Glidobactins	*B. cepacia*	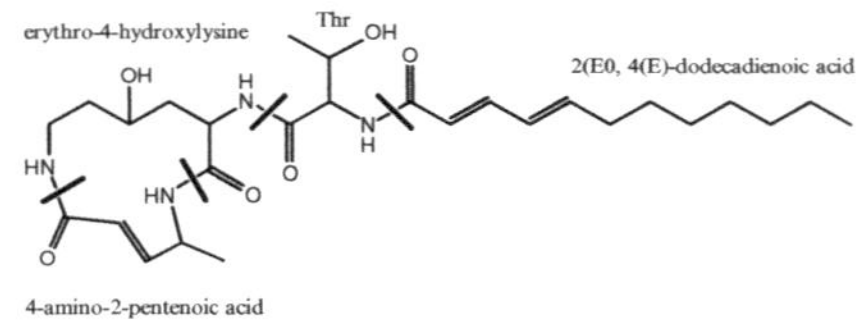	*Thielaviopsis basicola*	Shoji et al. (1990)
2-hydroxymethyl-chroman-4-one	*Burkholderia* sp.	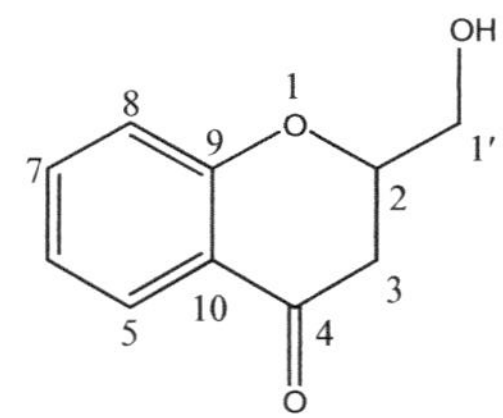	*P. ultimum*, *Phytophthora capsici*, *Sclerotinia sclerotiorum*	Kang et al. (2004)
4-hydroxy-2-alkylquinoline	*B. pseudomallei*, *B. thailandensis*, *B. ambifaria*	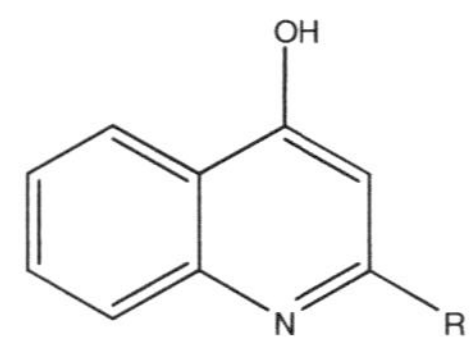	Antibacterial activity	Vial et al. (2008)
Lipopeptide-AFC-BC11	*B. cepacia* BC11	Unclear	*R. solani*, *P. ultimum*, *Colletotrichum* sp., *B. cinerea*, *Helminthosporium maydis*, *Fusarium* sp., *Rhizopus stolonifer*, *Rhodotorula glutinis*, *Sclerotium rolfsii*, *Scopulariopsis brevicaulis*	Kang et al. (1998)

Occidiofungin A (glycopeptide)	*B. contaminans*	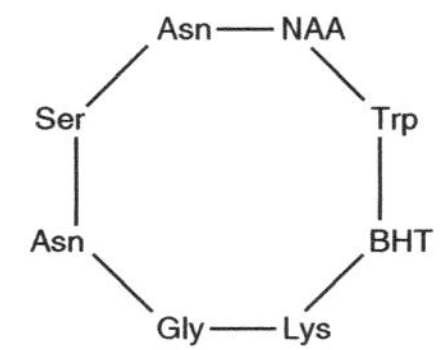	A broad range of fungal pathogens	Lu et al. (2009)
Phenazine	*B. cepacia*	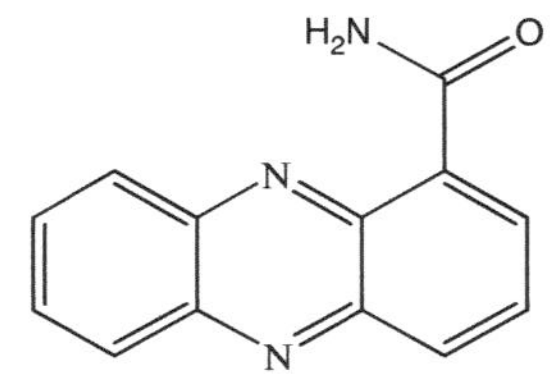	A broad range of fungal pathogens	Cartwright et al. (1995)
Pyrrolnitrin	*B. cepacia* NB-1 *B. cepacia* 5.5B *B.cepacia* *B. cepacia* K87 *B. pyrrocinia*	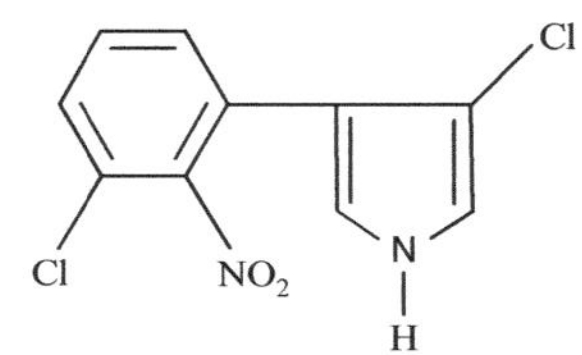	*R. solani*, *B. cinerea*, *C. gloeosporioides*, *Pyricularia oryzae*	Hwang et al. (2002)
Pyoluteorin	*Burkholderia* spp.	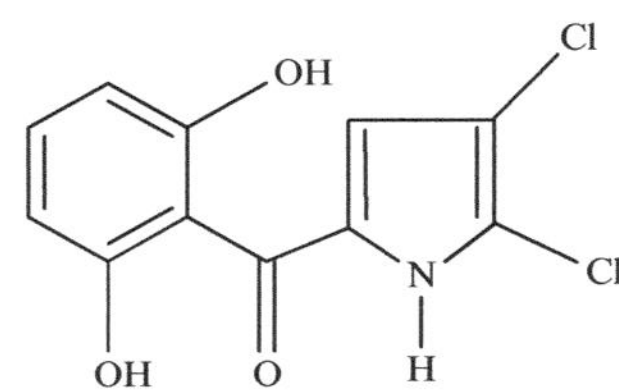	*Pythium* spp.	De Souza et al. (2003)

(*continued*)

Table 3 (continued)

Antifungals	Source	Structure	Pathogen	Reference
Quinolinones (pseudanes)	*P. cepacia* PC II			Moon et al. (1996)
Pyochelin (Siderophore)	*B. cepacia* complex	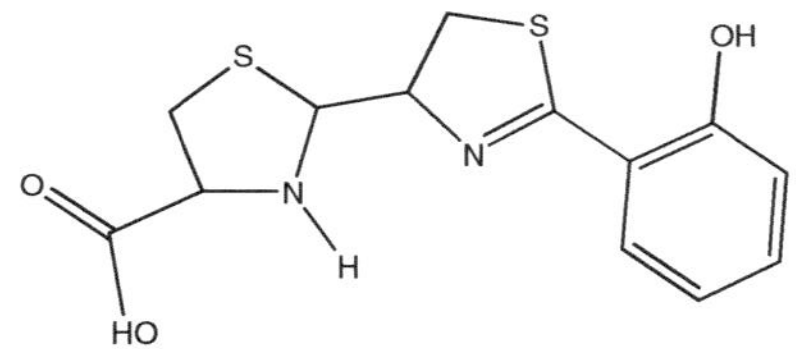 pyochelin	*Pythium* sp.	Thomas (2007)
Cepabactin (Siderophore)	*B. cepacia* complex	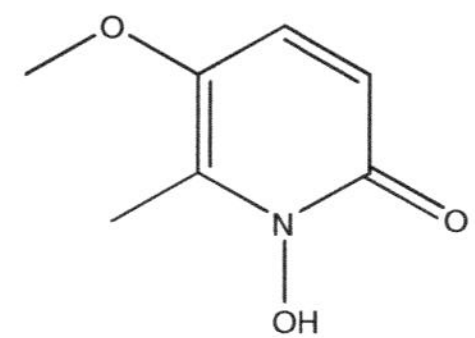 cepabactin	Antifungal activity	Thomas (2007)
Ornibactins (Siderophore)	*B. cepacia* complex	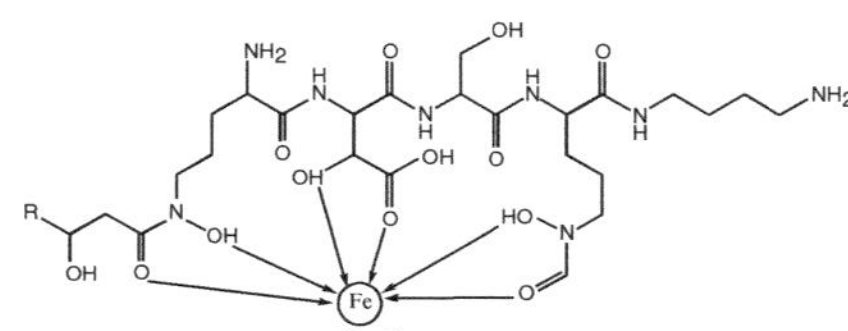 ornibactin	Antifungal activity	Thomas (2007)

(Raaijmakers et al. 2006). The antifungal activity has been studied for many different CLPs and for a wide variety of plant and human-pathogenic fungi and yeasts, including *R. solani*, *Phoma lingam*, *Podosphaera fusca*, *P. aphanidermatum*, *Alternaria brassicae*, *S. sclerotiorum*, *G. candidum*, *B. cinerea*, *Ophiostoma ulmi*, *Aspergillus* spp., *Fusarium* spp., *Penicillium digitatum*, *Cryptococcus neoformans*, *C. albicans*, and *C. glabrans*. In vitro studies showed that CLPs adversely affected mycelia of *R. solani* and *P. ultimum*, causing reduced growth and intracellular activity, hyphal swellings, increased branching, and rosette formation (Hansen et al. 2000; Thrane et al. 1999, 2000). The site of action on yeast cells was demonstrated to be cytoplasmic membrane. CLPs of iturin group can lyses spheroplasts of *S. cerevisiae* (Besson et al. 1984a, b). Moreover, a rapid leakage of potassium ions found in the presence of this antibiotics is directly associated to the killing effects. These results are consistent with a disruption of the structural integrity of the cytoplasmic membrane correlated to the loss of viability of the yeast cells.

Bacisubin secreted from *B. subtilis* strain B-916 is an antifungal protein and is strongly inhibited mycelial growth in *R. solani*, *M. grisease*, *S. sclerotiorum*, *A. oleracea*, *A. brassicae*, and *B. cinerea*, especially some species of *Alternaria* and *Botrytis*. The IC_{50} values of antifungal activity of bacisubin toward *A. brassicae*, *A. oleracea*, *R. solani*, and *B. cinerea* were as low as 0.055, 0.087, 4.01, and 2.74 mM, respectively. Bacisubin inhibited the growth of *R. solani* and induced increase in mycelial apex offshoot, distortion, tumescence, and rupture (Liu et al. 2007).

Apart from peptides, polyketides are the other dominant family of secondary metabolites having antimicrobial, immunosuppressive, antitumor, or other physiologically relevant bioactivities. Although polyketides are widespread secondary metabolites from bacteria, only a few (difficidin/oxydifficidin, bacillaene, and macrolactin) have been isolated and characterized from *Bacillus*. Difficidin has been shown to inhibit protein biosynthesis (Zweerink and Edison 1987), but the exact molecular target remains unknown. Patel et al. (1995) described that Bacillaene is a highly unstable inhibitor of prokaryotic protein synthesis with a partially characterized open-chain polyenic structure with the empirical formula C35H48O7. However, this formula was suspected to be incorrect, because the NRPS modules indicated the presence of two nitrogen atoms. Macrolactin is the third polyketide with macrolide-like structure; it is originally detected in an unclassified deep-sea marine bacterium and has been previously reported from several other *Bacillus* strains. The macrolactin carbon skeleton contains three separate diene structure elements in a 24-membered lactone ring. Until now, at least 17 macrolactins have been described and one of them, 7-O-malonyl-macrolactin A, has been recently reported as efficient against Gram-positive bacterial pathogens. A broad-spectrum antibiotic, 2,4-di-acetylphloroglucinol (2,4-DAPG), is a polyketide compound produced by many fluorescent pseudomonads, exhibits antifungal, antibacterial, antihelmenthic, and phytotoxic activities. Previous study has demonstrated that root-associated fluorescent *Pseudomonas* spp. with the capacity to produce 2,4-DAPG are the key components in biological control. It is synthesized by condensation of three molecules of acetyl CoA with one molecule of malonyl CoA to produce the precursor monoacetylphloroglucinol (MAPG), which is subsequently transacetylated to generate DAPG. The exact mechanism of DAPG action is still unclear, although it is known that disease suppression by this antifungal molecule is a result of interaction of specific root-associated microorganisms and the pathogen. This antibiotic also appears to cause ISR in plants. Plt is an aromatic polyketide antibiotic consisting of a resorcinol ring, which is derived through polyketide biosynthesis. This in turn is linked to a bichlorinated pyrrole moiety, whose biosynthesis remains unknown. Plt is produced by several *Pseudomonas* sp., including strains that suppress plant diseases caused by phytopathogenic fungi. Plt mainly inhibits the oomycetous fungi, such as *P. ultimum*.

Phzs are N-containing heterocyclic pigments synthesized by *Brevibacterium*, *Burkholderia*, *Pseudomonas*, and *Streptomyces*, and these compounds have been identified as virulence factors in a number of in vivo model systems. The Phz secreted by *Pseudomonas* are PCA, pyocyanin, 1-hydroxyphenazine (1-HP), and phenazine-1-carboxamide. Almost all Phz exhibits broad spectrum activity against bacteria and fungi. In addition to inhibiting fungal pathogenesis, Phz play an important role in microbial competition in rhizosphere, including survival and competence. The broad-spectrum activity exhibited by Phz compounds against fungi and other bacteria is not well understood. However, it is considered that Phz can accept electrons, yielding a relatively stable anion radical that readily undergoes redox cycle. It includes biosynthesis of Mn-containing superoxide dismutase (MnSOD) which causes enhanced production of O_2^- (superoxide radical). There is a distinct possibility that the antibiotic action of pyocyanin is actually a result of toxicity of O_2^- and H_2O_2 produced in increased amounts in its presence.

Prn was thought to inhibit bacterial growth by complexing with phospholipids of cell membranes. Although Prn inhibited the respiration of intact cells, the oxidative phosphorylation of mitochondria isolated from *C. utilis* was not inhibited. Tripathi and Gottlieb (1970) concluded that inhibition of electron transfer in yeast was the site of action of Prn. Previous study clearly shows that Prn inhibits respiration of fungal mitochondria and mammalian respiratory system (Wong and Airallb 1970). Complete inhibition of electron transport requires its higher concentration.

Most organisms require iron as an essential element in a variety of metabolic and informational cellular pathways. More than 100 enzymes acting in primary and secondary metabolism possess iron-containing cofactors such as iron-sulfur clusters or heme groups. The reversible Fe(II)/Fe(III) redox pair is best suited to catalyze a broad spectrum of redox reactions and to mediate electron chain transfer. Furthermore, several transcriptional and posttranscriptional regulators interact with iron to sense its intracellular level or the current status of oxidative stress to efficiently control the expression of a broad array of genes involved mainly in iron acquisition or reactive oxygen species protection. However, in most microbial habitats, Fe(II) is oxidized to Fe(III) either spontaneously by reacting with molecular oxygen or enzymatically during assimilation and circulation in host organism. In the environment, Fe(III) forms ferric oxide hydrate complexes ($Fe_2O_3{\cdot}nH_2O$) in the presence of oxygen and water at neutral to basic pH. These complexes are very stable, leading to a free Fe(III) concentration of 10^{-9}–10^{-18} M. Many microorganisms produce siderophores that bind iron and enhance microbial growth by solubilizing ferric iron and by accelerating iron transport.

Siderophores are themselves growth inhibitors of various phytopathogenic fungi, such as *P. parasitica*, *P. ultimum*, *F. oxysporum* veri *dianthi*, and *S. sclerotiorum*. Siderophores, whose chemical structures depend upon their producer microorganism, may provide iron (III) to some vegetable cells. These metabolites, due to their antagonistic capability against pathogenic microorganisms, could act as growth factors in plants.

4 Understanding Biosynthesis of Antifungal Metabolites at the Molecular Level

CLPs from *Pseudomonas* species are produced by nonribosormal peptide synthesis (NRPSs) via a thiotemplate process. NRPSs possess a modular structure and each module is a building block resulting in the stepwise incorporation and modification of one amino acid unit. Their substrates are not restricted to the usual proteinogenic amino acids but also can incorporate multiple nonproteinogenic D-amino acids, carboxy acids, or fatty acids. For CLP-producing *Pseudomonas* spp., a number of partial and complete sequences of NRPSs have been obtained over the past decade. Two of the best-characterized biosynthetic templates are the synthetase clusters for arthrofactin and syringomycin.

Bacteria of the *Bacillus* genus produce a wide variety of antibacterial and antifungal antibiotics. Some of these compounds, like subtilin, subtilosin A, TasA, and sublancin, are of ribosomal origin, but others, such as bacilysin, chlorotetain, mycobacillin, rhizocticins, bacillaene, difficidin, and lipopeptides belonging to the surfactin, iturin, and fengycin families, are formed by nonribosomal peptide synthetases and/or polyketide synthases (PKS)(Leclére et al. 2005). The model organism *B. subtilis* 168 and the plant root-colonizing *B. amyloliquefaciens* FZB42 produce a wide variety of antibacterial and antifungal antibiotics, and their gene clusters involved in antibiotics biosynthesis have been identified. In *B. amyloliquefaciens* FZB42 (Chen et al. 2007), the nine gene clusters (*srf*, *bmy*, *fen*, *nrs*, *dhb*, *bac*, *mln*, *bae*, *dfn*) direct the synthesis of bioactive peptides and polyketides by modularly organized megaenzymes defined as nonribosomal peptide synthetases (NRPSs) and PKS. Four gene clusters (*bmy*D, *pks*2, *pks*3, and *nrs*) are not found in *B. subtilis* 168. Except for the gene cluster encoding bacilysin synthesis, the functional activities of the remaining gene clusters depend on Sfp, an enzyme that transfers 4′-phosphopantetheine from coenzyme A to the carrier proteins of nascent peptide or polyketide chains.

Prn and Plt are broad-spectrum antibiotics produced by several strains of *Pseudomonas* and *Burkholderia* species. The *prn* operon has been completely sequenced; *prnABCD* spans 5.8 kb DNA which encodes Prn biosynthetic pathway in which four ORFs (Hammer et al. 1997), *prnA*, *prnB*, *prnC*, and *prnD*, are involved. All four ORFs are located on a single transcriptional unit. The four genes encode proteins which are identical in size. Among these, *prnA* gene product catalyses the chlorination of L-trp to 7 chloro-L-trp. The *prnC* gene product chlorinates it at the 3-position to form an amino pyrrolnitrin. The *prnD* gene product catalyses the oxidation of aminopyrrolnitrin to a nitro group to form pyrrolnitrin. The organization of *prn* genes in the operon is identical to the order in which the reactions are catalyzed in the biosynthetic pathway. Phz nucleus is formed by the symmetrical condensation of two molecules of chorismic acid, and the amide nitrogen of glutamine serves as the immediate source of N in the heterocyclic nucleus. PCA is the first phenazine formed, which gets converted to PCA and acts as the key intermediate in the synthesis of other phenazine in *Pseudomonas* and *Burkholderia* species. Phz compounds reported previously are pyocyanin, 1-hydroxyphenazine, and phenazine-1-carboxamide. Seven genes, *phzABCDEFG*, are involved in the synthesis of PCA (Mavrodl et al. 2001). These are localized within a 6.8 kb fragment in *P. fluorescens* 2-79. The phenazine biosynthetic loci in *P. fluorescens* 2-79, *P. aeruginosa* PAO1, and *P. chlororaphis* PCL 1394 are highly conserved. Each *phz* locus contains a set of seven gene core operons, regulated in a cell density-dependent manner by homologues of LuxI and LuxR.

The pathways of biosynthesis of siderophores from members of *Pseudomonas* and *Burkholderia* have been investigated clearly, and genes involved in these siderophores have been identified as shown in Table 4.

5 Concluding Remarks and Further Perspectives

PGPR can promote plant growth directly or indirectly. Some PGPR can directly cause plant growth promotion by producing and secreting plant growth regulators such as auxins, gibberellins, and cytokinins. Other PGPR affect plant growth by indirect mechanisms that involve suppression of bacterial, fungal, and nematode pathogens. To date, PGPR represent a variety of bacterial species from more than 20 genera, but the mechanism is yet unclear. At present, most research focus on the PGPR as biocontrol agent; however, in fact biological control bacteria can not only produce inhibitory metabolites but also produce some metabolites with growth promoting activity. Although the chemical structure and biosynthesis genes of many antifungal compounds are known, its regulatory mechanisms and the relationship between growth-promoting activity and antifungal activity do not understand. Furthermore, particular bacteria can produce a variety of antibiotics simultaneously, and some of them are considered as signal molecules that can regulate plant growth or pathogenicity of pathogenic fungal (bacteria). The solution of these problems will contribute to the extensive application of PGPR in agriculture and development of agricultural biotechnology.

Table 4 Genes involved in the biosynthesis of anbiotics of *Bacillu*, *Pseudomonas*, and *Burkholderia*

Antibiotics	Species/strain	Gene/Protein information	GenBank accession no.	References
Viscosin	*P. fluorescens* PfA7B	Nonribosomal peptide synthetases	No sequence available	Braun et al. (2001)
Massetolide A	*P. fluorescens* R1SS101	Nonribosomal peptide synthetases	AY303770; AY303771	De Souza et al. (2003)
Amphisin	*Pseudomonas* sp. DSS73	*amsY*, peptide synthetase	AJ416154	Koch et al. (2002)
Arthrofactin	*Pseudomonas* sp. MIS38	*arfA B C*; nonribosomal peptide synthetases	AB107223	Roongsawang et al. (2003)
Tolaasin	*P. tolaasii*	TL1, TL2, TL3; high molecular weight protein	No sequence available	Rainey et al. (1993)
Tolaasin	*Pseudomonas* NZ17	Homology to syringomycin synthetase	No sequence available	Godfrey et al. (2001)
Syringopeptin	*P. syringae* pv. *syringae* B728a; *P. syringae* pv. *Syringae* B301D	Syringopeptin synthetase genes; *sypABC*; syringopeptin synthetase; *pseABC*; tripartite	CP000075; AF286216	Feil et al. (2005); Scholz-Schroeder et al. (2001, 2003); Kang and Gross 2005
Syringomycin	*P. syringae* pv. *syringae* B728a; *P. syringae* pv. *Syringae* B301D	Syrinogmycin synthetase genes; *syrE*, *syrB1*, *syrC*, *syrB2*, *syrD*, putative ABCtransporter protein pseABC	CP000075, AF047828, U25130, M97223	Feil et al. (2005); Guenzi et al. (1998); Scholz-Schroeder et al. (2001); Zhang et al. (1995); Vaillancourt et al. (2005);
Putisolvin	*P. putida* PCL1445	*psoA*, putisolvin synthetase	DQ151887	Kuiper et al. (2004); Dubern et al. (2005)
Orphan	*P. fluorescens* Pf-5	*ofaA*, *ofaB ofaC*	YP259252.2, YP259253.1, YP_259254.1	Loper and Gross (2007)
2,4-Diaacetyl phloroglucinol	*P. fluorescens* Q2-87 *P. fluorescens* F113	*phlABCD*, *phlF*	U41818, AF129856	Dwivedi and Johri (2003); Delany et al. (2000)
Pyoluteorin	*P. fluorescens* Pf-5	*pltABCDEFG*	PFL2784-PFL2800	Paulsen et al. (2005)
Pyrrolntin	*P. fluorescens*	*prnABCD*	PFL3599-PFL3627	Paulsen et al. (2005)
Phenazines	*P. fluorescens* 2-79 *P. aeruginosa PAO1* *P. chlororaphis PCL 1394*	*phzABCDEFG*		Dwivedi and Johri (2003)

(*continued*)

Table 4 (continued)

Antibiotics	Species/strain	Gene/Protein information	GenBank accession no.	References
Pyochelin	*P. fluorescens* Pf-5	*pchABCDEF*	PFL3473-PFL3504	Paulsen et al. (2005)
Pyoverdine	*P. aeruginosa PAO1*	*pvdADEFIJ*	PA 2386-2388, PA2396-2402, PA	Lamont and Martin (2003)
Hydrogen cyanide	*P. fluorescens* Pf-5 *P. fluorescens* CHA0	*hcnABC*	PFL2577-PFL2579	Paulsen et al. (2005)
Lipopeptide-AFC-BC11	*B. cepacia* BC11	*afcABCD*	AF076477	Kang et al. (1998)
Pyrrolntin	*B. cepacia* LT4-12-W,	*prnABCD*	AF161183	Hammer et al. (1997)
Glidobactin	*Burkholderia* sp.	*Glb ABCDEFGH*	ZP_00417086, NP_929149, NP_929145, ZP_01144206 YP_111276, YP_111275, YP_111274, YP_111273,	Schellenberg et al. (2007)
Ornibactin	*B. cenocepacia*	*OrbGKL*, *pvdAF*	BCAL1690, BCAL 1698, BCAL 1699, BCAL 1701, BCAL 1702	Thomas (2007)
4-hydroxy-2-alkylquinoline	*B. pseudomallei*, *B. thailandensis*, *B. ambifaria*	*pqsABCDE*	BPSS00481-BPSS00485 BTH II1935-BTH II1931	Vial et al. (2008)
Subtilin	*Bacillus subtilis* ATCC 6633	*spa BCD*	M83944	Chung et al. (1992)
Surfactin	*B. amyloliquefaciens* FZB42 *B. subtilis 168*	*srfABCD*, *aat*, *334*, *ycx*, *CycxD*, *sfd*, *yczE; srfAA*, *AB*, *AD*, *ycxABCD*		Chen et al. (2007)
Bacillomycin D	*B. amyloliquefaciens* FZB42	*bmyBCAD*		Chen et al. (2007)
Fengycin	*B. amyloliquefaciens* FZB42 *B. subtilis 168*	*fenABCDE; PpsABCDE*		Chen et al. (2007)
Bacillibactin	*B. amyloliquefaciens* FZB42 *B. subtilis 168*	*dhbABCDEF*		Chen et al. (2007)

Bacilysin	*B. amyloliquefaciens* FZB42 *B. subtilis 168*	*BacABCDE, ywfG; ywfCDEFG*		Chen et al. (2007)
Macrolactin	*B. amyloliquefaciens* FZB42	*mlnABCDEFGHI*		Chen et al. (2007)
Bacillaene	*B. amyloliquefaciens* FZB42 *B. subtilis 168*	*baeBSDE, acpK, baeGHIJLMNRS; pksBCDE, acpK, pksGHIJLMNRS*		Chen et al. (2007)
Difficidin	*B. amyloliquefaciens* FZB42	*dfnAYXBCDEFGHIJKLM*		Chen et al. (2007)
Iturin A	*Bacillus subtilis* RB14	*ituABCD*	AB050629	
Zwittermicin A	*Bacillus cereus* UW85	*zmaDERG, zmaHI, zmaJL, zmaM, zmaNP, zmaS, zmaT, zmaU, zmaV*	FJ430564	Kevany et al. (2009)

References

Ajit NS, Verma R, Shanmugam V (2006) Extracellular chitinases of fluorescent Pseudomonads antifungal to *Fusarium oxysporum* f. sp. *dianthi* causing carnation wilt. Curr Microbiol 52:310–316

Arima K, Imanaka H, Kousaka M, Fukuda A, Tamura C (1964) Pyrrolnitrin, a new antibiotic substance, produced by *Pseudomonas*. Agric Biol Chem 28:575–576

Azevedo EC, Rios EM, Fukushima K, Campos-Takaki GM (1993) Bacitracin production by a new strain of *Bacillus subtilis*. Extraction, purification, and characterization. Appl Biochem Biotechnol 42(1):1–7

Bangera MG, Thomashow LS (1999) Identification and characterization of a gene cluster for synthesis of the polyketide antibiotic 2,4-diacetylphloroglucinol from Pseudomonas fluorescens Q2–87. J Bacteriol 181:3155–3163

Becker JO, Cook RJ (1988) Role of siderophores in suppression of *Pythium* species and production of increased growth response of wheat by fluorescent pseudomonads. Phytopathology 78:778–782

Besson F, Peypoux F, Michel G, Delcambe L (1984a) Identification of antibiotics of iturin group in various strains of *Bacillus subtilis* on the cell membrane of *Saccharomyces cerevisiae*. J Antibiot 37:172–177

Besson F, Peypoux F, Quentin MJ, Michel G (1984b) Action of antifungal peptidolipids from *Bacillus subtilis* on the cell membrane of *Saccharomyces cerevisiae*. J Antibiot 37:172–177

Bonsall RF, Weller DM, Tomashow LS (1999) Quantification of 2,4-diacetylphloroglucinol produced by fluorescent Pseudomonas spp. in vitro and in the rhizosphere of wheat. Appl Environ Microbiol 63:951–955

Bottone EJ, Peluso RW (2003) Production by *Bacillus pumilus* (MSH) of an antifungal compound that is active against *Mucoraceae* and *Aspergillus* species: preliminary report. J Med Microbiol 52:69–74

Braun PG, Hildebrand PD, Ells TC, Kobayashi DY (2001) Evidence and characterization of a gene cluster required for the production of viscosin, a lipopeptide biosurfactant, by a strain of *Pseudomonas fluorescens*. Can J Microbiol 47:294–301

Budzikiewicz H (1993) Secondary metabolites from fluorescent pseudomonads. FEMS Microbiol Rev 104:209–228

Cartwright DK, Chilton WS, Benson DM (1995) Pyrrolnitrin and phenazine production by *Pseudomonas cepacia*, strain 5.5B., a biocontrol agent of *Rhizoctonia solani*. Appl Microbiol Biotech 43:211–216

Chan WC, Bycroft BW, Leyland ML, Lian LY, Roberts GC (1993) A novel post-translational modification of the peptide antibiotic subtilin: isolation and characterization of a natural variant from *Bacillus subtilis* ATCC 6633. Biochem J 291:23–27

Chang WT, Chen YC, Jao CL (2007) Antifungal activity and enhancement of plant growth by *Bacillus cereus* grown on shellfish chitin wastes. Bioresour Technol 98:1224–1230

Chen XH, Koumoutsi A, Scholz R, Eisenreich A, Schneider K, Heinemeyer I, Morgenstern B, Voss B, Hess WR, Reva O, Junge H, Voigt B, Jungblut PR, Vater J, Süssmuth R, Liesegang H, Strittmatter A, Gottschalk G, Borriss R (2007) Comparative analysis of the complete genome sequence of the plant growth-promoting bacterium. Bacillus amyloliquefaciens FZB42. Nat Biotechnol 25:1007–1014

Chen SX, Fei ZW, He MT, Shen P, Xie ZX (2008) Different response of fluorescent and non-fluorescent siderophores from *Pseudomonas* to ferric iron. Microbiology 35(10):572–1576

Chen XH, Koumoutsi A, Scholz R, Schneider K, Vater J, Süssmuth R, Piel J, Borriss R (2009) Genome analysis of *Bacillus amyloliquefaciens* FZB42 reveals its potential for biocontrol of plant pathogens. J Biotech 140:27–37

Chung YJ, Steen MT, Hansen JN (1992) The subtilin gene of Bacillus subtilis ATCC 6633 is encoded in an operon that contains a homolog of the hemolysin B transport protein. J Bacteriol 174:1417–1422

Cox CD, Rinehart KL, Moore ML, Cook JC (1981) Pyochelin: novel structure of an iron-chelating growth promoter for Pseudomonas aeruginosa. Proc Natl Acad Sci USA 78:4256–4260

De Souza JT, De Boer M, De Waard P, Van Beek TA, Raaijmakers JM (2003) Biochemical, genetic, and zoosporicidal properties of cyclic lipopeptide surfactants produced by *Pseudomonas fluorescens*. Appl Environ Microbiol 69:7161–7172

Delany I, Sheehan MM, Fenton A, Bardin S, Aarons S, O'Gara F (2000) Regulation of production of the antifungal metabolite 2,4-diacetylphloroglucinol in Pseudomonas fluorescens F113: genetic analysis of phlF as a transcriptional repressor. Microbiology 146:537–546

Denning GM, Iyer SS, Reszka KJ, O'Malley Y, Rasmussen GT, Britigan BE (2003) Phenazine-1-carboxylic acid, a secondary metabolite of *Pseudomonas aeruginosa*, alters expression of immunomodulatory proteins by human airway epithelial cells. Am J Physiol 285:584–592

Dubern JF, Lagendijk EL, Lugtenberg BJJ, Bloemberg GV (2005) The heat shock genes dnaK, dnaJ, and grpE are involved in regulation of putisolvin biosynthesis in *Pseudomonas putida* PCL1445. J Bacteriol 187:5967–5976

Duitman EH, Hamoen LW, Rembold M, Venema G, Seitz H, Saenger W, Bernhard F, Reinhardt R, Schmidt M, Ullrich C, Stein T, Leenders F, Vater J (1999) The mycosubtilin synthetase of *Bacillus subtilis* ATCC6633: a multifunctional hybrid between a peptide synthetase, an amino transferase, and a fatty acid synthase. Proc Natl Acad Sci USA 96:13294–13299

Dwivedi D, Johri BN (2003) Antifungals from fluorescent pseudomonads: biosynthesis and regulation. Curr Sci 85:1693–1703

Emmert EAB, Handelsman J (1999) Biocontrol of plant disease: a (Gram-) positive perspective. FEMS Microbiol Lett 171:1–9

Eshita SM, Roberto NH (1995) Bacillomycin Lc, a new antibiotic of the iturin group: Isolation, structures and antifungal activities of the congeners. J Antibiot 48:1240–1247

Feil H, Feil WS, Chain P, Larimer F, Dibartolo G, Copeland A, Lykidis A, Trong S, Nolan M, Goltsman E, Thiel J, Malfatti S, Loper JE, Lapidus A, Detter JC, Land M, Richardson PM, Kyrpides NC, Ivanova N, Lindow SE (2005) Comparison of the complete genome sequences of *Pseudomonas syringae* pv. Syringae B728a and pv. tomato DC3000. Proc Natl Acad Sci USA 102:11064–11069

Fenton AM, Stephens PM, Crowley J, O'Callaghan M, O'Gara F (1992) Exploitation of gene(s) involved in 2,4-diacetylphloroglucinol biosynthesis to confer a new biocontrol capability to a Pseudomonas strain. Appl Environ Microbiol 58:3873–3878

Fukumori F, Sashihara N, Kudo T, Horikoshi K (1986) Nucleotides sequences of two cellulose genes from alkalophilic *Bacillus* sp. Strain N-4 and their strong homology. J Bacteriol 168:479–485

Godfrey SAC, Marshall JW, Klena JD (2001) Genetic characterization of *Pseudomonas*' NZ17 – a novel pathogen that results in a brown blotch disease of *Agaricus bisporus*. J Appl Microbiol 91:412–420

Guenzi E, Galli G, Grgurina I, Gross DC, Grandi G (1998) Characterization of the syringomycin synthetase gene cluster A link between prokaryotic and eukaryotic peptide synthetases. J Biol Chem 273:32857–32863

Haavik HI, Froyshov O (1975) Function of peptide antibiotics in producer organisms. Nature 254:79–82

Hammer PE, Hill DS, Lam ST, Van Pee KH, Ligon JM (1997) Four genes from *Pseudomonas fluorescens* that encode the biosynthesis of pyrrolnitrin. Appl Environ Microbiol 63:2147–2154

Hansen M, Thrane C, Olsson S, Sorensen J (2000) Confocal imaging of living fungal hyphae challenged with the fungal antagonist viscosinamide. Mycologia 92:216–221

Hathout Y, Ho YP, Ryzhov V, Demirev P, Kurstakins CF (2000) A new class of lipopeptides isolated from *Bacillus thuringiensis*. J Nat Prod 63:1492–1496

Henriksen A, Anthoni U, Nielsen TH, Sorensen J, Christophersen C, Gajhede M (2000) Cyclic lipoundecapeptide tensin from *Pseudomonas fluorescens* strain. Acta Crystal 56:113–115

Hofemeister J, Conrad B, Adler B, Hofemeister B, Feesche J, Kucheryava N, Steinborn G, Franke P, Grammel N, Zwintscher A, Leenders F, Hitzeroth G, Vater J (2004) Genetic analysis of the biosynthesis of nonribosomal peptide- and polyketide-like antibiotics, iron uptake and biofilm formation by *Bacillus subtilis* A1/3. Mol Genet Genomics 272:363–378

Howell CR, Stipanovic RD (1980) Suppression of *Pythium ultimum* induced damping-off of cotton seedlings by *Pseudomonas fluorescens* and its antibiotic, pyoluteorin. Phytopathology 70:712–715

Hsieh FC, Lin TC, Meng M, Kao SS (2008) Comparing methods for identifying *Bacillus* Strains capable of producing the antifungal lipopeptide Iturin A. Curr Microbiol 56:1–5

Huang CJ, Wang TK, Chung SC, Chen CY (2005) Identification of an antifungal chitinase from a Potential Biocontrol Agent, *Bacillus cereus* 28-9. J Biochem Mol Bio 38:82–88

Hwang J, Chilton WS, Benson DM (2002) Pyrrolnitrin production by *Burkholderia cepacia* and biocontrol of *Rhizoctonia* stem rot of poinsettia. Biol Control 25:56–63

Idris HA, Labuschagne N, Korsten L (2008) Suppression of *Pythium ultimum* root rot of sorghum by rhizobacterial isolates from Ethiopia and South Africa. Biol Control 45:72–84

Kajimura Y, Sugiyama M, Kaneda M (1995) Bacillopeptins, new cyclic lipopeptide antibiotics from *Bacillus subtilis* FR-2. J Antibiot 48:1095–1103

Kang H, Gross DC (2005) Characterization of a resistance-nodulation-cell division transporter system associated with the *syr-syp* genomic island of *Pseudomonas syringae* pv. *syringae*. Appl Environ Microbiol 71:5056–5065

Kang Y, Carlson R, Tharpe W, Schell MA (1998) Characterization of genes involved in biosynthesis of a novel antibiotic from *Burkholderia cepacia* BC11 and their role in biological control of *Rhizoctonia solani*. Appl Environ Microbiol 64(10):3939–3947

Kang JG, Shin SY, Kim MJ, Bajpai V, Maheshwari DK, Kang SC (2004) Isolation and anti-fungal activities of 2-Hydroxymethyl-chroman-4-one produced by *Burkholderia* sp. MSSP J Antibiot 57:726–731

Keel C, Wirthner PH, Oberhansli TH, Voisard C, Burger U, Haas D, Defago G (1990) Pseudomonads as antagonists of plant pathogens in the rhizosphere: role of the antibiotic 2,4-diacetylphloroglucinol in the suppression of black root rot of tobacco. Symbiosis 9:327–342

Keel C, Weller DM, Natsch A, Défago G, Cook RJ, Thomashow LS (1996) Conservation of the 2,4-diacetylphloroglucinol biosynthesis locus among fluorescent *Pseudomonas* strains from diverse geographic locations. Appl Environ Microbiol 62:552–563

Kevany BM, Rasko DA, Thomas MG (2009) Characterization of the complete Zwittermicin A biosynthesis gene cluster from *Bacillus cereus*. Appl Environ Microbiol 75:1144–1155

Kildea S, Ransbotyn V, Khan MR, Fagan B, Leonard G, Mullins E, Doohan FM (2008) *Bacillus megaterium* shows potential for the biocontrol of septoria tritici blotch of wheat. Biol Control 47:37–45

Kim EJ (2003) Iron deficiency lenads to inhibition of oxygen transfer and enhanced formation of virulence factors in cultures of *Pseudomonas aeruginosa* PAO1. Microbiology 149:2627–2634

Kim PI, Bai H, Bai D, Chae H, Chung S, Kim Y, Park R, Chi YT (2004) Purification and characterization of a lipopeptide produced by *Bacillus thuringiensis CMB26*. J Appl Microbiol 97:942–949

Kim KM, Lee JY, Kim CK, Kang JS (2009) Isolation and characterization of surfactin produced by *Bacillus polyfermenticus* KJS-2. Arch Pharm Res 32(5):711–715

Kirinuki T, Ichiba T, Katayama K (1984) General survey of action site of altericidins on metabolism of *Alternaria kikuchiana* and *Ustilago maydis*. J Pesticide Sci 9(4):601–610

Koch B, Nielsen TH, Sorensen D, Andersen JB, Christophersen C, Molin GM, Sorensen J, Nybroe O (2002) Lipopeptide production in Pseudomonas sp. strain DSS73 is regulated by components of sugar beet seed exudate via the Gac two-component regulatory system. Appl Environ Microbiol 68:4509–4516

Kuiper I, Lagendijk EL, Pickford R, Derrick JP, Lamers GE, Thomas-Oates JE, Lugtenberg BJ Bloemberg GV (2004) Characterization of two *Pseudomonas putida* lipopeptide biosurfactants, putisolvin I and II, which inhibit biofilm formation and break down existing biofilms. Mol Microbiol 51:97–113

Lamont IL, Martin LW (2003) Identification and characterization of novel pyoverdine synthesis genes in *Pseudomonas aeruginosa*. Microbiology 149:833–842

Lebbadi M, Galvez A, Maqueda M, Martinez-Bueno M, Valdivia E (1994) Fungicin M4: a narrow spectrum peptide antibiotic from *Bacillus licheniformis* M-4. J Appl Bacteriol 77:49–53

Leclére V, Béchet M, Adam A, Guez JS, Wathelet B, Ongena M, Thonart P, Gancel F, Chollet-Imbert M, Jacques P (2005) Mycosubtilin overproduction by *Bacillus subtilis* BBG100 enhances the organism's antagonistic and biocontrol activities. Appl Environ Microbiol 71:4577–4584

Lee CH, Kim SH, Hyun BC, Suh JW, Yon CS, Kim CO, Lim YH Kim CS (1994) Cepacidine A, a novel antifungal antibiotic produced by *Pseudomonas cepacia* I. Taxonomy, production, isolation and biological activity. J Antibiot 47:1402–1405

Lee YK, Kim JS, Chung HY, Jang YS, Jang BI (2001) Two-dimensional electrophoresis analysis of antifungal activity related proteins in *Bacillus licheniformis* DM3 Korean. J Environ Agric 22:203–209

Levy E, Gough FJ, Berlin KD, Guiana PW Smith JT (1992) Inhibition of *Septoria tritici* and other phytopathogenic fungi and bacteria by *Pseudomonas fluorescens* and its antibiotics. Plant Pathol 41:335–341

Ligon JM, Hill DS, Hammer PE, Torkewitz NR, Hofmann D, Kempf HJ, van Pée KH (2000) Natural products with antifungal activity from *Pseudomonas* biocontrol bacteria. Pest Manag Sci 56:688–695

Liu YF, Chen ZY, Zhang J, Zhou M, Song FP, Lu F, Liu YZ (2007) Bacisubin, an antifungal protein with ribonuclease and hemagglutinating activities from *Bacillus subtilis* strain B-916. Peptides 28:553–559

Liu HM, Yan A, Zhang XH, Xu YQ (2008) Phenazine-1-carboxylic acid biosynthesis in *Pseudomonas chlororaphis* GP72 is positively regulated by the sigma factor RpoN World. J Microbiol Biotechnol 24:1961–1966

Lodewyckx C, Vangronsveld J, Porteous F, Moore ERB, Taghavi S, Mezgeay M, van der Lelie D (2002) Endophytic bacteria and their potential applications. Crit Rev Plant Sci 21: 583–606

Loper JE, Gross H (2007) Genomic analysis of antifungal metabolite production by *Pseudomonas fluorescens* Pf-5. Eur J Plant Pathol 119:265–278

Lu SE, Novak J, Austin FW, Gu GY, Ellis D, Kirk M, Wilson-Stanford S, Tonelli M, Smith L (2009) Occidiofungin, a unique antifungal glycopeptide produced by a strain of *Burkholderia* contaminans. Biochemistry 48:8312–8321

Majumder S, Mukhopadhyay NK, Ghosh SK, Bose SK (1988) Genetic analysis of the mycobacillin biosynthetic pathway in *Bacillus subtilis* B3. J Gen Microbiol 134(5):1147–1153

Maldonado MC, Corona J, Gordillo MA, Navarro AR (2009) Isolation and partial characterization of antifungal metabolites produced by *Bacillus* sp. IBA Curr Microbiol 59:646–650

Maurhofer M, Keel C, Haas D, Défago G (1995) Influence of plant species on disease suppression by *Pseudomonas fluorescens* strain CHA0 with enhanced antibiotic production. Plant Pathol 44:40–50

Mavrodl DV, Bonsall RF, Delaney SM, Soule MJ, Phillips G, Thomashow LS (2001) Functional analysis of genes for biosynthesis of pyocyanin and phenazine-1-carboxamide from *Pseudomonas aeruginosa* PAO1. J Bacteriol 183:6454–6465

Mondal SN, Hyakumachi M (2000) Soil factors affecting carbon loss and pathogenicity of oospores of *Pythium aphanidermatum*. Soil Biol Biochem 32:111–118

Moon SS, Kang PM, Park KS, Kim CH (1996) Plant growth promoting and fungicidal 4-quinolinones from *Pseudomonas cepacia*. Phytochem 42:365–368

Nair JR, Narasimman G, Sekar V (2004) Cloning and partial characterization of zwittermicin A resistance gene cluster from *Bacillus thuringiensis* subsp. *kurstaki* strain HD1. J Appl Microbiol 97:495–503

Nakano M, Marahiel MA, Zuber P (1988) Identification of a genetic locus required for biosynthesis of the lipopeptide antibiotic surfactin in *Bacillus subtilis*. J Bacteriol 170:5662–5668

Omura S, Iwai Y, Masuma R, Hayashi M, Furusato T, Takagaki T (1980) A new peptide antibiotic, alboleutin. J Antibiot 33(7):758–759

Ongena M, Jacques P (2007) *Bacillus lipopeptides*: versatile weapons for plant disease biocontrol. Cell 16:115–125

Ozcengiz G, Alaeddinoglu NG (1991) Bacilysin production by Bacillus subtilis: effects of bacilysin, pH and temperature. Folia Microbiol 36:522–526

Parker WL, Rathnum ML, Seiner V, Trejo WH, Principe PA, Sykes RB (1984) Cepacin A and cepacin B, two new antibiotics produced by Pseudomonas cepacia. J Antibiot 37:431–440

Patel PS, Huang S, Fisher S, Pirnik D, Aklonis C, Dean L, Meyers E, Fernandes P, Mayer F (1995) Bacillaene, a novel inhibitor of Procaryotic protein synthesis produced by *Bacillus subtilis*: production, taxonomy, isolation, physico-chemical characterization and biological activity. J Antibiot 48(9):997–1003

Paulsen IT, Press CM, Ravel J, Kobayashi DY, Myers GS, Mavrodi DV, DeBoy RT, Seshadri R, Ren Q, Madupu R, Dodson RJ, Durkin AS, Brinkac LM, Daugherty SC, Sullivan SA, Rosovitz MJ, Gwinn ML, Zhou L, Schneider DJ, Cartinhour SW, Nelson WC, Weidman J, Watkins K, Tran K, Khouri H, Pierson EA, Pierson LS III, Thomashow LS, Loper JE (2005) Complete genome sequence of the plant commensal *Pseudomonas fluorescens* Pf-5. Nat Biotech 23:873–878

Peypoux F, Guinand G, Michel G, Delcambeb L, Das C, Lederer E (1978) Structure of iturin A, a peptidolipid antibiotic from Bacillus subtilis. Biochemistry 17:3992–3996

Phister TG, O'Sullivan DJ, McKay LL (2004) Identification of bacilysin, chlorotetaine, and iturin a produced by *Bacillus* sp. strain CS93 isolated from pozol, a Mexican fermented maize dough. Appl Environ Microbiol 70(1):631–634

Raaijmakers JM, Vlami M, de Souza JT (2002) Antibiotic production by bacterial biocontrol agents. Anton Van Leeuwen 81:537–547

Raaijmakers JM, de Bruijn I, de Kock MJD (2006) Cyclic Lipopeptide production by plant-associated Pseudomonas spp.: diversity, activity, biosynthesis, and regulation. Molecular Plant Microbe Interact 16(7):699–710

Rainey PB, Brodey CL, Johnstone K (1993) Identification of a gene-cluster encoding three high-molecular-weight proteins, which is required for synthesis of tolaasin by the mushroom pathogen *Pseudomonas tolaasii*. Mol Microbiol 8:643–652

Rajavel M, Mitra A, Gopal B (2009) Role of *Bacillus subtilis* BacB in the synthesis of Bacilysin. J Biol Chem 284:31882–31892

Ramette A, Frapolli M, Défago G, Moënne-Loccoz Y (2003) Phylogeny of HCN synthase-encoding hcnBC. Eur J Plant Pathol 119:265–278

Rapp C, Jung KM, Loeffler W (1988) Rhizocticins – new phosphono-oligopeptides with antifungal activity. Liebigs Ann Chem 7:655–661

Roncero MI (1983) Gene controlling xylan utilization by *Bacillus subtilis*. J Bacteriol 156: 257–263

Roongsawang N, Hase K, Haruki M, Imanaka T, Morikawa M, Kanaya S (2003) Cloning and characterization of the gene cluster encoding arthrofactin synthetase from *Pseudomonas* sp. MIS38. Chem Biol 10:869–880

Schellenberg B, Bigler L, Dudler R (2007) Identification of genes involved in the biosynthesis of the cytotoxic compound glidobactin from a soil bacterium. Environ Microbiol 9(7):640–1650

Schneider K, Chen XH, Vater J, Franke P, Graeme N, Borriss R, Süssmuth RD (2007) Macrolactin is the polyketide biosynthesis product of the pks2 cluster of *Bacillus amyloliquefaciens* FZB42. J Nat Prod 70:1417–1423

Scholz-Schroeder BK, Hutchison ML, Grgurina I, Gross DC (2001) The contribution of syringopeptin and syringomycin to virulence of *Pseudomonas syringae* pv. *syringae* strain B301D on the basis of *sypA* and *syrB1* biosynthesis mutant analysis. Mol Plant Microbe Interact 14:336–348

Scholz-Schroeder BK, Soule JD, Gross DC (2003) The *sypA*, *sypB* and *sypC* synthetase genes encode twenty-two modules involved in the nonribosomal peptide synthesis of syringopeptin by *Pseudomonas syringae* pv. *syringae* B301D. Mol Plant Microbe Interact 16:271–280

Shibazaki M, Sugawara T, Nagai K, Shimizu Y, Yamaguchi H, Suzuki K (1996) YM-47522, a novel antifungal antibiotic produced by *Bacillus* sp. Taxonomy, fermentation, isolation and biological properties. J Antibiot 49:340–344

Shoji J, Hinoo H, Kato T, Hattori T, Hirooka K, Tawara K (1990) Isolation of cepafungins I, II and III from *Pseudomonas species*. J Antibiot 43:783–787

Silo-Suh LA, Stabb EV, Raffel S, Handelsman J (1998) Target range of zwittermicin A an aminopolyol antibiotic from *Bacillus cereus*. Curr Microbiol 37:6–11

Sorensen D, Nielsen TH, Christophersen C, Sorensen J, Gajhede M (2001) Cyclic lipoundecapeptide amphisin from *Pseudomonas* sp. strain DSS73. Acta Crystallogr 57:1123–1124

Steijl H, Niemann GJ, Boon JJ (1999) Changes in chemical composition related to fungal infection and induced resistance in carnation and radish investigated by pyrolysis mass spectrometry. Physiol Mol Plant Pathol 55:297–311

Sugawara T, Shibazaki M, Nakahara H, Suzuki K (1996) YM47522, a novel antifungal antibiotic produced by *Bacillus* sp. structure and relative stereochemistry. J Antibiot 49:340–344

Takeda R (1958) Pseudomonas pigments. I. Pyoluteorin, a new chlorine-containing pigment produced by *Pseudomonas aeruginosa*. Hako Kogaku Zasshi 36:281–290

Tamehiro N, Okamoto-Hosoya Y, Okamoto S, Ubukata M, Hamada M, Naganawa H, Ochi K (2002) Bacilysocin, a novel phospholipid antibiotic produced by *Bacillus subtilis* 168. Antimicrob Agents Chemother 46(2):315–332

Tawara S, Matsumoto S, Hirose T, Matsumoto Y, Nakamoto S, Mitsuno M, Kamimura T (1989) In vitro antifungal synergism between pyrrolnitrin and clotrimazole. Jpn J Med Mycol 30: 202–210

Thomas MS (2007) Iron acquisition mechanisms of the *Burkholderia cepacia* complex. Biometals 20:431–452

Thomashow LS, Weller DM (1988) Role of a phenazine antibiotic from *Pseudomonas fluorescens* in biological control of *Gaeumannomyces graminis* var. *tritici*. J Bacteriol 170:3499–3508

Thrane C, Olsson S, Nielsen TH, Sorensen J (1999) Vital fluorescent stains for detection of stress in *Pythium ultimum* and *Rhizoctonia solani* challenged with viscosinamide from *Pseudomonas fluorescens* DR54. FEMS Microbiol Ecol 30:11–23

Thrane C, Harder NT, Neiendam NM, Sorensen J, Olsson S (2000) Viscosinamide-producing *Pseudomonas fluorescens* DR54 exerts a biocontrol effect on *Pythium ultimum* in sugar beet rhizosphere. FEMS Microbiol Ecol 33:139–146

Toshima H, Maru K, Saito M, Ichihara A (1999) Study on Fungitoxic 3-Amino-2-piperidinone-containing lipids: total syntheses of cepaciamides A and B. Tetrahedron Lett 40:939–942

Tripathi RK, Gottlieb D (1969) Mechanism of action of the antifungal antibiotic pyrrolnitrin. J Bacteriol 100:310–318

Tsuge K, Akiyama T, Shoda M (2001) Cloning, sequencing, and characterization of the iturin A operon. J Bacteriol 183:6265–6273

Tsuge K, Matsui K, Itaya M (2007) Production of the non-ribosomal peptide plipastatin in *Bacillus subtilis* regulated by three relevant gene blocks assembled in a single movable DNA segment. J Biotechnol 129(4):592–603

Vaillancourt FH, Yin J, Walsh CT (2005) From the cover: $SyrB_2$ in syringomycin E biosynthesis is a nonheme FeII {alpha}-ketoglutarate- and O_2-dependent halogenase. Proc Natl Acad Sci USA 102:10111–10116

Vanittanakoma N, Loeffler W (1986) Fengycin – a novel antifungal lipopeptide antibiotic produced by *Bacillus subtilis* F-29-3. J Antibiot 39:888–901

Vial L, Lépine F, Milot S, Groleau M, Dekimpe V, Woods DE, Eric D (2008) *Burkholderia pseudomallei*, *B. thailandensis*, and *B. ambifaria* produce 4-hydroxy-2-alkylquinoline analogues with a methyl group at the three position that Is required for quorum-sensing regulation. J Bacteriol 190:5339–5352

Vincent MN, Harrison LA, Brackin JM, Kovacevich PA, Mukerji P, Weller DM, Pierson EA (1991) Genetic analysis of the antifungal activity of a soilborne *Pseudomonas aureofaciens* strain. Appl Environ Microbiol 57:2928–2934

Wang J, Liu J, Chen H, Yao J (2007) Characterization of *Fusarium graminearum* inhibitory lipopeptide from *Bacillus subtilis* IB. Appl Microbiol Biotechnol 76:889–894

Whipps JM (1997) Developments in the biological control of soilborne plant pathogens. Adv Bot Res 26:1–134

Wilson KE, Flor JE, Schwartz RE, Joshua H, Smith JL, Pelak BA, Liesch JM, Hensens D (1987) Difficidin and oxydifficidin: novel broad spectrum antibacterial antibiotics produced by *Bacillus subtilis*. 2. Isolation and physicochemical characterization. J Antibiot 40:1682–1691

Woeng CA, TFC BGV, van der Bij AJ, van der Drift KMGM, Schripsema J, Kroon B (1998) Biocontrol by phenazine-1-carboxamide-producing *Pseudomonas chlororaphis* PCL1391 of tomato root rot caused by *Fusarium oxysporum* f. sp. radicis-lycopersici. Mol Plant Microbe Interact 16:525–535

Wong DT, Airallb JM (1970) The mode of action of antifungal agents: effect of pyrrolnitrin on mitochondrial electron transport. J Antibiot 23:55–61

Xiao L, Xie CC, Cai J, Lin ZJ, Chen YH (2009) Identification and characterization of a chitinase-produced *Bacillus* showing significant antifungal activity. Curr Microbiol 58:528–533

Yang CY, Hoa YC, Pang JC, Huang SS, Tschen JS (2009) Cloning and expression of an antifungal chitinase gene of a novel *Bacillus subtilis* isolate from Taiwan potato field. Bioresour Technol 100:1454–1458

Zhang JH, Quigley NB, Gross DC (1995) Analysis of the *syrB* and *syrC* genes of *Pseudomonas syringae* pv. *syringae* indicates that syringomycin is synthesized by a thiotemplate mechanism. J Bacteriol 177:4009–4020

Zhang B, Xie C, Yang X (2008) A novel small antifungal peptide from *Bacillus* strain B-TL2 isolated from tobacco stems. Peptides 29:350–355

Zhao ZZ, Wang QS, Wang KM, Brian K, Liu CG, YC GU (2010) Study of the antifungal activity of *Bacillus vallismortis* ZZ185 in vitro and identification of its antifungal components. Bioresour Technol 101:292–297

Zweerink MM, Edison A (1987) Difficidin and oxydifficidin: novel broad spectrum antibacterial antibiotics produced by *Bacillus subtilis* III. Mode of action of difficidin. J Antibiot 40:1691–1692

Role of Plant Growth Promoting Rhizobacteria in Biocontrol of Plant Diseases and Sustainable Agriculture

Mohd. Sayeed Akhtar and Zaki A. Siddiqui

Contents

1 Introduction 158
2 Mechanisms of Disease Suppression 159
3 Interactions of PGPR with Pathogens 160
3.1 Antibiotic Production 160
3.2 Enzyme Production 161
3.3 Siderophores Production 161
4 Interactions of PGPR with Plants 176
4.1 Induced Resistance 177
4.2 Root Colonization 178
4.3 Genetic Variations in the Host 179
5 Interactions of PGPR in the Rhizosphere 179
5.1 Interactions with the Microbial Community 180
5.2 Interactions of PGPR Strains 180
6 A Practical Control System Using PGPR 181
7 Conclusion 182
References 182

Abstract The management of plant diseases in the sustainable agriculture has become a challenge for plant pathologist. Increasing knowledge and growing concern of pesticide applications on environment have aroused interest in alternative methods of plant protection. Plant growth promoting rhizobacteria (PGPR) are the important group of microorganisms, which play a major role in the biocontrol of plant pathogens. PGPR can profoundly improve seed germination, root development, and water uptake by plants. These rhizobacteria stimulate plant growth directly by producing growth hormones and improving nutrient uptake or indirectly

Z.A. Siddiqui (✉) and M.S. Akhtar
Department of Botany, Aligarh Muslim University, University Aligarh 202002, India
e-mail: zaki_63@yahoo.co.in

D.K. Maheshwari (ed.), *Plant Growth and Health Promoting Bacteria*,
Microbiology Monographs 18, DOI 10.1007/978-3-642-13612-2_7,

by changing microbial balance in favor of beneficial microorganisms in the rhizosphere and can suppress a broad spectrum of bacterial, fungal, nematode, and even some viral diseases. Although significant control of plant pathogens has been demonstrated by PGPR in laboratory and greenhouse studies, results in the field trials have been inconsistent. Recent progress in our understanding of their diversity, colonizing ability, and mechanisms of action, formulation, and their application may facilitate their development as reliable biocontrol agents against plant pathogens. Use of PGPR has become a common practice in many regions of the world, and greater application of PGPR is possible for sustainable agriculture in near future.

1 Introduction

Management of plant diseases has become a challenge for the plant pathologist for sustainable agriculture. Increasing knowledge and health hazards associated with the applications of pesticides have aroused interest in alternative methods of plant protection. One of the best methods that may be used by plant pathologists is Biocontrol. Out of different organisms used for biocontrol, rhizosphere microorganisms may provide a front line defense against pathogen attack and are ideal for use as biocontrol agents (Weller 1988; Siddiqui 2006). Biocontrol involves harnessing of disease-suppressive microorganisms to improve plant health (Handelsman and Stabb 1996). Disease suppression by biocontrol agents is the manifestation of interactions among the plant, the pathogen, the biocontrol agent, the microbial community on and around the plant, and the physical environment. Among the wide range of beneficial microorganisms, plant growth promoting rhizobacteria (PGPR) play a vital role in the management of plant diseases (Kloepper and Schroth 1978; Glick 1995; Siddiqui 2006). PGPR are free-living bacteria that may impart beneficial effects on plants. PGPR inhabit the rhizosphere, the volume of soil under the immediate influence of the plant root system, and favors the establishment of a large amount of active microbial population. Plants release metabolically active cells from their roots and deposit as much as 20% of the carbon allocated to roots in the rhizosphere, suggesting a highly evolved relationship between the plant and rhizosphere microorganisms (Handelsman and Stabb 1996), and the dynamic nature of the rhizosphere creates interactions that lead to biocontrol of diseases (Rovira 1965, 1969, 1991; Hawes 1991; Waisel et al. 1991).

Biocontrol of plant diseases is particularly complex because diseases mostly occur in the dynamic environment at the interface of the plant root as well as in the aerial parts of plants. PGPR enhance seedling emergence, colonize roots, and stimulate overall plant growth. PGPR also improve seed germination, root development, mineral nutrition and water uptake/utilization. They can also suppress diseases of plants. Numerous recent reviews present comprehensively the variety of microbial biocontrol agents (Weller 1988; Handelsman and Stabb 1996; Siddiqui and Mahmood 1995a, 1996, 1999; Whipps 2001; Weller et al. 2002; Bakker et al.

2003; Compant et al. 2005; Siddiqui 2006). This chapter focuses on the potentiality of PGPR, mechanisms involved in the biocontrol of plant diseases to understand the behavior and interaction of mycorhizospheric organisms with PGPR. This understanding will facilitate the application of PGPR for the biocontrol of plant diseases under field conditions.

2 Mechanisms of Disease Suppression

There are different ways by which PGPR can affect the plant growth directly: by fixing atmospheric nitrogen, synthesizing several plant hormones and enzymes, and solubilizing minerals that can modulate plant hormone levels.

A particular plant growth promoting bacterium may possess one or more of these mechanisms (Compant et al. 2005; Siddiqui 2006). The indirect promotion of plant growth occurs when PGPR lessen or prevent the deleterious effects of one more phytopathogenic organism by producing siderophores that limit the available iron to the pathogen, producing antibiotics that kill the pathogen, antibiosis, and inducing systemic resistance in plant (Fig. 1). PGPR also cause cell wall structural modifications and biochemical/physiological changes leading to the synthesis of proteins and chemicals involved in plant defense mechanisms. PGPR has been successfully used for the biocontrol of nematode, fungal, bacterial, and viral diseases of plants in different parts of the world (Tables 1–4). Some of the biocontrol mechanisms that have been dealt and will be discussed as follows:

- Interactions of PGPR with pathogens
- Interactions of PGPR with plants
- Interactions of PGPR in the rhizosphere

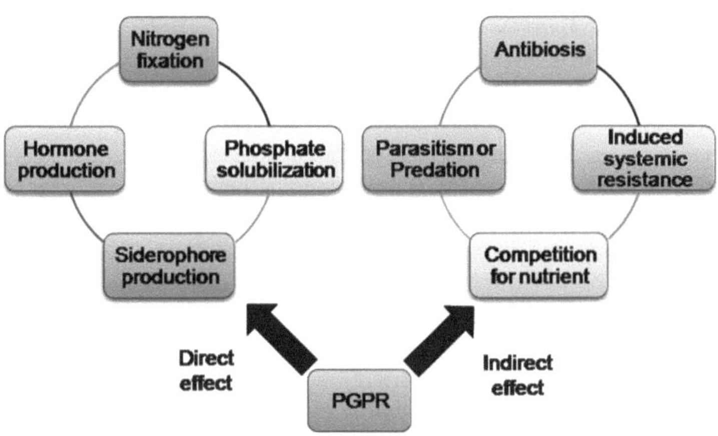

Fig. 1 Some direct or indirect effects of plant growth promoting rhizobacteria on the plant growth

3 Interactions of PGPR with Pathogens

During interaction process of PGPR with phytopathogens, the former produce certain antibiotics, cell wall degrading enzymes, siderophore, etc., and release of such metabolites decides the fate of the pathogen.

3.1 Antibiotic Production

One of the most effective mechanisms to prevent proliferation of phytopathogens is the synthesis of antibiotics by PGPR. There are numerous reports of the production of antifungal metabolites by bacteria in vitro that may also have activity in vivo. A variety of metabolites such as amphisin, butyrolactones, 2,4-diacetylphloroglucinol (DAPG), cyclic lipopeptide, HCN, kanosamine, oligomycin A, oomycin A, phenazine-1-carboxylic acid (PCA), pyoluterin (Plt), pyrrolnitrin (Pln), tensin, tropolone, viscosinamide, xanthobaccin, and zwittermycin A are produced by PGPR (Defago 1993; Milner et al. 1996; Keel and Defago 1997; Whipps 1997; Kang et al. 1998; Kim et al. 1999; Nakayama et al. 1999; Thrane et al. 1999; Nielsen et al. 2002; Raaijmakers et al. 2002; de Souza et al. 2003; Compant et al. 2005). To demonstrate a role for antibiotics in biocontrol, mutants lacking production of antibiotics or antibiotics over-producing mutants have been used (Bonsall et al. 1997; Chin-A-Woeng et al. 1998; Nowak-Thompson et al. 1999). Alternatively, the use of reporter genes or probes to demonstrate the production of antibiotics in the rhizosphere is becoming more common (Kraus and Loper 1995; Raaijmakers et al. 1997; Chin-A-Woeng et al. 1998). Indeed, isolation and characterization of genes or gene clusters responsible for antibiotic production have now been achieved (Kraus and Loper 1995; Bangera and Thomashow 1996; Hammer et al. 1997; Kang et al. 1998; Nowak-Thompson et al. 1999). Significantly, both Phl and PCA have been isolated from the rhizosphere of wheat following introduction of biocontrol strains of *Pseudomonas* (Thomashow et al. 1990; Bonsall et al. 1997; Raaijmakers et al. 1999), confirming that such antibiotics are produced in vivo. Further, Ph1 production in the rhizosphere of wheat was strongly related to the density of the bacterial population present and the ability to colonize roots (Raaijmakers et al. 1999). PCA from *Pseudomonas aureofaciens* has even been used as a direct field treatment for the control of *Sclerotinia homeocarpa* on creeping bent grass (Powell et al. 2000).

The first antibiotics clearly implicated in biocontrol by fluorescent pseudomonads were the phenazine derivatives (Handelsman and Stabb 1996). *P. fluorescens* strain 2-79 and *P. aureofaciens* strain 30-84 contribute to disease suppression of take-all of wheat (Weller and Cook 1983; Brisbane and Rovira 1988). *P. fluorescens* strain CHA0 produces hydrogen cyanide, 2,4-diacetylphloroglucinol, and pyoluteorin, which directly interferes with the growth of various pathogens and contributes to the disease suppression (Voisard et al. 1989; Keel et al. 1992; Maurhofer et al. 1994b; Duffy and Defago 1999). Furthermore, a quantitative relationship between antibiotic production and disease suppression is suggested by the enhanced production of 2, 4-diacetylphloroglucinol and pyoluteorin accomplished by adding extra copies of a 22-kb fragment of DNA that improves suppression of *Pythium* on cucumber (Maurhofer et al. 1992). Antibiotic DAPG has been shown to act as the inducing agent in CHA0-mediated induced systemic resistance (ISR) in tomato against root-knot nematode *M. javanica* (Siddiqui and Shaukat 2003) and suggest that more antibiotics may be capable of eliciting ISR in plants. The role of individual antibiotic compound in suppression of root pathogens has been clearly established using mutation analysis and molecular genetic tools, and purified antibiotics compounds viz. DAPG overproducing mutant of *P. fluorescens* offered a better protection against the take all of wheat and bacterial wilt of tomato (Hongyou et al. 2005).

3.2 *Enzyme Production*

Biocontrol of *Phytophthora cinnamomi* causing root rot of *Banksia grandis* was obtained using a cellulase-producing isolate of *Micromonospora carbonacea* (El-Tarabily et al. 1996) and *Phytophthora fragariae* var. *rubi* causing raspberry root rot that was suppressed by the application of actinomycete isolates selected for the production of β-1,3-, β -1,4-, and β -1,6-glucanases (Valois et al. 1996). Chitinolytic enzymes produced by both *Bacillus cereus* and *Pantoea* (*Enterobacter*) *agglomerans* also appear to be involved in the biocontrol of *Rhizoctonia solani* (Chernin et al. 1995, 1997; Pleban et al. 1997). Tn*5* mutants of *E. agglomerans* deficient in chitinolytic activity were unable to protect cotton, and the expression of the *chiA* gene for endochitinase in *E. coli* allowed the transformed strain to inhibit *R. solani* on cotton seedlings. Similar techniques involving *Tn*5 insertion mutants and subsequent complementation demonstrated that biocontrol of *Pythium ultimum* in the rhizosphere of sugar beet by *Stenotrophomonas maltophila* was due to the production of extracellular protease (Dunne et al. 1997). The incidence of plant disease caused by the phytopathogenic fungi *R. solani*, *Sclerotium rolfsii*, and *P. ultimum* was reduced by using a β-1, 3-glucanse producing strain of *Pseudomonas cepacia*, which was able to degrade the fungal mycelia. Many of the bacterial enzymes that can lyse fungal cells, including chitinases and β-1, 3-glucanse, are encoded by a single gene.

3.3 *Siderophores Production*

Iron is an important micronutrient used by bacteria and it is essential for their metabolism. In the soil, it is unavailable for direct assimilation by microorganisms because ferric iron (FeIII), which predominates in nature, is only sparingly soluble and too low in concentration to support microbial growth (Rachid and Ahmed 2005). To survive, soil microorganisms synthesize and secrete low-molecular-

Table 1 Effects of PGPR on plant parasitic nematodes

Nematode	PGPR	Effect	References
Meloidogyne incognita			
M. incognita	*Bacillus thuringiensis*	Prevented *M. incognita* from forming galls on tomato	Ignoffo and Dropkin (1977)
M. incognita	244 isolates	Only 125 bacterial isolates imparted positive effect on tomato and cucumber, rarely on both and negative effect on nematodes.	Zavaleta-Mejia and VanGundy (1982)
M. incognita	*Serratia marcescens*	Bacterium produced a volatile metabolite and was nematoxic.	Zavaleta-Mejia (1985)
M. incognita	354 isolates	*Pseudomonas fluorescens* (strains JOB204, JOB 209) and *Bacillus* (JOB203) were most effective and clover plants treated with these bacteria had fewer galls and large roots.	Becker et al. (1988)
M. incognita	*Bacillus licheniformis*, *Pseudomonas mendocina*	*B. licheniformis* caused greater reduction in nematode multiplication than *P. mendocina* on tomato.	Siddiqui and Husain (1991)
M. incognita	*Bacillus licheniformis*, *Alcaligenes faecalis*	*B. licheniformis* caused a greater reduction in nematode multiplication than *A. faecalis* on chickpea.	Siddiqui and Mahmood (1992)
M. incognita race3	*B. subtilis*	*B. subtilis* reduced nematode multiplication and improved growth of chickpea.	Siddiqui and Mahmood (1993)
M. incognita	*B. subtilis*	Reduced nematode population on tomato.	Gautam et al. (1995)
M. incognita	*B. subtilis*	Seed treatment with bacteria reduced nematode multiplication on chickpea.	Siddiqui and Mahmood (1995b)
M. incognita	Endophytic bacterial stains	Reduced galling of cotton roots by root-knot nematode.	Hallmann et al. (1997)
M. incognita	*P. fluorescens* PRS9, *B. polymyxa*	Reduced the galling and nematode population on tomato.	Khan and Akram (2000)
M. incognita	*B. subtilis*, *Azotobacter chroococcum* *Azospirillium lipoferum*	Reduced the no. of galls per root system, egg-mass production and nematode population on mung bean.	Khan and Kounsar (2000)
M. incognita	*P. aeruginosa*	Reduced galling on tomato.	Siddiqui and Haque (2001)
M. incognita	*P. fluorescens* (strains GRP3 and PRS9)	GRP3 strain was better in reducing galling and nematode multiplication than PRS9.	Siddiqui et al. (2001)
M. incognita	*Pseudomonas fluorescens*	Seed treatment significantly reduced the galling on okra.	Devi and Dutta (2002)
M. incognita	*P. fluorescens*, *A. brasilense*, *Azotobacter chroococcum*, Microphos	Best management of *M. incognita* was obtained when Microphos culture (mixture of *P. straita*, *Paenibacillus. polymyxa* and *Aspergillus niger*) was used with *A. chroococcum* and *A. brasilense* on brinjal.	Siddiqui et al. (2002)

M. incognita	*A. chroochoccum, Azospirillum* sp.	*Azotobacter* was better in reducing galling than *Azospirillum* sp. in okra.	Sharma and Mishra (2003)
M. incognita	*P. fluorescens, A. chroococcum*	Greater biocontrol of *M. incognita* was observed when *P. fluorescens* was used with the straw of *Zea mays.*	Siddiqui and Mahmood (2003)
M. incognita	*P. fluorescens, A. chroococcum, Azospirillum brasilense*	*P. fluorescens* was better at improving tomato growth and reducing galling and nematode multiplication than *A. chroococcum* or *A. brasilense.*	Siddiqui (2004)
M. incognita	*Pseudomonas straita*	Reduced reproduction of *M. incognita* on pea.	Siddiqui and Singh (2005a)
M. incognita, H. cajani	*Bacillus* and fluorescent pseudomonads isolates	Four isolates of *Pseudomonas* and 2 of *Bacllus* (Pa70, Pf18, Pa116, Pa324, B18, and B160) were considered potentially useful for the biocontrol of nematodes.	Siddiqui et al. (2005)
M. incognita	*Bacillus* consortium (Strains INR7, T4 and IN 937b)	*Bacillus* spp. had no significant differences over untreated control in term growth of micro- propagated papaya but the inoculation of *Bacillus* with AM fungi reduced the development of *M. incognita* in plants.	Jaizme-Vega et al. (2006)
M. incognita	20 isolates of fluorescent pseudomonads	Out of 20, four isolates of fluorescent pseudomonads (Pf604, Pf605, Pf611 and Pa616) have inhibitory effect against the hatching and penetration of nematodes.	Siddiqui and Shakeel (2006)
M. incognita	*P. putida, Paenibacillus polymyxa*	Inoculation of plants with *P. putida* most effectively reduces galling and nematode multiplication than *P. polymyxa* on chickpea.	Akhtar and Siddiqui (2007)
M. incognita	*P. aeruginosa*	*P. aeruginosa* caused a significant reduction in galling and nematode multiplication on chickpea	Siddiqui and Akhtar (2007)
M. incognita, H. cajani	*Bacillus* isolates	Isolate B 615 and B 603 were found more promising for the control of nematodes.	Siddiqui and Shakeel (2007)
M. incognita	*Bacillus* and *Pseudomonas* isolates	Out of 18 isolates, B 28 was best in improving tomato growth of *M. incognita* inoculated plants.	Siddiqui et al. (2007b)
M. incognita, H. cajani	*Bacillus* and *Pseudomonas* isolates	Combined use of Pa324 with B18 provided better biocontrol of nematodes than use of either of them on pigeon pea.	Siddiqui et al. (2008)
M. incognita	*P. straita, Rhizobium* sp	Inoculation of *Rhizobium* caused a greater increase in chickpea growth than caused by *P. straita.*	Akhtar and Siddiqui (2008a)
M. incognita	*Pseudomonas alcaligenes, B. pumilus*	*P. alcaligenes* caused a greater increase in shoot dry in plants inoculated with nematodes than did *B. pumilus* on chickpea.	Akhtar and Siddiqui (2008b)
M. incognita	*Paenibacillus polymyxa* GBR-1	Culture filtrate of *Paenibacillus polymyxa* GBR-1 under in vitro significantly reduced egg hatch and caused substantial mortality of *M. incognita* juveniles.	Khan et al. (2008)

(continued)

Table 1 (continued)

Nematode	PGPR	Effect	References
M. incognita	*Pseudomonas putida*	Use of composted manure with *P. putida* was more beneficial for tomato growth than the use of urea with bacterium.	Siddiqui and Akhtar (2008a)
M. incognita	*P. putida*	Use of *P. putida* caused 39% reduction in galling and nematode multiplication on tomato.	Siddiqui and Akhtar (2008b)
M. incognita	*P. putida*	Combined use of neem leaf litter with *P. putida* plus *G. intraradices* was best in improving growth of nematode infected tomato.	Siddiqui and Akhtar (2008c)
M. incognita	*P. putida*, *P. alcaligenes*, *Pseudomonas* isolate Pa 28	*P. putida* caused greatest reduction in galling and nematode multiplication followed by *P. alcaligenes* and Pa 28.	Akhtar and Siddiqui (2009)
M. incognita	*A. chroococcum*, *B. subtilis*, *P. putida*	Highest increase in the growth of nematode inoculated plants was observed when *P. putida* was used with cattle manure on tomato.	Siddiqui and Futai (2009)
M. incognita	*B. subtilis*, *P. polymyxa*, *Burkholderia cepacia*	The greatest increase in growth of nematode inoculated plants and reduction in nematode galling was observed when *P. polymyxa* was used with *P. lilacinus* on tomato.	Siddiqui and Akhtar (2009a)
M. incognita	*B. cepacia*, *B. subtilis*	Application of *B. cepacia* caused 36% increase in shoot dry mass of nematode inoculated plants followed by *B. subtilis* (32%) on tomato.	Siddiqui and Akhtar (2009b)
M. incognita	10 isolates of *Pseudomonas* and *Bacillus*	Fluorescent Pseudomonads isolates (Pf1, Pa2, Pa3, Pa4, and Pf5) caused greater inhibitory effect on the hatching and penetration of *M. incognita* than *Bacillus* isolates (B1, B2, B3,B4 and B5) on pea.	Siddiqui et al. (2009)
M. javanica			
M. javanica	*B. cereus*	Inhibited penetration of nematodes on tomato roots.	Oka et al. (1993)
M. javanica	*P. fluorescens*	Reduced nematode multiplication and morphometrics of *M. javanica* females on tomato in different soil.	Siddiqui and Mahmood (1998)
M. javanica	*B. subtilis*	Greatest growth of tomato and high reduction in nematode multiplication occurred when ammonium sulphate was used with *B. subtilis* and *G. mosseae*.	Siddiqui and Mahmood (2000)
M. javanica	*P. aeruginosa*	Reduced the galling and nematode population on tomato.	Siddiqui et al. (2000)
M. javanica	*P. fluorescens*, *Azotobacter chroococcum*, *Azospirillum brasilense*	Use of *P. fluorescens* with *Glomus mosseae* was better at improving chickpea growth and reducing galling and nematode multiplication than other treatments.	Siddiqui and Mahmood (2001)

M. javanica	*P. aeruginosa*, *P. fluorescens*	Bare root dip or soil drench treatment reduced nematode penetration into tomato roots.	Siddiqui and Shaukat (2002)
M. javanica	*P. fluorescens* CHA0	Use of *P. fluorescens* with ammonium molybdate reduced the nematode penetration in mung bean.	Hamid et al. (2003)
M. javanica	*Brevibacillus brevis* or *B. subtilis*	Use of B7 strain as seed dressing was found to be most effective in reducing nematode population on mung bean.	Li et al. (2005)
M. javanica	*P. putida*, *P. alcaligenes*, *P. polymyxa*, *B. pumilus*	Use of all PGPR strains reduced the galling and nematode reproduction in lentil but *P. putida* was found best in reducing galling and nematode reproduction.	Siddiqui et al. (2007a)
M. javanica	*P. fluorescens* EPS291 and EPS817	Both the isolates significantly increased the plant growth and reduced nematode reproduction in micropropagated banana.	Rodriguez-Romero et al. (2008)
M. javanica	*P. putida* *P. alcaligenes*	Individually all the PGPR strains significantly reduced the disease severity in chickpea.	Siddiqui and Akhtar (2009c)
M. exigua			
M. exigua	*Paenibacillus macerans*	Culture filtrate under in vitro condition showed potential against the root-knot nematode *M. exigua* juveniles.	Oliveira et al. (2009)
Globodera pallida			
Globodera pallida	Number of bacterial isolates	Seed treatment reduced nematode penetration in potato roots.	Racke and Sikora (1985)
Globodera pallida	*Agrobacterium radiobacter*	Reduced nematode infection by 40% when sprayed on seed pieces of potato.	Sikora et al. (1989)
Globodera pallida	*B. sphaericus*, *A. radiobacter*	Rhizobacteria systemically induced resistance against potato cyst nematode.	Hasky-Günther et al. (1998)
Heterodera cajani			
H. cajani	*B. subtilis*	Bacteria reduced nematode multiplication on pigeon pea	Siddiqui and Mahmood (1995c)
M. incognita, *H. cajani*, *H. zeae*, *H. avenae*	*B. subtilis*, *B. cereus*, *B. pumilus*, *Pseudomonas* sp.	Most effective isolates against all tested species were *B. subtilis* and *B. pumilus*. The non-cellular extract exhibited high larvicidal properties.	Gokte and Swarup (1988)
H. cajani	*P. fluorescens*	Reduced multiplication of *H. cajani* on pigeon pea.	Siddiqui et al. (1998)

(*continued*)

Table 1 (continued)

Nematode	PGPR	Effect	References
Heterodera schachtii			
H. schachtii	290 isolates	Eight isolates were antagonistic to *H. schachtii*, 3 isolates were identified as *P. fluorescens.*	Oostendorp and Sikora (1989a)
H. schachtii	8 isolates	Nematode penetration was reduced by 6 of 8 isolates tested.	Oostendorp and Sikora (1989b)
Other nematodes species			
Rotylenchulus reniformis, Meloidogyne spp.	*B. subtilis*	Reduced nematode reproduction and galling on cotton, tomato, peanut, and sugar beet.	Sikora (1988)
Criconemella xenoplax	*Pseudomonas aureofaciens*	One strain inhibited nematode multiplication in greenhouse test.	Westcott and Kluepfel (1992)
Criconemella xenoplax	*Pseudomonas aureofaciens*	Bacteria suppressed population of ring nematode.	Kluepfel et al. (1993)
Panagrellus sp.	*P. fluorescens*	Bacteria cultivated on plate count broth reduced nematodes up to 57.4%.	Weidenborner and Kunz (1993)
C. elegans, R. reniformis, P. penetrans	*Bacillus thuringiensis*	Isolate 371 of bacterium reduced nematode populations on tomato and strawberry.	Zuckerman et al. (1993)
R. reniformis	*Pseudomonas solanacearum*	Slight inhibition of nematode activity on aubergine roots.	Kermarrec et al. (1994)
R.. similis, Meloidogyne spp.	*P. putida, P. fluorescens*	Inhibited invasion of *R. similis* and *Meloidogyne* spp. in banana, maize, and tomato.	Aalten et al. (1998)
Heterodera cruciferae	Fluorescent pseudomonads	Growth and hatching of nematode eggs were inhibited	Aksoy and Mennan (2004)

Table 2 Effects of PGPR on fungal diseases of plants

Fungus	PGPR	Effect	References
Gaeumannomyces sp.			
Gaeumannomyces graminis var. *tritici*	*P. fluorescens*	Strains of *P. fluorescens* may be involved in the suppression of *G. graminis var. tritici*	Cook and Rovira (1976)
G. graminis	*Pseudomonas* spp. (fluorescent strains)	27% yield increase due to biocontrol of bacteria in winter wheat under field conditions.	De Freitas and Germida (1990)
G. graminis var. *tritici*	*P. aureofaciens Q2-87*	Inhibition of fungus was demonstrated both in vitro and in vivo.	Harrison et al. (1993)
G. graminis, Rhizoctonia solani	*Bacillus subtilis, B. cereus* isolates, *P. corrugata*	*Bacillus* isolate A47 and *B. subtilis* B908 reduced the take-all disease in sodic acid soil while *B. subtils* B931 was more effective in reducing Rhizoctonia root-rot in calcareous sandy loam soil of wheat.	Maarten et al. (1998)
Pythium spp.			
Pythium sp.	*P. fluorescens*	In *Pythium* contaminated sites, significant increases were observed in plant height, number of heads and grain yield of winter wheat.	Weller and Cook (1986)
P. ultimum P17	Fluorescent pseudomonads	Significantly suppressed root-rot disease on tulip.	Emma (1990)
P. aphanidermatum	*P. corrugate, P. aureofaciens*	Induced systemic resistance in cucumber roots.	Chen et al. (1999)
Pythium sp.	*P. putida, B. subtilis, E. aerogenes, E.agglomerans, B. cereus*	Most strains increased root length of cucumber in *Pythium*-infected plants *in vitro*.	Uthede et al. (1999)
P. aphanidermatum, F. o. f. sp. *cucurbitacearum*	*B. subtilis, P. putida*	Growth and yield of lettuce and cucumber were increased and disease severity reduced.	Amer and Utkhede (2000)
P. aphanidermatum, Aspergillus sp. *F. oxysporum* f. sp. *ciceri, Rhizoctonia solani*	*Pseudomonas* isolates	Two strains MRS23 and CRP55P have shown antifungal activity.	Goel et al. (2002)
P. aphanidermatum	*P. fluorescens, P. putida*	*P. fluorescens* isolate Pf1 was effective in reducing the damping-off incidence in tomato and hot pepper.	Ramamoorthy et al. (2002a)

(*continued*)

Table 2 (continued)

Fungus	PGPR	Effect	References
P. aphanidermatum OP4	Fluorescent *Pseudomonas* (CH31, CH1)	Suppressed the root-rot disease on cucumber.	Moulin et al. (1996)
Fusarium spp.			
Fusarium sp.	*P. fluorescens*	Observed induced resistance and phytoalexin accumulation in carnation.	Van Peer et al. (1991)
F. udum	*B. subtilis*	Increased shoot dry weight and reduced wilt of pigeon pea.	Siddiqui and Mahmood (1995c)
F. oxysporum f. sp. *raphani* *A. brassicicola*, *F. oxysporum*	*P. fluorescens*	Protected radish plants through induction of systemic resistance against these pathogens.	Hoffland et al. (1996)
F. culmorum	*P. chlororaphis 2E3, O6*	Strong inhibition of the fungus on spring wheat in the field.	Kropp et al. (1996)
F. udum	*B. subtilis*	Seed treatment with *B. subtilis* significantly reduced the incidence of wilt of pigeon pea.	Podile and Laxmi (1998)
F. udum	*P. fluorescens*	Wilt incidence was reduced in pigeon pea.	Siddiqui et al. (1998)
F. oxysporum f. sp. *lycopersici*	*P. fluorescens* PRS9, *B. polymyxa*	Reduced the wilting index and rhizosphere population of fungus on tomato.	Khan and Akram (2000)
F. oxysporum *M. phaseolina* *F. solani* *R. solani*	*P. aeruginosa*	Significantly suppressed growth of root infecting fungi on tomato.	Siddiqui et al. (2000)
F. oxysporum f. sp. *lycopersici*	*B. subtilis*, *P. fluorescens*, *Aspergillus awamori*, *A. niger*, *P. digitatum*	Use of all the PSM increased the yield and also reduced the rhizospheric population of wilt fungus by 23–49% on tomato.	Khan and Khan (2001)
F. oxysporum	Fluorescent Pseudomonads isolates	All the five isolates have shown the antifungal activity against the pathogen on pea.	Kumar et al. (2001)
F. udum, *Aspergillus niger*	*B. subtilis* AF1	AF1 supplemented with chitin or chitin material showed better control of crown rot and wilt diseases of ground nut and pigeon pea.	Manjula and Podile (2001)

F. moniliformae, *F. graminearum*, *Macrophomina phaseolina*	*Pseodomonas* sp. EM85 *Bacillus* sp. (MR-11(2), MRF)	All these isolates had the ability to suppress the diseases caused by *F. moniliforme*, *F. graminearum* and *M. phaseolina* on maize.	Pal et al (2001)
F. solani f. sp. *phaseoli*, *R. solani*, *F. oxysporum*	*B. subtilis* GBO3, MBI600	Seed treatment with these isolates caused increase in biomass and decrease in disease severity in glasshouse on bean.	Estevez de Jansen et al. (2002)
F. oxysporum f. sp. *ciceri*, *Aspergillus* sp., *P. aphanidermatum*, *R. solani*	*Pseudomonas* isolates	Two strains MRS23 and CRP55P have shown antifungal activity.	Goel et al. (2002)
F. oxysporum, *R. solani*	*P. fluorescens*	Out of 40 strains, 18 strains showed strong antifungal activity.	Kumar et al. (2002)
F. oxysporum f. sp. *lycopersici*	*P. fluorescens* Pf1	Pf1 protected tomato plants from wilt disease.	Ramamoorthy et al. (2002b)
F. udum, *F. oxysporum* f. sp. *ciceris*	*P. aeruginosa PNA 1*	*P. aeruginosa* protected pigeon pea and chickpea from *Fusarium* wilt.	Anjaiah et al. (2003)
F. chlamydosporium	*P. fluorescens*	Reduced the severity of disease on *Coleus*.	Boby and Bagyaraj (2003)
F. oxysporum f. sp. *melonis*	*P. putida*	Control on muskmelon achieved by seed treatment of *P. putida* strain 30 was 63% and 46–50% for strain 180.	Bora et al. (2004)
F. udum	*Bacillus* and fluorescent pseudomonads isolates	Four isolates, namely Pa116, P324, B18 and B160, have shown antifungal activity.	Siddiqui et al. (2005)
F. oxysporum f. sp. *ciceri*	*Paenibacillus lentimorbus* NRRL B-30488	Seed treatment with B-30488 caused greater mortality in non-bacterized seedlings compared to bacterized seedlings of chickpea.	Dasgupta et al. (2005)
F. udum	Fluorescent *Pseudomonas*	Four isolates, namely Pf604, Pf605, Pf611 and Pa616 have shown antifungal activity but isolate Pf605 reduced the wilt disease index of pigeon pea under pot condition.	Siddiqui and Shakeel (2005)
F. oxysporum f. sp. *lycopersici*	Fluorescent Pseudomonads	Significantly reduced the disease severity on tomato but the results were more pronounced when applied in combination with *T. harzianum*.	Yigit and Dikilitas (2007)
F. oxysporum f. sp. *radicis-lycopersici*	*Bacillus subtilis* EU07	Inoculation of *B. subtilis* (EU07) reduced the disease incidence up to 75% on tomato.	Baysal et al. (2009)

(*continued*)

Table 2 (continued)

Fungus	PGPR	Effect	References
Rhizoctonia spp.			
R. solani	*P. cepacia* R55, R85 *P. putida* R104	Increase of 62–78% of dry weight of winter wheat grown in *R. solani* infected soil.	De Freitas and Germida (1991)
R. solani	*B. subtilis* RB14	Antibiotics production by *B. subtilis* RB14 suppressed the damping off disease of tomato in vitro and in pot conditions.	Asaka and Shoda (1996)
R. solani	*Bacillus megaterium* (B153-2-2)	Seed treatment significantly reduced damage caused by *R. solani* on soybean in different soil.	Zheng and Sinclair (2000)
R. solani	*Pseudomonas fluorescens*	Mixture of 3 strains reduced disease and promoted growth of rice.	Nandakumar et al. (2001)
R. solani	*Bacillus subtilis*, *Burkholderia cepacia*	Combination of *B. subtilis* RB14-C with *B. cepacia* BY can lead to greater damping-off suppression on tomato than by these strains separately.	Szczech and Shoda (2004)
R. solani	*P. fluorescens* A6RI	Increased the growth of pathogen inoculated plants.	Berta et al. (2005)
	Fluorescent pseudomonads	Out of 103 isolates, only 52 isolates showed antifungal activity against the *R. solani*, in vitro condition.	Ahmadzadeh and Tehrani (2009)
Macrophomina phaseolina			
M. phaseolina	*B. licheniformis*, *A. faecalis*	Reduced root-rot disease of chickpea.	Siddiqui and Mahmood (1992)
M. phaseolina	*B. subtilis*	*B. subtilis* was superior to *P. lilacinus* for the management of *M. phaseolina* on chickpea.	Siddiqui and Mahmood (1993)
M. phaseolina	*B. subtilis*	*B. subtilis* resulted in greater shoot dry weight of chickpea than with any fungal filtrate.	Siddiqui and Mahmood (1995b)
M. phaseolina	*P. fluorescens* 4-92	*P. fluorescens* increased disease resistance by 33% in chickpea.	Srivastava et al. (2001)
M. phaseolina	Fluorescent *Pseudomonas* GRC_2	Seed bacterization with *Pseudomonas* isolates GRC_2 strain reduced the charcoal rot disease of peanut in *M. phaseolina* infested soil.	Gupta et al. (2002)
M. phaseolina	*P. fluorescens*	Seed treatment with *P. fluorescens* and neem cake reduced the root rot indices on green gram.	Begum and Kumar (2005)
M. phaseolina			

	Pseudomonas alcaligenes *Bacillus pumilus*	*P. alcaligenes* caused greater reduction against the root-rot disease than *B. pumilus* on chickpea.	Akhtar and Siddiqui (2008b)
Colletotrichum spp.			
Colletotrichum orbiculare	*P. putida*, *S. marcescens*, *Flavomonas oryzihabitans*, *B. pumilus*	PGPR mediated ISR was operative under field conditions against naturally occurring anthracnose of cucumber.	Wei et al. (1996)
Colletotrichum orbiculare	*B. pumilis*, *B. subtilis*, *Curtobacterium flaccumfaciens*	Mixture of these PGPR strains as seed treatment caused disease reduction on cucumber.	Raupach and Kloepper (1998)
Colletotrichum orbiculare	*S. marcescens* 90-166	Seed treatment suppressed anthracnose of cucumber.	Press et al. (2001)
Colletotrichum capsici	*Pseudomonas fluorescens*	Increased accumulation of enzymes involved in phenyl propanoid pathway and PR-proteins in hot pepper.	Ramamoorthy and Samiyappan (2001)
Colletotrichum lindemuthianum	*Pseudomonas aeruginosa*, *Pseudomonas fluorescens*	*P. aeruginosa* induced resistance only in resistant interactions while *P. fluorescens* induced resistance in susceptible and moderately resistant interactions on bean.	Bigirimana and Hofte (2002)
Colletotrichum falcatum	*P. fluorescens*	Induced systemic resistance against red rot of sugarcane.	Viswanathan and Samiyappan (2002)
Colletotrichum gloeosporioides	*P. fluorescens* FP7	Suppressed the anthracnose pathogen on mango leading to improved yield attributes.	Vivekananthan et al. (2004)
Other fungi			
Sclerotium rolfsii, *Fusarium* sp.	*P. putida* *P. fluorescens* *P. alcaligenes*	Reduced the incidence of disease caused by *S. rolfsii* in bean, and fusarium wilt of cotton and tomato.	Gamliel and Katan (1993)
Verticillium dahliae	*Pseudomonas PsJN*	Reduced disease incidence in tomato.	Sharma and Nowak (1998)

(*continued*)

Table 2 (continued)

Fungus	PGPR	Effect	References
Cronartium quercuum f. sp. *fusiforme*	*B. pumilus* SE34 *S. marcescens* 90-166	Two bacterial isolates out of 8, significantly reduced number of galls and induced systemic resistance against fusiform rust on Loblolly pine.	Enebak and Carey (2000)
Botrytis cineria	*Pseudomonas* PsJn	PsJn inhibits growth of *B. cineria* by disrupting cellular membrane and cell death.	Barka et al. (2002)
Curvularia lunata	*Bacillus* species BC121	Showed high antagonistic activity against *C. lunata*.	Basha and Ulaganathan (2002)
Cnaphalocrocis medinalis	*P. fluorescens* strains Pf1, FP7	Mixture of two strains performed better than the individual strains in reducing sheath blight of rice.	Radja Commare et al. (2002)
Phytophthora infestans	*P. fluorescens* 89B61 *B. pumilus* SE34	Elicited systemic protection against late blight of tomato and reduced disease severity.	Yan et al. (2002)
Sclerospora graminicola	*Bacillus pumilus*	Out of 7 PGPR strains, maximum vigor index resulted from treatment with strain INR7 followed by IN937b.	Niranjan Raj et al. (2003)
S. graminicola	*Pseudomonas fluorescens*	The isolates offered protection ranging from 20 to 75% against downy mildew to pearl millet.	Niranjan-Raj et al. (2004)
Alternaria triticina	*P. fluorescens* *A. chroococcum*	*P. fluorescens* caused greater reduction in *A. triticina* infected leaf area than *A. chroococcum.*	Siddiqui and Singh (2005b)
A. triticina	*Bacillus* and Fluorescent Pseudomonads	Out of 6 isolates, B28 was found best in improving plant growth and also caused reduction in percent leaf infected area of wheat.	Siddiqui (2007)
Exobasidium vexans	*Pseudomonas* and *Bacillus*	Seed treatment with PGPR strains reduced the disease severity on tea under field condition.	Saravanakumar et al. (2007)

Table 3 Effects of PGPR on bacterial diseases of plants

Pathogenic bacteria	PGPR	Effect	References
Xanthomonas compestris pv. *citri*	*P. fluorescens*	Control of citrus canker by siderophore production.	Unnamalai and Gnanamanickam (1984)
Erwinia carotovora	*P. putida* W4P63	Increased yield of Rosset Burbank potato and suppressed soft rot potential of tubers.	Xu and Gross (1986)
E. amylovora	*P. fluorescens* A506	Reduction in the population size of *E. amylovora* in pear flowers with *P. fluorescens* was due to competition.	Wilson and Lindow (1993)
P. syringae pv. *tomato*	*P. fluorescens* WCS417	*P. fluorescens* protected radish through induction of systemic resistance against a virulent bacterial leaf pathogen.	Hoffland et al. (1996)
P. solanacearum	*P. fluorescens* M29 and M40	Isolate M40 reduced tomato wilt significantly.	Kim and Misaghi (1996)
P. syringae pv. *lachrymans*	*P. putida*, *S. marcescens*, *Flavomonas oryzihabitans*, *B. pumilus*	PGPR strains caused significant protection against pathogen on cucumber.	Wei et al. (1996)
E. amylovora	*P. fluorescens* A506	Strain A506 and antibiotics acted additively in the control of frost and fire blight disease.	Lindow et al. (1996)
P. syringae pv. *lachrymans*, *Erwinia tracheiphila*	*B. pumilis*, *B. subtilis*, *Curtobacterium flaccumfaciens*	Seed treatment of strains mixture caused reduction in angular spot and wilt of cucumber.	Raupach and Kloepper (1998)
Ralstonia solanacearum	Fluorescent pseudomonads	All three strains suppressed wilt of tomato and increased yield.	Jagadeesh et al. (2001)
Xanthomonas oryzae pv. *oryzae*	*P. fluorescens*	Showed resistance to the rice bacterial blight pathogen.	Vidhyasekaran et al. (2001)
P. syringae pv. *tomato*	*Azospirillium brasilense*	Prevented bacterial speck disease development and improved tomato growth.	Bashan and Bashan (2002)
R. solanacearum	*Serratia* J2, *Pseudomonas*, *Bacillus* BB11	All the three strains suppress wilt of tomato and increase yield.	Guo et al. (2004)

(*continued*)

Table 3 (continued)

Pathogenic bacteria	PGPR	Effect	References
Xanthomonas compestris pv. *compestris*	*B. cereus, B. lentimorbus, B. pumilus*	Incidence and severity of black rot of cabbage were reduced when antagonists were applied.	Massomo et al. (2004)
R. solanacearum	*Bacillis* and *Pseudomonas* isolates	Out of 120, six isolates (PFMRI, BS-DFS, PF9, PF20, BC, and BS-wly) having antagonistic activity against the bacterial wilt on potato in vitro condition.	Aliye et al. (2008)

Table 4 Effects of PGPR on viral diseases of plants

PGPR	Viruses	Effects	References
Tobacco mosaic virus	*B. uniflagellatus*	Cultures and extracts from cultures reduced numbers of lesions from TMV.	Mann (1969)
Tobacco necrosis virus	*P. fluorescens* CHA0	Reduction in TNV leaf necrosis in *P. fluorescens* treated tobacco plants.	Maurhofer et al. (1994a)
Cucumber mosaic virus	*P. fluorescens, Serratia marcescens*	Treatment of cucumber or tomato plants with PGPR induced systemic resistance against CMV.	Raupach et al. (1996)
Tomato mottle virus	*B. amyloliquefaciens, B. subtilis, B. pumilus*	Disease severity ratings were significantly less in all PGPR powder based treatments.	Murphy and Zehnder (2000)
Cucumber mosaic cucumo virus (CCMV)	*B. amyloliquefaciens, B. subtilis, B. pumilus*	PGPR mediated ISR occurred against CCMV following mechanical inoculation on tomato.	Zehnder et al. (2000)
Pepper mild mottle virus (PMMoV)	*Bacillus amyloliquefaciens*	*Bacillus* induced systemic resistance against PMMoV in tobacco via salicyclic acid and jasmonic acid dependent pathways.	Ahn et al. (2002)

weight iron-binding compounds (400–1,000 daltons) known as siderophores. Siderophores bind FeIII with a very high affinity (Whipps 2001). The bacterium that originally synthesized the siderophores takes up the iron siderophore complex by using a receptor that is specific to the complex and is located in the outer cell membrane of the bacterium. Once inside the cell, the iron is released and is then available to support the microbial growth. PGPR can prevent the proliferation of fungal and other pathogens by producing siderophores that bind most of the FeIII in the area around the plant root. The resulting lack of iron prevents pathogens from proliferating in this immediate vicinity (Loper and Henkels 1999; Siddiqui 2006). The siderophores synthesis in bacteria is generally regulated by iron sensitive fur proteins, global regulators (GasS and GasA), the sigma factors (RpoS, PvdS, and Fpv1), quorum sensitive autoinducers (*N*-acyl homoserine lactone), and many site-specific recombinase (Cornelis and Matthijis 2002; Ravel and Cornelis 2003; Compant et al. 2005). A myriad of environmental factors can also modulate the siderophore synthesis, pH, iron level and forms of iron ions, presence of trace elements, and an adequate supply of C, N, and P (Duffy and Defago 1999). Microbial siderophores vary widely in overall structure but most contain hydroxamate and catechol groups, which are involved in chelating the ferric ion (Neilands 1995).

Suppression of soil borne plant pathogens by siderophore producing pseudomonads was observed (Bakker et al. 1987; Becker and Cook 1988; Loper 1988), and the wild type siderophore producing strain was more effective in suppressing disease compared with non-siderophore-producing mutants. Siderophore production is an important feature for the suppression of plant pathogens and promotion of plant growth. In another study, siderophore producing mutant *P. putida* was most effective than the wild type in suppression of Fusarium wilt of tomato (Vandenburgh and Gonzalez 1984), while a siderophore-deficient mutant of *P. aeruginosa* lost its biocontrol activity (Buysens et al. 1994). Fluorescent siderophores production was observed as a mechanism of biocontrol of bacterial wilt disease in the fluorescent pseudomonads RBL 101 and RSI 125 (Jagadeesh et al. 2001), while Akhtar and Siddiqui (2009) reported that siderophore producing Pseudomonads strains significantly reduced the root-rot disease in chickpea. Press et al. (2001) reported the catechol siderophore biosynthesis gene in *Serratia marcescens* 90–166 associated with induced resistance in cucumber against anthracnose, while *P. fluorescens* inhibited the growth of *Fusarium culmorum* in vitro (Kurek and Jaroszuk-Scisel 2003).

The capacity to utilize siderophores is important for the growth of bacteria in the rhizosphere (Jurkevitch et al. 1992) and on the plant surface (Loper and Buyer 1991). Specific siderophore producing *Pseudomonas* strains rapidly colonized roots of several crops and resulted in increased yield (Schroth and Hancock 1982). Enhanced plant growth caused by pseudomonad strains was often accompanied by the reduction in pathogen populations on the roots. There is convincing evidence to support a direct role of siderophore mediated iron competition in the biocontrol activity exhibited by such isolates (Leong 1986; Loper and Buyer 1991). The antagonism depends on the amount of iron available in the medium; siderophores produced by a biocontrol agent and sensitivity of target pathogens (Kloepper et al. 1980; Weger et al. 1988). Production of ALS 84 and siderophores contributed to the biocontrol of crown gall by *Agrobacterium rhizogenes* K84 especially under conditions of iron limitation (Penyalver et al. 2001).

Iron nutrition of the plant influences the rhizosphere microbial community structure (Yang and Crowley 2000), and the role of the pyoverdine siderophores produced by many *Pseudomonas* species has been clearly demonstrated in the control of *Pythium* and *Fusarium* species (Loper and Buyer 1991; Duijff et al. 1993). Pseudomonads also produce two other siderophores, pyochelin and its precursor salicylic acid. Pyochelin is thought to contribute to the protection of tomato plants from *Pythium* by *P. aeruginosa* 7NSK2 (Buysens et al. 1996). Different environmental factors can also influence the quantity of siderophores produced (Duffy and Defago 1999).

4 Interactions of PGPR with Plants

Inoculation with PGPR imparts resistance in various plant species against a variety of pathogens including bacteria, viruses, and fungi. And apart from inducing certain morphological changes in the plant itself, it also generates accumulation of phenolics and increases the levels of certain enzymes.

4.1 Induced Resistance

Use of selected PGPR strains was shown to trigger a plant-mediated resistance in above ground plant parts (Van Peer et al. 1991; Wei et al. 1991). This type of resistance is often referred as ISR and has been demonstrated in many plant species including bean, carnation, cucumber, radish, tobacco, tomato, and *Arabidopsis thaliana* (van Loon et al. 1998). Rhizobacteria-mediated ISR resembles phenotypically with classic pathogen induced resistance, in which noninfected parts of a previously pathogen infected plant become more resistant to further infection. This form of resistance is referred as systemic acquired resistance (SAR) (Ross 1961). The difference between ISR and SAR is that ISR is induced by nonpathogenic rhizobacteria, while SAR is induced systemically after inoculation with necrotizing pathogens. Moreover, ISR is independent of salicyclic acid but involves jasmonic acid and ethylene signaling, while SAR requires salicyclic acid as a signaling molecule in plants. ISR is accompanied by the expression of sets of genes distinct from the PR genes whereas SAR is accompanied by the induction of pathogenesis related proteins. Both ISR and SAR are effective against a broad spectrum of plant pathogens (Kuc 1982; van Loon et al. 1998).

The effectiveness of ISR and SAR to a range of viral, bacterial, fungal, and oomycete pathogens was tested on *Arabidopsis*. *Arabidopsis thaliana* L. has many features favoring its use as a model in studies of PGPR (O-Callaghan et al. 2000). In this model system, the nonpathogenic rhizobacterial strain *P. fluorescens* WCS417r was used as the inducing agent (Pieterse et al. 1996) to trigger ISR in several plant species (Van Peer et al. 1991; Leeman et al. 1995; Duijff et al. 1998; Bigirimana and Hofte 2002). Colonization of *Arabidopsis* roots by *P. fluorescens* WCS417r protected the plants against different plant pathogens (Pieterse et al. 1996; Van Wees et al. 1997; Ton et al. 2002). Protection against different pathogens was expressed both in reduction in disease symptoms and inhibition of pathogen growth. Since rhizobacteria were spatially separated from pathogens, the mode of disease suppression in the plants is through ISR. The ability to develop ISR appears to depend on the host/rhizobacterium combination (Pieterse et al. 2002) and suggests that specific recognition between the plant and the ISR-inducing rhizobacterium is required for the induction of ISR. Several bacterial components as potential inducers of ISR are involved including outer membrane lipopolysaccharides and iron regulated siderophores (Leeman et al. 1995; van Loon et al. 1998).

Changes that have been observed in plant roots exhibiting ISR include the following: (1) strengthening of epidermal and cortical cell walls and deposition of newly formed barriers beyond infection sites including callose, lignin, and phenolics (Benhamou et al. 1996a, b, c, 2000; Duijff et al. 1997; Jetiyanan et al. 1997; M'Piga et al. 1997); (2) increased levels of enzymes such as chitinase, peroxidase, polyphenol oxidase, and phenylalanine ammonia lyase (M'Piga et al. 1997; Chen et al. 2000); (3) enhanced phytoalexin production (Van Peer et al. 1991; Ongena et al. 1999); and (4) enhanced expression of stress-related genes (Timmusk and Wagner 1999). However, not all of these biochemical changes are found in all

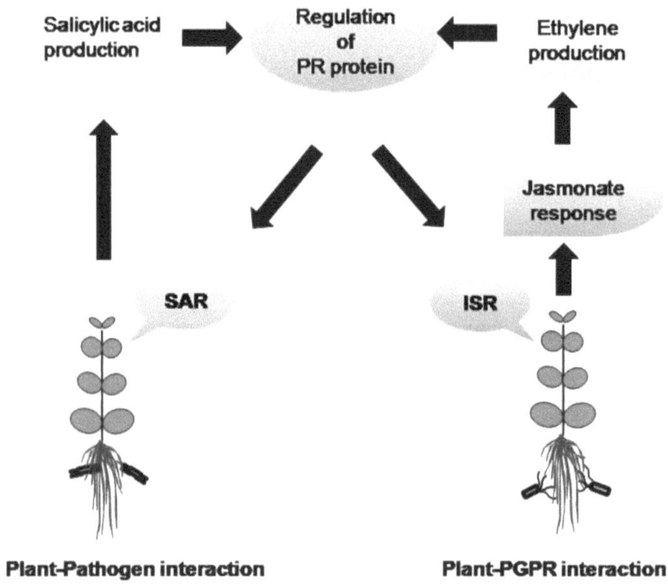

Fig. 2 Signaling pathway in plants responsible for the disease resistance in plants

bacterial–plant combinations (Steijl et al. 1999). Protection from diseases by biocontrol and its consistency in the field are generally not sufficient to compete with conventional methods of disease control. Combined use of antagonistic microorganisms with different mechanisms of action may improve efficacy and consistency of biocontrol agents (De Boer et al. 1999). Moreover, combination of ISR and SAR that results in an enhanced level of protection against specific bacterial pathogens (Van Wees et al. 2000) offers great potential to integrate both forms of induced resistance in agricultural practices. Induced resistance appears to be more useful for the management of viral diseases of plants where other management strategies are not generally successful (Fig. 2).

4.2 Root Colonization

Rhizosphere colonization is important not only as the first step in pathogenesis of soil borne microorganisms but also is crucial in the application of microorganisms for beneficial purposes (Lugtenberg et al. 2001). PGPR generally improves plant growth by colonizing the root system and pre-empting the establishment of, or suppressing deleterious rhizosphere microorganisms (Schroth and Hancock 1982). PGPR must be able to compete with the indigenous microorganisms and efficiently colonize the rhizosphere of the plants to be protected. Colonization is widely believed to be essential for biocontrol (Weller 1983; Parke 1991), and a biocontrol agent should grow and colonize the root surface. The ineffectiveness of PGPR in the field has often been attributed to their inability to colonize plant roots (Benizri et al. 2001; Bloemberg and Lugtenberg 2001). Colonization or even initial population size of the biocontrol agent has been significantly correlated with disease suppression (Parke 1990; Bull et al. 1991).

Cell surface characteristics influence the attachment of bacteria to roots, which may be necessary for colonization (Vesper 1987; Anderson et al. 1988). Certain mutants that affect accumulation of secondary metabolites also influence colonization of roots in the field (Mazzola et al. 1992; Carroll et al. 1995). Analysis of mutants indicates that prototrophy for amino acids and vitamin B1, rapid growth rate, utilization of organic acids and lipopolysaccharide properties contribute to colonization (Lugtenberg et al. 1996). A variety of bacterial traits and specific genes contribute to colonization but only few have been identified (Benizri et al. 2001; Lugtenberg et al. 2001). These include motility, chemotaxis to seed and use specific components of root exudates, production of pili or fimbriae, production of specific cell surface components, ability of protein secretion, and quorum sensing (Lugtenberg et al. 2001). Competition of introduced bacteria with indigenous microorganisms already present in the soil and rhizosphere of the developing plant is another important aspect for root colonization.

4.3 Genetic Variations in the Host

Plants vary in their ability to support and respond to beneficial microorganisms (Handelsman and Stabb 1996). The ability to support certain biocontrol agents varies among plant species and among cultivars. Some plants appear to attract and support biocontrol agents, which are antagonistic to pathogens (Neal et al. 1973; Azad et al. 1985). Legumes vary in their response to *P. polymyxa* (Chanway et al. 1988), and *Bacillus* isolates from wheat roots enhanced growth of wheat in a cultivar-specific manner (Chanway et al. 1988). Plant species vary in their ability to induce genes for pyoluteorin biosynthesis in *P. fluorescencs* (Kraus and Loper 1995) probably because of variation in composition of root exudates among species. Moreover, different cultivars vary in terms of survival or disease incidence in the presence of a pathogen and biocontrol agent (Liu et al. 1995; King and Parke 1996). Strains of *P. fluorescens* that overproduce pyoluteorin and 2,4-diacetyl-phloroglucinol provide superior disease suppression compared with the parent strain in some host–pathogen combinations and not others, and effect correlate with host, and not pathogen, besides sensitivity to antibiotics (Maurhofer et al. 1995).

5 Interactions of PGPR in the Rhizosphere

Soil being a sink of microorganisms, therefore, influences the ability of introduced PGPR strain to interact with microbial community comprising beneficial and deleterious rhizospheric microorganisms.

5.1 Interactions with the Microbial Community

Many biocontrol agents suppress disease effectively in the laboratory but fail to do so in the field. These biocontrol agents may be affected by indigenous soil microbial communities and may also influence the community into which they are introduced. Certain fluorescent pseudomonads displace resident microflora in some cases reducing populations of deleterious microorganisms (Yuen and Schroth 1986). Manipulation of introduced PGPR populations may lead to enhanced suppression of other soil borne plant pathogens. Limited induced soil suppressiveness can also be achieved through stimulation in microbial community structure and function by several cultural practices (Kloepper et al. 1999). This may include the application of organic manures and plant straw (Siddiqui and Mahmood 2003; Siddiqui 2004; Siddiqui and Akhtar 2008c), inclusion of antagonistic plants in cropping systems and other integrated pest management approaches.

5.2 Interactions of PGPR Strains

In general, a single biocontrol agent is used for biocontrol of plant disease against a single pathogen (Wilson and Backman 1999). On the one hand, this may sometimes account for the inconsistent performance by the biocontrol agent, because a single agent is not active in all soil environments or against all pathogens that attack the host plant. On the other hand, mixtures of biocontrol agents with different plant colonization patterns may be useful for the biocontrol of different plant pathogens via different mechanisms of disease suppression. Moreover, mixtures of biocontrol agents with taxonomically different organisms that require different optimum temperature, pH, and moisture conditions may colonize roots more aggressively, improve plant growth and efficacy of biocontrol. Naturally occurring biocontrol results from mixtures of biocontrol agents rather than from high populations of a single organism. The greater suppression and enhanced consistency against multiple cucumber pathogens was observed using strain mixtures of PGPR (Raupach and Kloepper 1998).

Incompatibility of the coinoculants may sometimes arise and thus inhibit each other as well as the target pathogens (Leeman et al. 1996). Thus, an important prerequisite for successful development of strain mixtures appears to be the compatibility of the coinoculated microorganisms (Baker 1990; De Boer et al. 1997). A biocontrol product composed of a mixture of strains is more costly than a product composed of single strain due to increased costs of production and registration of such product. However, greater emphasis on the development of mixtures of biocontrol agents is needed, because they may better adapt to the environmental changes that occur throughout the growing season and protect against a broader range of pathogens. Mixtures of microorganisms may increase the genetic diversity of biocontrol systems that persist longer in the rhizosphere and utilize a wider array of biocontrol mechanisms (Pierson and Weller 1994). Multiple organisms may enhance the level and consistency of biocontrol by a more stable rhizosphere community and effectiveness over a wide range of environmental conditions. In particular, combination of fungi and bacteria may provide protection at different times, under different conditions, and occupy different or complementary niches.

6 A Practical Control System Using PGPR

Selection of effective strains of bacteria is of prime importance for the biocontrol of plant pathogens. Isolation of bacteria from pathogen suppressive soils may increase the chances of finding effective strains (Cook and Baker 1983). The suppressive soil becomes apparent where the severity or incidence of disease is lower than the expected when compared with that in the surrounding soil (Cook and Baker 1983). To obtain effective strains, the isolation of bacteria should be conducted from the same environment in which they will be used (Weller et al. 1985). The ability to colonize roots and resistance against antibiotics are other parameters necessary to screen the effective strains (Siddiqui et al. 2005). Screening of biocontrol agents by a seedling bioassay chamber is required to determine the compatibility of an antagonist with the microflora of a field soil (Randhawa and Schaad 1985). Selection of field-effective strains can also be facilitated by a greenhouse assay. The important considerations in the development of the assays in the greenhouse are the inoculum potential of the pathogen (Weller et al. 1985), environmental conditions, and dose of the bacterium (Xu and Gross 1986). Many factors such as temperature, soil moisture, and soil texture influence the survival and establishment of bacteria. Formulation and application methods are often of paramount importance in effecting biocontrol (Papavizas and Lumsden 1980).

PGPR have great potential in the biocontrol of plant pathogens but the use of these rhizobacteria by farmers in the field is still lacking. The most obvious reasons for the limited use are the limited numbers of PGPR formulations available and inconsistent performance of these formulations. Mixtures of different strains are required to overcome inconsistency in the biocontrol performance. These mixtures of rhizobacteria may be used as seed treatment, which may be useful in reducing the quantity of bacterial inoculum required. This will facilitate systemic spread of the bacterial inoculum along the surface of the developing root system, and their antagonistic activity on the root surface during the early root infection by the pathogens. Rhizobacteria suspensions or formulations can also be mixed with organic manures in large vessels. They can be stored at 30–35°C for 5–10 days, mixing each day with water to keep them moist (Siddiqui and Mahmood 1999). Within 10 days, bacteria will attain high populations and this organic manure can be used at planting or after planting for the biocontrol of plant pathogens and better plant growth in the field.

7 Conclusion

Revelations about the mechanisms of PGPR action open new door to design strategies for improving the efficacy of biocontrol agents (Wang et al. 2000; Morrissey et al. 2004). Numerous studies have indicated that PGPR have great potential in the biocontrol of plant pathogens but most of the studies have been conducted in sterilized soil and in pots. There is an urgent need to conduct studies under field conditions. Colonization of root by PGPR is also important to increase their potential as biocontrol agents. Studies on the physical and chemical factors of soil, which affect root colonization, are needed. Moreover, use of mixture of effective strains of PGPR is advisable compared with use of single strain. The application of organic amendments with effective strains of PGPR is recommended because organic materials encourage the growth of organisms that compete with or destroy pathogens (Siddiqui and Mahmood 1999; Siddiqui and Akhtar 2008a, c). PGPR may also be used with fungal biocontrol agents and with arbuscular mycorrhizal fungi for greater beneficial effects. The absence of commercial interest in the biocontrol of plant pathogens by PGPR is also a major obstacle to progress. It is hoped that the future will see greater use of PGPR for the biocontrol of plant pathogens.

References

Aalten PM, Vitour D, Blanvillain D, Gowen SR, Sutra L (1998) Effect of rhizosphere fluorescent *Pseudomonas* strains on plant parasitic nematodes *Radopholus similis* and *Meloidogyne* spp. Lett Appl Microbiol 27:357–361

Ahmadzadeh M, Tehrani AS (2009) Evaluation of fluorescent pseudomonads for plant growth promotion, antifungal activity against *Rhizoctonia solani* on common bean, and biocontrol potential. Biol Control 48:101–107

Ahn IP, Park K, Kim CH (2002) Rhizobacteria-induced resistance perturbs viral disease progress and triggers defense related gene expression. Mol Cells 13:302–308

Akhtar MS, Siddiqui ZA (2007) Biocontrol of a chickpea root-rot disease complex with *Glomus intraradices*, *Pseudomonas putida* and *Paenibacillus polymyxa*. Austral Plant Pathol 36:175–180

Akhtar MS, Siddiqui ZA (2008a) Biocontrol of a root-rot disease complex of chickpea by *Glomus intraradices*, *Rhizobium* sp. and *Pseudomonas straita*. Crop Prot 23:410–417

Akhtar MS, Siddiqui ZA (2008b) *Glomus intraradices*, *Pseudomonas alcaligenes*, *Bacillus pumilus* as effective biocontrol agents for the root-rot disease complex of chickpea (*Cicer arietinum* L.). J General Plant Pathol 74:53–60

Akhtar MS, Siddiqui ZA (2009) Use of plant growth promoting rhizobacteria for the biocontrol of root-rot disease complex of chickpea. Austral Plant Pathol 38:44–50

Aksoy HM, Mennan S (2004) Biological control of *Heterodera cruciferae* (Tylenchida: Heteroderidae) Franklin 1945 with fluorescent *Pseudomonas* spp. J Phytopathol 152:514–518

Aliye N, Fininsa C, Hiskias Y (2008) Evaluation of rhizosphere bacterial antagonists for their potential to bioprotect potato (*Solanum tuberosum*) against bacterial wilt (*Ralstonia solanacearum*). Biol Control 47:282–288

Amer GA, Utkhede RS (2000) Development of formulations of biological agents for management of root rots of lettuce and cucumber. Can J Microbiol 46:809–816

Anderson AJ, Hablbzadegah-Tarl P, Tepper CS (1988) Molecular studies on the role of a root surface agglutinin in adherence and colonization by *Pseudomonas putida*. Appl Environ Microbiol 54:375–380

Anjaiah V, Cornelis P, Koedam N (2003) Effect of genotype and root colonization in biological control of Fusarium wilts in pigeon pea and chickpea by *Pseudomonas aeruginosa* PNA1. Can J Microbiol 49:85–91

Asaka O, Shoda M (1996) Biocontrol of *Rhizoctonia solani* damping off of tomato with *Bacillus subtilis* RB14. Appl Environ Microbiol 62:4081–4085

Azad HR, Davis JR, Kado CI (1985) Relationships between rhizoplane and rhizosphere bacteria and Verticillium wilt resistance in potato. Arch Microbiol 140:347–351

Baker R (1990) An overview of current and future strategies and models for biological control. In: Hornby D (ed) Biological control of soil-borne plant pathogens. C.A.B International, Wallingford, UK, pp 375–388

Bakker PAHM, Bakker AW, Marugg JD, Weisbeek PJ, Schippers B (1987) Bioassay for studying the role of siderophores in potato growth stimulation by *Pseudomonas* spp. in short potato rotations. Soil Biol Biochem 19:443–449

Bakker PAHM, Ran LX, Pieterse CMJ, van Loon LC (2003) Understanding the involvement of rhizobacteria-mediated induction of systemic resistance in biocontrol of plant diseases. Can J Plant Pathol 25:5–9

Bangera MG, Thomashow LS (1996) Characterization of a genomic locus required for synthesis of the antibiotic 2, 4-diacetylphloroglucinol by the biological control agent *Pseudomonas fluorescens* Q2-87. Mol Plant-Microbe Interact 9:83–90

Barka EA, Gognies S, Nowak J, Audran J, Belarbi A (2002) Inhibitory effect of endophyte bacteria on *Botrytis cineria* and its influence to promote the grapevine growth. Biol Control 24:135–142

Basha S, Ulaganathan K (2002) Antagonism of *Bacillus* species (strain BC121) towards *Curvularia lunata*. Curr Sci 82:1457–1463

Bashan Y, Bashan LE (2002) Protection of tomato seedlings against infection by *Pseudomonas syringae* pv. *tomato* by using the plant growth-promoting bacterium *Azospirillum brasilense*. Appl Environ Microbiol 68:2637–2643

Baysal O, Caliskan M, Yesilova O (2009) An inhibitory effect of a new *Bacillus subtilis* strain (EU07) against *Fusarium oxysporum* f. sp. *radicis-lycopersici*. Physiol Mol Plant Pathol 73:25–32

Becker JO, Cook RJ (1988) Role of siderophores in suppression of *Pythium* species and production of increased-growth response of wheat by fluorescent pseudomonads. Phytopathology 78:778–782

Becker JO, Zavaleta-Mejia E, Colbert SF, Schroth MN, Weinhold AR, Hancock JG, VanGundy SD (1988) Effects of rhizobacteria on root-knot nematodes and gall formation. Phytopathology 78:1466–1469

Begum MZ, Kumar MS (2005) Management of disease complex involving *Heterodera cajani* Koshy, 1967 and *Macrophomina phaseolina* (Tassi) Goid on greengram (*Vigna radiata* L. Wilczek). Indian J Nematol 35:192–194

Benhamou N, Bélanger RR, Paulitz TC (1996a) Pre-inoculation of Ri T-DNA-transformed pea roots with *Pseudomonas fluorescens* inhibits colonization by *Pythium ultimum* Trow: an ultra structural and cytochemical study. Planta 199:105–117

Benhamou N, Belanger RR, Paulitz TC (1996b) Induction of differential host responses by *Pseudomonas fluorescens* in Ri T-DNA-transformed pea roots after challenge with *Fusarium oxysporum* f. sp. *pisi* and *Pythium ultimum*. Phytopathology 86:1174–1185

Benhamou N, Gagne S, Quere DL, Dehbi L (2000) Bacterial-mediated induced resistance in cucumber: beneficial effect of the endophytic bacterium *Serratia plymuthica* on the protection against infection by *Pythium ultimum*. Phytopathology 90:45–56

Benhamou N, Kloepper JW, Quadt-Hallman A, Tuzun S (1996c) Induction of defense-related ultra structural modifications in pea root tissues inoculated with endophytic bacteria. Plant Physiol 112:919–929

Benizri E, Baudoin E, Guckert A (2001) Root colonization by inoculated plant growth promoting rhizobacteria. Biocont Sci Technol 11:557–574

Berta G, Sampo S, Gamalero E, Musasa N, Lemanceau P (2005) Suppression of Rhizoctonia root-rot of tomato by *Glomus mosseae* BEG 12 and *Pseudomonas fluorescens* A6RI is associated with their effect on the pathogen growth and on the root morphogenesis. Eur J Plant Pathol 111:279–288

Bigirimana J, Hofte M (2002) Induction of systemic resistance to *Colletotrichum lindemuthianum* in bean by a benzothiadiazole derivative and rhizobacteria. Phytoparasitica 30:159–168

Bloemberg GV, Lugtenberg BJJ (2001) Molecular basis of plant growth promotion and biocontrol by rhizobacteria. Curr Opin Plant Biol 4:343–350

Boby VU, Bagyaraj DJ (2003) Biological control of root-rot of *Coleus forskohlii* Briq., using microbial inoculatnts. World J Microbiol Biotechnol 19:175–180

Bonsall RF, Weller DM, Thomashow LS (1997) Quantification of 2, 4-diacetylphloroglucinol produced by fluorescent Pseudomonas spp. in vitro and in the rhizosphere of wheat. Appl Environ Microbiol 63:951–955

Bora T, Ozaktan H, Gore E, Aslan E (2004) Biological control of *Fusarium oxysporum* f. sp. *melonis* by wettable powder formulations of two strains of *Pseudomonas putida*. J Phytopathol 152:471–475

Brisbane PG, Rovira AD (1988) Mechanisms of inhibition of *Gaeumannomyces graminis* var. *tritici* by fluorescent pseudomonads. Plant Pathol 37:104–111

Bull CT, Weller DM, Thomashow LS (1991) Relationship between root colonization and suppression of *Gaeumannomyces graminis* var. *tritici* by *Pseudomonas fluorescens* strain 2-79. Phytopathology 81:954–959

Buysens S, Heungens K, Poppe J, Hofte M (1996) Involvement of pyochelin and pyoverdin in suppression of *Pythium*-induced damping-off of tomato by *Pseudomonas aeruginosa* 7NSK2. Appl Environ Microbiol 62:865–871

Buysens S, Poppe J, Hofte M (1994) Role of siderophores in the plant stimulation and antagonism by *Pseudomonas aeruginosa* 7NSK2. In: Ryder MH, Stephens PM, Bowen GD (eds) Improving plant productivity with rhizosphere bacteria. Commonwealth Scientific and Industrial Research Organization, Adelaide, Australia, pp 139–141

Carroll H, Moenne-Loccoz Y, Dawling DN, Gara FO (1995) Mutational disruption of the biosynthesis genes coding for the antifungal metabolite 2, 4 diacetylphloroglucinol does not influence the ecological fitness of Pseudomonas fluorescens F113 in the rhizosphere of sugar beets. Appl Environ Microbiol 61:3002–3007

Chanway CP, Nelson LM, Holl FB (1988) Cultivar-specific growth promotion of spring wheat (*Triticum aestivum* L.) by co-existent *Bacillus* species. Can J Microbiol 34:925–929

Chen C, Belanger RR, Benhamou N, Paulitz TC (1999) Role of salicyclic acid in systemic resistance induced by *Pseudomonas* spp. against *Pythium aphanidermatum* in cucumber roots. Eur J Plant Pathol 105:477–486

Chen C, Bélanger RR, Benhamou N, Paulitz TC (2000) Defense enzymes induced in cucumber roots by treatment with plant growth-promoting rhizobacteria (PGPR) and *Pythium aphanidermatum*. Physiol Mol Plant Pathol 56:13–23

Chernin L, Ismailov Z, Haran S, Chet I (1995) Chitinolytic *Enterobacter agglomerans* antagonistic to fungal plant pathogens. Appl Environ Microbiol 61:1720–1726

Chernin LS, de la Fuente L, Sobolev V, Haran S, Vorgias CE, Oppenheim AB, Chet I (1997) Molecular cloning, structural analysis and expression in *Escherichia coli* of a chitinase gene from *Enterobacter agglomerans*. Appl Environ Microbiol 63:834–839

Chin-A-Woeng TFC, Bloemberg GV, van der Bij AJ (1998) Biocontrol by phenazine-1-carboxamide-producing *Pseudomonas chlororaphis* PCL 1391 of tomato root rot caused by *Fusarium oxysporum* f. sp. *radicis-lycopersici*. Mol Plant-Microbe Interact 11:1069–1077

Compant S, Duffy B, Nowak J, Clement C, Barka EA (2005) Use of plant growth promoting bacteria for biocontrol of plant diseases: principles, mechanisms of action and future prospects. Appl Environ Microbiol 71:4951–4959

Cook RJ, Baker KF (1983) The nature and practice of biological control of plant pathogens. APS, St. Paul, MN, p 539
Cook RJ, Rovira AD (1976) Role of bacteria in the biological control of *Gaeumannomyces graminis* by suppressive soils. Soil Biol Biochem 8:269–273
Cornelis P, Matthijis S (2002) Diversity of siderophore-mediated iron uptake systems in fluorescent pseudomonads: not only pyoverdines. Environ Microbiol 4:787–798
Dasgupta SM, Khan N, Nautiyal CS (2006) Biologic control ability of plant growth promoting *Paenibacillus lentimorbus* NRRL B-30488 isolated from milk. Curr Microbiol 53:502–505
De Boer M, van der Sluis I, van Loon LC, Bakker PAHM (1997) In vitro compatibility between fluorescent *Pseudomonas* spp strains can increase effectivity of *Fusarium* wilt control by combinations of these strains. In: Ogoshi A, Kobayashi K, Homma Y, Kodama F, Kondo N, Akino S (eds) Plant growth-promoting rhizobacteria - present status and future prospects. Fourth Proceedings international workshop on plant growth-promoting rhizobacteria. Nakanishi Printing, Sappora, Japan, pp 380–382
De Boer M, Van der Sluis I, vanLoon LC, Bakker PAHM (1999) Combining fluorescent *Pseudomonas* spp. strains to enhance suppression of *Fusarium* wilt of radish. Eur J Plant Pathol 105:201–210
De Freitas JR, Germida JJ (1990) Plant growth promoting rhizobacteria for winter wheat. Can J Microbiol 36:265–272
De Freitas JR, Germida JJ (1991) *Pseudomonas cepacia* and *Pseudomonas putida* as winter wheat inoculants for biocontrol of *Rhizoctonia solani*. Can J Microbiol 37:780–784
de Souza JT, de Boer M, de Waard P, van Beek TA, Raaijmakers JM (2003) Biochemical, genetic, and zoosporicidal properties of cyclic lipopeptide surfactants produced by *Pseudomonas fluorescens*. Appl Environ Microbiol 69:7161–7172
Defago G (1993) 2, 4-Diacetylphloroglucinol, a promising compound in biocontrol. Plant Pathol 42:311–312
Devi SL, Dutta U (2002) Effect of *Pseudomonas fluorescens* on root-knot nematode (*Meloidogyne incognita*) of okra plant. Indian J Nematol 32:215–216
Duffy BK, Defago G (1999) Environmental factors modulating antibiotic and siderophore biosynthesis by *Pseudomonas fluorescens* biocontrol strains. Appl Environ Microbiol 65:2429–2438
Duijff BJ, Gianinazzi-Pearson V, Lemanceau P (1997) Involvement of the outer membrane lipopolysaccharides in the endophytic colonization of tomato roots by biocontrol *Pseudomonas fluorescens* strain WCS417r. New Phytol 135:325–334
Duijff BJ, Meijer JW, Bakker PAHM, Schippers B (1993) Siderophore-mediated competition for iron and induced resistance in the suppression of *Fusarium* wilt of carnation by fluorescent *Pseudomonas* spp. Neth J Plant Pathol 99:277–289
Duijff BJ, Pouhair D, Olivian C, Alabouvette C, Lemanceau P (1998) Implication of systemic induced resistance in the suppression of fusarium wilt of tomato by *Pseudomonas fluorescens*WCS417r and by nonpathogenic *Fusarium oxysporum* Fo47. Eur J Plant Pathol 104:903–910
Dunne C, Crowley JJ, Moënne-Loccoz Y, Dowling DN, de Bruijn FJ, O'Gara F (1997) Biological control of *Pythium ultimum* by *Stenotrophomonas maltophilia* W81 is mediated by an extracellular proteolytic activity. Microbiol 143:3921–3931
El-Tarabily KA, Sykes ML, Kurtböke ID, Hardy Gest J, Barbosa AM, Dekker RFH (1996) Synergistic effects of a cellulase-producing *Micromonospora carbonacea* and an antibiotic-producing *Streptomyces violascens* on the suppression of *Phytophthora cinnamomi* root rot of *Banksia grandis*. Can J Bot 74:618–624
Emma A (1990) Flourescent *Pseudomonas* isolate E11.3 a biocontrol agent for *Pythium* root-rot in tulips. Eur J Plant Pathol 96:261–272
Enebak SA, Carey WA (2000) Evidence for induced systemic protection to *Fusarium* rust in Loblolly pine by plant growth promoting rhizosphere. Plant Dis 84:306–308
Estevez de Jansen C, Percicha JA, Grahamb PH (2002) Integrated management strategies of bean root rot with *Bacillus subtilis* and *Rhizobium* in Minnesota. Field Crop Res 74:107–115

Gamliel A, Katan J (1993) Suppression of major and minor pathogens by fluorescent pseudomonads in solarized and nonsolarized soils. Phytopathology 83:68–75

Gautam A, Siddiqui ZA, Mahmood I (1995) Integrated management of *Meloidogyne incognita* on tomato. Nematol Medit 23:245–247

Glick B (1995) The enhancement of plant growth by free-living bacteria. Can J Microbiol 41:109–117

Goel AK, Sindhu SS, Dadarwal KR (2002) Stimulation of nodulation and plant growth of chickpea (*Cicer arietinum* L.) by *Pseudomonas* spp. antagonistic to fungal pathogens. Biol Fertil Soils 36:391–396

Gokte N, Swarup G (1988) On the potential of some bacterial biocides against root-knot and cyst nematodes. Indian J Nematol 18:152–153

Guo J, Ying Qi H, Guo Y, Ge H, Gong L, Zhang L, Sun P (2004) Biocontrol of tomato wilt by plant growth promoting rhizobacteria. Biol Control 29:66–72

Gupta CP, Dubey RC, Maheshwari DK (2002) Plant growth enhancement and suppression of *Macrophomina phaseolina* causing charcoal rot of peanut by *fluorescent Pseudomonas*. Biol Fertil Soils 35:399–405

Hallmann J, Quadt-Hallmann A, Mahafee WF, Kloepper JW (1997) Bacterial endophytes in agricultural crops. Can J Microbiol 43:895–914

Hamid M, Siddiqui IA, Siddiqui SS (2003) Improvement of *Pseudomonas fluorescens* CHA0 biocontrol activity against root-knot nematode by the addition of ammonium molybdate. Letters Appl Microbiol 36:239–244

Hammer PE, Hill DS, Lam ST, van Pée KH, Ligon JM (1997) Four genes from *Pseudomonas fluorescens* that encode the biosynthesis of pyrrolnitrin. Appl Environ Microbiol 63:2147–2154

Handelsman J, Stabb EV (1996) Biocontrol of soil borne plant pathogens. Plant Cell 8:1855–1869

Harrison LA, Letendre L, Kovacevich P, Pierson E, Weller DM (1993) Purification of an antibiotic effective against *Gaeumannomyces graminis* var. *tritici* produced by a biocontrol agent *Pseudomonas aureofaciens*. Soil Biol Biochem 25:215–221

Hasky-Günther K, Hoffmann-Hergarten S, Sikora RA (1998) Resistance against the potato cyst nematode *Globodera pallida* systemically induced by the rhizobacteria *Agrobacterium radiobacter* (G12) and *Bacillus sphaericus* (B43). Fundam Appl Nematol 21:511–517

Hawes MC (1991) Living plant cells released from the root cap: a regulator of microbial populations in the rhizosphere? In: Keister DL, Cregan PB (eds) The rhizosphere and plant growth. Kluwer Academic Publishers, Boston, MA, pp 51–59

Hoffland E, Hakulinen J, van Pelt JA (1996) Comparison of systemic resistance induced by avirulent and nonpathogenic *Pseudomonas* species. Phytopathology 86:757–762

Hongyou Z, Hailei W, Xili L, Ye W, Liqun Z, Wenhua T (2005) Improving biocontrol activity by *Pseudomonas fluorescens* through chromosomal integration of 2, 4-diacetyl-phloroglucinol biosynthesis genes. Chinese Sci Bull 50:775–781

Ignoffo CM, Dropkin VH (1977) Deleterious effects of the thermo stable toxin of *Bacillus thuringiensis* on species of soil inhabiting, mycophagous and plant parasitic nematodes. J Krans Ent Soc 50:394–395

Jagadeesh KS, Kulkarni JH, Krisharaj PU (2001) Evaluation of role of fluorescent siderophore in the biological control of bacterial wilt in tomato using Tn*5* mutants of fluorescent *Pseudomonas* sp. Curr Sci 81:882–883

Jaizme-Vega MC, Rodriguez-Romero A, Nunez ABL (2006) Effect of the combined inoculation of arbuscular mycorrhizal fungi and plant growth promoting rhizobacteria on papaya (*Carica papaya* L.) infected with root-knot nematode *Meloidogyne incognita*. Fruits 61:151–162

Jetiyanan K, Tuzun S, Kloepper JW (1997) Lignification, peroxidase and superoxidase dismutases as early plant defense reactions associated with PGPR-mediated induced systemic resistance. In: Ogoshi A, Kobayashi K, Homma Y, Kodama F, Kondo N, Akino S (eds) Plant growth-promoting rhizobacteria: present status and future prospects. Sapporo, Japan, pp 265–268

Jurkevitch E, Hadar Y, Chen Y (1992) Differential siderophore utilization and iron uptake by soil and rhizosphere bacteria. Appl Environ Microbiol 58:119–124
Kang Y, Carlson R, Tharpe W, Schell MA (1998) Characterization of genes involved in biosynthesis of a novel antibiotic from *Burkholderia cepacia* BC11 and their role in biological control of *Rhizoctonia solani*. Appl Environ Microbiol 64:3939–3947
Keel C, Defago G (1997) Interactions between beneficial soil bacteria and root pathogens: mechanisms and ecological impact. In: Gange AC, Brown VK (eds) Multitrophic interactions in terrestrial system. Oxford, Blackwell Science, pp 27–47
Keel C, Schnider U, Maurhofer M, Voisard C, Laville J, Burger U, Wirthner P, Haas O, Defago G (1992) Suppression of root diseases by *Pseudomonas fluorescens* CHA0: importance of the bacterial secondary metabolite 2, 4-diacetylphloroglucinol. Mol Plant-Microbe Interact 5:4–13
Kermarrec A, Jacqua G, Anais J (1994) Effect of *Fusarium solani* and *Pseudomonas solanacearum* on the infestation of aubergine with the plant parasitic nematode, *Rotylenchulus reniformis*. Nematologica 40:152–154
Khan MR, Akram M (2000) Effects of certain antagonistic fungi and rhizobacteria on wilt disease complex of tomato caused by *Meloidogyne incognita* and *Fusarium oxysporum* f. sp. *lycopersici*. Nematol Medit 28:139–144
Khan MR, Khan SM (2001) Biomanagement of *Fusarium* wilt of tomato by the soil application of certain phosphate solubilizing microorganisms. Int Pest Manag 47:227–231
Khan MR, Kounsar K (2000) Effect of seed treatment with certain bacteria on the growth of mungbean and reproduction of *Meloidogyne incognita*. Nematol Medit 28:221–226
Khan Z, Kim SG, Jeon YH, Khan HU, Son SH, Kim YH (2008) A plant growth promoting rhizobacterium, *Paenibacillus polymyxa* strain GBR-1, suppresses root-knot nematode. Bioresource Technol 99:3016–3023
Kim BS, Moon SS, Hwang BK (1999) Isolation, identification and antifungal activity of a macrolide antibiotic, oligomycin A, produced by *Streptomyces libani*. Can J Bot 77:850–858
Kim DH, Misaghi IJ (1996) Biocontrol performance of two isolates of *Pseudomonas fluorescens* in modified soil atmosphere. Phytopathology 86:1238–1241
King EB, Parke JL (1996) Population density of the biocontrol agent *Burkholderia cepacia* AMMDRl on four pea cultivars. Soil Biol Biochem 28:307–312
Kloepper JW, Schroth MN (1978) Plant growth-promoting rhizobacteria in radish. In: Proc 4th International conf plant pathogenic bacteria. Gilbert-Clarey, Tours, France, pp 879–882
Kloepper JW, Leong J, Teintze M, Schroth MN (1980) Enhanced plant growth by plant growth promoting rhizobacteria. Nature 286:885–886
Kloepper JW, Rodriguez-Kabana R, Zehnder GW, Murphy JF, Sikora E, Fernadez C (1999) Plant root- bacterial interactions in biological control of soil borne diseases and potential extension to systemic and foliar diseases. Austral Plant Pathol 28:21–26
Kluepfel DA, McInnis TM, Zehr EA (1993) Involvement of root colonizing bacteria in peach orchard soils suppressive of the nematode *Criconemella xenoplax*. Phytopathology 83:1240–1245
Kraus J, Loper JE (1995) Characterization of a genomic region required for production of the antibiotic pyoluteorin by the biological control agent *Pseudomonas fluorescens* Pf-5. Appl Environ Microbiol 61:849–854
Kropp BR, Thomas E, Pounder JI, Anderson AJ (1996) Increased emergence of spring wheat after inoculation with *Pseudomonas chlororaphis* isolate 2E3 under field and laboratory conditions. Biol Fert Soils 23:200–206
Kuc J (1982) Induced immunity to plant disease. Bioscience 32:854–860
Kumar BSD, Berggren I, Martensson AM (2001) Potential for improving pea production by co-inoculation with fluorescent *Pseudomonas* and *Rhizobium*. Plant Soil 229:25–34
Kumar NR, Arasu VT, Gunasekaran P (2002) Genotyping of antifungal compounds producing plant growth-promoting rhizobacteria *Pseudomonas fluorescens*. Curr Sci 82:1463–1466
Kurek E, Jaroszuk-Scisel J (2003) Rye (*Secale cereale*) growth promotion by *Pseudomonas fluorescens* strains and their interactions with *Fusarium culmorum* under various soil conditions. Biol Control 26:48–56

Leeman M, Den Ouden FM, Van Pelt JA, Dirkx FPM, Steijl H, Bakker PAHM, Schippers B (1996) Iron availability affects induction of systemic resistance to *Fusarium* wilt of radish by *Pseudomonas fluorescens*. Phytopathology 86:149–155

Leeman M, Van Pelt JA, Den Ouden FM, Heinsbroek M, Bakker PAHM, Schippers B (1995) Induction of systemic resistance against *Fusarium* wilt of radish by lipopolysaccharides of *Pseudomonas fluorescens*. Phytopathology 85:1021–1027

Leong J (1986) Siderophores: their biochemistry and possible role in the biocontrol of plant pathogens. Annu Rev Phytopathol 24:187–209

Li B, Xie GL, Soad A, Coosemans J (2005) Suppression of *Meloidogyne javanica* by antagonistic and plant growth promoting rhizobacteria. J Zhejing Uni Sci 6:496–501

Lindow SE, McGourty G, Elkins R (1996) Interaction of antibiotics with *Pseudomonas fluorescens* strain A506 in the control of fire blight and frost injury to pear. Phytopathology 86:841–849

Liu L, Kloepper JW, Tuzun S (1995) Induction of systemic resistance in cucumber by plant growth-promoting rhizobacteria: duration of protection and effect of host resistance on protection and root colonization. Phytopathology 85:1064–1068

Loper JE (1988) Role of fluorescent siderophore production in biological control of *Pythium ultimum* by a *Pseudomonas fluorescens* strain. Phytopathology 78:166–172

Loper JE, Buyer LS (1991) Siderophores in microbial interactions on plant surfaces. Mol Plant-Microbe Interact 4:5–13

Loper JE, Henkels MD (1999) Utilization of heterologous siderophores enhances levels of iron available to *Pseudomonas putida* in the rhizosphere. Appl Environ Microbiol 65:5357–5363

Lugtenberg BJJ, Dekkers L, Bloemberg GV (2001) Molecular determinants of rhizosphere colonization by *Pseudomonas*. Annu Rev Phytopathol 39:461–490

Lugtenberg BJJ, van der Bij A, Bloemberg G, Chin-A-Woeng T, Dekkers L, Kravchenko L, Mulders I, Phoelich C, Simons M, Spaink H, Tikhonovich I, de Weger L, Wijffelman C (1996) Molecular basis of rhizosphere colonization by *Pseudomonas* bacteria. In: Stacey G, Mullin B, Gresshoff PM (eds) Biology of plant-microbe interactions. ISPMB, St. Paul, Minnesota, pp 433–440

M'Piga P, Belanger RR, Paulitz TC, Benhamou N (1997) Increased resistance to *Fusarium oxysporum* f. sp. *radicis-lycopersici* in tomato plants treated with the endophytic bacterium *Pseudomonas fluorescens* strain 63-28. Physiol Mol Plant Pathol 50:301–320

Maarten HR, Zhinong Y, Teri ET, Albert D, Rovira WT, Raymond LC (1998) Use of strains of *Bacillus* isolates in china to suppress take all and Rhizoctonia root rot and promote seedling growth of glasshouse grown wheat in Australia soils. Soil Biol Biochem 1:19–29

Manjula K, Podile AR (2001) Chitin-supplemented formulations improve biocontrol and plant growth promoting efficiency of *Bacillus subtilis* AF1. Can J Microbiol 47:618–625

Mann EW (1969) Inhibition of tobacco mosaic virus by a bacterial extract. Phytopathology 59:658–662

Massomo SMS, Mortensen CN, Mabagala RB, Newman MA, Hockenhul J (2004) Biological control of black rot (*Xanthomonas campestris* pv. *campestris*) of cabbage in Tanzania with *Bacillus* strains. J Phytopathol 152:98–102

Maurhofer M, Hase C, Meuwly P, Metraux JP, Defago G (1994a) Induction of systemic resistance of tobacco to tobacco necrosis virus by root colonizing *Pseudomonas fluorescens* strain CHA0: influence of the gacA gene and of pyoverdine production. Phytopathology 84:139–146

Maurhofer M, Keel C, Haas D, Defago G (1994b) Pyoluteorin production by *Pseudomonas fluorescens* strain CHA0 is involved in the suppression of *Pythium* damping-off of cress but not of cucumber. Eur J Plant Pathol 100:221–232

Maurhofer M, Keel C, Haas D, Defago G (1995) Influence of plant species on disease suppression by *Pseudomonas fluorescens* strain CHA0 with enhanced antibiotic production. Plant Pathol 44:40–50

Maurhofer M, Keel C, Schnider U, Voisard C, Haas D, Defago G (1992) Influence of enhanced antibiotic production in *Pseudomonas fluorescens* strain CHA0 on its disease suppressive capacity. Phytopathology 82:190–195

Mazzola M, Cook RJ, Thomashow LS, Weller DM, Pierson LS (1992) Contribution of phenazine antibiotic biosynthesis to the ecological competence of fluorescent pseudomonads in soil habitats. Appl Environ Microbiol 58:2616–2624

Milner JL, Silo-Suh L, Lee JC, He H, Clardy J, Handelsman J (1996) Production of kanosamine by *Bacillus cereus* UW85. Appl Environ Microbiol 62:3061–3065

Morrissey JP, Dow JM, Mark GL, O'Gara F (2004) Are microbes at the root of a solution to world food production? EMBO Rep 5:922–926

Moulin F, Lemanceau P, Alabouvette C (1996) Suppression of *Pythium* root-rot of cucumber by a Fluorescent *Pseudomonas* is related to reduced root colonization by *Pythium aphanidermatum*. J Phytopathol 144:125–129

Murphy JF, Zehnder GW (2000) Plant growth-promoting rhizobacterial mediated protection in tomato against tomato mottle virus. Plant Dis 84:779–784

Nakayama T, Homma Y, Hashidoko Y, Mizutani J, Tahara S (1999) Possible role of xanthobaccins produced by *Stenotrophomonas* sp. strain SB-K88 in suppression of sugar beet damping-off disease. Appl Environ Microbiol 65:4334–4339

Nandakumar R, Viswanathan R, Babu S, Shella J, Raghuchander T, Samiyappan R (2001) A new bio-formulation containing plant growth promoting rhizobacterial mixture for the management of sheath blight and enhanced grain yield in rice. Biocontrol 46:493–510

Neal JL Jr, Larson RI, Atkinson TG (1973) Changes in rhizosphere populations of selected physiological groups of bacteria related to substitution of specific pairs of chromosomes in spring wheat. Plant Soil 39:209–212

Neilands JB (1995) Siderophores: structure and function of microbial iron transport compounds. J Biol Chem 270:26723–26726

Nielsen TH, Sørensen D, Tobiasen C, Andersen JB, Christeophersen C, Givskov M, Sorensen J (2002) Antibiotic and biosurfactant properties of cyclic lipopeptides produced by fluorescent *Pseudomonas* spp. from the sugar beet rhizosphere. Appl Environ Microbiol 68:3416–3423

Niranjan Raj S, Chaluvaraju G, Amruthesh KN, Shetty HS, Reddy MS, Kloepper JW (2003) Induction of growth promotion and resistance against downy mildew on pearl millet (*Pennisetum glaucum*) by rhizobacteria. Plant Dis 87:340–345

Niranjan-Raj S, Shetty NP, Shetty HS (2004) Seed bio-priming with *Pseudomonas fluorescens* isolates enhances growth of pearl millet plants and induces resistance against downy mildew. Inter J Pest Manag 50:41–48

Nowak-Thompson B, Chaney N, Wing JS, Gould SJ, Loper JE (1999) Characterization of the pyoluteorin biosynthetic gene cluster of *Pseudomonas fluorescens* Pf-5. J Bacteriol 181:2166–2174

O-Callaghan KJ, Dixon RA, Cocking EC (2000) *Arabidopsis thaliana*: a model for studies of colonization by non pathogenic and plant growth promoting bacteria. Aust J Plant Physiol 28:975–982

Oka Y, Chet I, Spiege IY (1993) Control of root-knot nematode *Meloidogyne javanica* by *Bacillus cereus*. Biocon Sci Technol 3:115–126

Oliveira DF, Carvalho HWP, Nunes AS, Silva Geraldo H, Campos VP, Júnior HMS, Cavalheiro AJ (2009) The activity of amino acids produced by *Paenibacillus macerans* and from commercial sources against the root-knot nematode *Meloidogyne exigua*. Eur J Plant Pathol 124:57–63

Ongena M, Daayf F, Jacques P, Thonart P, Benhamou N, Paulitz TC, Cornelis P, Koedam N, Belanger RR (1999) Protection of cucumber against *Pythium* root rot by fluorescent pseudomonads: predominant role of induced resistance over siderophores and antibiosis. Plant Pathol 48:66–76

Oostendorp M, Sikora RA (1989a) Seed treatment with antagonistic rhizobacteria for the suppression of *Heterodera schachtii* early root infection of sugarbeet. Rev Nematol 12:77–83

Oostendorp M, Sikora RA (1989b) In-vitro relation ships between rhizosphere bacteria and *Heterodera schachtii*. Rev Nematol 13:269–274

Pal KK, Tilak KVBR, Saxena AK, Dey R, Singh CS (2001) Suppression of maize root diseases caused by *Macrophomina phaseolina, Fusarium moniliformae* and *Fusarium graminearum* by growth promoting rhizobacteria. Microbiol Res 56:209–223
Papavizas GC, Lumsden RD (1980) Biological control of soil borne fungal propagules. Annu Rev Phytopathol 18:389–413
Parkè JL (1990) Population dynamics of *Pseudomonas cepacia* in the pea spermosphere in relation to biocontrol of *Pythium*. Phytopathology 80:1307–1311
Parke JL (1991) Root colonization by indigenous and introduced microorganisms. In: Keister DL, Cregan PB (eds) The rhizosphere and plant growth. Kluwer Academic Publishers, Boston, MA, pp 33–42
Penyalver R, Oger P, Lopez MM, Farrand SK (2001) Iron binding compounds from *Agrobacterium* spp.: biological control stains *Agrobacterium rhizogenes* K84 produce a hydroxamate siderophore. Appl Environ Microbiol 67:654–664
Pierson EA, Weller DM (1994) Use of mixtures of fluorescent pseudomonads to suppress take-all and improve the growth of wheat. Phytopathology 84:940–947
Pieterse CMJ, VanWees SCM, Hoffland E, Van Pelt JA, van Loon LC (1996) Systemic resistance in *Arabidopsis* induced by biocontrol bacteria is independent of salicyclic acid accumulation and pathogenesis-related gene expression. Plant Cell 8:1225–1237
Pieterse CMJ, VanWees SCM, Ton J, VanPelt JA, van Loon LC (2002) Signalling in rhizobacteria-induced systemic resistance in *Arabidopsis thaliana*. Plant Biol 4:535–544
Pleban S, Chernin L, Chet I (1997) Chitinolytic activity of an endophytic strain of *Bacillus cereus*. Lett Appl Microbiol 25:284–288
Podile AR, Laxmi VDV (1998) Seed bacterization with *Bacillus subtilis* AF1 increases Phenylalanine ammonialyase and reduces the incidence of Fusarial wilt in pigeonpea. J Phytopathol 146:255–259
Powell JF, Vargas JM, Nair MG, Detweiler AR, Chandra A (2000) Management of dollar spot on creeping bentgrass with metabolites of *Pseudomonas aureofaciens* (TX-1). Plant Dis 84:19–24
Press CM, Loper JE, Kloepper JW (2001) Role of iron in rhizobacteria-mediated induced systemic resistance of cucumber. Phytopathology 91:593–598
Raaijmakers JM, Bonsall RF, Weller DM (1999) Effect of population density of *Pseudomonas fluorescens* on production of 2, 4-diacetylphloroglucinol in the rhizosphere of wheat. Phytopathology 89:470–475
Raaijmakers JM, Vlami M, de Souza JT (2002) Antibiotic production by bacterial biocontrol agents. Antonie van Leeuwenhoek 81:537–547
Raaijmakers JM, Weller DM, Thomashow LS (1997) Frequency of antibiotic-producing *Pseudomonas* spp. in natural environments. Appl Environ Microbiol 63:881–887
Rachid D, Ahmed B (2005) Effect of iron and growth inhibitors on siderophores production by *Pseudomonas fluorescens*. Afr J Biotechnol 4:697–702
Racke J, Sikora RA (1985) Einfluss von Rhizospha-rebakterien auf *Rhizoctonia solani* und den Befall der kartefftrovte Hanja mit *Globodera pallida*. Vortr. Pflanzenzucht, 9:21–28 Status Seminar Grunback 2–21 April
Radja Commare R, Nandkumar R, Kandan A, Suresh S, Bharathi M, Raguchander T, Samiyappan R (2002) *Pseudomonas fluorescens* based bioformulation for the management of sheath blight disease and leaffolder insect in rice. Crop Prot 21:671–677
Ramamoorthy V, Samiyappan R (2001) Induction of defense related genes in *Pseudomonas fluorescens* treated chili plants in response to infection by *Colletotrichum capsici*. J Mycol Plant Pathol 31:146–155
Ramamoorthy V, Raghuchander T, Samiyappan R (2002a) Enhancing resistance of tomato and hot pepper to *Pythium* diseases by seed treatment with fluorescent pseudomonads. Eur J Plant Pathol 108:429–441
Ramamoorthy V, Raghuchander T, Samiyappan R (2002b) Induction of defense related proteins in tomato roots treated with *Pseudomonas fluorescens* Pf1 and *Fusarium oxysporum* f. sp. *lycopersici*. Plant Soil 239:55–68

Randhawa PS, Schaad NW (1985) A seedling bioassay chamber for determining bacterial colonization and antagonism on plant roots. Phytopathology 75:254–259
Raupach GS, Kloepper JW (1998) Mixture of plant growth-promoting rhizobacteria enhance biological control of multiple cucumber pathogens. Phytopathology 88:1158–1164
Raupach GS, Liu L, Murphy JF, Tuzun S, Kloepper JW (1996) Induced systemic resistance of cucumber and tomato against cucumber mosaic virus using plant growth promoting rhizobacteria. Plant Dis 80:891–894
Ravel J, Cornelis P (2003) Genomics of pyoverdine-mediated iron uptake in pseudomonads. Trends Microbiol 11:195–200
Rodriguez-Romero AS, Badosa E, Montesinos E, Jaizme-Vega MC (2008) Growth promotion and biological control of root-knot nematodes in micropropagated banana during the nursery stage by treatment with specific bacterial strains. Ann Appl Biol 152:41–48
Ross AF (1961) Systemic acquired resistance induced by localized virus infections in plants. Virology 14:340–358
Rovira AD (1965) Interactions between plant roots and soil microorganisms. Annu Rev Microbiol 19:241–266
Rovira AD (1969) Plant root exudates. Bot Rev 35:35–57
Rovira AD (1991) Rhizosphere research-85 years of progress and frustration. In: Keister DL, Cregan PB (eds) The rhizosphere and plant growth. Kluwer Academic Publishers, Boston, MA, pp 3–13
Saravanakumar D, Vijayakumar C, Kumar N, Samiyappan R (2007) PGPR-induced defense responses in the tea plant against blister blight disease. Crop Prot 26:556–565
Schroth MN, Hancock JG (1982) Disease suppressive soil and root colonizing bacteria. Science 216:1376–1381
Sharma HKP, Mishra SD (2003) Effect of plant growth promoter microbes on root-knot nematode *Meloidogyne incognita* on okra. Curr Nematol 14:57–60
Sharma V, Nowak J (1998) Enhancement of Verticillium wilt resistance in tomato transplants by *in vitro* co-culture of seedlings with a plant growth promoting rhizobacterium (*Pseudomonas* sp. Strain PsJN). Can J Microbiol 44:528–536
Siddiqui IA, Shaukat SS (2003) Suppression of root-knot disease by *Pseudomonas fluorescens* CHA0 in tomato: importance of bacterial secondary metabolite, 2, 4-diacetylpholoroglucinol. Soil Biol Biochem 35:615–1623
Siddiqui IA, Haque SE (2001) Suppression of the root rot-root knot disease complex by *Pseudomonas aeruginosa* in tomato: The influence of inoculums density, nematode populations, moisture and other plant associated bacteria. Plant Soil 237:81–89
Siddiqui IA, Shaukat SS (2002) Rhizobacteria-mediated induction of systemic resistance (ISR) in tomato against *Meloidogyne javanica*. J Phytopathol 150:469–473
Siddiqui IA, Qureshi SA, Sultana V, Haque SE, Ghaffar A (2000) Biological control of root rot-root knot disease complex in tomato. Plant Soil 227:163–169
Siddiqui S, Siddiqui ZA, Iqbal A (2005) Evaluation of fluorescent pseudomonads and *Bacillus* isolates for the biocontrol of wilt disease complex of pigeonpea. World J Microbiol Biotechnol 21:729–732
Siddiqui ZA (2004) Effects of plant growth promoting bacteria and composted organic fertilizers on the reproduction of *Meloidogyne incognita* and tomato growth. Bioresource Technol 95:223–227
Siddiqui ZA (2006) PGPR: prospective biocontrol agents of plant pathogens. In: Siddiqui ZA (ed) PGPR: biocontrol and biofertilization. Springer, The Netherlands, pp 111–142
Siddiqui ZA (2007) Biocontrol of *Alternaria triticina* by plant growth promoting rhizobacteria on wheat. Arch Phytopathol Plant Protec 40:301–308
Siddiqui ZA, Akhtar MS (2007) Biocontrol of a chickpea root-rot disease complex with phosphate solubilizing microorganisms. J Plant Pathol 89:31–42

Siddiqui ZA, Akhtar MS (2008a) Effects of fertilizers, AM fungus and plant growth promoting rhizobacterium on the growth of tomato and on the reproduction of root-knot nematode *Meloidogyne incognita*. J Plant Interac 3:263–271
Siddiqui ZA, Akhtar MS (2008b) Synergistic effects of antagonistic fungi and a plant growth promoting rhizobacterium, an arbuscular mycorrhizal fungus, or composted cow manure on the populations of *Meloidogyne incognita* and growth of tomato. Biocon Sci Technol 18:279–290
Siddiqui ZA, Akhtar MS (2008c) Effects of organic wastes, *Glomus intraradices* and *Pseudomonas putida* on the growth of tomato and on the reproduction of root-knot nematode *Meloidogyne incognita*. Phytoparasitica 36:460–471
Siddiqui ZA, Akhtar MS (2009a) Effects of antagonistic fungi and plant growth promoting rhizobacteria on growth of tomato and reproduction of root-knot nematode *Meloidogyne incognita*. Austral Plant Pathol 38:22–28
Siddiqui ZA, Akhtar MS (2009b) Effects of antagonistic fungi, plant growth promoting rhizobacteria and arbuscular mycorrhizal fungi alone and in combination on the reproduction of *Meloidogyne incognita* and growth of tomato. J Gen Plant Pathol 75:144–153
Siddiqui ZA, Akhtar MS (2009c) Effect of plant growth promoting rhizobacteria, nematode parasitic fungi and root-nodule bacterium on root-knot nematodes *Meloidogyne javanica* and growth of chickpea. Biocont Sci Technol 19:511–521
Siddiqui ZA, Futai K (2009) Biocontrol of *Meloidogyne incognita* using antagonistic fungi, plant growth-promoting rhizobacteria and cattle manure on tomato. Pest Manag Sci 65:943–948
Siddiqui ZA, Husain SI (1991) Studies on the biological control of root-knot nematode. Curr Nematol 2:5–6
Siddiqui ZA, Mahmood I (1992) Biological control of root-rot disease complex of chickpea caused by *Meloidogyne incognita* race 3 and *Macrophomina phaseolina*. Nematol Medit 20:199–202
Siddiqui ZA, Mahmood I (1993) Biological control of *Meloidogyne incognita* race-3 and *Macrophomina phaseolina* by *Paecilomyces lilacinus* and *Bacillus subtilis* alone and in combination of chickpea. Fundam Appl Nematol 16:215–18
Siddiqui ZA, Mahmood I (1995a) Role of plant symbionts in nematode management. A review. Bioresource Technol 54:217–26
Siddiqui ZA, Mahmood I (1995b) Management of *Meloidogyne incognita* race 3 and *Macrophomina phaseolina* by fungus culture filtrates and *Bacillus subtilis* on chickpea. Fundam Appl Nematol 18:71–76
Siddiqui ZA, Mahmood I (1995c) Biological control of *Heterodera cajani* and *Fusarium udum* by *Bacillus subtilis, Bradyrhizobium japonicum* and *Glomus fasciculatum* on pigeonpea. Fundam Appl Nematol 18:559–556
Siddiqui ZA, Mahmood I (1996) Biological control of plant parasitic nematodes by fungi. A Review. Bioresource Technol 58:229–239
Siddiqui ZA, Mahmood I (1998) Effect of a plant growth promoting bacterium, an AM fungus and soil types on the mophometrics and reproduction of *Meloidogyne javanica* on tomato. Appl Soil Ecol 8:77–84
Siddiqui ZA, Mahmood I (1999) Role of bacteria in the management of plant parasitic nematodes. A review. Bioresource Technol 69:167–179
Siddiqui ZA, Mahmood I (2000) Effects of *Bacillus subtilis, Glomus mosseae* and ammonium sulphate on the development of *Meloidogyne javanica* and on growth of tomato. Thai J Agri Sci 33:29–35
Siddiqui ZA, Mahmood I (2001) Effects of rhizobacteria and root symbionts on the reproduction of *Meloidogyne javanica* and growth of chickpea. Bioresource Technol 79:41–45
Siddiqui ZA, Mahmood I (2003) Effects of plant straws and plant growth promoting bacteria on the reproduction of *Meloidogyne incognita* and growth of tomato. Boil Agri Hort 21:53–62
Siddiqui ZA, Shakeel U (2006) Use of fluorescent pseudomonads isolates for the biocontrol of wilt disease complex in pigeonpea in green house and under pot condition. Plant Pathol J 5:99–105

Siddiqui ZA, Shakeel U (2007) Screening of *Bacillus* isolates for potential biocontrol of the wilt disease complex of pigeon pea (*Cajanus cajan*) under green house and small-scale field condition. J Plant Pathol 89:179–183
Siddiqui ZA, Singh LP (2005a) Effects of fly ash, Pseudomonas striata and Rhizobium sp. on the reproduction of nematode Meloidogyne incognita and on the growth and transpiration of pea. J Environ Biol 26:117–122
Siddiqui ZA, Singh LP (2005b) Effects of fly ash and soil microorganisms on the plant growth, photosynthetic pigments and leaf blight of wheat. J Plant Dis Protec 112:146–155
Siddiqui ZA, Baghel G, Akhtar MS (2007a) Biocontrol of *Meloidogyne javanica* by *Rhizobium* and plant growth-promoting rhizobacteria on lentil. World J Microbiol Biotechnol 23:435–441
Siddiqui ZA, Iqbal A, Mahmood I (2001) Effects of *Pseudomonas fluorescens* and fertilizers on the reproduction of *Meloidogyne incognita* and growth of tomato. Appl Soil Ecol 16:179–185
Siddiqui ZA, Khan S, Mahmood I (2002) Use of Rhizobacteria for the management of *Meloidogyne incognita* on *Solanum melongena*. Thai J Agri Sci 35:1–8
Siddiqui ZA, Mahmood I, Hayat S (1998) Biocontrol of *Heterodera cajani* and *Fusarium udum* on pigeonpea using *Glomus mosseae, Paecilomyces lilacinus* and *Pseudomonas fluorescens*. Thai J Agri Sci 31:310–321
Siddiqui ZA, Qureshi A, Akhtar MS (2009) Biocontrol of root-knot nematode *Meloidogyne incognita* by *Pseudomonas* and *Bacillus* isolates on *Pisum sativum*. Arch Phytopathol Plant Protec 42:1154–1164. doi:10.1080/03235400701650890
Siddiqui ZA, Shakeel U, Siddiqui S (2008) Biocontrol of wilt disease complex of pigeonpea by fluorescent pseudomonads and *Bacillus* spp. under pot and field conditions. Acta Phytopathol et Entom Hung 43:79–94
Siddiqui ZA, Sharma B, Siddiqui S (2007b) Evaluation of *Bacillus* and *Pseudomonas* isolates for the biocontrol of *Meloidogyne incognita* on tomato. Acta Phytopathol et Entom Hung 42:25–34
Sikora RA (1988) Interrelationship between plant health promoting bacteria, plant parasitic nematodes and soil microorganisms. Med Fac Landbouww Rijksuniv Gent 53:867–878
Sikora RA, Racke J, Bodenstein F (1989) Influence of plant health promoting bacteria antagonistic to *Globodera pallida* and *Heterodera schachtii* on soil borne fungal and bacterial plant pathogens of potato and sugarbeet. J Nematol 21:588
Srivastava AK, Singh T, Jana TK, Arora DK (2001) Induced resistance and control of charcoal rot in *Cicer arietinum* (chickpea) by *Pseudomonas fluorescens*. Can J Bot 79:787–795
Steijl H, Niemann GJ, Boon JJ (1999) Changes in chemical composition related to fungal infection and induced resistance in carnation and radish investigated by pyrolysis mass spectrometry. Physiol Mol Plant Pathol 55:297–311
Szczech M, Shoda M (2004) Biocontrol of rhizoctonia damping-off of tomato by *Bacillus subtilis* combined with *Bukholderia cepacia*. J Phytopathol 152:549–556
Thomashow LS, Weller DM, Bonsall RF, Pierson LSP (1990) Production of the antibiotic phenazine-1-carboxylic acid by fluorescent *Pseudomonas* species in the rhizosphere of wheat. Appl Environ Microbiol 56:908–912
Thrane C, Olsson S, Nielsen TH, Sörensen J (1999) Vital fluorescent stains for detection of stress in *Pythium ultimum* and *Rhizoctonia solani* challenged with viscosinamide from *Pseudomonas fluorescens* DR54. FEMS Microbiol Ecol 30:11–23
Timmusk S, Wagner EGH (1999) The plant-growth-promoting rhizobacterium *Paenibacillus polymyxa* induces changes in *Arabidopsis thaliana* gene expression: a possible connection between biotic and abiotic stress responses. Mol Plant-Microbe Interac 12:951–959
Ton J, Van Pelt JA, van Loon LC, Pieterse CMJ (2002) Differential effectiveness of salicylate-dependent and jasmonate/ethylene-dependent induced resistance in *Arabidopsis*. Mol Plant-Microbe Interac 15:27–34
Unnamalai N, Gnanamanickam SS (1984) *Pseudomonas fluorescens* is an antagonist to *Xanthomonas citri* (Hasse) Dye, the incitant of citrus canker. Curr Sci 53:703–704

Uthede RS, Koch CA, Menzies JG (1999) Rhizobacterial growth and yield promotion of cucumber plants inoculated with *Pythium aphanidermatum*. Can J Plant Pathol 21:265–271

Valois D, Fayad K, Barbasubiye T, Garon M, Déry C, Brzezinski R, Beaulieu C (1996) Glucanolytic actinomycetes antagonistic to *Phytophthora fragariae* var. *rubi*, the causal agent of raspberry root rot. Appl Environ Microbiol 62:1630–1635

van Loon LC, Bakker PAHM, Pieterse CMJ (1998) Systemic resistance induced by rhizosphere bacteria. Annu Rev Phytopathol 36:453–483

Van Peer R, Niemann GJ, Schippers B (1991) Induced resistance and phytoalexin accumulation in biological control of *Fusarium* wilt of carnation by *Pseudomonas* sp. strain WCS417r. Phytopathology 81:728–733

Van Wees SCM, De Swart EAM, VanPelt JA, van Loon LC, Pieterse CMJ (2000) Enhancement of induced disease resistance by simultaneous activation of salicylate and jasmonate dependent defense pathways in *Arabidopsis thaliana*. Proc Natl Acad Sci USA 97:8711–8716

Van Wees SCM, Pieterse CMJ, Trijssenaar A, Westende YAM, Hartog F, van Loon LC (1997) Differential induction of systemic resistance in *Arabidopsis* by biocontrol bacteria. Mol Plant-Microbe Interac 10:716–724

Vandenburgh PA, Gonzalez CF (1984) Methods for protecting the growth of plants by employing mutants siderophore producing strains of *Pseudomonas putida*. U.S. Patent No. 4479936

Vesper SJ (1987) Production of pili (fimbriae) by *Pseudomonas fluorescens* and correlation with attachment to corn roots. Appl Environ Microbiol 53:1397–1405

Vidhyasekaran P, Kamala N, Ramanathan A, Rajappan K, Paranidharan V, Velazhahan R (2001) Induction of systemic resistance by *Pseudomonas fluorescens* Pf1 against *Xanthomonas oryzae* pv. *oryzae* in rice leaves. Phytoparasitica 29:155–166

Viswanathan R, Samiyappan R (2002) Induced systemic resistance by fluorescent pseudomonads against red rot disease of sugarcane caused by *Colletotrichum falcatum*. Crop Prot 21:1–10

Vivekananthan R, Ravi M, Ramanathan A, Samiyappan R (2004) Lytic enzymes induced by *Pseudomonas fluorescens* and other biocontrol organisms mediate defence against the anthracnose pathogen in mango. World J Microbiol Biotechol 20:235–244

Voisard C, Keel C, Haas D, Defago G (1989) Cyanide pmduction by *Pseudomonas fluorescens* helps suppress black root rot of tobacco under gnotobiotic conditions. EMBO J 8:351–358

Waisel Y, Eshel A, Katkafl U (1991) Plant roots: the hidden half. Marcel Dekker, New York, NY

Wang C, Knill E, Glick BR, Defago G (2000) Effect of transferring 1-aminocyclopropane-1-carboxylic acid (ACC) deaminase genes into *Pseudomonas fluorescens* strain CHA0 and its *gacA* derivative CHA96 on their growth-promoting and disease-suppressive capacities. Can J Microbiol 46:898–907

Weger LA, Arendonk JJCM, Recourt K, Hofstad GAJM, Weisbeek PJ, Lugtenberg B (1988) Siderophore-mediated uptake of Fe^{3+} by the plant growth-stimulating *Pseudomonas putida* strain WCS358 and by other rhizosphere microorganisms. J Bacteriol 170:4693–4698

Wei G, Kloepper JW, Tuzun S (1991) Induction of systemic resistance of cucumber to *Colletotrichum orbiculare* by select strains of plant growth promoting rhizobacteria. Phytopathology 81:1508–1512

Wei G, Kloepper JW, Tuzun S (1996) Induced systemic resistance to cucumber diseases and increased plant growth by plant growth promoting rhizobacteria under field conditions. Phytopathology 86:221–224

Weidenborner M, Kunz B (1993) Infuence of fermentation conditions on nematicidal activity of *Pseudomonas fluorescens*. Zeitschrift fur Pfleanzenkrankheiten und pflanzenschuts 100:90–94

Weller DM (1983) Colonization of wheat roots by a fluorescent pseudomonad suppressive to take-all. Phytopathology 73:1548–1553

Weller DM (1988) Biological control of soilborne plant pathogens in the rhizosphere with bacteria. Annu Rev Phytopathol 26:379–407

Weller DM, Cook RJ (1983) Suppression of take-all of wheat by seed treatments with fluorescent pseudomonads. Phytopathology 73:463–469

Weller DM, Cook RJ (1986) Increased growth of wheat by seed treatment with fluorescent pseudomonads, and implications of *Pythium* control. Can J Plant Pathol 8:328–334

Weller DM, Raaijmakers JM, McSpadden Gardener BB, Thomashow LS (2002) Microbial populations responsible for specific soil suppressiveness to plant pathogens. Annu Rev Phytopathol 40:309–348

Weller DM, Zhang BX, Cook RJ (1985) Application of a rapid screening test for selection of bacteria suppressive to take-all of wheat. Plant Dis 68:710–713

Westcott SW, Kluepfel DA (1992) Inhibition of *Criconemella xenoplax* egg hatch by a strain of *Pseudomonas aureofaciens*. J Nematol 24:626

Whipps JM (1997) Developments in the biological control of soil-borne plant pathogens. Adv Bot Res 26:1–134

Whipps JM (2001) Microbial interactions and biocontrol in the rhizosphere. J Exp Bot 52:487–511

Wilson M, Lindow SE (1993) Interactions between the biological control agent *Pseudomonas fluorescens* A506 and *Erwinia amylovora* in pear blossoms. Phytopathology 83:117–123

Wilson M, Backman PA (1999) Biological control of plant pathogens. In: Ruberson JR (ed) Handbook of pest management. Marcel Dekker, New York, NY, pp 309–335

Xu GW, Gross DC (1986) Field evaluation of the interactions among fluorescent pseudomonads, *Erwinia carotovora* and potato yields. Phytopathology 76:423–430

Yan Z, Reddy MS, Ryu CM, McInroy JA, Wilson MA, Kloepper JW (2002) Induced systemic protection against tomato late blight elicited by plant growth-promoting rhizobacteria. Phytopathology 92:1329–1333

Yang CH, Crowley DE (2000) Rhizosphere microbial community structure in relation to root location and plant iron nutritional status. Appl Environ Microbiol 66:345–351

Yigit F, Dikilitas M (2007) Control of *Fusarium* wilt of tomato by combination of fluorescent *Pseudomonas*, non pathogenic *Fusarium* and *Trichoderma harzianum* T-22 in green house condition. Plant Pathol J 6:159–163

Yuen GY, Schroth MN (1986) Interactions of *Pseudomonas fluorescens* strain E6 with ornamental plants and its effect on the composition of root-colonizing microflora. Phytopathology 76:176–180

Zavaleta-Mejia E (1985) The effect of soil bacteria on *Meloidogyne incognita* (Kofoid & White) Chitwood infection. Dissertation abstract Int Sci Eng 46:108

Zavaleta-Mejia E, VanGundy SD (1982) Effects of rhizobacteria on *Meloidogyne* infection. J Nematol 14:475–476

Zehnder GW, Yao C, Murphy JF, Sikora ER, Kloepper JW (2000) Induction of resistance to tomato against cucumber mosaic cucumo virus by plant growth promoting rhizobacteria. Biocontrol 45:127–137

Zheng XY, Sinclair JB (2000) The effects of traits of *Bacillus megaterium* on seed and root colonization and their correlation with the suppression of rhizoctonia root-rot of soybean. Biocontrol 45:223–243

Zuckerman BM, Dicklow MB, Acosta N (1993) A strain of *Bacillus thuringiensis* for the control of plant parasitic nematodes. Biocon Sci Technol 3:41–46

Potential of Bacilli for Biocontrol and Its Exploitation in Sustainable Agriculture

Olga Susana Correa and Marcelo Abel Soria

Contents

1 Introduction 198
2 Mechanism Involved in Microbial Biological Control 199
3 Sustainability of Plant Disease Control Using Bacilli 201
4 Causes That Restrict the Adoption of Biological Control 204
5 Conclusions and Future Considerations 205
References 206

Abstract Plant diseases are caused mainly by fungi, bacteria, viruses, and nematodes, and their control is necessary to feed an increasing population. Control of plant diseases often rely on chemical pesticides, which have contributed to improvements in crop productivity and quality over the past years. However, the intensive use of agrochemical pesticides results in soil and groundwater pollution. Consequently, there are worldwide efforts to develop other alternatives to chemical pesticides for controlling plant diseases. Among them, the use of microorganisms and their products, referred as biological control, are regarded as promissory alternatives to reduce the use of chemical products. Different *Bacillus* species excrete peptides and lipopeptides to the culture medium, such as fungicine, iturin, bacillomicine and others, that have antifungal antibacterial and surfactant activity. In addition, these species produce spores that are resistant to heat and desiccation, which allows the preparation of more stable and durable formulations. A variety of biological control products based on *Bacillus* species are available for agronomical use; but in order to translate these developments into a broader and more effective use, a greater understanding of the complex interactions among plants, microorganisms, and the environment is required. This chapter describes some mechanisms of

---

O. Susana Correa (✉) and M. Abel Soria
Microbiología Agrícola y Ambiental, Facultad de Agronomía, Universidad de Buenos Aires, Av. San Martín 4453, C1417DSE Buenos Aires, Argentina
e-mail: correa@agro.uba.ar

D.K. Maheshwari (ed.), *Plant Growth and Health Promoting Bacteria*,
Microbiology Monographs 18, DOI 10.1007/978-3-642-13612-2_8,

biocontrol exhibited by species of *Bacillus*, the current status of research and application of biological control using *Bacillus* species, constraints to microbial biocontrol implementation, and briefly outlines the future directions that might lead to the development of more diverse and effective biological controls for plant diseases.

1 Introduction

Agricultural production in the twenty-first century faces the challenge of increasing food production without negatively affecting the environment. The effective control of diseases is an essential component in every crop production system. To this end, resistant plant cultivars, cultural practices, and chemical applications are routinely used to provide disease control. Among these, the use of resistant cultivars and the careful management of cultural practices have the least aggressive effect on the environment. However, not every disease has a corresponding resistant or tolerant plant cultivar and through natural selection, the pathogens frequently overcome the resistance present in current cultivars in a few years (Cook 1993; Howarth 1991; Rusell 1995). Besides, cultural practices are not always economically or technologically feasible. Since World War II, numerous synthetic pesticides have been developed and successfully used for the control of crop pests and diseases. On the other hand, chemical pesticides also lose their effectiveness because of the development of genetic resistance in pathogen populations or they are banned or its use restricted by new regulations. In addition, available chemical pesticides are often expensive and also have adverse effects on human beings and the environment (Gupta 2004; Bortoli et al. 2009). So, in order to keep the pace of increase food demands, we need to search new solutions to control plant disease problems by alternative methods that result in effective control with minimum impact on humans, animals, and the environment.

Despite the synthetic pesticides dominating the phytosanitary market worldwide, their irrational selection and misuse have determined a decline in their use since 2000, thus increasing the need for new strategies of phytopathogen control. Therefore, biological control appears to constitute an appropriate alternative for controlling diseases in an environmentally friendly manner. Biological control can be defined as the use of one organism to reduce the population density of another organism and thus can include the control of animals, weeds, and diseases (Bale et al. 2008). The use of beneficial microorganisms for controlling plant diseases represents an environmentally friendly alternative to chemical pesticides, and can be used where conventional pesticides should be avoided because of residue concerns or in organic farming. Moreover, biological control can be applied together with chemicals in order to reduce the doses of chemicals and pathogen resistance, and as part of an integrated pest management (IPM) schema. The final goal is to minimize the use of synthetic pesticides.

Another undesirable effect of an excessive use of agrochemicals for management of plant diseases is their detrimental impact on the microbial biodiversity of the agroecosystems. Many of the chemical pesticides kill not only the target species of pathogen but also other non-harmful or beneficial organisms (Hanazato 2001). Two examples of beneficial microorganisms affected by chemical product meant to control pathogens are the nitrogen-fixing symbiotic bacteria and the fungi that form mycorrhizal associations with plants. In addition, during the past few years, there is a growing and widespread concern about the use of non-sustainable technologies for food production (Allen et al. 2008; Saifi and Drake 2008).

The use of bacteria as biocontrol agents has been extensively studied (Expert and Digat 1995; Asaka and Shoda 1996; Podile and Prakash 1996; Kim et al. 1997; Mao et al. 1997; Singh et al. 1998; de Vrije et al. 2001). Most of the bacterial biopesticides belong to the genera *Agrobacterium*, *Bacillus*, and *Pseudomonas* (Adesemoye et al. 2008). Currently, the contribution of biocontrol to plant health management is small but it is expected that it will increase in the next years (Ongera and Jacques 2008).

The process of developing biological control begins with in vitro and in vivo screenings that continues with the study of mechanisms of control such as competition, antibiosis, and induced systemic resistance. The next stage, the production of large amounts of efficient biomass at a low cost, requires studies of microbial physiology and the use of biotechnological processes. Adequate formulations and application methods have to be designed to ensure that the microbial biomass will attain a high level of biocontrol activity (Schisler et al. 2004). The legal registration procedures is usually the hardest part, it is a time-consuming and expensive process that must prove the effectiveness of the product and also that it does not entail any significant adverse effect on human health and the environment.

2 Mechanism Involved in Microbial Biological Control

The main mechanisms by which microbial biocontrol agents (MBCAs) can control other microorganisms are direct competition for space and nutrients, antibiosis or toxin production, predation or parasitism, and induced host resistance (Compant et al. 2005). Most MBCAs exhibit only one of these mechanisms, whereas some can use more than one. The molecular bases of biological control activity are diverse. Several biochemical pathways and gene regulatory networks are involved in the different processes that lead to pathogen control.

There are variations in the range of antibiotics between species and even within species, at the level of strains (Nagórska et al. 2007). This variability enhances the effectiveness of the use of *Bacillus* as a biocontrol tool, since the greater the spectrum of antifungals released the more difficult it becomes to the pathogen to adapt by natural selection. It also has another practical implication, because the life span of a product in the market can be expected to extend for several years before the target organism can develop genetic resistance. The same reasoning applies to biosurfactant production: there are differences across strains in types and relative amounts of compounds produced, all of which can have differential activities against a diverse set of targets. Some strains have good surfactant and poor antibiotic biosynthetic activities and vice versa; since it has been postulated that both types of compounds act synergistically, it makes sense to combine both types of strains in the same product.

The genus *Bacillus* has several bacterial species that produce lipopeptides with biological actives for inhibiting plant pathogens (Ongera and Jacques 2008). These molecules have antagonistic activity against bacteria, fungi, and oomycetes. In Fig. 1 a clear antagonisms effect in a dual Petri dish culture is showed. *Bacillus amyloliquefaciens* strain BNM122 has showed high antagonistic activity both in vitro and in vivo against several fungi that cause plant diseases (Souto et al. 2004). The antagonistic fungal activity exhibited by strain BNM122 was related to the coproduction of iturin, which has antifungal activity and surfactin, which has surfactant properties (Souto et al. 2004).

Most of the biological activity of these compounds is related to their effect on the lipids of the cell membrane, where they can promote, depending on concentration, irreversible pore formation in the double layer of phospholipids (Fig. 2).

These antifungal peptides inhibit the growth of a large number of fungi, including *Aspergillus*, *Penicillium*, and *Fusarium* species (Munimbazi and Bullerman

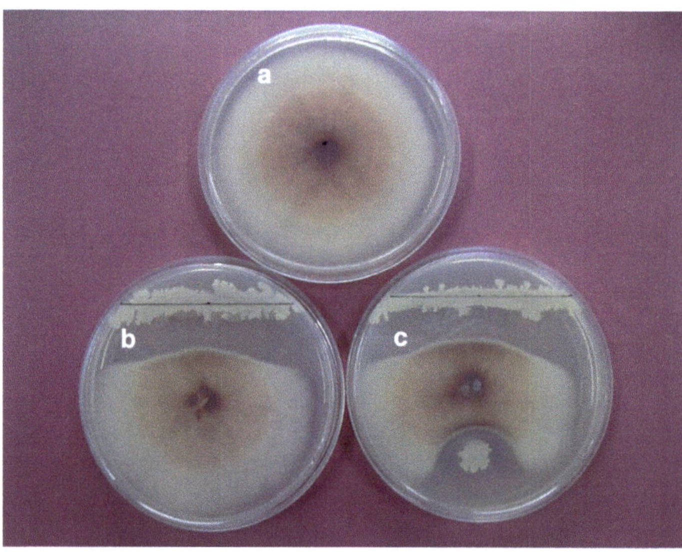

Fig. 1 In vitro antagonism of *Bacillus amyloliquefaciens* strain BNM122 against *Fusarium oxysporum* in dual culture on Petri dishes with potato dextrose agar medium. An inoculum of *F. oxysporum* was placed in the *middle* of the plates. (**a**) Growth of *F. oxysporum* with no bacterial inhibition; (**b**) Fungal growth inhibition by strain BNM122 streaked on one edge of the plate; (**c**) Fungal growth inhibition by strain BNM122 streaked (at the *top*) and spotted (at the *bottom*) on plate edges. Clear zones of fungal growth inhibition are observed toward the growth of BNM122

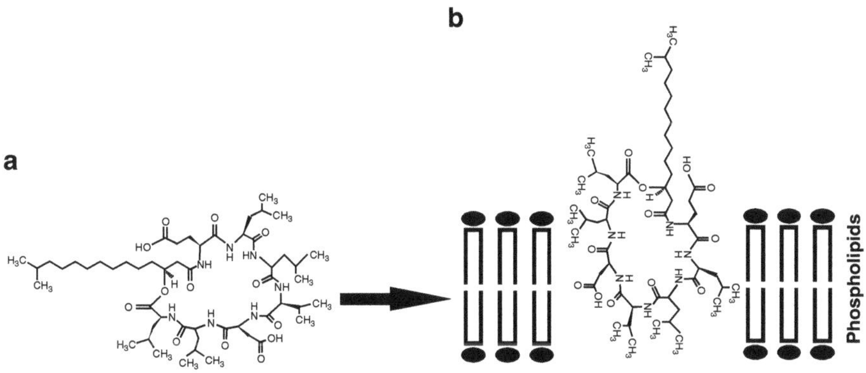

Fig. 2 Chemical structure of surfactin (**a**) and its action mode on fungal plasma membrane (**b**)

1998), as well as some yeasts (Thimon et al. 1995). In addition to their antagonistic activity against pathogens, *Bacillus* lipopeptides also have others more specific and important functions. Several of these compounds are involved in plant tissue colonization and in inducing plant resistance to phytopathogens (Ongera and Jacques 2008), whereas others like surfactin and mycosubtilin are important for bacterial surface motility and as wetting agents, reducing surface tension (Leclère 2006).

In the past few years, *Bacillus* species have also been proposed as biological control of plant parasitic nematodes belonging to the genera *Meloidogyne*, *Heterodea*, and *Rotylenchulus* (Tian et al. 2007). Nematodes cause great crop losses and are one of the most important agricultural pests. They are difficult to control because they inhabit the soil and attack the underground parts of the plants. Despite there are several chemical nematicides that are effective and easy to apply, they are being withdrawn from the market because of concerns regarding public and environmental safety. Hydrolytic enzymes such as proteases that degrade nematode cuticle are the main mechanisms involved in nematode biocontrol by *Bacillus* species (Lian et al. 2007). In Table 1, some of these useful substances and the *Bacillus* species that produce them are consigned.

3 Sustainability of Plant Disease Control Using Bacilli

Biological control products based on *Bacillus* species have huge potential in systems of IPM in order to reduce environmental contamination and to obtain safe and healthy foods. Cultural control, crop rotation, chemical pesticides, resistant host, and biocontrol agents are all part of IMP. Probably, one of the best known examples of integrated management is the application of fungicides and biocontrol agents for seed treatments. In the USA, almost all cotton planted is protected with Kodiak, a *B. subtilis* GB03 product, and fungicides. Biocontrol agents and fungicides in

Table 1 Bacterial species, the lipopeptide synthesized and main mechanism of action

Bacteria	Lipopeptide	Action as	References
B. subtilis	Bacillomycin	Antifungal	Peypoux et al. (1981)
B. subtilis and *B. amyloliquefaciens*	Fengycin	Antifungal/surfactant	Koumoutsi et al. (2004)
B. subtilis and *B. amyloliquefaciens*	Iturin	Antifungal	Delcambe (1965)
B. subtilis	Mycosubtilin	Antifungal/surfactant	Peypoux and Michel (1976)
B. cereus	Kanosamine, Zwittermycin A	Antifungal	Emmert and Handelsman (1999)
B. licheniformis, *B. coagulans*, *B. pumilis*	Lichenysin	Antifungal/surfactant	Huszcza and Burczyk (2006)

combination provide effective control of plant pathogens, which in many cases are not controlled by the available fungicides alone (Brannen and Kenney 1997). However, the potential use of lower fungicide doses when combined with a *Bacillus*-based biocontrol product has not been well explored yet, although some studies did report increases in disease control and reduction in chemical pesticide doses when bacilli-based products and chemicals are combined. In this sense, Cook et al. (2002) reported significant increases in winter wheat yield when *Bacillus* sp. strain L-324-92 was used in association with difeconazole plus mefoxam.

The integration of biological control *Bacillus*-based products with disease-resistant host plants should also be considered as part of IPM (Jacobsen et al. 2004). Published studies have demonstrated that the protective effect of *B. mycoides* Bm J against *Cercospora* leaf spot and that the control of *Fusarium* wilt with *B. subtilis* GB03 were more effective when the more resistant plant cultivar was used (Jacobsen et al. 2002; Hervas et al. 1998). These results highlight the importance of integrating several tools to gain stability in disease management programs. Also, mixtures of organisms with different modes of action may enhance the spectrum of activity, but unfortunately, there is limited knowledge and understanding of the interactions of such mixtures (Fravel 2005).

One of the main topics of discussion about the practical applications of biocontrol agents is how effective they are in real field applications. Since biological products can be very sensitive to environmental conditions, such as temperature, humidity, and sunlight exposure among other factors, they tend to be less stable than their traditional chemical counterparts. Ojiambo and Scherm (2006) conducted a statistical meta-analysis of 53 reports published between 2000 and 2005 that accounted for 149 combinations of target organism, biocontrol agent, plant host, and cultural treatments. These authors found that after normalization the range of observed results was quite wide, going from cases in which the biocontrol agent potentiated the pathogen to highly effective biocontrol. However, overall, the effect of biocontrol agents was positive and statistically significant. At a finer level of aggregation, they discovered that there were no differences in the effects observed between field or greenhouse conditions; soilborne or aerial diseases or when

Table 2 *Bacillus* species and strains; pathogens and diseases controlled, and host plants

Antagonistic species	Pathogen disease	Host plant	References
B. subtilis ZJY-116	*Fusarium* head blight	Wheat and barley	Zhang et al. (2005)
B. subtilis 6051	*Pseudomonas syringae* pv. tomato	*Arabidopsis*	Bais et al. (2004)
B. subtilis M4	Damping-off	Tomato, bean	Ongera et al. (2005)
B. subtilis RC8	*Fusarium verticillioides*	Maize	Cavaglieri et al. (2004)
B. subtilis AF1	Wilt in pigeon pea	Pigeon pea	Manjura and Podile (2001)
B. amyloliquefaciens MET0908	Anthracnose	Watermelon	Kim and Chung (2004)
B. amyloliquefaciens RC-2	Mulberry anthracnose	Mulberry	Hiradate et al. (2002)
B. amyloliquefaciens B94	*Rhizoctonia solani*–Damping-off	Soybean	Yu et al. (2002)
B. amyloliquefaciens BNM122	*Rhizoctonia solani*–Damping-off	Soybean	Souto et al. (2004)
B. cereus UW85	*Phytophthora megasperma*	Alfalfa	Handelsman et al. (1990)
B. mycoides BacJ	Cercospora leaf spot	Sugar beet	Bargabus et al. (2002)

considering the disease intensity. Interestingly, they did find a significant difference regarding host lifestyle: biocontrol agents were more effective controlling pathogens attacking annual plants compared to perennials. Some examples of bacilli reported as biological control agents of different plant diseases are shown in Table 2. Also, *Bacillus*-based products seemed less effective to other biocontrol species. However, a more detailed analysis showed that many experiments used commercial products that included strains of *Bacillus* species, probably due to its wide availability, even for conditions when they would not be recommended, such as treatment of aerial pathogens. A further test removing entries with potential misuse was performed, and it showed that *Bacillus*-based products were as effective as other biocontrol organisms. Another interesting conclusion from this meta-analysis study was that there were no differences between bacterial or fungal biocontrol agents, and that they could effectively control both bacterial and fungal targets. However, there was a significant difference between r- and K-strategists agents. Irrespective of whether they were fungal or bacterial, r-type organisms achieved greater controlling effects. Microorganisms with an r-strategy can divide very fast under favorable conditions and reach a high population size, a requisite for disease suppression.

Although *Bacillus*-based products are more effective controlling soil-borne pathogens compared to aerial targets, there is an interest among researchers in finding and overcoming the factors hindering the development of foliar formulations. The discoveries in this field can mutually benefit with those related to the use of microorganisms as biopesticides. The work of Ojiambo and Scherm (2006) found that one or two spray applications can be enough for controlling an aerial microbial target, a convenient rate of application from an economical and management point of view. However, Gan-Mor and Matthews (2003) pointed out that the configuration and handling of field equipment requires special attention in the case of biopesticide, because the optimal settings and adequate procedures are different from those recommended for chemical products. These observations suggest that some failures in controlling aerial pathogens could be caused by improper handling and not only by strictly biological factors. Clearly, the successful deployment of biological control agents in the field needs specialized training of farm personnel.

4 Causes That Restrict the Adoption of Biological Control

Despite the advantages of biological control for a sustainable agriculture, few products are commercially available. There are many reasons for this; one is the difficulty to obtain a formulation with a good shelf live, others are the lack of knowledge about their modes of action, and the differences among regulatory policies in different countries that complicate the inscription process. Sporulating Gram positive bacteria, like those belonging to the genus *Bacillus*, offer a solution to the formulation problem. Their spores have high resistance to dryness which constitutes and advantage for the production of certain classes of commercial products. The spores can be formulated as dry powders and can be stored for a long period of time without loss of concentration and effectiveness (Emmert and Handelsman 1999).

Another limitation to the extensive use of biocontrol agents is that most of the research reports focus only on the control of the target pathogen without further investigations on their impact on the agroecosystem and the environment in general. Some exceptions are the studies made with *B. cereus* UW85, a biocontrol agent of *Phytophthora* damping-off and root rot of soybean in the USA (Osburn et al. 1995). For this organism, researchers have studied the basis for disease biocontrol, the interaction with the plant, with the pathogen, and also the impact of strain UW85 on soil microbial communities (Handelsman et al. 1990; Silo-Suh et al. 1994; He et al. 1994; Gilbert et al. 1993; Halverson et al. 1993; Milner et al. 1995). In the same research line, Souto et al. (2004) and Correa et al. (2009) have studied the mechanisms of action of *B. amyloliquefaciens* strain BNM122 and their impact on the microbial community of soybean rhizosphere. Using culture-dependent and -independent methods, the authors demonstrate that this bacterium have a lower impact on soil microbial communities and non-target microorganisms than that exhibited when a chemical fungicide was applied (Fig. 3).

Soybean plants, whose seeds had been inoculated with *B. amyloliquefaciens* strain BNM 122 or treated with fungicides did not show differences in plant growth (mg pl^{-1}) and nodulation (nodules per plant) but a significant reduction was observed in the mycorrhizal symbiosis (Fig. 3). The important reduction observed in this beneficial non-target fungal symbiosis can be attributed to the wide spectrum

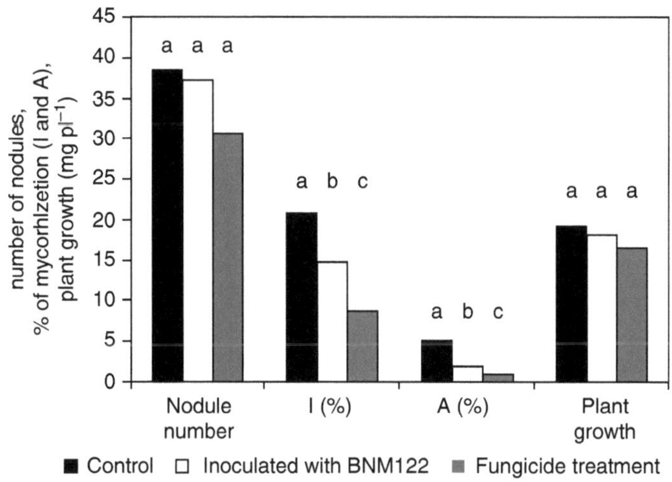

Fig. 3 Effect of seed inoculation with strain BNM122 on soybean nodule number, percentage of intensity and abundance (*I%* and *A%*) of mycorrhization, and plant growth (mg pl^{-1}). Plants were grown in pots, in greenhouse, under natural condition of light and temperature. Before sowing, all seeds were inoculated with the nodulating nitrogen-fixation bacteria *Bradyrhizobium japonicum* Treatments were non-inoculated seeds (control), inoculated with the biocontrol strain BNM122, and seeds treated with a mixture of chemical fungicides (carbendazim and thiram). Different *letters* indicate significant differences ($p < 0.05$) among treatments. Adapted from Correa et al. (2009)

of action displayed by both treatments. However, it is worth pointing out the lesser negative effect on root mycorrhization exhibited by the *Bacillus* biocontrol agent.

5 Conclusions and Future Considerations

Although scientists have been working for more than 50 years on biological control and IPM systems, the commercial importance of the business grows slowly and biological control products represent less than 2% of the plant protection market worldwide (Kiewnick 2007). If our objective is to deploy environmentally sounder alternatives to plant disease control, research in biological control should be well supported and funded. We have antecedents of how well some bacterial biocontrol agents have performed in the past. The success of *B. thuringiensis* and *B. sphaericus*, two larvicides, that have been successfully used to replace DDT and malathion. Both have proved to be extremely effective and to pose lower human health and environmental risk and to be useful for resistance management (Grisolia et al. 2009). All in all, it is important to bear in mind that these products do not have the efficacy of chemical counterparts; although in many occasions biological control is a valuable complement of chemical protection. For these reasons, the use of biological control pesticides is expected to increase in the coming years, especially in developing countries, together with the application of integrated disease management schemas. Despite the great interest in the exploitation of MBCAs, there are still barriers that impede their adoption in agriculture. Among them, the main obstacles are the long time for product registration and the slow adoption of new developments in a market more accustomed to chemical alternatives (Marrone 2007).

In the near future, the greater sustainability and lesser environmental risks associated with MBCAs will be factors balancing their lower performance and will become drivers in the search of better and diversified products, to enhance competitiveness and greater market penetration (Bailey et al. 2010). This will require government support such as tax benefits and incentives. Also of great importance is to increase the level of biocontrol-related education, to ensure the availability of detailed extension information and the training of distributors and farmers.

References

Adesemoye AO, Obini M, Ugoji EO (2008) Comparison of plant growth-promotion with *Pseudomonas aeruginosa* and *Bacillus subtilis* in three vegetables. Braz J Microbiol 39:423–426

Allen VG, Brown CP, Segarra E, Green CJ, Wheeler TA, Acosta-Martinez V, Zobeck TM (2008) In search of sustainable agricultural systems for the Llano Estacado of the U.S. Southern High Plains. Agric Eco Environ 124:3–12

Asaka O, Shoda M (1996) Biocontrol of *Rhizoctonia solani* damping-off of tomato with *Bacillus subtilis* RB14. Appl Environ Microbiol 62:4081–4085

Bailey KL, Boyetchko SM, Längle T (2010) Social and economic drivers shaping the future of biological control: a Canadian perspective on the factors affecting the development and use of microbial biopesticides. Biol Control 52:221–229

Bais HP, Fall R, Vivanco JM (2004) Biocontrol of *Bacillus subtilis* against infection of Arabidopsis roots by *Pseudomonas syringae* is facilitated by biofilm formation and surfactin production. Plant Physiol 134:307–319

Bale JS, van Lenteren JC, Bigler F (2008) Biological control and sustainable food production. Philos Trans R Soc B 363:761–776

Bargabus RL, Zidack NK, Sherwood JW, Jacobsen BJ (2002) Characterization of systemic resistance in sugar beet elicited by a non-pathogenic, phyllosphere colonizing Bacillus mycoides, biological control agent. Physiol Mol Plant Pathol 61:289–298

Bortoli GM, Azevedo MB, Silva LB (2009) Cytogenetic biomonitoring of Brazilian workers exposed to pesticides: micronucleus analysis in buccal epithelial cells of soybean growers. Mutat Res 675:1–4

Brannen PM, Kenney DS (1997) Kodiak-A successful biological-control product for suppression of soil-borne plant pathogens of cotton. J Ind Microbiol Biotechnol 19:165–171

Cavaglieri L, Passone A, Etcheverry M (2004) Screening procedures for selecting rhizobacteria with biocontrol effects upon *Fusarium verticillioides* growht and fumonisin B1 production. Res Microbiol 155:747–754

Compant S, Duffy B, Nowak J, Clément C, Barka EA (2005) Use of plant growth-promoting bacteria for biocontrol of plant diseases: principles, mechanisms of action, and future prospects. Appl Environ Microbiol 71:49–51

Cook RJ (1993) Making greater use of introduced microorganisms for biocontrol of plant pathogens. Annu Rev Phytopathol 31:53–80

Cook RJ, Weller DM, El-Banna AY, Vakoch D, Zhang J (2002) Yield responses of direct-seeded wheat to rhizobacteria and fungicide seed treatment. Plant Dis 86:780–784

Correa OS, Montecchia MS, Berti MF, Fernández Ferrari MC, Pucheu NL, Kerber NL, García AF (2009) *Bacillus amyloliquefaciens* BNM122, a potential microbial biocontrol agent applied on soybean seeds, causes a minor impact on soil microorganisms. Appl Soil Ecol 41:185–194

de Vrije T, Antoine N, Buitelaar RM, Bruckner S, Dissevelet M, Durand A, Gerlagh M, Jones EE (2001) The fungal biocontrol agent *Coniothyrium minitans*: production by solid-state fermentation, application and marketing. Appl Microbiol Biotechnol 56:58–68

Delcambe L (1965) Iturine. I. Preparation, purification et poids moléculaire. Bull Soc Chim Belg 74:315–328

Emmert EAB, Handelsman J (1999) Biocontrol of plant disease: a (Gram-) positive prespective. FEMS Microbiol Lett 171:1–9

Expert JM, Digat B (1995) Biocontrol of *Sclerotinia* wilt of sunflower by *Pseudomonas fluorescens* and *Pseudomonas putida* strains. Can J Microbiol 41:685–691

Fravel DR (2005) Commercialization and implementation of biocontrol. Annu Rev Phytopathol 43:337–359

Gan-Mor S, Matthews GA (2003) Recent developments in sprayers for application of biopesticides-and overview. Biosys Engin 84:119–125

Gilbert GS, Parke JL, Clayton MK, Handelsman J (1993) Effects of an introduced bacterium on bacterial communities on roots. Ecology 74:840–854

Grisolia CK, Oliveira-Filho EC, Ramos FR, Lopes MC, Freitas Muniz DH, Monnerat RG (2009) Acute toxicity and cytotoxicity of *Bacillus thuringiensis* and *Bacillus sphaericus* strains on fish and mouse bone marrow. Ecotoxicology 18:22–26

Gupta PK (2004) Pesticide exposure-Indian scene. Toxicology 198:83–90

Halverson LJ, Clayton MK, Handelsman J (1993) Population biology of *Bacillus cereus* UW85 in the rhizosphere of field-grown soybeans. Soil Biol Biochem 25:485–493

Hanazato T (2001) Pesticide effects on freshwater zooplankton: an ecological perspective. Environ Pollut 112:1–10

Handelsman J, Raffel S, Mester EH, Wunderlich L, Grau CR (1990) Biological control of damping-off of alfalfa seedlings with Bacillus cereus UW85. Appl Environ Microbiol 56:713–718

He H, Silo-Suh LA, Lethbridge BJ, Raffel SJ, Clardy J, Handelsman J (1994) Zwittermicin A, an antifungal and plant protection agent from *Bacillus cereus*. Tetrahedron Lett 35:2499–2502

Hervas A, Landa B, Datnoff LE, Jimenez-Díaz RM (1998) Effect of commercial and indigenous microorganisms on *Fusarium* wilt development in chickpea. Biol Control 13:166–176

Hiradate S, Yoshida S, Sugie H, Yada H, Fujii Y (2002) Mulberry anthracnose antagonists (iturins) produced by *Bacillus amyloliquefaciens* RC-2. Phytochemistry 61:639–698

Howarth FG (1991) Environmental impacts of classical biological control. Annu Rev Entomol 36:485–509

Huszcza E, Burczyk B (2006) Surfactin isoforms from *Bacillus coagulans*. Z Naturforsch C 61:727–733

Jacobsen BJ, Larson B, Zidack NK, Ansley J, Eckhoff JLA, Bergman J (2002) Integrated management of Cercospora leaf spot. Sugarbeet Res Ext Re 33:235–240

Jacobsen BL, Zidack NK, Larson BJ (2004) The role of *Bacillus*-based biological control agents in integrated pest management system: plant diseases. Phytopathology 94:1272–1275

Kiewnick S (2007) Practicalities of developing and registering microbial biological control agents. CAB Rev 2, No. 13 www.cababstractsplus.org/cabreviews

Kim PI, Chung KC (2004) production of an antifungal protein for control of *Colletotrichum lagenarium* by *Bacillus amyloliquefaciens* MET0908. FEMS Microbiol Lett 234:177–183

Kim D, Cook RJ, Weller D (1997) *Bacillus* sp. L324-92 for biological control of three root diseases of wheat grown with reduced tillage. Phytopathology 87:551–558

Koumoutsi A, Chen XH, Henne A, Liesegang H, Gabriele H, Franke P, Vater J, Borris R (2004) structural and functional characterization of gene clusters directing nonribosomal synthesis of bioactive lipopeptides in *Bacillus amyloliquefaciens* strain FZB42. J Bact 186:1084–1096

Leclère V (2006) The lipopeptides mycosubtilin and surfactin enhance spreading of *Bacillus subtilis* by their surface-active properties. Arch Microbiol 186:475–483
Lian LH, Tian BY, Xiong R, Zhu MZ, Xu J, Zhang DQ (2007) Proteases from Bacillus: a new insight into the mechanism of action for rhizobacterial suppression of nematode populations. Lett Appl Microbiol 45:262–269
Manjura K, Podile AR (2001) Chitin-supplemented formulations improve biocontrol and plant growth promoting efficiency of Bacillus subtilis AF 1. Can J Microbiol 47:618–625
Mao W, Lewis JA, Hebbar PK, Lumsden RD (1997) Seed treatment with fungal or a bacterial antagonist for reducing corn damping-off caused by species of *Pythium* and *Fusarium*. Plant Dis 81:450–454
Marrone PG (2007) Barriers to adoption of biological control agents and biological pesticides. CAB Rev. 2, No. 51 www.cabasbstractsplus.org/cabreviews
Milner JL, Lethbridge RSJ, BJ HJ (1995) Culture conditions that influence accumulation of zwittermicin A by *Bacillus cereus* UW85. Appl Microbiol Biotechnol 43:685–691
Munimbazi C, Bullerman LB (1998) Isolation and partial characterization of antifungal metabolites of *Bacillus pumilus*. J Appl Microbiol 84:959–968
Nagórska K, Bikowski M, Obuchowski M (2007) Multicellular behaviour and production of a wide variety of toxic substances usage of *Bacillus subtilis* as a powerful biocontrol agent. Acta Biochim Pol 54:495–508
Ojiambo PS, Scherm H (2006) Biological and application-oriented factors influencing plant disease suppression by biological control: a meta-analytical review. Phytopathology 96:1168–1174
Ongera M, Jacques P (2008) *Bacillus* lipopeptides: versatile weapons for plant disease biocontrol. Trends Microbiol 16:115–125
Ongera M, Jacques P, Toure Y, Destain J, Jabrane A, Thonart P (2005) Involvement of fengycin-type lipopeptides in the multifaceted biocontrol potential of *Bacillus subtilis*. Appl Microbiol Biotechnol 69:29–38
Osburn RM, Milner JL, Oplinger ES, Smith RS, Handelsman J (1995) Effect of *Bacillus cereus* UW85 on the yield of soybean at two field sites in Wiscosin. Plant Dis 79:551–556
Peypoux F, Michel G (1976) Structure de la mycosubtiline, antibiotique isole de *Bacillus subtilis*. Eur J Biochem 63:391–398
Peypoux F, Besson F, Michel G, Delcambe L (1981) Structure of Bacillomycin D, a new antibiotic of the Iturin group. Eur J Biochem 118:323–327
Podile AR, Prakash AP (1996) Lysis and biological control of *Aspergillus niger* by *Bacillus subtilis* AF1. Can J Microbiol 42:533–538
Rusell PE (1995) Fungicide resistance: occurrence and management. J Agric Sci 124:317–323
Saifi B, Drake L (2008) A coevolutionary model for promoting agricultural sustainability. Ecol Econ 65:24–34
Schisler DA, Slininger PJ, Behle RW, Jackson MA (2004) Formulation of *Bacillus* sp. for biological control of plant diseases. Phytopathology 94:1267–1271
Silo-Suh LA, Lethbridge BJ, Raffel SJ, He H, Clardy J, Handelsman J (1994) Biological activities of two fungistatic antibiotics produced by *Bacillus cereus* UW85. Appl Environ Microbiol 60:2023–2030
Singh PP, Shin YC, Park CS, Ching YR (1998) Biological control of Fusarium wilt of cucumber by chitinolytic bacteria. Phytophatology 89:92–99
Souto GI, Correa OS, Montecchia MS, Kerber NL, Pucheu NL, Bachur M, García AF (2004) Genetic and functional characterization of a *Bacillus* sp. strain excreting surfactin and antifungal metabolites partially identified as iturin like compounds. J Appl Microbiol 97:1247–1256
Thimon L, Peypoux F, Wallach J, Michel G (1995) Effect of lipopeptide antibiotic, iturin A, on morphology and membrane ultrastructure of yeast cells. FEMS Microbiol Lett 128:101–106
Tian B, Yang J, Zhang K-Q (2007) Bacteria used in the biological control of plant-parasitic nematodes: populations, mechanisms of action, and future prospects. FEMS Microbiol Ecol 61:197–213

Yu GY, Sinclair JB, Hartman GL, Bertagnolli BL (2002) Production of iturin A by *Bacillus amyloliquefaciens* suppressing *Rhizoctonia solani*. Soil Biol Biochem 34:955–963

Zhang X, Zhang BX, Zhang Z, Shen WF, Yang CH, Yu JQ, Zhao YH (2005) Survival of the biocontrol agents Brevibacillus brevis ZJY-1 and Bacillus subtilis ZJK-116 on the spikes of barley in the field. J Zhejiang Univ Sci B 6:770–777

Plant Growth Promoting Rhizobacteria as Biocontrol Agents Against Soil-Borne Plant Diseases

Nico Labuschagne, T. Pretorius, and A.H. Idris

Contents

1 Introduction 212
1.1 Concepts and Definitions 213
1.2 Advantages and Disadvantages 214
2 Biocontrol of Soilborne Diseases by Means of PGPR with Emphasis on Cereal Crops 215
3 Modes of Action of PGPR as Biocontrol Agents 220
3.1 Production of Antifungal Metabolites 220
3.2 Induction of Systemic Resistance 223
3.3 Root Colonization and Rhizosphere Competence 224
4 Latest Advances and Future Prospects of PGPR as Biocontrol Agents in Plants 224
5 Conclusion 225
References 226

Abstract Soil-borne diseases are responsible for major crop losses worldwide. Alternatives to the use of synthetic chemicals for disease control are increasingly being sought due to among other reasons, the detrimental effects of these compounds on the environment. In this chapter, biological control of soil-borne plant diseases by means of plant growth promoting rhizobacteria (PGPR) is reviewed with emphasis on cereals. The concepts and definitions of PGPR, biocontrol agents, biopesticides, biofertilizers, and soil inoculants are discussed and overlap between these categories are illustrated. Advantages and disadvantages of the use of PGPR

N. Labuschagne (✉) and T. Pretorius
Department of Microbiology and Plant Pathology, University of Pretoria, Pretoria, South Africa
e-mail: nl@up.ac.za, t.pretorius@tuks.co.za

A.H. Idris
ARC-Plant Protection Research Institute, Private bag X134, Queenswood 0121, 0001, Pretoria, South Africa
e-mail: HassenA@arc.agric.za

D.K. Maheshwari (ed.), *Plant Growth and Health Promoting Bacteria*,
Microbiology Monographs 18, DOI 10.1007/978-3-642-13612-2_9,

as biocontrol agents are mentioned. Biocontrol of soil-borne diseases of crops is discussed and illustrated by means of specific examples of effective application of growth promoting rhizobacteria for control of soil-borne pathogens on cereals such as wheat and sorghum. The modes of action of PGPR with biocontrol activity is discussed with reference to the production of antibiotics, siderophores, and cell wall degrading enzymes as well as induction of systemic resistance, root colonization efficacy, and rhizosphere competence.

1 Introduction

Research in the area of plant growth-promoting rhizobacteria (PGPR) has opened up a fascinating world of remarkable diversity not only in terms of the rhizobacteria but also in terms of the multifaceted beneficial plant–microbe interactions and effects involved. These interactions and effects encompass both enhancement of plant growth directly and indirectly through biological control of plant pathogens.

From the volume of scientific publications appearing on the topic of biocontrol by means of PGPR, it is evident that this is an active and growing field of science. Some of the reasons for the sustained interest in PGPR and also biocontrol by means of PGPR include the following:

(a) Huge crop losses sustained due to diseases including soilborne diseases
(b) The increase in production costs, especially fertilizer costs
(c) The global trend toward the use of more environmentally friendly production methods

Huge amounts of money are being spent on application of synthetic pesticides to control soilborne diseases worldwide. The application of rhizobacteria that colonize the roots of crop plants and suppress soilborne diseases is becoming an alternative choice to the use of chemical fungicides because of increased environmental and health concerns as mentioned earlier (Raupach and Kloepper 1998; Walsh et al. 2001; Kobayashi et al. 2002). The use of PGPR as soil inoculants for control of soilborne diseases, therefore, constitutes a viable biological alternative.

Rhizobacteria with biocontrol efficacy often provide long-term protection from soilborne pathogens at the root surface because they are often rhizosphere competent, that is, they have the capacity to rapidly colonize the rhizosphere and spread down the root from a single seed treatment or drench application into the soil (Rangarajan et al. 2003; Whipps 2007).

The literature on PGPR is voluminous and in the last 10 years there have been more than 26 reviews, including some chapters in books, on the topic of biocontrol by means of rhizobacteria. However, many of these reviews did not only deal with PGPR as biocontrol agents but also discussed micro-organisms other than PGPR (Avis et al. 2008; Compant et al. 2005; Fravel 2005; Lucy et al. 2004; Pielach et al.

2008; Preston 2004; Raaijmakers et al. 2008; Whipps 2001; Zahir et al. 2004). The review by Lucy et al. (2004) gives an extensive summary of examples of free-living PGPR tested on various crop types.

The current chapter is focused on biological control of soilborne diseases and mechanism of biocontrol agents (PGPR) with special emphasis on soilborne diseases of cereals.

1.1 Concepts and Definitions

At the outset, it is necessary to clarify some key concepts and definitions. In the literature on PGPR and biocontrol agents (BCA), different definitions and classifications are being used. PGPR as a group of free-living bacteria occupying the rhizophere and rhizoplane of plants finds itself in between a number of different groupings. Some authors consider PGPR and BCAs to be separate groups while others consider BCAs to be a subgroup of PGPR. Bashan and Holguin (1998), for example, proposed the division of PGPR into two classifications namely "biocontrol-plant growth promoting bacteria" and "plant growth promoting bacteria." Clearly there is overlap between these groupings. For example, a PGPR can have plant growth enhancing activity as its primary effect and as its secondary effect, it reduces the disease by enabling the plant to outgrow and thereby "escape" the disease. However, there are many specific examples of PGPR with direct biocontrol activity as will be discussed later in this chapter.

In addition to PGPR and BCAs, there are also the classifications of "biofertilizers," "biopesticides," "biofungicides," and "soil inoculants." Depending on the definition one ascribes to, some PGPR can be classified as biofertilizers (purely enhancing plant growth), biocontrol agents (suppressing or controlling plant disease), and biopesticides (controlling plant pests). "Soil inoculants" are mostly used as a general term for biological products (microbials), which are applied to the soil.

We define PGPR as a group of free-living rhizosphere occupying bacteria that enhances plant growth and can also be classified as biocontrol agents, biofertilizers, or biopesticides, depending on their activities/abilities. We concur with the definition of Menn and Hall (1999) for biopesticides as microbials or products derived from microbes, plants, and other biological entities, applied for control of plant pests. Furthermore, we concur with the definition for biofertilizers proposed by Vessey (2003) as a substance that contains living microorganisms, which, when applied to seed, plant surfaces, or soil, colonizes the rhizosphere or the interior of the plant and promotes growth by increasing the supply and availability of primary nutrients to the host plant. In the current review, we define biocontrol agents more specifically as microbials capable of suppressing or controlling plant diseases. The relationships and overlap between some of these groupings are illustrated in the proposed model in Fig.1.

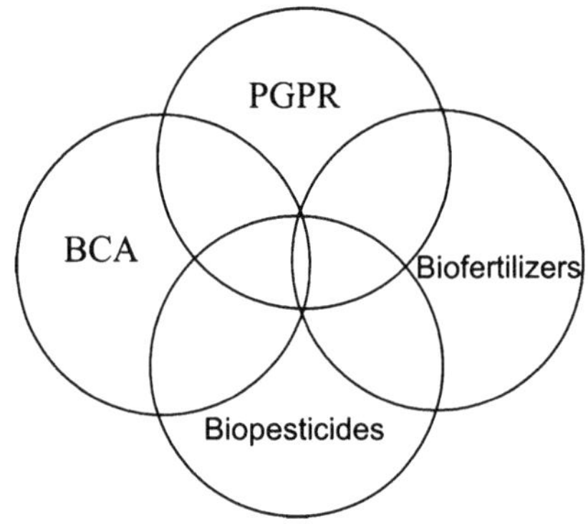

Fig. 1 Proposed model illustrating the relationships/overlaps between PGPR, BCAs, biopesticides, and biofertilizers

1.2 Advantages and Disadvantages

PGPR as biocontrol agents have certain advantages over conventional chemical control compounds. Firstly, PGPR are beneficial, naturally occurring micro-organisms, which are environmentally friendly and nontoxic. Secondly, from an ecological perspective, their application is sustainable (long term). Another advantage of PGPR is the fact that they possess a diverse range of modes of action including antibiosis, production of siderophores, cell wall degrading enzymes, bio-surfactants and volatiles, and also induces systemic resistance in plants. The fact that some PGPR by definition directly enhances the growth of plants is an additional advantage.

There are also, however, certain disadvantages to the use of PGPR based BCAs compared with conventional chemical control compounds. Firstly, being live micro-organisms, they are more sensitive to environmental conditions such as temperatures, soil conditions desiccation, etc. Shelf life of commercial PGPR or BCAs in general is shorter than that of the chemical pesticides or fungicides. Secondly, and probably the most important disadvantage, is the fact that up to now, efficacy of PGPR and BCA in general has been inconsistent under field conditions. Many scientific publications report effective biocontrol under environmentally controlled conditions in vitro or in greenhouses, but much fewer data exist regarding efficacy under field conditions. However, this does not detract from the fact that PGPR as BCAs is constantly becoming more effective as researchers are gaining more knowledge on the factors and mechanisms involved in biological control of plant diseases by means of PGPR and the factors that play a role in biocontrol of plant diseases.

Another area for application of PGPR as BCAs is that of formulation and application of the commercial product. Formulating a live micro-organism into a commercial product in such a way that it remains viable and that it can be applied by growers on a large scale is evidently more difficult than formulating a chemical pesticide.

We concur with the view of various other authors that biological products, be they BCAs or biopesticides, should not be seen as replacements for chemical pesticides on agricultural crops, but rather as important components of an integrated disease control program.

2 Biocontrol of Soilborne Diseases by Means of PGPR with Emphasis on Cereal Crops

The importance of soilborne diseases is indirectly illustrated by the fact that soil fumigation often results in increases in production of between 7 and 100%, for example, in wheat (Cook 1992), although other factors apart from disease control are also involved in this phenomenon. Soilborne diseases affect all crops and encompass the whole spectrum of plant pathogens including fungi, bacteria, and nematodes. Several groups of soilborne fungi attack most of the economically important crops causing infection resulting in huge yield losses (Gohel et al. 2006).

Cereals are as much affected by soilborne diseases as any other crop. Crown rot of wheat and barley in the Pacific Northwest in the US, for example, caused by a complex consisting of mainly *Fusarium* spp, can cause yield losses of up to 35% in commercial fields (Smiley et al. 2005). The economic as well as socioeconomic importance of cereals, such as wheat (*Triticum aestivum* L), rice (*Oryza sativa*), and maize (*Zea mays* L), which the most important crops worldwide, makes control of cereal root diseases a priority.

There are many examples of effective control of soilborne diseases by means of PGPR (Whipps 2001; Lucy et al. 2004), and many bacterial strains have been shown to have potential for development as biocontrol agents on cereals (Table 1). The biocontrol potential of *Bacillus* spp. as important agents to combat root and soilborne pathogens has been reported in many crops including chickpea (Landa et al. 1997). Several *Bacillus* spp. isolated from the rhizosphere of chickpea had shown antagonistic activity against fusarium wilt caused by *Fusarium oxysporum*. Similarly, several strains of *Bacillus* spp. isolated from the rhizosphere of sorghum in Ethiopia and wild grass spp. in South Africa have shown effective biocontrol of the root and crown rot pathogens *F. oxysporum* and *Pythium ultimum*, respectively, in sorghum under greenhouse conditions (Figs. 2 and 3) (Idris et al. 2007, 2008). Effective control of crown and root rot of wheat, caused by *F. oxysporum*, has been achieved with a strain of *Paenibacillus alvei* in South Africa (Labuschagne and Idris, unpublished data) (Fig. 4). Apart from disease control, this strain has also been demonstrated to induce about 40% increase in wheat shoot mass in the absence of pathogens. On the basis of this and other data, *P. alvei* strain has been included together with another PGPR strain in a commercial product, which is being marketed as a soil inoculant in South Africa under the trade name BacUp®.

Several commercial biocontrol products are currently available on cereals and a variety of other crops (Coping 2001; McSpadden and Fravel 2002) and new products are constantly entering in the market. Although there are several PGPR products available as soil inoculants on cereals, most of these are marketed as biofertilizers and not as biocontrol agents (Ryder et al. 1999).

Table 1 Examples of biocontrol of fungal plant pathogens on cereal crops by means of rhizobacterial application

Biocontrol agent	Plant pathogen	Crop	Mode of action of biocontrol agent	Reference
Pseudomonas fluorescens	*Microconidium nivale/ Fusarium nivale*	Wheat	Growth promotion, siderophore production, in vitro antibiosis	Amein et al. (2008)
Acinetobacter, Pseudomonas, Staphylococcus, Bacillus, Enterobacter, Pantoae, Alcallgenes	*Fusarium oxysporum, F. culmorum, F. solani, Pythium ultimum, Alternaria alternate, Botrytis cinerea, Phytophtora cryptogea*	Wheat	Antagonism and growth promotion	Egamberdieva et al. (2008)
Pseudomonas fluorescens	*Fusarium culmorum*	Rye	Fe(III) chelating compounds (including siderophores)	Kurek and Jaroszuk-Scisel (2003)
Bacillus sp. L324-92	*Gaeumannomyces graminis* var *tritici*, Rhizoctonia root rot, R. *solani* AG8, Pythium root rot, *Pythium irregulare P. ultimum.*	Wheat	Not specified	Kim et al. (1997)
Bacillus subtilis and *B. cereus*	Take all (*G. graminis* var *tritici*) Rhizoctonia root rot (R. *solani* AG8)	Wheat	Growth promotion	Ryder et al. (1999)
Bacillus spp, *Pseudomonas fluorescens*	*G. graminis, R. solani, R. oryzae, P. ultimum,*	Wheat	Not specified	Cook et al. (2002)
Bacillus subtilis CE1	*Fusarium verticilloides*	Maize	Not specified	Cavaglieri et al. (2005)
Pseudomonas chlororaphis	*Macrophomina phaseolina* (charcoal rot of sorghum)	Sorghum	Extracellular antibiotics, production of volatiles, siderophores, effective root colonization	Das et al. (2008)
	Fusarium oxysporum	Sorghum		Idris et al. (2007)

Bacillus stearothermophilus, *B. cereus*, *B. licheniformis*, *B. circulans*, *Chromobacterium violaceum*			Antibiotic production, chitinolytic enzymes, efficient root colonization	
Bacillus cereus, *Brevibacterium laterosporus*, *Pseudomonas fluorescens*, *Serratia marcescens*	*Pythium ultimum*	Sorghum	Antibiotic production, siderophores, induction of systemic resistance	Idris et al. (2008)
Pseudomonas fluorescens MKB 100 and MKB 249, *P. frederiksbergensis* 202, *Pseudomonas* spp. MKB 158	*Fusarium culmorum*	Wheat and barley	Induced resistance, antibiotic production, pathogenesis related proteins (induced resistance) in wheat	Khan et al. (2006)

[a]Additional examples of biocontrol agents on other crops can be found in the reviews by Whipps (2001) and Lucy et al. (2004), the latter review including a comprehensive table of isolates reported to be effective against soilborne pathogens of several crops

Fig. 2 Greenhouse experiment illustrating the efficacy of three *Bacillus* strains for biocontrol of root and crown rot of sorghum caused by *F. oxysporum*. All plants treated with *F. oxysporum* alone died (Control a, *far right*) whereas 100% of the plants inoculated with both the pathogen and Bacillus isolates KBE5-7, NAE5-7, and KBE9-1 survived, showing no symptoms of infection. (Adopted from Idris et al. 2007)

Fig. 3 Example of suppression of *P. ultimum* root rot in 4-week-old sorghum seedlings by bacterial strains isolated from the rhizosphere of wild grasses in South Africa. (**a**) Plants inoculated with *P. ultimum* and treated with rhizobacterial isolates. (**b**) Control plants that were treated only with *P. ultimum* developed visible root rot and necrotic leaves. (Adopted from Idris et al. 2007)

However, most putative biocontrol agents fail in the field. Factors that can affect biocontrol efficacy in the field include effects of the environment, ecological factors, and difficulties in production, formulation, and delivery of BCAs. Suggested solutions to overcome these constraints include combination of BCAs with

Fig. 4 Effective control of *F. oxysporum* crown and root rot of wheat with a strain of *Paenibacillus alvei* in the greenhouse (Labuschagne N and Idris A H, unpublished data). Plants on the *left*: inoculated with *F. oxysporum* and treated with *Paenibacillus alvei*. Plants in the *middle*: pathogen free (uninoculated) and untreated. Plants on the *right*: inoculated with *F. oxysporum* only

chemical pesticides/fungicides, modification of agronomic practices, application of BCA mixtures, and genetic manipulation. The efficacy of biocontrol is also affected by the screening and sourcing protocol used in the development of BCAs (Spadaro and Gullino 2005; Fravel 2005). It has been shown that root colonization is an important aspect, which has a determinative impact on biocontrol efficacy (Van Bruggen et al. 2008).

The outcome (i.e., success) of a biocontrol agent treatment depends on the following:

1. The method of inoculation/application
2. The physiological state of the BCA
3. The concentration and dosage of the BCA
4. The presence or absence of nutrients
5. The presence or absence of adjuvants such as adhering or protective agents (Knudsen et al. 1997)
6. The media used for BCA production
7. The volume of treatment (Levenfors et al. 2008)
8. The plant type and cultivar. Both plant and cultivar specificity has been observed for some BCAs (Khan et al. 2006)

Other indirect factors include the effect of fungi on BCA colonization as reported for wheat roots (Mazzola and Cook 1991) and host plant–BCA interaction (Lugtenberg et al. 2002). Consideration should also be given to the effect of the

BCA application on the microbial ecology and occurrence of phenomena such as disease replacement where a particular root disease is controlled but another takes its' place (Kim et al. 1997).

3 Modes of Action of PGPR as Biocontrol Agents

For successful and sustainable biocontrol under field conditions, it is imperative that the mode of action of the BCA strains being used is known. The mode of action involved will be a determining factor in the type of disease control strategy to be implemented.

3.1 Production of Antifungal Metabolites

PGPR including those associated with cereal crops produce various types of antifungal metabolites capable of reducing or suppressing infection by pathogenic fungi in several crops (Ongena et al. 1999; Bloemberg and Lugtenberg 2001; Raaijmakers et al. 2002).

3.1.1 Antibiotics

Antibiosis is an attractive and a highly effective mode of action of rhizobacteria in the suppression of soilborne infections in a number of crops (Handelsman and Stab 1996). Most biocontrol strains of PGPR produce one or several groups of antibiotics, which inhibit fungal pathogens (Haas and Defago 2005). Antibiotics produced by these biocontrol PGPR reduce or suppress soilborne infections of cereal crops including wheat, rice, maize, chickpea, and barley (Raaijmakers et al. 2002). Some of these antibiotics cause membrane damage to pathogens such as *Pythium* spp. and inhibit zoospores formation (de Souza et al. 2003). Others such as the phenazines inhibit electron transport in disease causing organisms and also act by damaging lipids and other macromolecules (Haas and Defago 2005).

Genetic analysis of many biocontrol strains of *Pseudomonas* indicated that there is a positive correlation between disease suppression and antibiotic production (Vincent et al. 1991). It was demonstrated that with increasing populations of *Pseudomonas* spp., which produce the antibiotic 2,4-diacetylphloroglucinol (2,4-DAPG), there was a rapid decline in take-all disease in wheat caused by the fungus *Gaeumanomyces graminis* var. *tritici* (Raaijmakers and Weller 1998; de Souza et al. 2003). The production of phenazine-1-carboxylic acid (PCA), another group of antibiotics by *Pseudomonas fluorescens* and *Pseudomonas aureofaciens* strains, has also been described elsewhere. Bacterization of wheat seeds with *P. fluorescens* strains 2–79 producing the antibiotic PCA resulted in significant suppression of take-all in about 60% of field trials (Weller 2007).

There is a growing list of reports of *Bacillus* spp. as biocontrol agents in various crops. Kim et al. (1997), for instance, isolated and discovered a potential biocontrol strain, *Bacillus* sp. L324-92, with a broad spectrum inhibitory activity against take-all, root rot caused by *Rhizoctonia. solani*, *Pythium irregulare*, and *Pythium ultimum*. In other experiment (El-Meleigi et al. 2007), treatment of spring wheat seeds with antibiotic producing strains of *Bacillus* spp. has been reported as a powerful tool to control root rot causing fungal pathogens in dry land fields. According to this work, application of *Paenibacillus polymyxa* to wheat seeds suppressed infection by root rot pathogens *Fusarium graminearum* and *Cochliobolus sativum*.

The potential uses of antibiotic producing PGPR as biocontrol agents have been reported in many other cereals including maize, sorghum, rice, and chickpea. In maize for instance, *Fusarium verticilloides*, causing root rot and yield loss, has been significantly suppressed by the application of *Bacillus amyloliquifaciens* as seed treatment (Pereira et al. 2009). Von der Weid et al. (2005) recently described *Paenibacillus brasilensis* PB177, a new strain isolated from the rhizosphere of maize in Brazil that produces antimicrobial substances suggesting that it could be a potential biocontrol agent in the rhizosphere of maize.

In another biocontrol experiments, Idris et al. (2007, 2008) demonstrated the bio-control of *F. oxysporum* and *Pythium ultimum* on sorghum with *Bacillus* spp. (mentioned under point 2 earlier in this chapter). It was demonstrated that the bacterial strains produce antimicrobial metabolites, possibly antibiotics, which suppressed the growth of the fungal pathogens in vitro (Fig. 5).

3.1.2 Siderophores

Biocontrol PGPRs also exert their antagonistic activity against plant pathogens by means of secretion of siderophores. These low molecular weight compounds

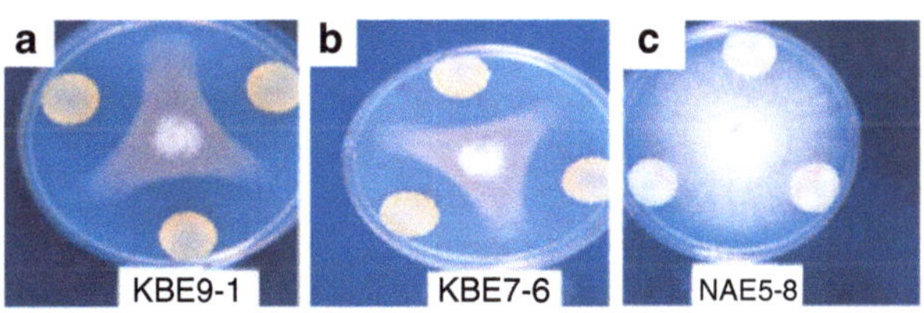

Fig. 5 Dual culture assay for screening of biocontrol agents based on the production of antifungal metabolites in agar plates. *Bacillus* strains KBE9-1, KBE7-6, and NAE5-8 were inoculated in three equidistant positions at the margin of Potato Dextrose Agar (PDA) plates with *F. oxysporum* agar block placed in the centre. The growth of the fungal mycelium was significantly inhibited by strains KBE9-1 (**a**) and KBE7-6 (**b**) showing prominent inhibition zones on the plates. Strain NAE5-8 does not produce any inhibition zones, indicating the absence of antifungal metabolites (**c**). (Adopted from Hassen 2007, PhD thesis)

(400–1, 500 Da) preferentially chelate iron (Fe^{+++}) and transport it into the cell across the cell membrane (Neilands 1995; Wandersman and Delepelaire 2004). The siderophores bind most of the Fe^{+3} in the rhizosphere and effectively prevent the proliferation of fungal pathogens by depriving them of available iron (Kloepper et al. 1980; O' Sulivan and O' Gara 1992). Suppression of the pathogens arises because iron deficiency causes growth inhibition, decrease in nucleic acid synthesis, inhibition of sporulation, and causes changes in cell morphology (Mathiyazhagan et al. 2004).

Among the biocontrol rhizobacteria, the fluorescent *Pseudomonas* spp. are efficient competitors for iron (Fe^{+3}) in the rhizosphere of various crops producing two major types of siderophores: the fluorescent pigmented pyoverdins (pseudobactins) (Lemanceau et al. 1993) and the nonfluorescent siderophore called pyochelins (Leeman et al. 1996). Siderophores produced by certain strains of the *P. fluorescens-putida* group are responsible for enhanced plant growth and biocontrol and are most often associated with fungal suppression in the rhizosphere of several crops (Battu and Reddy 2009). According to these workers, siderophore mediated the suppression of rice fungal pathogens *R. solani* and *Pyricularia oryze* in an in-vitro assay on Kings-B medium. Earlier, Becker and Cook (1988) reported the role of siderophores produced by *Pseudomonas* strain B324 in the suppression of *Pythium* root rot of wheat. Mutants deficient in pyoverdins production are less effective than parental strains in suppression of fungal pathogens (Loper and Henkels 1999). It is thus believed that siderophore production is another important mechanism by which some strains of bacteria protect plants against root pathogens.

3.1.3 Cell Wall Degrading Enzymes

One of the major mechanisms used by biocontrol agents to control soilborne pathogens involves the production of cell wall degrading enzymes (Chet et al. 1990; Kobayashi et al. 2002). Cell wall degrading enzymes such as β-1, 3-glucanase, chitinase, cellulase, and protease secreted by biocontrol strains of PGPR exert a direct inhibitory effect on the hyphal growth of fungal pathogens. Chitinase and β-1,3-glucanase degrade chitin, an insoluble linear polymer of β-1,4-N-acetylglucoseamine, which is the major component of the fungal cell wall.

The β-1, 3-glucanase synthesized by strains of *Paenibacillus* and *Streptomyces* spp. lyse fungal cell walls of pathogenic *F. oxysporum*. In a similar manner, *Bacillus cepacia* synthesizes β-1,3-glucanase, which destroys the cell walls of the soilborne pathogens *R. solani, P. ultimum*, and *S. rolfsi* (Compant et al. 2005). Potential biocontrol agents with chitinolytic activities include *B. licheniformis*, *B. cereus*, *B. circulans*, and *B. thuringiensis* (Sadfi et al. 2001). Among the Gram-negative bacteria, *Serratia marcescens*, *Enterobacter agglomerans*, *Pseudomonas aeruginosa*, and *P. fluorescens* have been found to have chitinolytic activities (Nelson and Sorenson 1999).

Cell wall degrading enzymes of rhizobacteria affect the structural integrity of the walls of the target pathogen (Budi et al. 2000). Someya et al. (2000) studied the chitinolytic and antifungal activities of a potent biocontrol strain of *S. marcescens* B2 against the soilborne pathogens *R. solani* and *F. oxysporum*. The mycelia of the fungal pathogens coinoculated with this strain showed various abnormalities such as partial swelling in the hyphae and at the tip, hyphal curling or bursting of the hyphal tip. Examples of protection from phytopathogenic infection as a result of the activity of cell wall degrading enzymes include control of *Sclerotium rolfsii* and *F. oxysporum* on beans (Felse and Panda 1999).

3.2 Induction of Systemic Resistance

Induced systemic resistance (ISR) is the state of defensive capacity developed by the plant when stimulated by diverse agents including rhizobacteria (van Loon et al. 1998). Once resistance is induced in plants, it will result in nonspecific protection against pathogenic fungi, bacteria, and viruses (Silva et al. 2004). The mode of action of disease suppression by nonpathogenic rhizosphere bacteria should be distinguished from pathogen induced systemic acquired resistance (SAR) (Bakker et al. 2003). Colonization of the plant root system by rhizobacteria can indirectly lead to reduced pathogen attack through induction of systemic resistance (Kloepper and Beauchamp 1992). PGPR elicit ISR in plants by increasing the physical and mechanical strength of the cell wall as well as changing the physiological and biochemical reactions of the host. This results in the synthesis of defense chemicals such as chitinase, peroxidase, and pathogenesis-related proteins (Ramamoorthy et al. 2001; Nandakumar et al. 2001; Silva et al. 2004).

In rice, *P. fluorescens* strains showed inhibitory effect on the mycelial growth of *R. solani* by inducing resistance in the plant (Radjacommare et al. 2004). The bacteria induced resistance against the sheath blight fungus by activating chitinase genes in rice (Nandakumar et al. 2001). Another biocontrol PGPR, *S. marcescens* strain B2, which inhibits several soil borne pathogens including *F. oxysporum* under greenhouse conditions, could not inhibit the same pathogens in a dual culture assay indicating that this is due to the induction of systemic resistance (Someya et al. 2002).

In beans, *P. aeroginosa* ISR against infection by *Collehotricum lindemuthianum* (Bigirimana and Hofte 2002). Benhamou et al. (1996) investigated ISR in *Pisum sativum* and found that pea roots inoculated with *P. fluorescens* strain 63–28 produced more chitinase at the site of penetration by *F. oxysporum* f. sp. *pisi*. Several strains of *Bacillus* spp. also have the capacity to induce systemic resistance in various crops against a wide range of pathogens. *Bacillus subtilis* AF1 isolated from soils suppressive to pigeon pea (*Cajanus cajan*) wilt caused by *Fusarium* sp. caused lysis of *Aspergillus niger* by stimulating the production of phenylalanine ammonia lyase and peroxidase by the plant thereby eliciting induction of systemic resistance (Kloepper et al. 2004).

Similar to other modes of action, rhizobacterial-mediated ISR can be an important additional means of environmentally friendly plant disease control (van Loon et al. 1998).

3.3 Root Colonization and Rhizosphere Competence

Root colonization is an important prerequisite for bacteria to be considered as true PGPRs, and it is commonly believed that a biocontrol agent should colonize the rhizosphere and the surface of the plant it protects (Silva et al. 2003; Handelsman and Stab 1996; Benizri et al. 2001). Therefore, any given PGPR is often ineffective as a biocontrol agent against root disease if it does not colonize the roots efficiently (Montealegre et al. 2003).

Pseudomonas and *Bacillus* spp. are the most important root colonizing PGPR in various crops. Several members of this group have widespread distribution in the soil, are efficient colonizers of the rhizosphere, and produce various types of metabolites inhibitory to a wide range of pathogens in plants (Rangarajan et al. 2003). Many other root colonizing strains of PGPR have also been found to have antifungal properties toward a number of pathogens in soil.

However, for many of the potential biocontrol strains including *Pseudomonas* and *Bacillus* spp., biological control of soilborne diseases is often inconsistent. One of the major factors associated with this inconsistency is insufficient root colonization by introduced bacteria (Bloemberg and Lugtenberg 2001). Correlation of poor biocontrol performance of a biocontrol agent with inefficient root colonization has been confirmed by means of mutants of *Pseudomonas* strains, which had lost their biocontrol activity. In this regard, it is essential to understand the bacterial traits that contribute to root colonization.

It is now possible to detect and enumerate microorganisms in situ on plant surfaces using molecular techniques. In the study of root colonization of bacteria in situ, one of the approaches was the use of marker genes such as the *gfp* gene encoding the green fluorescent protein (GFP). GFP transformed bacteria can be monitored and visualized using confocal laser scanning microscopy (CLSM) (Bloemberg and Lugtenberg 2001). Apart from visualizing root colonization, the GFP technique can also be used to study the colonization patterns of different biocontrol agents.

4 Latest Advances and Future Prospects of PGPR as Biocontrol Agents in Plants

With the advancement and innovations of current biotechnological research over the past ten years, there is now vastly improved knowledge on the beneficial effects of both biocontrol and growth enhancing PGPR. Several strategies have so far been exploited to increase the efficacy of biocontrol strains to develop them for widespread use in agriculture. Because of their metabolic versatility, excellent root colonization capability, and their capacity to produce a wide range of antifungal metabolites, intense biotechnological research is being done on the soil borne fluorescent Pseudomonads (Walsh et al. 2001). For example, the antifungal metabolite 2,4-diacetylphloroglucinol (2,4 DAPG) is an important metabolite produced by these biocontrol strains. In this regard, the development of sensitive in situ detection methods of 2,4-DAPG helped to understand the relationship between effective BCA pseudomonads and suppressive soils in the suppression of take-all disease caused by *Gaeumanomyces graminis* var. *tritici* (Raaijmakers et al. 1999; Walsh et al. 2001).

Improving the biocontrol efficacy of potential rhizobacteria by means of genetic modifications involves, for instance, the construction of strains that produce increased levels of antimicrobial and growth enhancing metabolites (Walsh et al. 2001). By transforming *P. fluorescens* CHAO with the gene coding for 1-aminocyclopropane-1-carboxylic acid deaminase, for instance, the plant growth promotion and biocontrol capacity of this strain have been increased (Wang et al. 2000). Novel perspectives are emerging regarding biocontrol and optimizing the application of biocontrol strains for future use.

The identification of *P. fluorescens* genes associated with root colonization and that are specifically expressed in the rhizosphere (*rhi* genes) by means of in-vivo expression technology (IVET) is another important innovation (Bloemberg and Lugtenberg 2001). Many such root colonizing genes and traits from *P. fluorescens* have been identified and used to improve root colonization patterns of wild type *Pseudomonas* strains (Lugtenberg and Dekkers 1999) In some biocontrol PGPR, efficient root colonization is linked to a site-specific recombinase gene, and transfer of this gene from a rhizosphere competent *P. fluorescens* strain to a noncompetent strain improved its root colonization ability (Compant et al. 2005).

5 Conclusion

Two important principles pointed out by Baker and Cook (1974) should be borne in mind. First, there is no one system by which biological control works, each relationship is unique. Second, analysis of the microorganisms involved, as well as their relationships and interactions on biochemical/molecular level, becomes a means of perfecting the result obtained and is not a necessary precursor to attempting biological control.

In conclusion, as there are numerous examples of effective biocontrol candidates, the future challenge is not to prove that biocontrol is possible, but to improve efficacy and durability of biocontrol in the field. This will only be achieved through a better understanding of the biocontrol mechanisms, plant–microbe interactions and processes as well as microbial ecology in the soil and rhizosphere. The necessary molecular tools for studying these processes and interactions are already available. If this is achieved, the efficacy of biocontrol could conceivably be improved through application of this knowledge to develop improved screening protocols, formulation, and application procedures as well as new innovative integrated disease management practices.

References

Amein T, Omer Z, Welch C (2008) Application and evaluation of *Pseudomonas* strains for biocontrol of wheat seedling blight. Crop Prot 27:532–536

Avis TJ, Gravel V, Antoun H, Tweddell RJ (2008) Multifaceted beneficial effects of rhizosphere microorganisms on plant health and productivity. Soil Biol Biochem 40:1733–1740

Baker KF, Cook JR (1974) Biological control of plant pathogens. WH Freeman and Company, San Francisco

Bakker PAHM, Ran LX, Cm P, van Loon CC (2003) Understanding the involvement of rhizobacteria mediated induction of systemic resistance in biocontrol of plant pests. Can J Plant Pathol 25:5–9

Bashan Y, Holguin G (1998) Proposal for the division of plant growth promoting rhizobacteria into two classifications: biocontrol-PGPB (plant growth promoting bacteria) and PGPB. Soil Biol Biochem 30(9):1225–1228

Battu PR, Reddy MS (2009) Siderophore mediated antibiosis of rhizobacterial fluorescent *Pseudomonads* against rice fungal pathogens. Int J Pharm Tech Res 1:227–229

Becker JO, Cook RJ (1988) Role of siderophore in suppression of *Pythium* species and production of increased growth response of wheat by fluorescent *Pseudomonas*. Phytopathology 78:778–782

Benhamou NR, Belanger R, Paulitz TC (1996) Induction of differential host responses by *Pseudomonas fluorescens* in Ri T-DNA transformed pea roots after challenge with *Fusarium oxysporum* f. sp. pisi and *Pythuim ultimum*. Phytopathology 86:114–178

Benizri E, Baudoin E, Guckert A (2001) Root colonization by inoculated plant growth promoting rhizobacteria. Biocontrol Sci Technol 11:557–574

Bigirimana J, Hofte M (2002) Induction of systemic resistance to *Colletotrichum lindemuthianum* in bean by a benzothiadiazole derivative and rhizobacteria. Phytoparasitica 30:159–168

Bloemberg GV, Lugtenberg BJJ (2001) Molecular basis of plant growth promotion and biocontrol by rhizobacteria. Curr Op Plant Biol 4:343–350

Budi SW, van Tuinen D, Arnould C, Dumas-Gaudot E, Gianinazzi-Pearson V, Gianinazzi S (2000) Hydrolytic enzyme activity of *Paenibacillus* sp. strain B2 and effects of the antagonistic bacterium on cell integrity of two soil borne pathogenic bacteria. Appl Soil Ecol 15:191–199

Cavaglieri I, Orlando J, Rodriguez MI, Chulze S, Etcheverry M (2005) Biocontrol of *Bacillus subtilis* against *Fusarium verticillioides* in vitro and at the Maize root level. Res Microbiol 156:748–754

Chet I, Ordentilich A, Shapira R, Oppenheim A (1990) Mechanisms of biocontrol of soilborne plant pathogens by rhizobacteria. Plant Soil 129:85–92

Compant S, Duffy B, Nowak J, Clément C, EA BI (2005) Use of plant growth-promoting bacteria for biocontrol of plant diseases: principles, mechanisms of action, and future prospects. Appl Environ Microbiol 71:4951–4959

Cook JR, Weller DM, El-Banna AY, Vakoch D, Zhang H (2002) Yield responses of direct-seeded wheat to rhizobacteria and fungicide seed treatments. Plant Dis 86(7):780–784

Cook RJ (1992) Wheat root health management and environmental concern. Can J Plant Pathol 14:76–85

Coping LG (ed) (2001) The biopesticides manual: a world compendium. British Crop Protection Council, Farnham, Surrey, UK

Das IK, Indira S, Annapurna A, Prabhakar SN (2008) Biocontrol of charcoal rot in sorghum by fluorescent Pseudomonads associated with the rhizosphere. Crop Prot 27:1407–1414

de Souza JT, Weller DM, Raaijmakers JM (2003) Frequency, diversity and activity of 2, 4-diacetylphloroglucinol producing fluorescent *Pseudomonas* spp. in Dutch take-all decline soils. Phytopathology 93:54–63

Egamberdieva D, Kamilova F, Validov S, Gafurova L, Kucharova Z, Lugtenberg B (2008) High incidence of plant growth-stimulating bacteria associated with the rhizosphere of wheat grown on salinated soil in Uzbekistan. Environ Microbiol 10(1):1–9. doi:10.1111/j.1462-2920.2007.01424.x

El-Meleigi MA, Hassen ZM, Ibrahim GH (2007) Biological control of common root rot of spring wheat by coating seeds with *Bacillus* or *Trichoderma* spp. JKUA: Met Environ Arid & Agric Sci 18:3–12

Felse AP, Panda T (1999) Production of microbial chitinases. Bioprocess Eng 23:127–134

Fravel DR (2005) Commercialization and implementation of biocontrol. Annu Rev Phytopathol 43:337–59

Gohel V, Singh A, Vimal A, Ashwini P, Chhatpar HS (2006) Bioprospecting and antifungal potential of chitinolytic microorganisms. African J Biotechnol 5(2):54–72

Haas D, Defago G (2005) Biological control of soil-borne pathogens by fluorescent pseudomonads. Natra Rev Microb. doi:10.1038/nrmicro1129

Handelsman J, Stab EV (1996) Biocontrol of soilborne plant pathogens. Plant C 8:1855–1869

Hassen AI (2007) Efficacy of rhizobacteria for growth promotion and biocontrol of *Pythium ultimum* and *Fusarium oxysporum* on sorghum in Ethiopia and South Africa. PhD Thesis, University of Pretoria, South Africa

Idris AH, Labuschagne N, Korsten L (2007) Screening rhizobacteria for biological control of *Fusarium* root and crown rot of sorghum in Ethiopia. Biol Control 40:97–106

Idris AH, Labuschagne N, Korsten L (2008) Suppression of *Pythium ultimum* root rot of sorghum by rhizobacterial isolates from Ethiopian and South Africa. Biol Control 45:72–84

Khan MR, Fischer S, Egan D, Doohan FM (2006) Biological control of *Fusarium* seedling blight disease of wheat and barley. Phytopathology 96:386–394

Kim DS, Cook RJ, Weller DM (1997) *Bacillus* sp. L324-92 for biological control of three root diseases of wheat grown reduced tillage. Phytopathology 87:551–558

Kloepper JW, Beauchamp CJ (1992) A review of issues related to measuring colonization of plant roots by bacteria. Canad J Microbiol 38:1219–1232

Kloepper JW, Leong J, Teintze M, Scroth MN (1980) Enhancing plant growth by siderophores produced by plant growth promoting rhizobacteria. Nature 286:885–886

Kloepper JW, Ryu CM, Zhang S (2004) Induced systemic resistance and promotion of plant growth by *Bacillus* spp. Phytopathology 94:1259–1266

Knudsen IMB, Hockenhull J, Jensen DF, Gerhardson B, Hokenberg M, Tahoven R, Teperi E, Sunderheim L, Hendriksen B (1997) Selection of biological control agents for controlling soil and seed-borne disease in the field. Eur J Plant Pathol 103:775–784

Kobayashi DY, Reedy RM, Bick JA, Oudemans PV (2002) Characterization of chitinase gene from *Stenotrophomonas maltophilia* strain 34S1 and its involvement in biological control. Appl Environ Microbiol 68(3):1047–1054

Kurek E, Jaroszuk-Scisel J (2003) Rye (*Secale cereale*) growth promotion by *Pseudomonas fluorescens* strains and their interactions with *Fusarium culmorum* under various soil conditions. Biol Control 26:48–56

Landa BB, Hervas A, Bethiol W, Jimenez-Diaz RM (1997) Antagonistic activity of bacteria from the chickpea rhizosphere against *Fusarium oxysporum* f.sp.ciceris. Phytoparasitica 25:305–318

Leeman M, den Ouden FM, Pelt JA, Dirik FPM, SteijL H, Bakker PAHM, Schippers B (1996) Iron availability affects induction of systemic resistance to *Fusarium* wilt of radish by *Pseudomonas fluorescens*. Phytopathology 86:149–155

Lemanceau P, Bakker PAHM, De Kogel WJ, Alabouvette C, Schippers B (1993) Antagonistic effect of non pathogenic *Fusarium oxysporum* Fo47 and pseudobactin 358 upon pathogenic *Fusarium oxysporum* f. sp. dianthi. Appl Environ Microbiol 59:74–82
Levenfors JP, Eberhard TH, Levenfors JJ, Gerhardson B, Hokeberg M (2008) Biological control of snow mould (*Microdochium nivale*) in winter cereals by *Pseudomonas brassicacearum*, MA250. Biocontrol 53:651–665
Loper JE, Henkels MD (1999) Utilization of heterologous siderophores enhances level of iron available to *Pseudomonas putida* in the rhizosphere. Appl Environ Microbiol 65:5357–5363
Lucy M, Reed E, Glick BR (2004) Applications of free living plant growth-promoting rhizobacteria. Antonie Leeuwenhoek 86:1–25
Lugtenberg BJ, Dekkers LC (1999) What makes *Pseudomonas* bacteria rhizosphere competent? Environ Microbiol 1(1):9–13
Lugtenberg BJJ, Chin-A-Woeng TFC, Bloemberg GV (2002) Microbe-plant interactions: principles and mechanisms. Antonie Leeuwenhoek 81:373–383
Mathiyazhagan S, Kavitha K, Nakkeerans S, Chandrasekar MK, Renukadevi P, Krishnamoorthy AS, Fernando WGD (2004) PGPR mediated management of stem blight of *Phyllanthus amarus* (Schum and Thonn) caused by *Corynespora cassiicola* (Berk and Curt) wei. Archives Phytopathol Plant Prot 37:183–199
Mazzola M, Cook RJ (1991) Effects of fungal root pathogens on the population dynamics of biocontrol strains of fluorescent pseudomonades in the wheat rhizosphere. Appl Environ Microbiol 57(8):2171–2178
McSpadden Gardener BB, Fravel DR (2002) Biological control of plant pathogens: research, commercialization and application in the USA. Online Plant Health Progress. doi:10.1094/PHP-2002-0510-01-RV (www.planthealthmanagementnetwork.org/php/)
Menn JJ, Hall FR (1999) Biopesticides present status and future prospects. In: Hall FR, Menn JJ (eds) Biopesticides use and delivery. Humana, Totowa, New Jersey, pp 1–10
Montealegre JR, Reyes R, Perez LM, Herrera R, Silva P, Besoain X (2003) Selection of bioantagonistic bacteria to be used in biological control of *Rhizoctonia solani* in tomato. Electron J Biotechnol 6:116–127
Nandakumar R, Babu S, Viswanathan R, Raguchander T, Samiyappan R (2001) Induction of systemic resistance in rice against sheath blight disease by *Pseudomonas fluorescens*. Soil Biol Biochem 33:603–612
Neilands J (1995) Siderophores: structure and function of microbial Iron transport compounds. J Biol Chem 45:26723–26726
Nelson MN, Sorenson J (1999) Chitinolytic activity of *Pseudomonas fluorescens* isolates from barley and sugar beet rhizosphere. FEMS Microbiol Ecol 30:217–227
O' Sulivan DJ, O' Gara F (1992) Traits of fluorescent *Pseudomonas* spp. involved in the suppression of plant root pathogens. Microbiol Rev 56:662–676
Ongena M, Daayf F, Jacques P, Thonart P, Benhamou N, Paulitz TC, Cornelis P, Koedam N, Belanger RR (1999) Protection of cucumber against *Pythium* root rot by fluorescent pseudomonads: predominant role of induced resistance over siderophores and antibiosis. Plant Pathol 48:66–76
Pereira P, Nesci A, Etcheverrg MG (2009) Efficacy of Bacterial seed treatments for the control of *Fusarium verticilliodes* in maize. Biocontrol 54:103–111
Pielach CA, Roberts DP, Kobayashi DY (2008) Metabolic behavior of bacterial biological control agents in soil and plant rhizospheres. Adv Appl Microbiol 65:199–215
Preston GM (2004) Plant perceptions of plant growth-promoting *Pseudomonas*. Phil Trans R Soc Lond B 359:907–918
Raaijmakers JM, Bonsal RF, Weller DM (1999) Effect of population density of *Psedomonas fluorescens* on production of 2, 4- diacetylphloroglucinol producing bacteria isolated from the maize rhizosphere. Appl Environ Microbiol 66:948–955

Raaijmakers JM, Paulitz TC, Steinberg C, Alabouvette C, Moënne-Loccoz Y (2008) The rhizosphere: a playground and battlefield for soilborne pathogens and beneficial microorganisms. Plant Soil. doi:10.1007/s11104-008-9568-6

Raaijmakers JM, Vlami M, de Souza JT (2002) Antibiotic production by bacterial biocontrol agent. Anton van Leeuwenhoek 81:537–547

Raaijmakers JM, Weller DM (1998) Natural plant protection by 2, 4-diacetylphloroglucinol producing *Pseudomonas* spp. in take all decline soils. Mol Plant-Microbe Interact 11:144–152

Radjacommare R, Kandan A, Nandakumar R, Samiyapan R (2004) Association of the hydrolytic enzyme chitinase against *Rhizoctonia solani* in rhizobacteria treated rice plants. J Phytopathol 152:365–370

Ramamoorthy V, Viswanathan R, Raguchander T, Prakasam V, Samayapan R (2001) Induction of systemic resistance by plant growth promoting rhizobacteria in crop plants against pests and diseases. Crop Prot 20:1–11

Rangarajan S, Saleena LM, Vasudevan P, Nair S (2003) Biological suppression of rice disease by *Pseudomonas* spp. under saline conditions. Plant Soil 251:73–82

Raupach GS, Kloepper JW (1998) Mixtures of plant growth promoting rhizobacteria enhance biological control of multiple cucumber pathogens. Phytopathology 88(11):1158–1163

Ryder MH, Yan Z, Terrace TE, Rovira AD, Tang W, Correll RL (1999) Use of *Bacillus* isolated in China to suppress take-all and rhizoctonia root rot, and promote seedling growth of glasshouse-grown wheat in Australian soils. Soil Biol Biochem 31:19–29

Sadfi N, Cherif M, Fliss I, Boudabbous A, Antoun H (2001) Evaluation of bacterial isolates from salty soils and *Bacillus thuringiensis* strains for the biocontrol of *Fusarium* dry rot of potato tubers. J Plant Pathol 83:101–118

Silva HSA, Da Silva RR, Mounter A (2003) Development of root colonization bioassay for rapid screening of rhizobacteria for potential biocontrol agents. J Phytopathol 151:42–46

Silva HSA, de Silva RR, Macagnan D, de Almeda Halfeld-Viera B, Pereira MCB, Mounteer A (2004) Rhizobacterial Induction of systemic resistance in tomato plants: non specific protection and increase in enzyme activities. Biol Control 29:288–295

Smiley RW, Gourlie JA, Easley SA, Patterson LM, Whittaker RG (2005) Crop damage estimates for crown rot of wheat and barley in the Pacific Northwest. Plant Dis 89:595–604

Someya N, Kataoka N, Komagata T, Hirayae K, Hibi T, Akutsu K (2000) Biological control of cyclamen soilborme disedses by *Serratia marcescens* strain B2. Plant Dis 84:334–340

Someya N, Nakajima M, Hibi T, Yamaguchi I, Akutsu K (2002) Induced resistance to rice blast by antagonistic bacterium *Serratia marcescens* strain B2. J Gen Plant Pathol 68:177–182

Spadaro D, Gullino ML (2005) Improving the efficacy of biocontrol agents against soilborne pathogens. Crop Prot 24:601–613

Van Bruggen AHC, Semenov AM, Zelenev VV, Semenov AV, Raaijmakers JM, Savler RJ, de Vos O (2008) Wave-like distribution pattern of *GFP*-marked *Pseudomonas Fluorescens* along roots of wheat plants grown in two soils. Microb Ecol 55:466–475

van Loon LC, Bakker PAHM, Pieterse CM (1998) Systemic resistance induced by rhizosphere bacteria. Ann Rev Phytopathol 36:453–483

Vessey JK (2003) Plant growth promoting rhizobacteria as biofertilizers. Plant Soil 255:571–586

Vincent MN, Harrison LA, Brackin JM, Kovacevich PA, Mukerji P, Weller DM, Pierson EA (1991) Genetic analysis of the antifungal activity of a soilborne *Pseudomonas aureofaciens* strain. Appl Environ Microbiol 57:2928–2934

Von der Weid I, Artursson V, Seldin L, Jansson JK (2005) Antifungal and root surface colonization properties of GFP-tagged *Paenibacillus brasilensis* PB177. World J Microbiol Plant Pathol 12:1591–1597

Walsh UF, Morrissey JP, O'Gara F (2001) Pseudomonas for biocontrol of phytopathogens: from functional genomics to commercial exploitation. Curr Opin Biotech 12:289–295

Wandersman C, Delepelaire P (2004) Bacterial iron sources: from siderophores to hemophores. Annu Rev Microbiol 58:611–47

Wang C, Knill E, Glick BR, Defago G (2000) Effect of transferring 1-aminocyclopropane-1-carboxylic acid (ACC) deaminase genes into Pseudomonas fluorescens strain CHAO and its gacA derivative CHA96 on their growth promoting and disease suppressive capacities. Can J Microbiol 46:898–907

Weller DM (2007) Pseudomonas biocontrol agents of soilborne pathogens: looking back over 30 years. Phytopathology 97(2):250–256

Whipps JM (2001) Microbial interactions and biocontrol in the rhizosphere. J Exp Bot 52(Spec Issue):487–511

Whipps JM (2007) Complex multitrophic interactions in the plant environment can affect disease biocontrol. In: Proceedings of the XIV international plant protection congress. vol 2, Glasgow, Scotland, UK, pp 432–433

Zahir ZA, Arshad M, Williams T, Jr F (2004) Plant growth promoting rhizobacteria: applications and perspectives in agriculture. Adv Agron 81:97–168

Sustainable Approaches for Biological Control of Fusarium Wilt in Pigeon Pea (*Cajanus cajan* L. Millspaugh)

Piyush Pandey, Abhinav Aeron, and D. K. Maheshwari

Contents

1 Introduction 232
2 The Universal Pathogen: *Fusarium* spp. 233
2.1 Fusarium Wilt of Pigeon Pea 233
2.2 Resistant Varieties of Pigeon Pea: An Effective Strategy for Wilt Control (?) 234
3 Plant Growth Promoting Rhizobacteria 236
3.1 Biocontrol Agents: A Sustainable Eco-Friendly Strategy for Pathogen Control 237
3.2 Biological Control of *Fusarium* Wilt 238
4 Conclusion 243
References 243

Abstract *Cajanus cajan* (Pigeon pea) is an important crop of Indian subcontinent and African countries, cultivated in the tropics and subtropics. *Fusarium* wilt is one of the major yield and growth-limiting factors of pigeon pea. Along with nematodes such as *Meloidogyne incognita* and *Heterodera cajani*, *F. udum* result in highly destructive wilt disease complex, which is a major constraint for the successful cultivation of pigeon pea. *F. udum* from the same or different geographical origin have shown that the fungus is highly variable in cultural characteristics and pathogenicity. Although development and use of resistant cultivars is effec tive, economical, and environmentally sound strategy for disease control, still variable

P. Pandey
Department of Biotechnology, Division of Life Sciences, S.B.S.P.G. Institute of Biomedical Sciences and Research, Balawala, Dehradun 248161, Uttarakhand, India
e-mail: piyushgkp@rediffmail.com

A. Aeron and D.K. Maheshwari (✉)
Department of Botany and Microbiology, Gurukul Kangri University, Haridwar 249404, Uttarakhand, India
e-mail: abhinavaeron@gmail.com; maheshwaridk@gmail.com

D.K. Maheshwari (ed.), *Plant Growth and Health Promoting Bacteria*,
Microbiology Monographs 18, DOI 10.1007/978-3-642-13612-2_10,

responses with cultivation conditions had been a matter of concern. For an eco-friendly and sustainable management of fusarium wilt, biological control with the application of PGPR offers a potential nonchemical means for disease management. Several strains of *Pseudomonas* and *Bacillus* have been widely reported as effective biocontrol agents for pigeon pea wilt, though combination of several organisms have been proved more effective in field conditions.

1 Introduction

C. cajan L. Millspaugh, a multipurpose species, is extensively used as food grain and green manure crop for soil fertility amelioration in cropping systems (Tobita et al. 1994; Adu-Gyamfi et al. 1996). It is an important pulse crop in India and is a major source of protein for most of the vegetarian population worldwide (Nene et al. 1996). In India, cultivation area of pigeon pea increased from 2.2 million hectare (1.7 metric tons) in 1950–1951 to 3.8 M ha (2.9 M tons) in 1996–1997, while the productivity dropped from 780 to 753 kg/ha in the same period. In Asia, between 1972 and 2003, pigeon pea recorded 57% increase in area (2.44–3.81 m ha) and 61% increase in production (1.72–2.77 m tons). Globally, pigeon pea area has been recorded an increase of 43% since 1970. It is currently grown on 4.3 m ha (Anonymous 2007). In India, it had a low growth rate of 0.8% in production between 1949–1950 and 2004 because of various biotic and abiotic stresses (Singh et al. 2005). Kenya stands next to India in annual pigeon pea production. Kenya dedicates 200,000 ha of cultivated land annually to pigeon pea cultivation (Odeny et al. 2009). With more than 150,000 ha under cultivation, mostly located in the dry regions of the Eastern part of the country, Kenya is the main producer of pigeon pea in East-Africa and the second highest producer in the world, after India (Johansen et al. 1993).

The wilt disease complex is a major constraint for the successful cultivation of pigeon pea in India, and therefore there is an urgent need to workout a suitable biocontrol of wilt disease complex of pigeon pea (Hasan 1984; Siddiqui and Mahmood 1996, 1999). However, it suffers major economic loss, associated with poor yield mainly due to wilt caused by fusarial infection. *Fusarium* wilt of pigeon pea causes a loss of several million US$ (Reddy et al. 1990), an estimated yield loss of US$36 million in India and $5 million in eastern Africa (Kannaiyan et al. 1984). Previous studies (Songa et al. 1991; Khonga and Hillocks 1996) highlighted *Fusarium* wilt as one of the most important and wide spread diseases in Kenya with wilt incidence estimated at 60% (Kannaiyan et al. 1984). *Fusarium udum* (Singh 1983) along with *Heterodera cajani* (Husain et al. 1989) were reported to induce wilting and cause destruction to the pigeon pea crop in certain states of northern India (Perveen et al. 1999). Since then, the two decades have witnessed some very effective work for its control, though a definite practical measure is yet to be adapted.

2 The Universal Pathogen: *Fusarium* spp.

Hundreds of species of *Fusarium* are known, which play multiplicity of role in the environment. *Fusarium* species are important pathogens invading seeds, seedlings, and older plants of almost all kinds of vegetables, flowers, and cereals, as well as many fruit and forest tree. Most of the interest in this fungus arises because of its ability to cause diseases of economically important plant hosts, but it is near ubiquity in soils worldwide and its ecological activities indicate a much more diverse role in nature (Alves-Santos et al. 1999).

Fusarium sp. come in contact with host surface and recognizes the host. Sometimes, macroconidia release an extra cellular material from their tip, which is involved in adhesion (Schuerger and Mitchell 1993). Plant recognizes pathogen when physical contact occurs between them. Some of the cell wall components act as elicitors in their recognition by host (Ren and West 1992). *Fusarium* spp. penetrate the cell wall of host by producing several hydrolytic enzymes. Production of cutinase is supposed to be of major importance (Lin and Kolattukudy 1980). Fusaria are known to produce extracellular polygalacturonase and/or pectate lyase, the pectin enzymes, to breach pectinaceous barrier (Peres-Artes and Tena 1989). These deadly pathogens are known to cause wilting in host plant by affecting xylem tissues.

2.1 Fusarium Wilt of Pigeon Pea

Wilt disease of pigeon pea was first reported in 1906 by E.J. Butler from the state of Bihar, India. He was unable to distinguish the pigeon pea wilt pathogen from *F. vasinfectum* that attack cotton and sesamum (Butler 1906). He reported that wilt disease of pigeon pea is responsible for 15.25% mortality of plants, and the wilting may rise to more than 50% in epidemic year. In 1940, Padwick studied cultural characteristics of *F. udum* and found that it differed from *F. vasinfectum* because it produced abundant spores in sporodochia, and these spores are strongly hooked at the apex and so he proposed the name *F. udum* Butler var. *cajani* for the wilt pathogen of pigeon pea. On the basis of the specific shape and prominent hook at the apex of macroconidia, Booth (1977) proposed the name *F. udum* Butler, which is now widely accepted.

The *F. udum* grows systematically in taproot, lateral root, collar, stem branches, leaflets, petioles, rachis, pedicel, and pod hull. It is mainly soil borne but in the tolerant cultivars also carried in seeds. The fungus can survive for 2–3 years in soil (Kannaiyan et al. 1984). The pioneering studies on ecology of *F. udum* revealed that the fungi in root regions of healthy and diseased *C. cajan* differed qualitatively and quantitatively, as *F. udum* was always recorded on the rhizoplane of wilted plants and about 90% of the total fungal population of the rhizosphere of wilted plants was *F. udum* (Upadhyay and Rai 1982). In an early report, Sarojini (1951) isolated several *F. udum* strains from Coimbatore (India) sick soil and compared the virulence with respect to micro nutrient requirements.

Fig. 1 (**a**) Wilted infected plant of pigeon pea against a green, healthy plant; (**b**) *Fusarium* infested field of pigeon pea with infected plants; (**c**, **d**) Pigeon pea plant infected with *F. udum*, with brown streak of on stem

Limited studies on variability in the wilt fungus *F. udum* have indicated that the fungus exhibits physiologic specialization (Shit and Sen Gupta 1978; Reddy and Raju 1993). *F. udum* shows great deal of variation in cultural and morphological characteristics (Booth 1977; Rai and Upadhyay 1982; Kiprop 2002). The high variation in cultural and morphological characteristics of these pathogens is supposed to be because of environmental conditions, age of isolates, subculturing, method of storage, and culturing conditions. Wide variation in virulence to different genotypes of pigeon pea among *F. udum* isolates has been suggested, mainly because of environmental conditions and inoculation techniques (Shit and Sen Gupta 1978; Kiprop 2002).

Kiprop et al. (2002) isolated 79 single-spore isolates of *F. udum*, the causal agent of wilt disease of pigeonpea, from Kenya, India, and Malawi and characterized according to their cultural characteristics, pathogenicity, and vegetative compatibility group (VCG). They observed that isolates exhibited high variation in pathogenicity on a wilt-susceptible pigeonpea variety, and in mycelial growth and sporulation on potato dextrose agar medium. Further, the 79 isolates were categorized into two virulence groups, two groups of radial mycelial growth, and four groups of sporulation. Further, 38 *F. udum* isolates from pigeon pea were tested for variability in VCG and amplified fragment length polymorphism (AFLP). All the isolates were placed in single VCG with two subgroups, and one AFLP with more than ten AFLP groups (Kiprop et al. 2005). The disease symptoms of fusarium infestation of pigeon pea are given in Fig.1.

2.2 *Resistant Varieties of Pigeon Pea: An Effective Strategy for Wilt Control (?)*

A lot of research has been conducted on *Fusarium* wilt since the 1930s, especially in India, yet the genetics of resistance to this disease remains to be understood

(Saxena 2008). Some of the reports available (Shaw 1936; Joshi 1957; Jain and Reddy 1995; Pandey et al. 1996; Singh et al. 1998) are conflicting and inconclusive regarding the genetics of this destructive disease (Odeny et al. 2009). Pal (1934) reported that resistance to wilt in pigeon pea was controlled by multiple factors while Shaw (1936) observed two complementary genes. Later studies by Pathak (1970) confirmed the presence of two complementary genes while Pawar and Mayee (1986) reported the control of this trait by a single dominant gene. Ten pigeon pea lines were developed for use in African countries, which were resistant or tolerant to *F. udum* early maturing, short in height, and high yielding (Kimani et al. 1994).

It was found that germplasm from Asia and Africa possess different genetic mechanisms for resistance to *Fusarium* wilt (Odeny et al. 2009), rendering it difficult to raise a resistant variety with consistent performance in field conditions. Singh et al. (2004) checked the combined effect of root knot nematode, *Meloidogyne javanica*, and wilt pathogen, *F. udum*, in ten wilt resistant/tolerant accessions of pigeon pea. They found that presence of *M. javanica* with *F. udum* increased wilting from 8 to 33% in KPL 44, 15 to 60% in AWR 74/15, 25 to 50% in ICP 8859 and ICPL 89049, and 15 to 50% in ICP 12745, and hence proposed serious concerns over these cultivars. However, in other five accessions, wilting was not increased much in presence of nematodes. The lowest root knot index was observed in KPL 43 (1.50) and GPS 33 (1.75). Further, reaction to fusarium wilt as well as agronomic performance of elite pigeon pea germplasm was evaluated in three different countries during the 2001/2002 cropping season using wilt-sick plots (Gwata et al. 2006). The genotype ICEAP 00040 consistently showed a high (<20.0%) level of resistance to the disease in all three countries. ICEAP 00068, a short duration but susceptible to fusarium wilt (Gwata et al. 2007) cultivar was used to develop elite germplasm by breeding with three long-duration genotypes that were either resistant (ICEAP 00040; ICEAP 00020) or moderately resistant (ICP 13076) to fusarium wilt.

RAPD has been used to tag wilt resistance in pigeon pea (Kotresh et al. 2006; Dhanasekar et al. 2010). ICP 8863 (ICRISAT 1993) and ICP 9145 (ICRISAT 1994) are popular wilt resistant varieties. During 1978–1983, 61 pigeon pea lines and cultivars were screened for *F. udum* wilt at 15 wilt-endemic locations in India, and lines ICP 4769, 8863, 9168, 10958, 11299, and cultivars C 11 (ICP 7118) and BDN 1 (ICP 7182) were found to be resistant in all the years of testing at most of the locations, suggesting stability and broad-based resistance (Nene et al. 1985). Isozymes variability among different pigeon pea cultivars for resistance against wilt caused by *F. udum* and to assess the genetic variability among the resistant and susceptible cultivars was reported (Prasad et al. 2003).

Marley and Hillocks (2007) suggested that mechanisms of resistance to fusarium wilt (*F. udum*) were mainly because of phytoalexin synthesis. Wilt-susceptible (Malawi local) and wilt-resistant (ICP 9145) plants were stem-inoculated with a spore suspension containing 2×10^6 conidia/ml of the pathogen. Four fungitoxic isoflavonoid phytoalexins – hydroxygenistein, genistein, cajanin, and cajanol – were isolated from plants, 15 days after inoculation. Cajanol was identified as the main antifungal compound. Still it is evident that environmental and cultivation practices affect the wilt susceptibility. Therefore, it may be advised to use biocontrol agents and other PGPR (described below) to enhance yield and minimize loss with wilt resistant varieties.

3 Plant Growth Promoting Rhizobacteria

The term "plant growth promoting rhizobacteria" (PGPR) was first defined by Kloepper and Schroth (1978), to include soil bacteria that colonize the roots of plants following inoculation onto seed and enhance plant growth. The definition was revised as beneficial free-living soil bacteria that enhance plant growth, referred to as PGPR (Kloepper et al. 1989) or yield increasing bacteria (YIB) (Tang 1994). The bacteria useful to plants were proposed to be characterized into two general types: bacteria forming a symbiotic relationship with the plant, and another the free-living ones found in the soil but are often found near, on, or even within the plant tissues (Kloepper et al. 1988; Frommel et al. 1991). The premier examples of plant growth enhancing agents occur in many genera including *Actinoplanes*, *Agrobacterium*, *Alcaligens*, *Amorphosporangium*, *Arthrobacter*, *Azotobacter*, *Bacillus*, *Burkholderia*, *Cellulomonas*, *Enterobacter*, *Erwinia*, *Flavobacterium*, *Gluconacetobacter*, *Micromonospora*, *Pseudomonas*, *Rhizobia*, *Serratia*, *Streptomyces*, *Xanthomonas* as stated by large number of microbiologists (Kloepper et al. 1989; Tang 1994; Weller and Thomashao 1994; Glick 1995; Glick et al. 1995, 1998, 1999; Lucy et al. 2004). Recently, a new PGPR *Delftia tsuruhatensis* HR4, having both nitrogen fixing and biocontrol activity, was reported (Han et al. 2005). Further, Burelle et al. (2006) reported beneficial effect of PGPR and their application methods on bacterial survival, rhizosphere colonization, growth, yield, and selected indigenous rhizosphere microorganisms, without adversely affecting the beneficial indigenous microbial population.

PGPR had also been classified according to their beneficial effects (1) biofertilizers that fix nitrogen, subsequently used by the plant, thereby improving plant growth when the amount of nitrogen in the soil is limiting as observed by number of workers (Bartsev et al. 2004); (2) phytostimulators that can directly promote the growth of plants, usually by the production of hormones; and (3) biocontrol agents that are able to protect plants from infection by deleterious microorganisms (Bloemberg and Lugtenberg 2001). PGPR provide an effective and eco-friendly alternative to the agrochemicals, as they have been established to improve the yield and growth of crop plants (Vyas 2003; Vessey 2003).

Microbial inoculants as a source of biofertilizers have become a hope for most of the countries in relation with economical and environmental points of view (Ramamoorthy et al. 1994; Burelle et al. 2002; Singhal et al. 2003; Compant et al. 2005; Tilak et al. 2006a, b; Gahukar 2006).

3.1 Biocontrol Agents: A Sustainable Eco-Friendly Strategy for Pathogen Control

Interest in biocontrol has increased recently fuelled by public concerns over the use of chemicals in the environment in general, and the need to find alternatives to use of chemicals for disease control in particular (Whipps and Davies 2000; Whipps and Lumsden 2001). Seed treatment with fungicides does not protect the crop for long periods. Soil drenching with fungicides are not economical and they may establish imbalances in the microbial community unfavorable for activities of beneficial organisms (Jeyarajan et al. 1991). In addition, continuous use of the same fungicides for the same pathogen results in the development of resistant strains of the pathogen, besides polluting the environment (Pandey and Maheshwari 2007a). It is now widely recognized that biological control of plant pathogens using antagonistic fungi and bacteria is a distinct possibility for the future and can be successfully utilized especially within the framework of integrated disease management system (Muthamilan and Jeyarajan 1996).

The use of microorganisms for biological control instead of chemicals is a boon as it does not cause any harm to plant and free it from plant pathogens (Cook et al. 1995) and also exhibit their effect in stable form for long duration (Waage and Greathead 1988).

Studies on biological control of fusarium wilts have a long history (Alabouvette et al. 1998) as various disease suppressive mechanisms of biocontrol agents has been suggested, including siderophore-mediated competition for iron (Bakker et al. 1988; Raaijmakers et al. 1995), competition for substrate (Couteaudier and Alabouvette 1990), induction of systemic resistance (Van Peer et al. 1991; Van Loon 1997), and production of antibiotics (Chin-A-Woeng et al. 1998). Role of rhizobia in biocontrol of *Fusarium* has been suitably compiled (Deshwal et al. 2003).

There had been several mechanisms suggested for biocontrol mechanism including siderophore. The bacterial siderophores are known to sequester the limited supply of iron available in the rhizosphere making it unavailable to pathogenic fungi, thereby restricting their growth (O'Sullivan and O'Gara 1992). Recently, Bae et al. (2007) reported siderophore production from *Burkholderia gladioli* but based on their findings, they discarded its role in biocontrol activity.

Certain volatiles of bacterial origin including hydrogen cyanide (HCN), which is produced by many fluorescent pseudomonads in the exponential growth phase in media containing $FeCl_3$ or inorganic phosphate may also influence plant root pathogen (Voisard et al. 1989) and suppresses the diseases (Glick 1995). In one historical experiment, the effect of added cyanide was tested directly in the field, where "sick" soil was treated with $Ca(CN)_2$, which is a cheap water-soluble cyanide that is used by the mining industry and is known as "cyanogas." This treatment killed fungi en masse, significantly reduced "grey speck" disease of oats, and induced oat grain yield with no side effects on the fauna (Timonin 1947). Chanway et al. (1988) suggested that HCN produced by rhizobacteria form stable complexes with several divalent metal ions, and also cytochrome oxidase of many organisms is strongly inhibited by cyanide. Pandey et al. (2006) isolated a *Bacillus* sp., which produced HCN in vitro and reduced the radial growth of *F. udum*. Similarly, Siddiqui et al. (2008) isolated a *Pseudomonas* strain Pa324, a strong antagonist of *F. udum*, which had ability to produce HCN and siderophore in excessive amount.

Rhizobia are major biocontrol agents in natural and agricultural ecosystems. Tu (1979) suggested that rhizobia achieve this bioprotection by parasitizing the hyphal tips of the fungal pathogens and decreasing contact with the host plant cells. Different rhizobial strains were reported to successfully protect field-grown leguminous (soybean, mungbean) and nonleguminous (sunflower, okra) plants from infection by the root-borne pathogens including *Fusarium* species, irrespective to the mode of application including – seed dressing or soil drench (Ehteshamul-Haque and Ghaffar 1993). There had been substantial reports where rhizobia had been used for control of fusarium infections. Antoun et al. (1978) found 49 strains of *Sinorhizobium meliloti* that inhibited growth of *F. oxysporum* by up to 50%. Chakrabarty and Chakrabarty (1988) reported that presence of *R. meliloti* increased the production of phytoalexin 4-hydroxy-2, 3, 9-trimethoxypterocarpan, which inhibited *F. solani* f. sp. *pisi* affecting pea. Nautiyal (1997) screened the biological control activity of 256 rhizobial strains and noticed that *Rhizobium* NBRI9513 completely inhibited growth of *F. oxysporum*, *R. bactaticola*, and *Pythium* sp. in vitro condition. However, reports on rhizobial control of *F. udum* are still limited (Pandey and Maheshwari 2007a; Siddiqui and Shakeel 2009).

3.2 Biological Control of Fusarium *Wilt*

Biological control of fusarial wilt has attracted attention throughout the world. Currently, the idea of controlling soil-borne plant pathogens, including *Fusaria*, with chemical pesticides or fungicides is being challenged by the approach that biological control can have an important role in sustainable agriculture.

3.2.1 Use of *Bacillus* or *Pseudomonas*

Production of chitinases is an important attribute of biocontrol bacteria. Most of the fungi contain chitin (a homopolymer of β-1, 4 linked *N*-acetylglucosamine) in the cell wall, which ranges from 22 to 40% (Muzzarelli 1977). Therefore, formulations based on chitinases producing organisms offer potential biocontrol agents (Boller 1985). In a very early report, Mitchell and Alexander (1961) demonstrated biological control of *Fusarium* sp. and *Pythium sp.* by bacteria that degrades the cell wall of these plant pathogens. Biological control of *Fusarium* wilt of pigeon pea had been reported with chitinolytic activity of *Alcaligenes xylosoxydans* (Vaidya et al. 2001, 2003a). Further, they employed random mutagenesis through physical (UV, gamma radiation) and chemical agents (ethyl methane sulphonate

[EMS]) to obtain improved mutants for chitinase producing biocontrol strain of *A. xylosoxydans* (Vaidya et al. 2003b).

Bacillus is one of the most commonly found soil bacteria, which has been reported as excellent biocontrol agent by a number of workers (Dal-Soo et al. 1997; Bacon et al. 2001; Basha and Ulaganathan 2002; Chaurasia et al. 2005). *Bacillus* species as a group has been suggested to offer several advantages over other bacteria on protection against root pathogens because of their ability to form endospores and the broad spectrum activity of their antibiotics (Cavaglieri et al. 2005). *Bacillus brevis* inhibited the growth of pigeon pea pathogen – *F. oxysporum* f. sp. *udum* because of production of unknown antibiotic substance (Bapat and Shah 2000). Similarly, in vitro interaction of *F. udum* and a biocontrol strain of *Bacillus subtilis* AF 1 showed that the fungus forms chlamydospore-like structures and increases vacuolation, when both cultures are simultaneously inoculated into potato dextrose broth. Though, in their experiments, extracellular proteins of *B. subtilis* AF 1 reduced the growth of *F. udum* in proportion to the concentration of the protein precipitate, still formation of chlamydospore-like structures and vacuolated portions in mycelium of *F. udum* in the presence of AF 1 led the authors to conclude that *F. udum* has a mechanism to tolerate mycolytic activity (Harish et al. 1998). Recently, Siddiqui and Shakeel (2007) reported that two *Bacillus* strains (B615 and B603) had biocontrol potential against *F. udum*, in addition to inhibitory effect on the hatching and penetration of *H. cajani* and *Meloidogyne incognita* along with colonization of pigeon pea roots. In fact, the latter two nematodes cause serious damage in wilt disease complex of pigeon pea. Although restricted to pot trials, this work provides substantial evidence for PGPR to be used as broad spectrum control strategy of wilt disease in pigeon pea.

Anjaiah et al. (2003) found that *Pseudomonas aeruginosa* PNA1, an isolate from chickpea rhizosphere in India, protected pigeonpea from fusarium wilt disease. They also measured root colonization of pigeon pea using a *lacZ*-marked strain of PNA1, and observed tenfold lower root colonization of susceptible genotypes than that of moderately tolerant genotypes, indicating that this plant–bacteria interaction could be important for disease suppression in this plant. Further, strain PNA1 produced two phenazine antibiotics, phenazine-1-carboxylic acid and oxychlororaphin, in vitro, and its Tn*5* mutants (FM29 and FM13), which were deficient in phenazine production, caused a reduction or loss of wilt disease suppression in vivo, which suggest that phenazine production by PNA1 contributes to the biocontrol of fusarium wilt diseases in pigeon pea. The root nodulating bacterial isolate *Burkholderia* sp. MSSP (Pandey et al. 2007a) produces antibiotic 2-hydroxymethyl-chroman-4-one, because of which it shows antifungal properties against *F. udum* and many other phytopathogens (Kang et al. 2004).

In a similar kind of work, several pseudomonads were checked for biocontrol potential and strain Pf736 was found to cause greater increase in plant growth and higher reduction in nematode multiplication and wilting index followed by other Pa737, Pf718, and Pf719 pseudomonad strains (Siddiqui and Shakeel 2009). The use of these isolates along with *Rhizobium* (pigeon pea strain) further increased plant growth and reduced nematode multiplication and wilting index.

3.2.2 Use of Combination of Microorganisms

Combination of several organisms has been checked by many workers. *Bacillus subtilis*, *Bradyrhizobium japonicum*, and *Glomus fasciculatum* were used alone and in combination for the management of a wilt disease complex of pigeon pea caused by *H. cajani* and *F. udum* (Siddiqui and Mahmood 1995). Application of all the three management agents alone or in combination to plants inoculated with the pathogens increased shoot dry weight, number of nodules, phosphorus content, and reduced nematode multiplication and wilting index. Interestingly, in their experiments, another phenomenon in pigeon pea rhizosphere biology was identified as combined application of *G. fasciculatum* and *B. japonicum* increased root infection by *G. fasciculatum* whereas combined use with *B. subtilis* reduced mycorrhizal colonization.

In a related study, Siddiqui et al. (2008) experimented with six potential isolates of *Bacillus* and *Pseudomonas* under pot and field conditions for the biocontrol of wilt disease complex of pigeon pea. Under field condition, isolate Pa324 was best in reducing wilt disease complex followed by B18. Combined use of Pa324 with B18 provided better biocontrol of wilt disease complex than the use of either of them. Application of these isolates (Pa324 and B18) with *Rhizobium* sp. caused about 30% increase in yield under field condition and provided substantial protection against wilt disease complex of pigeon pea.

Jayalakshmi et al. (2003) also observed that the seed treatment with *Trichoderma viride* followed by *T. harzianum* was found to be effective in reducing the wilt disease incidence in pigeon pea by controlling *F. udum* effectively, when compared with individual treatments. Singh et al. (2002) checked *Aspergillus flavus*, *Aspergillus niger*, *Bacillus licheniformis* (strain-2042), *Gliocladium virens*, *Penicillium citrinum*, and *Trichoderma harzianum* for biological control of *F. udum*. They claimed that these were the most potent organisms in inhibiting the radial colony growth of the test pathogen. They observed maximum reduction of the wilt disease was with application of *G. virens* (50%) both in pots and in the fields, followed by *A. niger* (38%), *P. citrinum* (33%), and *T. harzianum* (28%), although the mechanism of biocontrol was not described.

Prasad et al. (2002) studied the efficacy of *T. harzianum* on various levels of *F. udum*. They applied *T. harzianum* as seed treatment (10 and 20 g/kg seed) and as a soil amendment (10 and 20 g/9 m^2) in field plots infested with the pathogen at three inocula levels (log 3.04, log 4.98, and log 5.34 colony-forming units (cfu)/g of soil). They observed that *Trichoderma* population increased to more than 10^8 cfu/g soil by 60 days in treated plots, whereas for seed treatments, fungal population reached a maximum of $10^{4.62}$ cfu/g soil within 45 days, and thereafter started to decline. However, even at the highest pathogen density (log 5.34), soil amendment with *T. harzianum* at 10 g gave about 30% disease reduction.

A novel mycolytic strain *Pantoea dispersa* was evaluated against *F. udum*, as a biocontrol agent in comparison with chemical fungicide Bavistin and antifungal biocontrol agent *Trichoderma* Monitor WP in both pot and field experiments (Maisuria et al. 2008). In the pot experiment, *P. dispersa* the treated pigeon pea

(T-15-15) seeds showed higher percentage of seed germination and decreased wilt incidence when compared with chemical fungicide; Bavistin and antifungal biocontrol agent *Trichoderma* Monitor WP treatments. Moreover, the root, shoot lengths, and growth were also found to be higher. The results of field study during three cropping seasons (2004/2007) suggested that the seed dressing by *P. dispersa* reduced wilt incidence (47%) during field trials, which was greater than Bavistin (41%) and *Trichoderma* Monitor WP (36%) treatments. Similarly, *T. harzianum* and *A. niger* were evaluated as biocontrol agents against *F. udum* in combination to two fungicides, Foltaf 80W (Captafol 80%) and Blue Copper-50, for the treatment of pigeon-pea wilt (Bhatnagar 1995). It was observed that the disease was more effectively controlled when biocontrol agents were applied with chemical fungicides, in comparison to the fungicides that were used alone.

More recently, Kumar et al. (2010) reported wilt disease management of *C. cajan* (L.) var. Manak by root nodulating *Sinorhizobium fredii* KCC5 and rhizospheric *Pseudomonas fluorescens* LPK2 amended with chemical fertilizers. Combinations of *S. fredii* KCC5 and *P. fuorescens* LPK2 with low dose of chemical fertilizers provided better disease management of wilt in *C. cajan*. The microbial combinations involving *S. fredii* KCC5 and *P. fuorescens* LPK2 reduced wilt disease, proved the most effective in reducing disease incidence due to *F. udum*.

3.2.3 Use of Bioformulations

Pandey and Maheshwari (2007a) formulated an effective bioformulation utilizing *Burkholderia* sp. MSSP, a known PGPR using green fluorescent protein (*gfp*) to monitor its population in carriers, including sugarcane – bagasse, sawdust, cocoa peat, rice husk, wheat bran, charcoal, rock phosphate; and paneer – whey. They concluded that whey and wheat bran proved to be efficient carrier materials for the bioformulation. Interestingly, viability of MSSP was also assessed in wheat bran and whey-based consortium, having three other bacterial strains, namely *Sinorhizobium meliloti* PP3, *Rhizobium leguminosarum* Pcc, and *Bacillus* sp. B1. Presence of other plant growth promoting bacteria did not have any detrimental effect on viability of MSSP. In fact, MSSP and PP3 strains were known to enhance seedling growth in mixed-species, coinoculated consortium (Pandey and Maheshwari 2007b), while *Bacillus* sp. B1 had biocontrol activity against *F. udum* (Pandey et al. 2006). Efficiency of wheat bran based multispecies consortium was studied on growth of pigeonpea in field conditions. Considerable increase in plant biomass, nodule number and weight, and number of pods was recorded when compared with individual trials, as well as control.

Similarly, seed treatment of groundnut and pigeonpea with peat based formulation of *B. subtilis* supplemented with 0.5% chitin or with 0.5% of sterilized *Aspergillus* mycelium controlled wilt of pigeon pea. It also increased growth promotion even in the presence of inoculum pressure (Manjula and Podile 2001). For formulation details, Nakkeeran et al. (2005) may be referred.

Treatment of pigeon pea seeds with talc based formulation of *P. fluorescens* (Pf1) effectively controlled fusarial wilt of pigeon pea under greenhouse and field conditions (Vidhyasekaran et al. 1997). More specifically, Seed treatment of pigeon pea with talc-based formulation of fluorescent pseudomonads at the rate of 4 g/kg of seed followed by soil application at the rate of 2.5 kg/ha at 0, 30, and 60 days after sowing controlled pigeonpea wilt incidence under field conditions.

Biocontrol agents *T. harzianum* and *P. fluorescens*, isolated from rhizosphere soil samples collected from various pigeon pea-growing fields, were immobilized in wheat bran, rice bran, paddy straw, and neem cake (Niranjana et al. 2009). It was found that boiled rice bran increased the growth of both biocontrol agents. Talc and sodium alginate formulations of mass-multiplied biocontrol agents were prepared and evaluated for their effects against fusarium wilt under greenhouse conditions. The fresh cultures of both biocontrol agents were found to increase seedling emergence and reduce fusarium wilt disease incidence when compared with the control and the formulations.

3.2.4 Others

A *Bacillus cereus* strain BS 03 and a *P. aeruginosa* strain RRLJ 04 were studied for their effect on induction of systemic resistance against *F. udum* wilt in pigeon pea, both individually and in combination with a rhizobial strain RH 2 (Dutta et al. 2008). They observed increased level of defense-related enzymes, viz., l-phenylalanine ammonia lyase (PAL), peroxidase (POX), and polyphenol oxidase (PPO), in coinoculated plants. Production of β-1, 3-glucanase and polymethyl galacturonase by the pathogen in culture medium was also sharply reduced in the presence of both the PGPR strains.

On the basis of results of an interesting set of experiments, Prasad et al. (2002) reported that “preinoculation” or “simultaneous inoculation” of pigeon pea seedlings with soilborne fungi nonpathogenic to pigeon pea, viz., *Fusarium oxysporum* f. sp.*niveum; F. oxysporum* f. sp.*ciceris; F. solani* f. sp. *pisi*; and *Cephalosporium sacchari*, before challenge inoculation with the pathogen *F. udum*, was effective in controlling wilt of pigeon pea to a great extent. Inoculation with the nonpathogens before the challenge inoculation was more effective than simultaneous inoculation and gave up to 81.6% protection.

In an unrelated work, fungitoxic effects of different plant extracts on *F. udum* was examined (Singh and Rai 2000). At 10% concentration of leaf extract from *Adenocallyma alliaceum*, the radial growth of *F. udum* was completely arrested. A leaf extract of *Citrus medica*, a root extract of *Asparagus adscendens*, rhizome extracts of *Curcuma longa* and *Zingiber officinale*, and a bulb extract of *Allium sativum* inhibited up to 100% growth at higher concentrations. The population of *F. udum* was found to be markedly reduced following treatments with plant powders. On the basis of these results, it may be proposed to evaluate that these plant materials as carrier for the formulation of PGPR amended bioinoculant for pigeon pea.

4 Conclusion

Pertaining to the economic importance of crop and extent of loss, the *Fusarium* wilt of pigeon pea has received considerable attention from scientists working in diverse field. Raising elite disease-resistant varieties or study of variability in virulence of *F. udum*, every attempt is toward understanding the disease complex for its management. Biological control had been proved very effective in lab as well as field conditions. Along with *Bacillus*, *Pseudomonas*, or *Trichoderma*, several other biocontrol agents (either monoculture or in combination) had been reported to be very effective by workers. Definite measure is required to popularize these technologies to completely replace the traditional use of fungicides.

Acknowledgments Financial support from UCOST, Dehradun, India is gratefully acknowledged.

References

Adu-Gyamfi JJ, Katayama K, Gayatri D, Rao TP, Oto O (1996) Improvement of fertilizer and nitrogen use efficiency in intercropping. In: Ito O, Johansen C, Adu-Gyamfi JJ, Katayama K, Rao JVDK, Rego TJ (eds) Dynamics of roots and nitrogen in cropping systems of the semi-arid tropics. Japan International Center for Agricultural Sciences (JIRCAS), JIRCAS International Agriculture Series No 3, Tsukuba, pp 493–506

Alabouvette C, Schipper SB, Lemanceau P, Bakker PAHM (1998) Biological control of *Fusarium* wilts. Toward development of commercial products. In: Boland GJ, Kuykendall LD (eds) Plant-microbe interactions and biological control. Marcel Dekker, New York, pp 15–30

Alves-Santos FM, Benito EP, Eslawa AP, Diaz-Minguez JM (1999) Appl Environ Microbiol 65:3335–3340

Anjaiah V, Cornelis P, Koedam N (2003) Effect of genotype and root colonization in biological control of fusarium wilts in pigeon pea and chick pea by *Pseudomonas aeruginosa* PNA1. Can J Microbiol 49:85–91

Anonymous (2007) http://www.icrisat.org/PigeonPea/PigeonPea.htm

Antoun H, Bordeleau LM, Gagnon C (1978) Antagonisme entre *Rhizobium meliloti* et *Fusarium oxysporum* en relation avec l'efficacite symbiotique. Can J of Plant Sci 58:75–78

Bacon CW, Yates IE, Hinton DM, Meredith F (2001) Biocontrol of *Fusarium moniliforme* in maize. Environ Health Persp 109:325–332

Bae Y-S, Park K, Choi OH (2007) Laboratory culture media-dependent biocontrol ability of *Burkholderia gladioli* strain B543. Plant Pathol J 23:161–165

Bakker PAHM, Weisbeek PJ, Schippers B (1988) Siderophore production by plant growth-promoting *Pseudomonas* spp. J Plant Nutr 11:925–933

Bapat S, Shah AK (2000) Biological control of fusarial wilt of pigeon pea by *Bacillus brevis*. Can J Microbiol 46:125–132

Bartsev A, Kobayashi H, Broughton WJ (2004) Rhizobial signals convert pathogens to symbionts at the legume interface. In: Gillins M, Holmes A (eds) Plant microbiology. BIOS Scientific Publishers, Taylor and Francis Group, London and New York, pp 19–31

Basha S, Ulaganathan K (2002) Antagonism of *Bacillus* species (strain BC121) towards *Curvularia lunata*. Curr Sci 82:1457–1463

Bhatnagar H (1995) Integrated use of biocontrol agents with fungicides to control wilt incidence in pigeon-pea. World J Microbiol Biotechnol 11:564–566

Bloemberg GV, Lugtenberg BJJ (2001) Molecular basis of plant growth promotion and biocontrol by rhizobacteria. Curr Opin Plant Biol 4:343–350
Boller T (1985). Induction of hydrolases as a defense reaction against pathogens. In: key JL, kosuge T (eds) Cellular and Molecular Biology of Plant Stress. Alan R Liss, Inc, New York, pp 247–262
Booth C (1977) The genus *Fusarium*. Common Wealth Mycological Institute, Kew, England
Burelle NK, Vavrina CS, Rosskopf EN, Shelby RA (2002) Field evaluation of plant growth promoting rhizobacteria amended transplant mixes and soil solarization for tomato and pepper production in Florida. Plant Soil 238:257–266
Burelle NK, Kloepper JW, Reddy MS (2006) Plant growth-promoting rhizobateria as transplant amendments and their effect on indigenous rhizosphere microorganism. Appl Soil Ecol 31:91–100
Butler EJ (1906) The wilt diseases of pigeonpea and pepper. Agric J of India 1:25–36
Cavaglieri L, Orlando J, Rodriguez MI, Chulze S, Etcheverry M (2005) Biocontrol of *Bacillus subtilis* against *Fusarium verticillioides in vitro* and at the maize root level. Res Microbiol 156:748–754
Chakrabarty U, Chakrabarty BN (1988) Interaction of *Rhizobium leguminasorum* and *Fusarium solani* f. sp. *pisi* on pea affecting disease development and phytoalexin production. Can J Bot 67:1698–1701
Chanway CP, Holl FB, Turkington R (1988) Genotypic coadaptation in plant growth promotion of forage species by *Bacillus polymyxa*. Plant Soil 106:281–284
Chaurasia B, Pandey A, Palini LMS, Trivedi P, Kumar B, Colvin N (2005) Diffusible and volatile compounds produced by an antagonistic *Bacillus subtilis* strain cause structural deformations in pathogenic fungi *in vitro*. Microbiol Res 160:75–81
Chin-A-Woeng TFC, Bloemberg GV, Van der Bij AJ, Van der Drift KMGM, Schripse-ma J, Kroon B, Scheffer RJ, Keel C et al (1998) Biocontrol by phenazine-1-carboxamide-producing *Pseudomonas chlororaphis* PCL1391 of tomato root rot caused by *Fusarium oxysporum* f. sp. radicis lycopersici. Mol Plant-Microbe Interact 11:1069–1077
Compant S, Duffy B, Nowak J, Clement C, Barka EA (2005) Use of plant growth promoting bacteria for biocontrol of plant diseases: principles, mechanisms of action and future prospects. Appl Environ Microbiol 71:4951–4959
Cook RJ, Thomasshow L, Weller DM, Fujimotto D, Mazzola M, Bangera G, Kim D (1995) Molecular mechanism of defense by rhizobacteria against root disease. Nat Acad Sci USA 92:4197–4201
Couteaudier Y, Alabouvette C (1990) Quantitative comparison of *Fusarium oxysporum* competitiveness in relation to carbon utilization. Microbiol Ecol 74:261–268
Dal-Soo K, Cook RJ, Weller DM (1997) *Bacillus* sp. L324-92 for biological control of three root diseases of wheat grown with reduced tillage. Biol Control 87:551–558
Deshwal VK, Pandey P, Kang SC, Maheshwari DK (2003) Rhizobia as a biocontrol agent against soil borne plant pathogenic fungi. Ind J Exp Biol 41:1160–1164
Dhanasekar P, Dhumal KN, Reddy KS (2010) Identification of RAPD markers linked to plant type gene in pigeonpea. Ind J Biotechnol 9:58–63
Dutta S, Mishra AK, Dileep Kumar BS (2008) Induction of systemic resistance against fusarial wilt in pigeon pea through interaction of plant growth promoting rhizobacteria and rhizobia. Soil Biol Biochem 40:452–461
Ehteshamul-Haque S, Ghaffar A (1993) Use of rhizobia in the control of root rot diseases of sunflower, okra, soybean and mungbean. J Phytopathol 138:157–163
Frommel MI, Nowak J, Lazarovits G (1991) Growth enhancement and developmental modifications of *in vitro* grown potato (*Solanum tuberosum* spp. *tuberosum*) as affected by a nonfluorescent *Pseudomonas* sp. Plant Physiol 96:928–936
Gahukar RT (2006) Potential and use of biofertilizers in India. Everyman's Science 40:354–361
Glick BR (1995) The enhancement of plant growth by free-living bacteria. Can J Microbiol 47:109–117

Glick BR, Karatuprovic DM, Newwll PC (1995) A novel procedure for rapid isolation of plant growth promoting pseudomonads. Can J Microbiol 41:533–536

Glick BR, Penrose DM, Li J (1998) A model for the lowering of plant ethylene concentrations by plant growth-promoting bacteria. J Theor Biol 190:63–68

Glick BR, Patten CL, Holguin G, Penrose DM (1999) Biochemical and genetic mechanisms used by PGPB. Imperial College Press, London, U.K

Gwata ET, Silim SN, Mgonja M (2006) Impact of a new source of resistance to fusarium wilt in pigeonpea. J Phytopathol 154:62–64

Gwata ET, Mligo JK, Silim SN (2007) Registration of pigeonpea cultivar Tumia. Crop Sci 47:436

Han JS, Cheng JH, Yoon TM, Song J, Rajkarnikar A, Kim WG, Yoo ID, Yang YY, Suh JW (2005) Biological control agent of common scab disease by antagonistic strain *Bacillus* sp. sunhua. J Appl Microbiol 99:213

Harish S, Manjula K, Podile AR (1998) *Fusarium udum* is resistant to the mycolytic activity of a biocontrol strain of *Bacillus subtilis* AF 1. Microb Ecol 25:385–390

Hasan A (1984) Synergism between *Heterodera cajani* and *Fusarium udum* attacking *Cajanus cajan*. Nematol Medit 12:159–162

Husain SL, Siddiqui ZA, Saddiqui MR (1989) Prevalence and geographical distribution of cyst forming nematodes in Uttar Pradesh, India. Indian J Nematol 19:108–114

ICRISAT (1993) Plant material description no. 44. International Crops Research Institute for the Semi-Arid Tropics, Patancheru, Andhra Pradesh, India

ICRISAT (1994) Plant material description no. 48. International Crops Research Institute for the Semi-Arid Tropics, Patancheru, Andhra Pradesh, India

Jain KC, Reddy MV (1995) Inheritance of resistance to *Fusarium* wilt in pigeonpea (*Cajanus cajan* L.). Ind J Genetics Plant Breeding 55:434–437

Jayalakshmi SK, Sreermula K, Benig VI (2003) Efficacy of *Trichoderma spp.* against pigeonpea wilt caused by *Fusarium udum*. J Biol Cont 17:75–78

Jeyarajan R, Ramakrishnan G, Sangeetha P (1991) Efficacy of *Trichoderma* as biocontrol agent for root rot disease of grain legumes. Petria 1:143

Johansen C, Silim SN, Singh L (1993) Towards a data base for pigeonpea in Africa. Int Pigeonpea Newslett 18:2–5

Joshi AR (1957) Genetics of resistance to disease and pests. Ind J Genet Plant Breed 17:305–317

Kang GK, Shin SY, Kim MJ, Bajpai V, Maheshwari DK, Kang SC (2004) Isolation and anti-fungal activities of 2-Hydroxymethyl-chroman-4-one produced by *Burkholderia* sp. MSSP. J Antibiot 57:726–731

Kannaiyan J, Nene YL, Reddy MV, Rajan JG, Raju TN (1984) Prevalence of pigeonpea diseases and associated crop losses in Asia, Africa and the Americas. Tropical Pest Management 30:62–71

Khonga EB, Hillocks RJ (1996) Soil borne diseases in maize-based cropping systems in southern Malawi: incidence and farmer's perceptions. Afr Plant Protect 2:131–138

Kimani PM, Nyende AB, Silim S (1994) Development of early maturing Fusarium wilt resistant pigeonpea cultivars. Afr Crop Sci J 2:35–41

Kiprop EK, Baudoin JP, Mwang'ombe AW, Kimani PM, Mergeai G, Maquet A (2002) Characterization of Kenyan isolates of *Fusarium udum* from Pigeonpea [*Cajanus cajan* (L.) Millsp.] by cultural characteristics, aggressiveness and AFLP analysis. J Phytopathol 150:517–527

Kiprop EK, Mwang'ombe AW, Baudoin JP, Kimani PM, Mergeai G (2002) Cultural characteristics, pathogenicity and vegetative compatibility of *Fusarium udum* isolates from pigeonpea (*Cajanus cajan* (L.) Millsp.) in Kenya. Eur J Plant Pathol 108:147–154

Kiprop EK, Mwang'ombe AW, Baudoin JP, Kimani PM, Mergeai G (2005) Genetic variability amongst *F. udum* isolates from pigeonpea. Afr Crop Sci J 13:163–172

Kloepper JW, Schroth MN (1978) Plant growth promoting rhizobacteria on radishes. Proceedings of the 4th International conference on pathogenic bacteria (Vol II), Station de Pathologie Végétale et Phytobacteriologie, INRA, Angers, France, pp 879–882

Kloepper JW, Hume DJ, Scher FM, Singeleton C, Tipping B, Laliberte M, Frauley K, Kutchaw T, Simonson C, Lifshitzr R, Zeleska I, Lee L (1988) Plant growth-promoting rhizobacteria (PGPR) on canola (rape seed). Plant Dis 72:42–46

Kloepper JW, Lifshitz R, Zablotowich RK (1989) Free living bacterial inocula for enhancing crop productivity. Trends Biotechnol 7:39–43

Kotresh H, Fakruddin B, Punnuri SM, Rajkumar BK, Thudi M et al (2006) Identification of two RAPD markers genetically linked to two recessive allele of a *Fusarium* wilt resistance gene in pigeon pea [*Cajanus cajan* (L.) Millsp.]. Euphytica 149:113–120

Kumar H, Bajpai VK, Dubey RC, Maheshwari DK, Chul KS (2010) Wilt disease management and enhancement of growth and yield of Cajanus cajan (L) var. Manak by bacterial combinations amended with chemical fertilizer. Crop Protect. doi:10.1016/j.cropro.2010.01.002

Lin TS, Kolattukudy PE (1980) Structural studies on cutinase, a glycoprotein containing novel amino acids and glucuronic acid amide at the N terminus. Eur J Biochem 106:341–351

Lucy M, Reed E, Glick R (2004) Application of free living plant growth promoting rhizobacteria. Antonie van Leeuwenhek 86:1–25

Maisuria VB, Gohel V, Mehta AN, Patel RR, Chhatpar HS (2008) Biological control of *Fusarium* wilt of pigeonpea by *Pantoea dispersa*, a field assessment. Ann Microbiol 58:411–419

Manjula K, Podile AR (2001) Chitin supplemented formulations improve biocontrol and plant growth promoting efficiency of *Bacillus subtilis* AF1. Can J Microbiol 47:618–625

Marley PS, Hillocks RJ (2007) The role of phytoalexins in resistance to fusarium wilt in pigeon pea (*Cajanus cajan*). Plant Pathol 42:212–218

Mitchell R, Alexander M (1961) The mycolytic phenomenon and biological control of *Fusarium* in soil. Nature 190:109–110

Muthamilan M, Jeyarajan R (1996) Integrated management of *Sclerotium* root rot of groundnut involving *T. harzianum*, *Rhizobium* and carbendazim. Ind J Mycol and Plant Pathol 26:204–209

Muzzarelli RAA (1977) Chitin. Pergamon, Oxford

Nakkeeran S, Dilantha Fernando WG, Siddiqui ZA (2005) Plant growth promoting rhizobacteria formulations and its scope in commercialization for the management of pests and diseases. In: Siddiqui ZA (ed) PGPR: biocontrol and biofertilization. Springer, Dordrecht, The Netherlands, pp 257–296

Nautiyal CS (1997) Rhizosphere competence of *Pseudomonas* sp. NBRI9926 and *Rhizobium* sp. NBRI9513 involved in the suppression of chickpea (*Cicer arietinum L.*) pathogenic fungus. Microb Ecol 23:145–158

Nene YL, Kannaiyan J, Reddy MV, Zote KK, Mahmood M, Hiremath RV, Shukla P, Kotasthane SR, Sengupta K, Jha DK, Haque MF, Grewal JS, Pal M (1985) Multilocational testing of pigeonpea for broad-based resistance to *Fusarium* wilt in India. Ind Phytopathol 40:33–36

Nene YL, Sheila VK, Sharma SB (1996) A world list of chickpea and pigeon pea pathogens, 5th edn. ICRISAT, Patencheru 502324, Andhra Pradesh, India

Niranjana SR, Lalitha S, Hariprasad P (2009) Mass multiplication and formulations of biocontrol agents for use against fusarium wilt of pigeonpea through seed treatment. International J Pest Management 55:317–324

O'Sullivan DJ, O'Gara F (1992) Traits of fluorescent *Pseudomonas* spp. involved in suppression of plant root pathogens. Microbial Rev 56:662–676

Odeny DA, Githiri SM, Kimani PM (2009) Inheritance of resistance to *Fusarium* wilt in pigeonpea {*Cajanus cajan* (L.) Millsp.}. J Animal Plant Sci 2:89–95

Pal BP (1934) Recent progress in plant breeding at Pusa. Agric Livest India 4:505–515

Pandey P, Maheshwari DK (2007a) Bioformulation of *Burkholderia* sp. MSSP with a multi-species consortium for growth promotion of Cajanus cajan. Can J Microbiol 53:213–222

Pandey P, Maheshwari DK (2007b) Two-species microbial consortium for growth promotion of *Cajanus cajan*. Curr Sci 92:1137–1142

Pandey RN, Pawar SE, Bhatia CR (1996) Inheritance of wilt resistance in pigeonpea. Indian J Genet Plant Breed 56:305–308

Pandey P, Saraf M, Dubey RC, Maheshwari DK (2006) Application of fusaria in agricultural and industrial biotechnology. In: Maheshwari DK, Dubey RC (eds) Biotechnological applications of microorganisms: a techno-commercial approach. I. K. International Publishing House, New Delhi, India, pp 199–212
Pathak GN (1970) Red gram. In: Pulse crops of Indian Council of Agricultural Research. New Delhi, India, pp 14–53
Pawar NB, Mayee CD (1986) Reaction of pigeonpea genotypes and their crosses to *Fusarium* wilt. Indian Phytopathol 39:70–74
Peres-Artes E, Tena M (1989) Pectic enzymes from two races of *Fusarium oxysporum* f. sp. *ciceri*. Enzyme production in culture and enzymatic activity on isolated chickpea cell walls. J Phytopathol 124:39–51
Perveen K, Haseeb A, Shukla PK (1999) Effect of *Meloidogyne incognita* and *Fusarium udum* on the disease development and growth of pigeonpea. Curr Nematol 10:33–40
Prasad RD, Rangeshwaran R, Hegde SV, Anuroop CP (2002) Effect of soil and seed application of *Trichoderma harzianum* on pigeonpea wilt caused by *Fusarium udum* under field conditions. Crop Protect 21:293–297
Prasad P, Eswara Reddy NP, Anandam RJ, Lakshmikantha Reddy G (2003) Isozymes variability among *Fusarium udum* resistant cultivars of pigeonpea (*Cajanus cajan* (L.) Millsp). Acta Physiologiae Plantarum 25:221–228
Raaijmakers JM, Leeman M, Van Oorschot MMP, van der Sluis L, Schippers B, Bakker PAHM (1995) Dose-response relationships in biological control of *Fusarium* wilt of radish by *Pseudomonas* spp. Phytopathol 85:1075–1081
Rai B, Upadhyay RS (1982) *Gibberella indica*: the perfect state of *Fusarium udum*. Mycologia 74:343
Ramamoorthy K, Vairavan K, Vijayalakshmi R, Jehangir KS (1994) Effect of phosphorus sources on growth and yield of pigeonpea. Ind J Pulses Res 7:84–85
Reddy MV, Raju TN (1993) Pathogenic variability in pigeonpea wilt pathogen *Fusarium udum*. In: Muralidharan K, Reddy CS (eds) Plant disease problems in Central India 1993. Proc Symp Central Zone. Indian Phytopath Soc Directorate of Rice Research, Hyderabad, India, pp 32–34
Reddy MV, Sharama SB, Nene YL (1990) Pigeonpea: diseases management. In: Nene YL, Hall SD, Sheila VK (eds) The pigeonpea. CAB International, Oxon, p 303
Ren YY, West CA (1992) Elicitation of Diterpene biosynthesis in rice (*Oryza sativa* L.) by chitin. Plant Physiol 99:1169–1178
Sarojini TS (1951) Soil conditions and root diseases, Part II. Fusarium udum disease of red gram [Cajanus cajan (Linn.) Millsp.]. Proc: Plant Sci 33:49–68
Saxena KB (2008) Genetic improvement of pigeonpea – a review. Trop Plant Biol 1:159–178
Schuerger AC, Mitchell DJ (1993) Influence of mucilage secreted by macroconidia of *Fusarium solani* f. sp. *phaseoli* on spore attachment to roots of *Vigna radiata* in hydroponic nutrient solution. Phytopathology 83:1162–1170
Shaw FLF (1936) Studies in Indian pulses: the inheritance of morphological characters and of wilt resistance in Arhar (*Cajanus indicus* Spreng). Ind J Agric Sci 6:139–187
Shit SK, Sen Gupta PK (1978) Possible existence of physiological races of *Fusarium oxysporum* f. sp. *udum*, the incitant of wilt of pigeonpea. Ind J Agric Sci 48:629–632
Siddiqui ZA, Mahmood I (1995) Biological control of *Heterodera cajani* and *Fusarium udum* by *Bacillus subtilis*, *Bradyrhizobium japonicum* and *Glomus fasciculatum* on pigeonpea. Fundam Appl Nematol 18:559–566
Siddiqui ZA, Mahmood I (1996) Effects of *Heterodera cajani*, *Meloidogyne incognita* on the wilt disease complex of pigeon pea. Indian J Nematol 26:102–104
Siddiqui ZA, Mahmood I (1999) The effect of inoculations of *Heterodera cajani*, *Meloidogyne incognita* with *Fusarium udum* and *Bradyrhizobium japonicum* on the wilt disease complex of pigeon pea. Ind Phytopathol 52:66–70

Siddiqui ZA, Shakeel U (2007) Screening of *Bacillus* isolates for potential biocontrol of the wilt disease complex of pigeon pea (*Cajanus cajan*) under greenhouse and small-scale field conditions. J Plant Pathol 89:179–183

Siddiqui ZA, Shakeel U (2009) Biocontrol of wilt disease complex of pigeon pea (*Cajanus cajan* (L.) Millsp.) by isolates of *Pseudomonas* spp. Afr J Plant Sci 3:01–12

Siddiqui ZA, Shakeel U, Siddiqui S (2008) Biocontrol of wilt disease complex of pigeonpea by fluorescent pseudomonads and Bacillus spp. under pot and field conditions. Acta Phytopathol Entomol Hung 43:77–92

Singh RS (1983) Wilt of pigeon pea. In: Singh RS (ed) Plant diseases, 51st edn. IBH Publishing, New Delhi, pp 412–417

Singh R, Rai B (2000) Antifungal potential of some higher plants against *Fusarium udum* causing wilt disease of *Cajanus cajan*. Microbios 102:165–173

Singh IP, Vishwa D, Chaudhary RC, Pandey DK (1998) Genetics of *Fusarium* wilt resistance in pigeonpea. In: National symposium on management of biotic and abiotic stresses in pulse crops, 26–28th June, IIPR Kanpur, India, p 15

Singh R, Singh BK, Upadhyay RS, Rai B, Lee YS (2002) Biological control of *Fusarium* wilt disease of pigeonpea. Plant Pathol J 18:279–283

Singh B, Ali SS, Askary Naimuddin TH (2004) Combined effect of *Fusarium udum* and *Meloidogyne javanica* on wilt resistant accessions of pigeon pea. Ann Plant Protect Sci 12:33–36

Singh R, Kumar V, Sharma S, Behl RK, Singh BP, Narula N (2005) Performance and persistence of green fluorescent protein (gfp) marked *Azotobacter chroococcum* in sterilized and unsterilized wheat rhizospheric soil. Chi J Appl Environ Biol 11:751–755

Singhal V, Pratibha M, Sengar RS (2003) Biofertilizer: boon for farmers. Indian Farming 4:11–12

Songa WA, Omanga P, Reddy MV (1991) Survey of pigeonpea wilt and other diseases in Machakos and Kitui districts of Kenya. Int Pigeonpea Newsl 14:25–26

Tang WH (1994) Yield increasing bacteria (YIB) and biological control of sheath blight of rice. In: Ryder MH, Stephens PM, Bowen GD (eds) Improving plant productivity with rhizosphere bacteria. Common wealth scientific and industrial research organization, Adelaide, Australia, pp 267–278

Tilak KVBR, Ranganayaki N, Manoharachari C (2006a) Synergistic effects of plant-growth promoting rhizobacteria and *Rhizobium* on nodulation and nitrogen fixation by pigeonpea (*Cajanus cajan*). Eur J Soil Sci 57:67–71

Tilak KVBR, Ranganayaki N, Pal KK, De R, Saxena AK, Nautiyal CS, Mittal S, Tripathi AK, Johri BN (2006b) Diversity of plant growth and soil health supporting bacteria. Curr Sci 89:136–150

Timonin MI (1947) Microflora of the rhizosphere in relation to the manganese-deficiency disease of oats. Soil Sci Soc Am Proc 11:284–292

Tobita S, Ito O, Matsunaga R, Rao TP, Rego TJ, Johansen C, Yoneyama T (1994) Field evaluation of nitrogen fixation and use of nitrogen fertilizer by sorghum/pigeon pea intercropping on an Alfisol in Indian semiarid tropics. Biol Fert Soils 17:241–248

Tu JC (1979) Evidence of differential tolerance among some root rot fungi to rhizobial parasitism *in vitro*. Physiol Plant Pathol 14:171–177

Upadhyay RS, Rai B (1982) Ecology of *Fusarium udum* causing wilt disease of pigeon pea: Population dynamics in the root region. Trans Brit Mycol Soc 78:209–220

Vaidya RJ, Shah IM, Vyas PR, Chhatpar HS (2001) Production of chitinase and its optimization from a novel isolate *Alcaligenes xylosoxydans*: potential in antifungal biocontrol. World J Microbiol Biotechnol 17:691–696

Vaidya RJ, Macmil SLA, Vyas P, Ghetiya LV, Thakor KJ, Chhatpar HS (2003a) Biological control of *Fusarium* wilt of pigeonpea *Cajanus cajan* (L.) Millsp. with chitinolytic *Alcaligenes xylosoxydans*. Ind J Exp Biol 41:1469–1472

Vaidya RJ, Vyas P, Chhatpar HS (2003b) Statistical optimization of medium components for the production of chitinase by *Alcaligenes xylosoxydans*. Enzyme Microb Technol 33:92–96

Van Loon LC (1997) Induced resistance in plants and the role of pathogenesis-related proteins. Eur J Plant Pathol 103:753–765

Van Peer R, Niemann GJ, Schippers B (1991) Induced resistance and phytoalexin accumulation in biological control of *Fusarium* wilt of carnation by *Pseudomonas* sp. strain WCS417r. Phytopathology 81:728 734

Vessey JK (2003) Plant growth promoting rhizobacteria as biofertilizers. Plant Soil 255:571–586

Vidhyasekaran P, Sethuraman K, Rajappan K, Vasumathi K (1997) Powder formulation of *Pseudomonas fluorescens* to control pigeonpea wilt. Biol Control 8:166–171

Voisard C, Keel C, Haas D, Défago G (1989) Cyanide production by *Pseudomonas fluorescens* helps suppress black root rot of tobacco under gnotobiotic conditions. EMBO J 8:351–358

Vyas SP (2003) Efficacy of biofertilizer on *Brassica* genotypes in arid Gujarat. Fertiliser News 48:49–51

Waage J, Greathead DJ (1988) Biological control-challenges and opportunities. In: Wood RKS, Way MJ (eds) Biological control of pests, pathogens, and weeds: developments and prospects. The Royal Society, London, pp 1–18

Weller DM, Thomashao LS (1994) Current challenges in introducing beneficial microorganisms into the rhizosphere. In: O'Gara F, Dowling DN, Boesten B (eds) Molecular ecology of rhizosphere microorganisms: biotechnology and the release of GMO's. Wiley-Vch, Weinheim, Germany, pp 1–18

Whipps JM, Davies KG (2000) Success in biological control of plant pathogens and nematodes by microorganisms. In: Gurr G, Wratten SD (eds) Measures of success in biological control. Kluwer, Dordrecht, The Netherlands, pp 231–269

Whipps JM, Lumsden RD (2001) Commercial use of fungi as plant disease biological control agents: status and prospects. In: Butt T, Jackson C, Magan N (eds) Fungal biocontrol agents – progress, problems and potential. CAB International, Wallingford, pp 36–43

Plant Growth Promoter Rhizobacteria in Plants Inhabiting Harsh Tropical Environments and Its Role in Agricultural Improvements

Suikinai Nobre Santos, Vanessa Nessner Kavamura, João Luiz da Silva, Itamar Soares de Melo, and Fernando Dini Andreote

Contents

1 Introduction 252
2 Soil Structure and Microbial Community 252
3 The Plant-Associated Bacteria 254
4 Plant Growth-Promoting Rhizobacteria 256
5 Bacterial Community in Harsh Tropical Ecosystems 258
5.1 The Adaptation of Microbes to Harsh Environments 258
5.2 The Brazilian Caatinga 261
5.3 The Mangrove: Ecosystem and the Interaction of Plant–Bacteria in this Niche .. 263
5.4 Plant Growth-Promoting Rhizobacteria in Harsh Environments 265
6 Agricultural Improvements: The Role of Plant Growth-Promoting Rhizobacteria 266
7 Conclusion 267
References 268

Abstract The importance of the interactions between plants and bacteria is well known for plant development and success of agriculture. A number of succeeded examples are reported in the literature for the improvement of plant yields and protection against pathogens and pests. However, some specific niches where these interactions are essential are still unexplored, like the environments where the agriculture is not practiced due to the harsh conditions found; mangroves and

---

S.N. Santos and V.N. Kavamura
Superior School of Agriculture "Luiz de Queiroz", University of São Paulo, Piracicaba, SP, Brazil
Laboratory of Environmental Microbiology, Embrapa Environment, Rodovia SP 340 – km 127.5, P.O. Box 13820-000, Jaguariúna, SP, Brazil

J.L. da Silva and I.S. de Melo
Laboratory of Environmental Microbiology, Embrapa Environment, Rodovia SP 340 – km 127.5, P.O. Box 13820-000, Jaguariúna, SP, Brazil

F.D. Andreote (✉)
Department of Soil Science, Surerior School of Agriculture "Luiz de Queiroz", University of Sao Paulo, Piracicaba, SP, Brazil
e-mail: fdandreo@gmail.com

D.K. Maheshwari (ed.), *Plant Growth and Health Promoting Bacteria*,
Microbiology Monographs 18, DOI 10.1007/978-3-642-13612-2_11,

the Brazilian semiarid caatinga. Digging into the bacterial diversity associates to plant growth promotion in such spots can help on the description of new species and features related to the plant growth-promoting rhizobacteria character under harsh tropical conditions. This chapter gives an overview of examples of such niches, where the bacterial community must be adapted to survive and support the plant development. Possible bacterial characteristics related to this ability will be discussed, as the production of biofilms and exopolysaccharides. Furthermore, the application of these biotechnological products will be evaluated and discussed allowing the reader to have a snapshot on this yet nonexplored biodiversity.

1 Introduction

Plant growth-promoting rhizobacteria, known as PGPR, are those with the ability in stimulating the plant development, acting in the nutritional and water supplementation, hormonal production, and plant protection against pathogens and pests. A number of studies have focused on the role of PGPR in a variety of crops cultivated all over the world. Although the plant species and cultivation techniques are variable in these studies, and regardless its importance, new approaches and new plant niches should be explored to enhance our knowledge about such interaction between plants and bacteria. Hence, to access the diversity of bacteria which are able to promote the plant growth in new environments can contribute in many fields; (1) the discovery of new microbial species and genotypes, which are capable of surviving in stressful environmental conditions; (2) the development of new technologies for the improvement of agricultural practices in soils where the availability of water and nutrients are low, the salinity is high and extreme temperatures are usual.

In this chapter, we explore some important features found in bacteria inhabiting harsh tropical conditions, where the practice of agriculture is not usual. The explored environments are the mangroves, where the salinity and the exposition to the sea effects are intense and the endogenous Brazilian caatinga, where the high temperatures and low water availability compose a harsh soil where plants have to develop. The bioprospection of PGPR in these niches might contribute in a better description of interactions going on under these conditions. It can also name new microbial species as the first coming candidates for the usage in program of plant protection by bacterial inoculation, leading to the safety and viable agricultural practices in lands now considered out of order for this activity.

2 Soil Structure and Microbial Community

Soils can be defined as the mineral layer used by plants to play their roles in the ecosystems (Paul and Clark 1996), subdivided into layers called horizons (Brock et al. 1994). However, if we consider the symbiotic relations, the soil can be defined as the superficial portion of the Earth which present the essential conditions for animal, plant, and microbial life. Although very heterogeneous and variable according to the depth, physical–chemical properties and location in distinct geographic regions, an average soil is formed by approximately 25% of air, 25% of water, and 50% of solids, divided in 45% minerals divided into sand, silt, and clay, and 5% of organic matter (4.5% inert organic matter and 0.5% of live organisms) (Stotzky 1972; Siqueira et al. 1994).

The microbial fraction in the soil (5%) is constituted by a diversity determined by the combination of the environmental conditions, which interact with the phenotypes of the microbes, resulting in higher populations to adapted genotypes and low populations to less adapted microbes. It also results in the way of life, as active microbial communities, or dormant cells, which can last for a period of inactivity of low dense populations. Concerning the niche occupied by microbes in soils, these species can live in association to clay particles, organic matter, in spaces between soil particles as well as in association to plants, colonizing the roots surfaces. But it is hard to assume that the life in soil is always easy and abundant in water and nutrients. The soil is a very stressful environment, where the competitiveness is constant or the niche occupation and nutrients uptaking. Such a balance and the importance of distinct features make the soil very densely inhabited and constituted by a wide diversity of microbial species (Dommergues et al. 1978; Siqueira et al. 1994).

Moreover, soils located in specific environments present even harsher conditions for the development of life, but do not limit its occupation by adapted species of microbes and plants. Such soils are deficient in nutrients and organic matter, present high acidity and salinity, extreme low or high capacity of cationic exchange, limiting its recovering and further usage as common soils. Some examples of these soils are those found in the caatinga and mangroves, which have extreme conditions of humidity – that is, low in caatinga and high in mangroves and limited amount of nutrients, due to the low content of organic matter in caatinga and the anoxic conditions in mangroves, limiting the organic matter processing (Stotzky 1972).

Considering the name of microbes that lives in soils, a number of studies are available; however, none of them have final conclusions about the main species inhabiting this niche. Consistent results have named bacterial groups found in soil, like *Proteobacteria*, *Actinobacteria*, and *Acidobacteria* as major soil inhabitants. Controversially, it is well known that the number of organisms in soil is immensurable, with amounts of 10^8 to 10^{10} microbial cells per gram of soil. Hence, a remark should be made for the density and diversity of bacteria in soils, possibly related to its essentiality to the maintenance of the functionality of this ecosystem, cycling nutrients, and harnessing the compounds degradation.

The quantity of bacterial cells in soils is dependent on environmental variables, like soil depth, pH, humidity, and temperature. Kuske et al. (2002) studying an arid soil, observed that the superficial layer from 0 to 10 cm, the bacterial counting was significantly higher than 20–30 cm. The amount of DNA extracted has also decreased with increase in depth. The culturability also decreases with depth increment (Sait et al. 2002).

3 The Plant-Associated Bacteria

In addition to microorganisms inhabiting soils, there is a group of microorganisms associated to plants that are extremely important to plant metabolism. They are found in synergism with plant roots and are called rhizosphere microorganisms. A wide diversity of bacteria can interact with plants, composing bacterial communities with important roles in plant development and health status (Hallmann et al. 1997). For a review of the bacterial communities associated with plants and how to assess them, please read Andreote et al. (2009). These interactions can vary according to the host plant in a process similar to those widely known for pathogenic microorganisms (Liu et al. 1995). Bacterial populations are distributed in the rhizosphere, epiphytic, and endophytic communities.

The rhizosphere (Fig. 1) was first defined by Hiltner in the beginning of the twentieth century as the volume of soil influenced by root plants (Hiltner 1904; Melo 2002), its extent varies with soil type and plant species (Campbell and Greaves 1990).

In rhizosphere, the quantities and types of substrate are different from those in the bulk soil and lead to colonization by different populations of bacteria, fungi, protozoa, and nematodes. Other physiochemical factors, which can be different in this region, are acidity, moisture and nutrients status, electrical conductivity, and

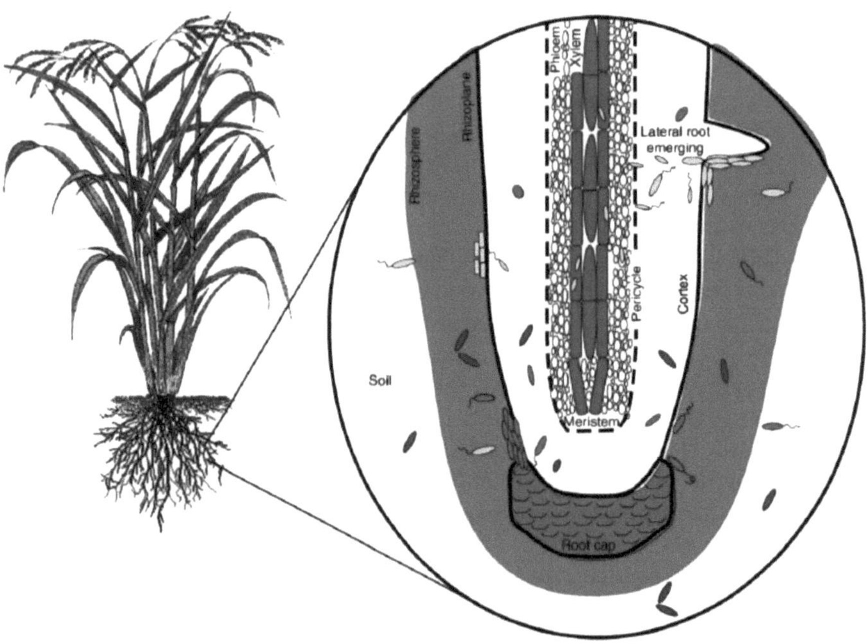

Fig. 1 Scheme of the rhizosphere system in detail and its role on the supplying of endophytes for plant colonization. Adapted and authorized
Source: Hardoim et al. (2008)

redox potential. The total rhizosphere environment is determined by an interacting trinity of the soil, the plant and the organisms associated with the roots (Campbell and Greaves 1990). A more recent definition state the rhizosphere as the soil compartment influenced by the root, including the root itself (Hartmann et al. 2008). In this way, roots associated bacteria became into the context.

Epiphytic and endophytic bacteria are characterized by the colonization of surface and inner tissues of plants, respectively. There is an ongoing discussion toward a better definition of these microorganisms; a commonly used definition of endophytes is those whose isolates form on surface-disinfected plant tissues (Hallmann et al. 1997). In addition to these definitions is the separation of endophytes according to their essentiality in niche occupations. In that case, the endophytic community is divided into "passenger" endophytes, i.e., bacteria that eventually invade internal plant tissues by stochastic events and "true" endophytes, those with adaptive traits enabling them to strictly live in association with the plant (Hardoim et al. 2008). Due to the novelty of this separation, and the problems involved in the methodological separation of these endophytic groups, we will consider in this review that the endophytic community is those bacteria that colonize inner tissues of healthy plants.

The cells in the rhizosphere, plant-surface or endophyte communities are variable. A superficial analysis of these communities could lead to the conclusion that there is a strict specificity for niche colonization. However, a more realistic scene is represented by the gradient of population distribution along plants. If a didactic approach is applied to explain bacterial communities associated with plants, it would divide these bacteria into distinct communities, with separation between epiphytic and endophytic communities in accordance with plant organs, such as roots, stems, and leaves. However, in nature the gradient of distribution will prevail over separation. It is important to note that bacteria in the rhizosphere are often similar to those in the endophytic community and on leaf surfaces. Concerning the role in plant growth promotion, both, the rhizosphere or endophytic bacteria can act supplying plants with their need and stimulating the plant development. It is remarkable that similar bacteria can be present in the rhizosphere and also colonizing inner tissues of the host plant (Hardoim et al. 2008) (Fig. 1). Chi et al. (2005) demonstrated that similar bacteria were distributed over the rice plant, from roots to leaves. However, the abundance of bacterial types along the different niches can differ, mainly due to differences in these niches in nutrient supply, atmospheric conditions, and competitiveness with other components of these communities (Rao et al. 2006). The behavior of these populations and how they colonize plants is determined by environmental conditions, like formation of biofilms that help bacteria fix to cell walls, avoiding the migration driven by sieve transportation. Similarly, in the parenchymatic region, being single-celled can enable better contact with cells and so better nutritional supply for the bacterium.

Among these bacteria, an important group is named rhizobacteria, which occupy diverse niches in the roots-plants system. Such bacteria have the ability to promote the plant growth by a diversity of mechanisms like the production of antibiotics, the nutritional supplying of the plants and by the induction of systemic resistance.

Major groups for the induction of systemic resistance are the genera *Bacillus* and *Pseudomonas*, while other bacteria like *Agrobacterium*, *Serratia*, *Enterobacter*, and *Rhizobium* can interfere in the recognition process between nematode and plants (Seldin et al. 1984; Melo 2002; Von der Weid et al. 2005; Tian et al. 2007).

In summary, the close-to-plant environment is the main niche for bacteria occupation, leading to the importance of these interactions for plant healthy. One can also consider that if these interactions are important ion common conditions of plants cultivation, it might be even more important or essential, when plants are developing under harsh environmental conditions.

4 Plant Growth-Promoting Rhizobacteria

The objective of this section is to give an overview about the PGPR, leaving the major responsibility of this task for other chapter, where experts are writing about it. PGPR were first defined by Kloepper and Schroth (1978) as being bacteria that colonize the roots of plants and help them in their growth and development (Zahir et al. 2003). This is achieved by several mechanisms such as nitrogen fixation, plant-growth hormone production, protection against diseases, and pathogens (Table 1). In this way, PGPR can be used as inoculants in assays of biofertilization, phytostimulation, and biocontrol (Bloemberg and Lugtenberg 2001) with application in agriculture, forests and environmental restoration (Lucy et al. 2004). Again, the strict division of tasks is merely didactic, considering that more than one mechanism can be present in one bacterial species.

The isolates from a sample can be examined for a wide array of traits associated with growth promotion. Cattelan et al. (1999) studied this by analyzing the siderophore, indoleacetic acid, chitinase, β-1,3-glucanase, 1-aminocyclopropane-1-carboxylate (ACC) deaminase, and cyanide production as well as phosphate solubilization of soil and rhizosphere isolates from soybean – *Glycine max*. After this screening, they have chosen 23 isolates positive for these traits and also tested their ability associated with biocontrol, bradyrhizobial inhibition, and rhizosphere competence. Ahmad et al. (2008) also screened some bacteria in vitro for the production of indoleacetic acid, ammonia, hydrogen cyanide, siderophore, phosphate solubilization, and antifungal activity.

The use of PGPR to inoculate plants can be convenient for reforestation purposes as shown by Requena et al. (1997). They have tested the ability of two arbuscular mycorrhizal fungi (AMF), one native and one exotic; two native *Rhizobium* bacteria and two PGPR, one exotic and one native in the combination of microbial inoculants. The native microorganisms were isolated from the rhizosphere of *Anthyllis cytisoides* and the other ones were obtained from existing collections. The native microorganisms were more effective inoculants than the exotic ones when biomass accumulation, nutrient uptake, and nitrogen fixation were evaluated. This report highlights the importance of previously knowledge of the microorganisms

Table 1 Examples of plant growth promotion features found in distinct bacterial species

Application	Plant growth-promoting rhizobacteria	References
Undescribed plant growth – promotion feature	Not-identified rhizobacteria: *PGB4*, *PGG2*, *Pseudomonas* sp., *Variovorax* sp., *Agrobacterium* sp., *Phyllobacterium* sp., *Bacillus firmus*, *B. mycoides*, *B. stearothermophilus*, *B. subtilis*, *B. subtilis/amyloliquefaciens*, *B. circulans*, *Brevibacillus brevis*, *Paenibacillus lautus and Stenotrophomona maltophilia*, *Pseudomonas alcaligenes PsA15*, *P. denitrificans PsD6*, *Bacillus polymyxa BcP26 and Mycobacterium phlei MbP1*, *Unindentified PGPR strains*, *Bacillus edaphicus*, *Pseudomonas putida*	Asghar et al. (2002), Ashrafuzzaman et al. (2009), Bertrand et al. (2001), Díaz et al. (2009), Egamberdiyeva (2009), Höflich (2004), Khalid et al. (2004), Sheng (2005), Trivedi and Pandey (2007)
Protection against drought stress	*Pseudomonas corrugata*, *Bacillus thuringiensis*	Kumar et al. (2007), Marulanda et al. (2006)
Indole acetic acid production	PGB4, PGG2, unindentified PGPR	Ashrafuzzaman et al. (2009), Khalid et al. (2004)
Phosphate solubilization	PGB4, PGG2, *Pseudomonas putida*	Ashrafuzzaman et al. (2009), Trivedi and Pandey (2007)
Nodulation	*Bacillus endophyticus*, *B. pumilus*, *B. subtilis*, *Paenibacillus lautus*, *P. macerans*, *P. polymyxa*, *Bacillus* sp.	Figueiredo et al. (2008), Camacho et al. (2001)
Nutrient uptake	Strain YAS34, *Azospirillum* sp., and *Azotobacter* sp.,*Bacillus endophyticus*, *B. pumilus*, *B. subtilis*, *Paenibacillus lautus*, *P. macerans*, *P. polymyxa*, two PGPR (A2 and E)	Alami et al. (2000), Biari et al. (2008), Figueiredo et al. (2008), Requena et al. (1997)
Revegetation	Two PGPR (A2 and E)	Requena et al. (1997)
Promotion of soil aggregation	Strain YAS34, *Pseudomonas mendocina*	Alami et al. (2000), Kohler et al. (2006)

inhabiting the environments, as well as their physiological and genetic adaptation, so they can be useful for further applied researches.

The indigenous rhizosphere bacteria are able to break a great variety of contaminants, but not all of them are necessarily known as plant growth-promoting rhizobacteria (PGPR) (Lucy et al. 2004). However, some PGPR can help in the productivity of some culture in soils with low nutrient content or even in contaminated soils. It occurs mainly when the microbial communities are involved in the control and absorption of metals and nutrients by surrounding plants (Stout and Nüsslein 2005). This role can be extended to the use of the term biofertilizer to PGPR, once this term is related to increase in nitrogen fixation, nutrient availability, and root growth (Vessey 2003). The nitrogen fixation can be increased as showed by Zhang et al. (1996). They have coinoculated nine growth-promoting rhizobacteria with *Bradyrhizobium* to test their ability to reduce the negative effects at suboptimal root zone temperatures in *G. max* in the nodulation and nitrogen fixation. They have observed that in certain temperatures some strains increased the number of nodules as well as the amount of fixed nitrogen when coinoculated with *Bradyrhizobium japonicum*. The most stimulating ones were *Serratia proteamaculans* in 15°C and 17.5°C; *Aeromonas hydrophila* in 17.5°C; and *Serratia liquefaciens* in 25°C.

5 Bacterial Community in Harsh Tropical Ecosystems

Microorganisms in general and bacterial community in particular adapt to grow under abiotic stresses. Such bacteria have been isolated from extreme environments of low and high temperature, sodic, and acidic habitats. These bacteria have the ability to recognize physio-chemical environment based on genetic features.

5.1 The Adaptation of Microbes to Harsh Environments

The microorganisms are able to grow using different carbon and nitrogen sources and inhabit a wide variety of ecological niches, such as extreme semiarid environments, mangroves, and desert areas. The key for the microbial adaptability may be related to their capacity of expressing only the genes of enzymes and biochemical traits which are required to a maximum growth rate in the particular environment, like the soils where they are found. This is possible due to their ability to recognize chemical and physical composition of the environment. This ability is codified by a cluster of genes that are only expressed when it is necessary. Therefore, the well-succeeded growth of a microbial population reflects its adaptation grade to the physical and chemical composition of a particular environment (Dick 1992).

The biodiversity of microorganisms allows their survival in several habitats (Parkinson and Coleman 1991). Among microorganisms, the bacteria represent the group with the greatest physiological diversity, which provides major adaptability. That is why, among certain limits, it is possible to select organisms that tolerate several stressful factors such as high temperature and soil acidity like in semiarid and mangrove regions, extremely low temperatures, minimum levels of carbon source, and high solar incidence like in polar region (Parkinson and Coleman 1991). The existence of a microorganism in determined time and place results from its evolution, from the existence of favorable abiotic factors and from its diverse biological relationships with competitors, antagonists, and predators. With the soil environment modification, the adaptation capacity of a community varies in function of its genetic constitution and it is known as "biological buffer of the soil."

However, little is known about the impacts that the environmental changes may have over the soil microorganisms.

The microorganisms in environments with special characteristics, such as high salinity as present in mangroves and low water and organic nutrients rate as in semiarid climate, may present physiological and genetic mechanisms responsible for the survival capacity, even in environments considered extremes. These characteristics allow the discovery of compounds with bioprospection potential and industrial applicability. In the presence of this scenery, the rhizosphere of plants from extreme environments can be considered as a promissory source for bioprospection of new microorganisms and products that result from microbial metabolism, aiming a biotechnological application (Melo 2002). According to Idris et al. (2007b) and Kishore and Pande (2007), several members of the *Bacillus* genera, such as *B. cereus*, *B. subtilis*, and *B. licheniformis* as well as *Pseudomonas* spp., isolated from rhizospheric soil of semiarid regions, synthesize natural metabolites that act as biological control agents and growth promoters. The biofilm formation is a mechanism that can also accomplish its role in the protection of microbes to harsh environments, as will be discussed in the next section.

5.1.1 Biofilm Formation: The Role of Exopolysaccharides

Biofilm is defined by the community of microorganisms when attached to a surface (O'Toole et al. 2000) or associated with interfaces (Davey and O'Toole 2000). According to O'Toole et al. (2000), biofilms can comprise one single microbial species or several and can also be formed on a wide variety of biotic and abiotic surfaces, depending on a response to a specific environmental condition. The matrix where those microorganisms attach is made of a chain of polysaccharides that can be produced in the interior of the microorganism and then be eliminated to the exterior as exopolysaccharides. They are produced by a large variety of microorganisms (Sutherland 1998) and are accumulated in the surface of cells (Coronado et al. 1996) (Fig. 2).

The use of exopolysaccharides is being associated to a mechanism of adaptation of microorganisms to a wide variety of environmental conditions such as salinity, shifts in temperature, and water stress. They make possible the degradation of some substances, help the colonization, virulence, and survival of some phytopathogens in the host (Roper et al. 2007) and can also protect against stressful environmental conditions (Coronado et al. 1996). Iwabuchi et al. (2000) discovered that *Rhodococcus rodochrous* was able to produce an exopolysaccharide containing D-glucose, D-galactose, D-mannose, D-glucuronic acid, and lipids that contributed to the bacterial tolerance to the aromatic faction of crude oil. They have added some exopolysaccharide in sea water with high nutrient content and some aromatic fraction of crude oil, resulting in a growth promotion ability of indigenous bacteria and increase in the degradation of crude oil by bacteria (Iwabuchi et al. 2002).

Chang et al. (2007) suggested that a strain of *Pseudomonas putida* must produce an exopolysaccharide called alginate that influences the development and the

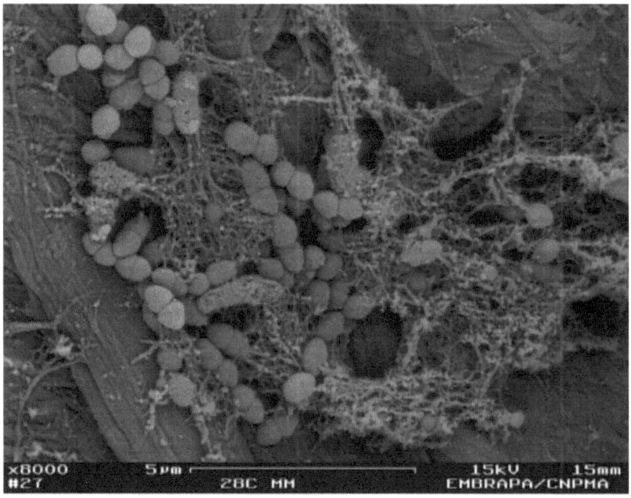

Fig. 2 Scanning electron microscopy showing some bacteria capable of producing exopolysaccharides. Note that cells are embedded in a net formed by a polymeric matrix
Source: EMBRAPA – Meio Ambiente 2009

physical–chemical properties of exopolysaccharide in response to a limited water supply. These responses should facilitate the maintenance of a hydrated environment, protecting the microorganisms against desiccation. Mutant bacteria without the gene responsible for alginate production displayed sensitivity to heat, paraquat, and hydrogen peroxide (Keith and Bender 1999). In the case of the terrestrial cyanobacterium *Nostoc commune*, exopolysaccharide production showed to be crucial to the stress tolerance during desiccation, freezing, and thawing (Tamaru et al. 2005).

Alami et al. (2000) studied the effects of a rhizobacterium capable of producing exopolysaccharide in the growth promotion of sunflower (*Helianthus annus*) under water stress. The inoculation of the strain modified the soil structure in the root system, acting against the negative effect of the lack of water in the growth. There is a factor called AlgU (AlgT) that controls the production of exopolysaccharide that is important for the adaptation of *Pseudomonas fluorescens* in dry environmental conditions (Schnider-Keel et al. 2001).

With this in mind, bacteria inoculated in the root system of plants can help in their survival in environments where water is a limiting factor. Marulanda et al. (2006) innoculated *Glomus intraradices and Bacillus thuringiensis* in *Retama sphaerocarpa*, observing an increase in the root growth of 201%, as well as higher water absorption. *Medicago sativa* plants were submitted to drought and an analysis of the involvement of the carbon metabolism and oxidative stress in the reduction of nitrogenase activity was performed. In a very severe drought, the activity of the enzyme was inhibited in 82% (Naya et al. 2007).

5.2 The Brazilian Caatinga

In Brazil, the semiarid region encompasses several minor-environments with singular conditions of climate, soil, and vegetation heterogeneity. This zone is located almost exclusively in the Northeast of the country, and is one of the five Brazilian geopolitical regions with nine states (Fig. 3). This large dry lands, stretching between 3 and 17°S and 35–45°W, covers approximately 8% of the Brazilian territory and occupies an area of about 900,000 km^2 (Giulietti 2006). The climate of this region is one of the most complex systems in the world (Giulietti et al. 2002), not only due to the size of this huge land and its diverse physiography, but also due to the conjunction of two major weather system, provided by the northeast and southeast trade winds, which create an enormous diversity and instability in rainfall patterns. These physical and climatic conditions provide the great diversity of vegetation types that characterize the semiarid region. The precipitation within the region varies from being extremely wet, with an annual rainfall of up to around 2,000 mm along the coast, to only 300–500 mm in the semiarid zone, where the rainfall is usually restricted to a few months during the year. It is indeed this factor of water availability, which is the controlling influence over the vegetation (Fig. 4) and fauna, as well as, to a great extent, human exploitation of natural resources, throughout the region.

The set of contrasting physical and climatic factors has combined to provide the diversity of vegetation types that characterize the semiarid region as a mosaic, reflecting the microlocal conditions particular to each region. Such change of rainfall regime can be modulated by the altitude, where the presence of hills or mountains provides a gradual range of condition of raining and temperature, modulating the landscapes compositions. From the coast to about 100–200 km inland, the vegetation is dominated by Atlantic Forest, with its lush, evergreen canopy leaves. Further inland, as rain is scarcer, the rain forest gives space to a forest in which the canopies semideciduous or leaves-free in dry season species are more abundant. More into the continent, the extreme dry conditions make the deciduous forest dominant, free of leaves and bleached due to the intensity of the sun. From these forests, the last two delimit the caatinga, excluding only the rain forest, which is more related to the coastline area. The structure of these forests can vary considerably from forest composed of often spiny trees, 6–10 m tall, deciduous or semideciduos, and often with a ground-layer of small deciduous shrubs and annual herbs, with predominance of Leguminosae, cacti, bromeliads, and Euphorbiaceae (Giulietti et al. 2006).

The caatinga biome covers about 735,000 km^2, and it is one of the most degraded vegetation in the semiarid region, with less than 1% of its area protected in permanent reservoirs. Recently, the Brazilian government put efforts on initiatives to better preserve its biodiversity. Areas of extreme biological interest were selected overlapping information of different groups of organisms and endogenous regions were proposed for the biome preservation (Giulietti et al. 2006). In addition, the scientific community has raised the question about the biotechnological

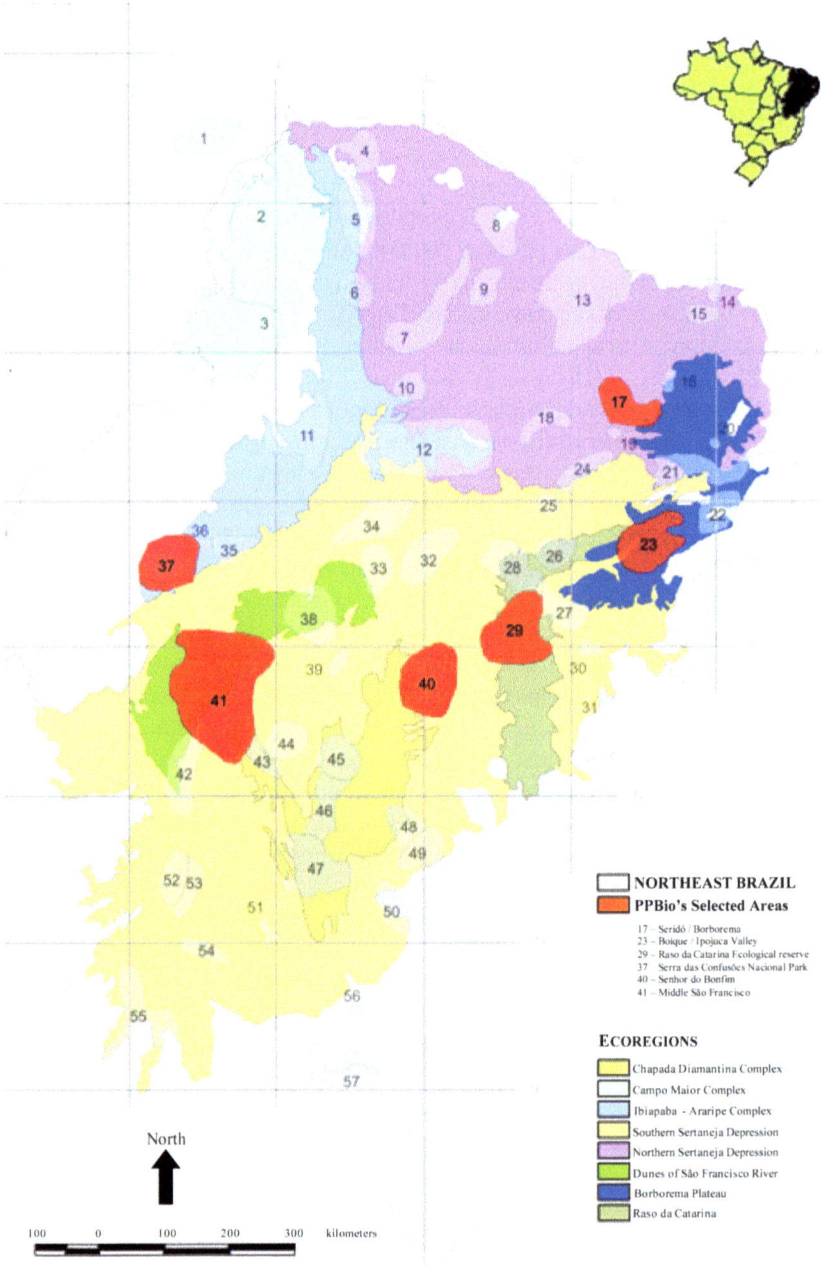

Fig. 3 Location of the semiarid climate in Brazil. Map of the Brazilian Northeast showing the ecoregions of the Caatinga Biome and the 57 priority areas for conservation (modified from Giulietti et al. 2006); with a remark to six (*red*) areas of Extreme Biological Importance

Fig. 4 Pictures from the Caatinga area in Northeast of Brazil. Typical vegetation and sight scene are shown, with a remark for plants from the Cactaceae family

potential found in species endogenous to the caatinga biome, where a remark is made for the microbial community, believed to be highly important in the biome functioning and maintenance.

A great example of such potential is represented by the bacteria associated to caatinga-plants. It is reasonable to consider that the plant life is not easy in such environmental conditions, and microbes associated to plants might have important roles in plant development. The first coming hypothesis for biotechnological possibilities is that, microbes associated to plants at caatinga might help the plant cultivation in dry agricultural lands, where water regime is not similar every year. Also, such bacteria can help the initial plant development and during the fruiting process, when the missing of water result in expressive decreasing of production. Other possibilities are still open to explored by enthusiastic scientists on the field of plant growth promotion rhizobacteria.

5.3 The Mangrove: Ecosystem and the Interaction of Plant–Bacteria in this Niche

Mangrove is a typical tropical ecosystem comprised of a coastal biome, located at the transition between the land and the sea (Sjõling et al. 2005; Zhou et al. 2006). This biome covers around 60–75% of the world's tropical and subtropical coastlines (Holguin et al. 2001) (Figs. 5 and 6), and it is characterized by the periodic tidal flooding which makes environmental factors such as salinity and nutrient availability highly variable (Crump et al. 2004; Holguin et al. 2006; Alongi 1989).

This constant change in environmental conditions makes the microbial diversity highly responsive, in order to adapt to these shifts, controlling and maintaining the

Fig. 5 The mangrove ecosystem shown in different distances. The remark should be made to the soil under the water film, responsible for the harsh conditions of anaerobiosis combined with the high salinity

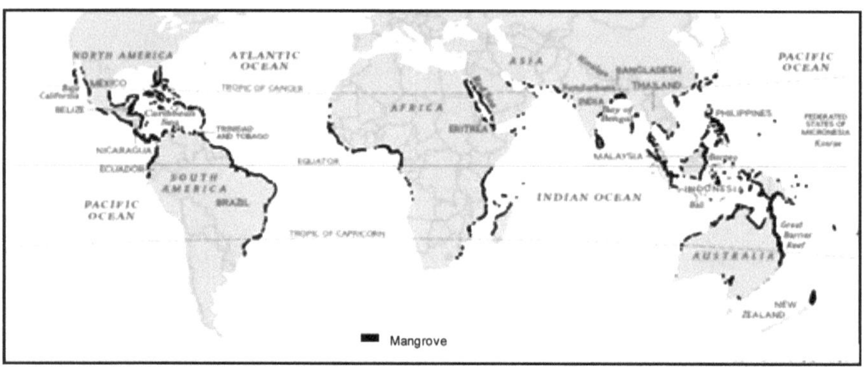

Fig. 6 The abundance and occurrence of mangrove coastline areas in the world represented by lines drawn in the color black

functioning of the mangrove (Holguin et al. 2006). However, a phylogenetic and functional description of microbial diversity in the mangrove ecosystem has not been addressed to the same extent as that of other environments (Zhou et al. 2006). A more thorough description of the bacterial diversity and distribution in a mangrove would improve our understanding of bacterial functionality and microbial interactions found in that ecosystem (Kathiresan et al. 2006 and Selvam 2006).

In the mangrove ecosystem, rhizobacteria are present and very important in the plant protection against the salinity. These bacteria are also able to fix nitrogen, supplying the plants with this essential nutrient (Holguin et al. 2001). The main bacterial groups and their role found as rhizobacteria in mangrove plants are affiliated to the genus *Vibrio*, *Listonella*, *Phyllobacterium*, the free-living nitrogen fixers; *Bacillus*, *Paenibacillus*, *Xanthobacter*, and *Vibrio* responsible for phosphate solubilization; and *Desulfovibrio*, *Desulfotomaculum*, *Desulfosarcina*, and *Desulfococcus* known as sulfate reducers.

The importance in the knowledge of plant-associated bacteria in mangroves resides on its biotechnological potential. The microbes in such conditions might help plants to survive in the mangrove, where the salinity and the depletion of oxygen exert a particular environment. A possible application to such microbes is the protection of plants against the salinity of soils, commonly found in newly available areas for agricultural practice.

5.4 Plant Growth-Promoting Rhizobacteria in Harsh Environments

Some PGPR are able to protect plants against abiotic stress such as drought and salinity. It was demonstrated the protection of soybean plants in saline soil by the inoculation with two PGPR *Bacillus subtilis* and *Bacillus megaterium* (Han and Lee 2005).

In semiarid ecosystems like those presented in Northeast of Brazil and in some Mediterranean regions, where the temperature is very high in the summer, the rainfalls are regularly scarce and the evaporation rate is considerably high. In cases where plants are submitted constantly to drought, it is interesting to apply rhizobacteria that could improve plant fitness reducing the growth inhibition ameliorating the drought. This was showed by Jaleel et al. (2007) who inoculated *P. fluorescens* in *Catharanthus roseus* and submitted to drought. There was an increase in fresh and dry weights in inoculated plants. Then, plants and microorganisms in such niches must be very well adapted to these conditions. The mechanisms by which PGPR tolerate some abiotic stress remain unknown. However, Yang et al. (2009) have proposed that PGPR are able to elicit the induced systemic tolerance (IST) of plants by physical and chemical changes that result in increased tolerance to abiotic stress. In soils with high salinity, the application of PGPR can alleviate salt stress on plant growth by the production of phytohormones and growth regulators (Egamberdiyeva 2009). The growth promotion can be attributed to the production of indole-3-acetic acid (IAA) in the case of *Azospirillum brasiliense*, which is triggered by some nutrient stresses as well as environmental fluctuations (Malhotra and Srivastava 2009). It has been demonstrated that the synthesis of auxins depend on tryptophan and it is also related to plant growth promotion, as showed by Idris et al. (2007a) for *Bacillus amyloliquefaciens*. Ethylene, a plant growth regulator, has also been established as a stress hormone. Under stress conditions the production of ethylene increases, affecting the root growth and the growth of the plant as a whole. The enzyme ACC deaminase is responsible for the regulation of ethylene biosynthesis. So, several studies have been focusing on the introduction of ACC deaminase genes into plants to regulate ethylene level in plants for optimum growth under stressed conditions (Saleem et al. 2007). In the case of mangrove ecosystems, although rich in organic matter, this niche is defective in nutrients, especially nitrogen and phosphorus (Vazquez et al. 2000). The supply of nutrients can be raised by the action of PGPR. This was shown by Biari et al. (2008) when they inoculated some PGPR in maize – *Zea mays* – seeds. They observed that the uptake of nutrients such as N, P, K, Fe, Zn, Mn, and Cu was significantly influenced by this inocula as well as the increment in plant height and shoot and seed dry weight.

6 Agricultural Improvements: The Role of Plant Growth-Promoting Rhizobacteria

The soil vegetation protects it against erosion and helps in the maintenance of the equilibrium between the factors related to its formation and degradation. The disruption of this relation provokes physical, chemical, and biological alterations that can cause a decay in the productivity and a degradation in the ecosystem. Tropical regions are considered to have the greater agricultural potential due to the abundance in light, water, and heat, essential to plant development. However, these areas are being more subjected to soil degradation, because of the high precipitation rates, weak soil structure, low organic matter content, and inadequate handling. The relation between the components of agriculture production and their environmental consequences depends on the production system and the type of exploration (Knoepp et al. 2000).

Due to the special features of PGPR presented in this chapter, Ashrafuzzaman et al. (2009) state that the use of these microorganisms is a very attractive option to replace the use of chemical fertilizers, pesticides, and other substances in agriculture. In this way, several researches have been conducted to improve growth and yield of many agricultural products by increasing nitrogen, phosphorus, and potassium availability, stimulating root nodulation and phytohormone production. PGPR has already been used as inoculum for rice, maize, wheat, soybean, and others. Phosphate is important for plant growth and in soils where this nutrient in unavailable, PGPR can be used for phosphate solubilization. In this way, Trivedi and Pandey (2007) immobilized cells of *P. putida* on sodium alginate beads to test the efficiency of wheat growth promotion by the solubilization of insoluble phosphate. The use of genetic manipulation by recombinant DNA methodology can also be useful for some agricultural purposes. Rodríguez et al. (2000) first reported the achievement of improved phosphate solubilizing ability from rhizobacterial strains by genetic construction using a gene involved in mineral phosphate solubilization. Potassium is also an important nutrient required for plant growth and development (Ashley et al. 2006). Sheng (2005) studied the ability of a potassium-releasing bacterial strain *Bacillus edaphicus* to promote cotton and rape growth in K-deficient soil. The strain was able to mobilize potassium as well as promote root and shoot growth. It is known that nitrogen is one of the most limiting nutrient for growth of plants (Stacey et al. 1992). The associative interaction of rhizobia to leguminous plants has evolved to help the nitrogen fixation, thus providing nitrogen to plant to develop its function. However, this association depends on the nodulation of the leguminous plants, allowing the rhizobia to penetrate plant tissues. In this way, PGPR can also help in this process. Inoculation of the common bean – *Phaseolus vulgaris* – with PGPR stimulated nodulation resulting in higher levels of nitrogen accumulation (Figueiredo et al. 2008).

When there is a concern about crop productivity, PGPR can be used for this purpose, by increasing growth rate and production of biomass. Ashrafuzzaman et al. (2009) tested the efficiency of ten isolates for rice growth enhancement, resulting in a significant increase in plant height, root length, production of root and shoot dry matter, and seed germination. One strain was also able to produce indoleacetic acid and solubilize phosphorus. The potential of growth promotion by rhizobacteria may be due to the expression of more than one trait, as it was observed by Dey et al. (2004). Inoculation of *Pseudomonas* spp. in *Arachis hypogea* displayed a suppression of phytopathogens, solubilization of tri-calcium phosphate, production of siderophore, and nodulation promotion, and all of these characteristics might have contributed to the enhancement of growth, yield, and nutrient uptake of peanut. Khalid et al. (2004) screened some PGPR for the potential of in vitro auxin production, applying four isolates in wheat seedlings, which resulted in increase in root and shoot elongation, root and shoot dry weight. *Pseudomonas corrugata* showed to be an appropriate inoculum for maize under rainfed conditions (Kumar et al. 2007).

When looking for an optimal crop production it is interesting to make a coinoculation of more than one PGPR as well as in combination with other rhizobia and fungi, i.e., Arbuscular Mycorrhizal Fungi (AMF), so they can act synergistically in the increase of plant growth (Gamalero et al. 2004). To examine this, Camacho et al. (2001) coinoculated a strain of *Bacillus* sp. with *Rhizobium tropici*, observing an increased bean nodulation.

7 Conclusion

This chapter gives an overview of typical harsh conditions found in specific environments exclusively found in tropical areas of the world. Exploiting these niches for PGPR functions may help on the development of biotechnological processes to save plants cultivated in common land against environmental harsh conditions. Also, an elusion could be made for the use of new land areas and for the maintenance of the agricultural yield under the climate changing process.

Acknowledgments We would like to thank the Brazilian Agricultural Research Corporation (Embrapa) and São Paulo Research Foundation (FAPESP) for the financial supporting of our ongoing research on the role of microbes in tropical environments.

References

Ahmad F, Ahmad I, Khan MS (2008) Screening of free-living rhizospheric bacteria for their multiple plant growth promoting activities. Microbiol Res 163:173–181

Alami Y, Achouak W, Marol C, Heulini T (2000) Rhizosphere soil aggregation and plant growth promotion of sunflowers by an exopolysaccharide-producing *Rhizobium* sp. strain Isolated from sunflower roots. Appl Environ Microbiol 66:3393–3398

Alongi DM, Boto KG, Tirendi F (1989) Effect of exported mangrove litter on bacterial productivity and dissolved organic carbon fluxes in adjacent tropical nearshore sediments. Mar Ecol Prog Ser 56:133–144

Andreote FD, Azevedo JL, Araújo WL (2009) Assessing the diversity of bacterial communities associated with plants. Braz J Microbiol 40:417–432

Ashley MK, Grant M, Grabov A (2006) Plant responses to potassium deficiencies: a role for potassium transport proteins. J Exp Bot 57:425–436

Ashrafuzzaman M, Hossen FA, Ismail MR, Hoque MA, Islam MZ, Shahidullah SM, Meon S (2009) Efficiency of plant growth-promoting rhizobacteria (PGPR) for the enhancement of rice growth. Afr J Biotechnol 8:1247–1252

Asghar HN, Zahir ZA, Arshad M, Khaliq A (2002) Relationship between in vitro production of auxins by rhizobacteria and their growth-promoting activities in *Brassica juncea* L. Biol Fertil Soils 35:231–237

Biari A, Gholami A, Rahmani HA (2008) Growth promotion and enhanced nutrient uptake of maize (*Zea mays* L.) by application of plant growth promoting rhizobacteria in arid region of Iran. J Biol Sci 8:1015–1020

Bloemberg GV, Lugtenberg BJJ (2001) Molecular basis of plant-growth promotion and biocontrol by rhizobacteria. Curr Opin Plant Biol 4:343–350

Brock TD, Madigan MT, Martinko JM, Parker J (1994) Biology of microorganisms, 7th edn. Prentice-Hall, New Jersey

Bertrand H, Nalin R, Bally R, Cleyet-Marel J-C (2001) Isolation and identification of the most efficient plant growth-promoting bacteria associated with canola (*Brassica napus*). Biol Fertil Soils 33:152–156

Camacho M, Santamaría C, Temprano F, Rodriguez-Navarro DN, Daza A (2001) Co-inoculation with *Bacillus* sp. CECT 450 improves nodulation in *Phaseolus vulgaris* L. Can J Microbiol 47:1058–1062

Campbell R, Greaves MP (1990) Anatomy and community structure of the rhizosphere. In: Lynch JM (ed) The rhizosphere. Wiley, New York, pp 11–34

Cattelan AJ, Hartel PG, Fuhrmann JJ (1999) Screening for plant growth-promoting rhizobacteria to promote early soybean growth. Soil Sci Soc Am J 63:1670–1680

Chang W-S, Mortel M, Nielsen L, Guzman GN, Li X, Halverson LJ (2007) Alginate production by *Pseudomonas putida* creates a hydrated microenvironment and contributes to biofilm architecture and stress tolerance under water-limiting conditions. J Bacteriol 189:8290–8299

Chi F, Shen SH, Cheng HP, Jing YX, Yanni YG, Dazzo FB (2005) Ascending migration of endophytic rhizobia, from roots to leaves, inside rice plants and assessment of benefits to rice growth physiology. Appl Environ Microbiol 71:7271–7278
Coronado C, Sánchez-Anddújar B, Palomares AJ (1996) Rhizobium extracellular structures in the symbiosis. World J Microbiol Biotechnol 12:127–136
Crump BC, Hopkinson CS, Sogin ML, Hobbie JE (2004) Microbial biogeography along an estuarine salinity gradient: combined influences of bacterial growth and residence time. Appl Environ Microbiol 70:494–1505
Davey ME, O'Toole GA (2000) Microbial biofilms: from ecology to molecular genetics. Microbiol Mol Biol Ver 64:847–867
Dey R, Pal KK, Bhatt DM, Chauhan SM (2004) Growth promotion and yield enhancement of peanut (*Arachis hypogea* L.) by application of plant growth-promoting rhizobacteria. Microbiol Res 159:371–394
Dick RPA (1992) A review: long-term effects of agricultural systems on soil biochemical and microbial paramaters. Agric Ecosyst Environ 40:25–30
Dommergues YR, Belser LW, Schmidt EL (1978) Limiting factors for microbial growth and activity in soil. Adv microb Ecol 2:49–104
Díaz K, Valiente C, Martínez M, Castillo M, Sanfuentes E (2009) Root promoting rhizobacteria in Eucalyptus globulus cuttings. World J Microbiol Biotechnol 25:867–873
Egamberdiyeva D, Höflich GH (2004) Effect of plant growth-promoting bacteria on growth and nutrient uptake of cotton and pea in a semi-arid region of Uzbekistan. J Arid Environ 56: 293–30
Egamberdiyeva D (2009) Alleviation of salt stress by plant growth regulators and IAA producing bacteria in wheat. Acta Physiol Plant 31:861–864
Figueiredo MVB, Martinez CR, Burity HA, Chanway CP (2008) Plant growth-promoting rhizobacteria for improving nodulation and nitrogen fixation in the common bean (*Phaseolus vulgaris* L.). World J Microbiol Biotechnol 24:187–1193
Gamalero E, Trotta A, Massa N, Copetta A, Martinotti MG, Berta G (2004) Impact of two fluorescent pseudomonads and an arbuscular mycorrhizal fungus on tomato plant growth, root architecture and P acquisition. Mycorrhiza 14:185–192
Giulietti NA, Harley RM, Queiroz LP, Rapini A (2006) Introduction: Top set the scene. In: Queiroz LP, Rapini A, Giulietti NA (eds) Towards greater knowledge of the Brazilian semi-arid biodiversity. Ministério Ciência e Tecnologia (MCT), Brasília, pp 15–19
Giulietti AM et al (2002) Flora. In: Pereira RM, Montenegro MM, Fonseca M (eds) Avaliação e ações prioritárias para a conservação da biodiversidade da Caatinga. Ministério do Meio Ambiente, SBF, Brasília, p 36
Giulietti NA, Harley RM, Queiroz LP, Rapini A (2006) Top set the scene. In: Queiroz LP, Rapini A, Giulietti NA (eds) Towards greater knowledge of the Brazilian semi-arid biodiversity. Ministério Ciênica e Tecnologia (MCT), Brasilia, pp 15–19
Hallmann J, QuadtHallmann A, Mahaffee WF, Kloepper JW (1997) Bacterial endophytes in agricultural crops. Can J Microbiol 43:895–914
Han HS, Lee KD (2005) Physiological responses of soybean – inoculation of *Bradyrhizobium japonicum* with PGPR in saline soil conditions. Res J Agric Biol Sci 1:216–221
Hardoim PR, van Overbeek LS, van Elsas JD (2008) Properties of bacterial endophytes and their proposed role in plant growth. Trends Microbiol 16:463–471
Hartmann A, Lemanceau P, Prosser JI (2008) Multitrophic interactions in the rhizosphere Rhizosphere microbiology: at the interface of many disciplines and expertises. FEMS Microbiol Ecol 65:179
Hiltner L (1904) Über neuere Erfahrungen und Probleme auf dem Gebiet der Bodenbakteriologie und unter besonderer Berücksichtigung der Gründung und Brache. Arb Deutsch Landwirt Ges 98:59–78

Holguin G, Vazquez P, Bashan Y (2001) The role of sediment microorganisms in the productivity, conservation, and rehabilitation of mangrove ecosystems: an overview. Biol Fertil Soils 33:265–278

Holguin G, Zamorano PG, Bashan LED, Mendoza R, Amador E, Bashan Y (2006) Mangrove health in an arid environment encroached by urban development – a case study. Sci Total Environ 363:260–274

Idris EE, Iglesias DJ, Talon M, Borriss R (2007a) Tryptophan-dependent production of indole-3-acetic acid (IAA) affects level of plant growth promotion by *Bacillus amyloliquefaciens* FZB42. Mol Plant Microbe Interact 20:619–626

Idris HA, Labuschagne N, Korsten L (2007b) Screening rhizobacterial for biological control of *Fusarium* root and crow rot of sorghum in Ethiopia. Biol Control 40:97–106

Iwabuchi N, Sunairi M, Anzai H, Nakajima M, Harayama S (2000) Relationships between colony morphotypes and oil tolerance in *Rhodococcus rhodochrous*. Appl Environ Microbiol 66:5073–5077

Iwabuchi N, Sunairi M, Urai M, Itoh C, Anzai H, Nakajima M, Harayama S (2002) Extracellular polysaccharides of *Rhodococcus rhodochrous* S-2 stimulate the degradation of aromatic components in crude oil by indigenous marine bacteria. Appl Environ Microbiol 68: 2337–2343

Jaleel CA, Manivannan P, Sankar B, Kishorekumar A, Gopi R, Somasundaram R, Panneerselvam R (2007) *Pseudomonas fluorescens* enhances biomass yield and ajmalicine production in *Catharanthus roseus* under water deficit stress. Colloids Surf B 60:7–11

Kathiresan K, Selvam M (2006) Evaluation of beneficial bacteria from mangrove soil. Bot Mar 49:86–88

Keith LMW, Bender CL (1999) AlgT (S22) controls alginate production and tolerance to environmental stress in *Pseudomonas syringae*. J Bacteriol 181:7176–7184

Khalid A, Arshad M, Zahir ZA (2004) Screening plant growth-promoting rhizobacteria for improving growth and yield of wheat. J Appl Microbiol 96:473–480

Kishore GK, Pande S (2007) Chitin-supplemented foliar application of chitinolytic *Bacillus cereus* reduces secerity of *Botrytis gray* mold disease in chickpea under controlled conditions. Lett Appl Microbiol 44:98–105

Kloepper JW, Schroth MN (1978) Plant growth-promoting rhizobacteria on radishes. In: Proceedings of the 4th international conference on plant pathogenic bacteria, Vol. 2. Station de Pathologie Vegetale et Phytobacteriologie, Anegrs. Anais. França, pp 879–882

Kohler J, Caravaca F, Carrasco L, Roldán A (2006) Contribution of *Pseudomonas mendocina* and *Glomus intraradices* to aggregate stabilization and promotion of biological fertility in rhizosphere soil of lettuce plants under field conditions. Soil Use Manag 22:298–304

Knoepp JD, Coleman DC, Crossley DA Jr, Clark JS (2000) Biological indices of soil quality: an ecosystem case study of their use. Fores Ecol Manage 138:357–368

Kumar B, Trivedi P, Pandey A (2007) *Pseudomonas corrugata*: a suitable bacterial inoculant for maize grown under rainfed conditions of Himalayan region. Soil Biol Biochem 39:3093–3100

Kuske CR, Ticknor LO, Miller ME, Dunbar JM, Davis JA, Barns SM, Belnap J (2002) Comparison of soil bacterial communities in rhizospheres of three plant species and in the interspaces in an arid grassland. Appl Environ Microbiol 68:1854–1863

Liu L, Kloepper JW, Tuzun S (1995) Induction of systemic resistance in cucumber against *Fusarium* wilt by plant growth promoting rhizobacteria. Phytopathology 85:695–698

Lucy M, Reed E, Glick BR (2004) Applications of free living plant growth-promoting rhizobacteria. Antonie Leeuwenhoek 86:1–25

Malhotra M, Srivastava S (2009) Stress-responsive indole-3-acetic acid biosynthesis by *Azospirillum brasilense* SM and its ability to modulate plant growth. Eur J Soil Biol 45:73–80

Marulanda A, Barea JM, Azcón R (2006) An indigenous drought-tolerant strain of *Glomus intraradices* associated with a native bacterium improves water transport and root development in *Retama sphaerocarpa*. Microb Ecol 52:670–678

Melo IS (2002) Recursos genéticos microbianos. In: Melo IS, Valadare-Inglis MC, Nass LL, Valois ACC (eds) Recursos Genéticos e Melhoramento – Microrganismos. EMBRAPA Meio Ambiente, Jaguariúna/SP, pp 2–48

Naya L, Ladrera R, Ramos J, González EM, Arrese-Igor C, Minchin FR, Becana M (2007) The response of carbon metabolism and antioxidant defenses of alfalfa nodules to drought stress and to the subsequent recovery of plants. Plant Physiol 144:1104–1114

O'Toole G, Kaplan HB, Kolter R (2000) Biofilm formation as microbial development. Annu Rev Microbiol 54:49–79

Parkinson D, Coleman DC (1991) Microbial communities, activity and biomass. Agric Ecosyt Environ 34(1):3–33

Paul EA, Clark FE (1996) Soil microbiology and biochemistry, 2nd edn. Academic Press, San Diego

Rao D, Webb JS, Kjelleberg S (2006) Microbial colonization and competition on the marine alga *Ulva australis*. Appl Environ Microbiol 72:5547–5555

Requena N, Jimenez I, Toro M, Barea JM (1997) Interactions between plant-growth-promoting rhizobacteria (PGPR), arbuscular mycorrhizal fungi and *Rhizobium* spp. in the rhizosphere of *Anthyllis cytisoides*, a model legume for revegetation in mediterranean semi-arid ecosystems. New Phytol 136:667–677

Rodríguez H, Gonzalez T, Selmán G (2000) Expression of a mineral phosphate solubilizing gene from *Erwinia herbicola* in two rhizobacterial strains. J Biotechnol 84:155–161

Roper MC, Greve LC, Labavitch JM, Kirkpatrick BC (2007) Detection and visualization of an exopolysaccharide produced by *Xylella fastdiosa* in vitro and in planta. Appl Environ Microbiol 73:7252–7258

Sait M, Hugenholtz P, Janssen PH (2002) Cultivation of globally distributed soil bacteria from phylogenetic lineages previously only detected in cultivation-independent surveys. Environ Microbiol 4:654–666

Saleem M, Arshad M, Hussain S, Bhatti AS (2007) Perspective of plant growth promoting rhizobacteria (PGPR) containing ACC deaminase in stress agriculture. J Ind Microbiol Biotechnol 34:635–648

Schnider-Keel U, Lejbølle KB, Baehler E, Haas D, Keel C (2001) The sigma factor AlgU (AlgT) controls exopolysaccharide production and tolerance towards desiccation and osmotic stress in the biocontrol agent *Pseudomonas fluorescens* CHA0. Appl Environ Microbiol 67:5683–5693

Seldin L, Van Elsas JD, Penido EGC (1984) *Bacillus azotofixans* sp. nov., a nitrogen-fixing species from Brazilian soils and grass roots. Int J Syst Bacteriol 34:451–456

Sheng XF (2005) Growth promotion and increased potassium uptake of cotton and rape by a potassium releasing strain of *Bacillus edaphicus*. Soil Biol Biochem 37:1918–1922

Siqueira JO, Moreira FM, Grisi BM, Hungria M, Araujo RS (1994) Microrganismos e processos biológicos do solo: perspectivas ambientais. In: Siqueira JO (ed) Empresa Brasileira de pesquisa Agropecuária. Centro nacional de pesquisa de Arroz e feijão, Centro Nacional de Pesquisa de Soja, Brasília DF, p 142

Sjõling S, Mohammed SM, Lyimoc TJ, Kyaruzib JJ (2005) Benthic bacterial diversity and nutrient processes in mangroves: impact of deforestation. Estuar Coast Shelf Sci 63:397–406

Stacey G, Burris RH, Evans HJ (1992) Biological nitrogen fixation. Chapman and Hall, New York

Stotzky G (1972) Activity, ecology, and population dynamics of microorganisms in soil. Crit Rev Microbiol 1:59–137

Stout LM, Nüsslein K (2005) Shifts in rhizoplane communities of aquatic plants after cadmium exposure. Appl Environ Microbiol 71:2484–2492

Sutherland IW (1998) Novel and established applications of microbial polysaccharides. Trends Biotechnol 16:41–46

Tamaru Y, Takani Y, Yoshida T, Sakamoto T (2005) Crucial role of extracellular polysaccharides in desiccation and freezing tolerance in the terrestrial cyanobacterium *Nostoc commune*. Appl Environ Microbiol 71:7327–7333

Tian B, Yang J, Zhang KQ (2007) Bacteria used in the biological control of plant-parasitic nematodes: populations, mechanism of action, and future prospects. FEMS Microbiol Ecol 61:197–213
Trivedi P, Pandey A (2007) Application of immobilized cells of *Pseudomonas putida* strain MTCC 6842 in alginate to solubilize phosphate in culture medium and soil. J Plant Nutr Soil Sci 170:629–631
Vazquez P, Holguin G, Puente ME, Lopez-Cortes A, Bashan Y (2000) Phosphate-solubilizing microorganisms associated with the rhizosphere of mangroves in a semiarid coastal lagoon. Biol Fertil Soils 30:460–468
Vessey JK (2003) Plant growth promoting rhizobacteria as biofertilizers. Plant Soil 255:571–586
Von Der Weid I, Artursson V, Seldin L, Jansson JK (2005) Antifungal and root surface colonization properties of GFP-tagged *Paenibacillus brasilensis* PB177. World J Microbiol Biotechnol 21:1591–1597
Yang J, Kloepper JW, Ryu C-M (2009) Rhizosphere bacteria help plants tolerate abiotic stress. Trends Plant Sci 14:1–4
Zahir ZA, Arshad M, Frankenberger WT Jr (2003) Plant growth promoting rhizobacteria: applications and perspectives in agriculture. Adv Agron 81:97–168
Zhang F, Dashti N, Hynes RK, Smith DL (1996) Plant growth promoting rhizobacteria and soybean [*Glycine max*(L.) Merr.] nodulation and nitrogen fixation at suboptimal root zone temperatures. Ann Bot 77:453–460
Zhou HW, Guo CL, Wong YS, Tam NFY (2006) Genetic diversity of dioxygenase genes in polycyclic aromatic hydrocarbon-degrading bacteria isolated from mangrove sediments. FEMS Microbiol Lett 262:148–157

Cold-Tolerant Agriculturally Important Microorganisms

Pankaj Kumar Mishra, Piyush Joshi, Shekhar Chandra Bisht, Jaideep Kumar Bisht, and Govindan Selvakumar

Contents

1 Introduction 274
2 Ecological Diversity of Cold-Tolerant Microorganisms 275
3 Cold Temperature Effects on Microbial Cells 275
4 Cold Tolerance Mechanisms in Microorganisms 276
4.1 Cell Membrane-Associated Changes 277
4.2 Role of Cryoprotectants in Cold Tolerance 278
4.3 Cold Acclimation Proteins 278
4.4 Cold-Shock Proteins 279
4.5 Role of Ice Nucleators and Antifreeze Proteins in Cold Tolerance 281
4.6 Cold-Adapted Enzymes 281
4.7 Role of RNA Degradosomes 282
4.8 Other Mechanisms of Cold Tolerance in Rhizobia 282
5 Agricultural Importance of Cold-Tolerant Microorganisms 282
5.1 Plant Growth Promotion by Cold-Tolerant Microbes 283
5.2 Diazotrophy Under Cold Temperature Conditions 286
5.3 Phosphate Solubilization by Cold-Tolerant Bacteria 287
5.4 Induction of Resistance to Cold Stress by PGPB 288
6 Industrial Potential of Psychrotolerant Microorganisms 289
7 Conclusion 290
References 290

Abstract Cold-tolerant microorganisms are endowed with the ability to grow at 0°C, though their growth optima lie in the mesophilic range. To overcome the stress induced by low temperatures they have evolved a variety of adaptive responses at

P.K. Mishra, P. Joshi, S.C. Bisht, and J.K. Bisht
Vivekananda Institute of Hill Agriculture (I.C.A.R.), Almora 263601, Uttarakhand, India
e-mail: misrapank12@rediffmail.com; joshinbri@rediffmail.com; shekhar_cbisht@yahoo.co.in; bishtjk@hotmail.com

G. Selvakumar (✉)
Indian Institute of Horticultural Research, Bangalore 560089, Karnataka, India
e-mail: gselva74@rediffmail.com

D.K. Maheshwari (ed.), *Plant Growth and Health Promoting Bacteria*, Microbiology Monographs 18, DOI 10.1007/978-3-642-13612-2_12,

the cellular and molecular levels. Multiple cell membrane modifications ensure that solute transport is not impaired at low temperatures. Other mechanisms include the synthesis of cold-shock proteins (Csps), cold acclimation proteins (Caps), cryoprotectants, ice nucleation factors, cold-adapted enzymes, and RNA degradosomes. The agricultural importance of such microbes stems from the fact that the world over temperate agro-ecosystems are characterized by low temperatures and short growing seasons that subject both plant and microbial life to cold temperature induced stress. Hence, there is a need to identify potential microbes that retain their functional traits under low temperature conditions. Such microbes can be profitably used as inoculants in agricultural production systems in the temperate regions of the world. This chapter deals with the cold tolerance/resistance mechanisms operating in microorganisms and the utility of cold-tolerant microbes in improving soil quality and productivity of agricultural crops.

1 Introduction

Among the various environmental stresses that microbes encounter due to their ubiquitous distribution on earth, cold temperature induced stress assumes paramount importance. This is due to the fact that most life processes are temperature-dependent and life almost comes to a standstill under suboptimal temperatures. Cold temperatures affect the cell interiors and a myriad of cellular processes, rendering microbial cells inactive or often irreversibly damaged. Since more than 80% of the earth's biosphere is exposed to temperatures below 5°C, throughout the year (Herbraud and Potier 1999), microorganisms capable of coping with low temperature stress have naturally evolved in several environments. Considering their ubiquity and dominance, cold-adapted microorganisms are widely regarded as the most successful colonizers of our planet (Russell 1990). During the past two decades, considerable research attention has been devoted to cold-adapted microorganisms driven by the realization that such microbes and their enzymes have a great potential for exploitation in biotechnology (Kottmier and Sullivan 1990). The agricultural importance of cold-tolerant microorganisms arises due to the fact that the cropping cycle in several parts of the world is subject to transient cold periods, which are deleterious to microbial processes such as symbiotic and asymbiotic nitrogen fixation, plant growth promotion, and disease suppression.

Life under low temperatures was identified by Forster (1887), who reported that microorganisms isolated from fish could grow at 0°C. Since then, a number of organisms particularly bacteria, yeasts, unicellular algae, and fungi have been reported to successfully colonize low temperature environments and contribute to nutrient cycling processes and primary biomass production. Following nearly two decades of debate over the term psychrophiles (named from the Greek word for "cold-loving"), the definition given by Morita in 1975 became widely accepted. He based his definitions of cold-adapted bacteria on their cardinal growth temperatures, viz. lower limit, optimum, and upper limit. Psychrophiles grow at or below zero (0°C) and have an optimum growth temperature at 15°C and an upper limit of 20°C. In contrast, psychrotolerant microbes (also called psychrotrophs) can also grow close to 0°C, and also grow at mesophilic temperatures with a growth optima usually above 30°C, hence they could be considered as cold-tolerant mesophiles (Morita 1975). Such organisms are much more widely distributed than psychrophiles and can be isolated from soils and waters in temperate regions, as well as from refrigerated food products. Though psychrotolerant organisms do grow at 0°C, they have highly extended lag periods, before the appearance of visible colonies on growth media under in vitro conditions.

2 Ecological Diversity of Cold-Tolerant Microorganisms

Generally, it is widely perceived that more extreme the environmental conditions of a niche, the lower the diversity of organisms. But most cold inhospitable environments are dominated by a variety of microorganisms, thereby making them the most versatile of all life forms. The lowest temperature limit for life seems to be around −20°C, which is the value reported for bacteria living in permafrost soil and in sea ice. Microbial activity at such temperatures is restricted to small amounts of unfrozen water inside the permafrost soil or the ice and brine channels. These contain high concentrations of salts, exopolymeric substances, and/or particulate matter, and fluid flow is maintained by concentration and temperature gradients (D'Amico et al. 2006). Cold-tolerant microorganisms are also widely encountered in refrigerated environments and have become a major cause of concern in the food processing and storage industry. In nature, the alpine soil environments are characterized by dramatic seasonal shift in physical and biochemical properties, due to intermittent snow cover and fluctuating sub freezing temperatures in winter and intense desiccating sunshine punctuated by infrequent rains during summer (Greenland and Losleben 2001). It is not uncommon to come across a wide variety of cold-tolerant microorganisms in the alpine and subalpine landscapes. Bacteria, archaea, and eukaryotes like yeast occur in cold environments. While bacteria dominate and are present in greater diversity than archaea in polar environments, archaea are wide spread in cold, deep ocean waters (Karner et al. 2001; Deming 2002). Morphological types encountered in cold environments include spore-formers, nonspore formers, and filamentous bacteria. Together they cover a wide range of metabolic types ranging from aerobes to anaerobes and include both heterotrophs and autotrophs.

3 Cold Temperature Effects on Microbial Cells

Temperature can influence the response of a microorganism either directly or indirectly. Direct effects include decreased growth rate, enzyme activities, alteration of cell composition, and differential nutritional requirements. Indirect effects are usually observed on the solubility of solute molecules, diffusion of nutrients, osmotic effects on membranes, and cell density (Herbert 1986). As temperature falls, the lag phase that precedes growth extends, leading to a decrease in the growth rate and the final cell number. During the lag phase that precedes growth in mesophiles, many physiological changes occur, including a decrease in the saturation of fatty acids and inhibition of DNA, RNA, and protein synthesis (Panoff et al. 1998).The effect of cold temperatures is largely felt on the solute transport system. The lipid bilayer which is the basic structure of the microbial membranes must have proper fluidity to maintain the cell permeability and movement of essential solutes. The functional state of this bilayer is a liquid–crystalline phase, but a decline in temperature induces a gel phase transition and a drastic loss of the membrane properties. A major difference between mesophiles and psychrotrophs is the ability to transport sugars into the cell at temperatures near 0°C (Wilkins 1973). The effect of the rapid cold shock on the membrane correlated with high rates of cell inactivation (90 and 70%) in *Escherichia coli* and *Bacillus subtilis*, respectively. Thus, membrane alternation seems to be the principal cause of cold-shock injury in *E. coli* and *B. subtilis* (Hoang et al. 2007).

In some bacteria, production of pigments and other enzymatic activities are enhanced at low temperatures, e.g., lipase and proteinase production by *Pseudomonas* and certain other genera occurs preferentially at low temperatures (Witter et al. 1966; Olson and Nottingham 1980).The prior temperature history of the cell has been found to be an important factor for the survival and growth of organisms because of its effects on the extent of lag phase before onset of growth (Dufrenne et al. 1997). A decrease in the poly-β-hydroxybutyrate (PHB) content of non-cold acclimated *Rhizobium* DDSS69 cultures was observed by Sardesai and Babu (2001a). Cold stress induces a shift in the carbon source utilization and enhances the susceptibility of bacteria to antibiotics (Ponder et al. 2005). *Vibrio cholerae* is known to enter the viable but nonculturable state in response to cold shock (Escalana et al. 2006). Since the agro-ecosystem is characterized by transient cold stress followed by warmer temperature regimes, we will focus mainly on the cold tolerance mechanisms of psychrotolerant microbes and their role in agriculture.

4 Cold Tolerance Mechanisms in Microorganisms

Unsaturation of fatty acids, reduction in the average fatty acid chain length, maintenance of membrane fluidity, synthesis of several cryoprotectant compounds, cold acclimation proteins (Caps), cold-shock proteins (Csps), ice nucleators and antifreeze proteins, cold-adapted enzymes, and RNA degradosomes are some of the cold tolerance mechanisms known to be active in microorganisms and are discussed herein.

4.1 Cell Membrane-Associated Changes

Since low temperatures primarily affect the lipid bi-layer of the bacterial cell rendering it impermeable to diffusion of solutes, the fluidity of the lipid bi-layer has to be maintained for the cell to function properly. It has been well established that microorganisms adjust their cell membrane constituents in accordance to their growth temperatures to ensure membrane functions such as solute transport (Russell et al. 1995; Mastronicolis et al. 1998). The most common changes in the cell membrane at cold temperature are the unsaturation of fatty acids, by desaturases situated in the membrane itself. In the anaerobic bacterium *Clostridium botulinum*, a decrease in temperature from 37 to 8°C leads to an increased level of unsaturation from 27 to 40% (Russell et al. 1995). This is achieved by an increase in the amount of branched fatty acid and reduction in the concentration of cyclic fatty acids and increases the monounsaturated straight chain fatty acids. The outcome of increasing fluidity may be attributed to the shortening of average fatty acid chain length owing to fewer carbon–carbon interactions between neighboring chains (Evans et al. 1998).

Other mechanisms include reduction in the average fatty acid chain length which is observed in the psychrophilic organism bacterium *Micrococcus cryophilus* (McGibbon and Russel 1983). A similar mechanism is also encountered in the yeast *Zygosaccharomyces bailii* at low temperature (Baleiras-Couto and Huis-In't-Veld 1995). An increase in the amount of branched fatty acids and reduction of the amount of cyclic fatty acids are observed in *Salmonella* spp. (Russell 1984) and *C. botulinum* (Evans et al. 1998). In *Listeria monocytogenes*, the major change that takes place as the temperature falls below optimum (e.g., 7°C) is the enhancement in amounts of $C_{15:0}$ at the expense of $C_{17:0}$. Such a reduction in fatty acid chain length reduces the melting temperature and aids in maintenance of the membrane fluidity at low temperatures. Moreover, there is also a small increase in $C_{18:1}$ which adds to fluidization of membrane at cold temperatures (Puettman et al. 1993; Annous et al. 1997). *B. subtilis* alters its membrane composition by enhancing the level of anteiso-branched fatty acid contents and decreasing the isobranched ones (Klein et al. 1999).

Maintenance of membrane fluidity is a major mode of survival of cold-adapted rhizobia since the symbiotic proteins (p Sym Nod), which are major determinants of nodule competitiveness, are membrane associated (Denarie et al. 1992). The induction of *nod* FE gene in cold-adapted *R. leguminosarum* bv. viciae was found to result in the de novo synthesis of phospholipids with specific polyunsaturated fatty acids (Geiger et al. 1993). Theberge et al. (1996) observed that the proportion of *cis*-vaccenic acid, the major unsaturated fatty acid increased by 30% as growth temperature of two cold-adapted *R. leguminosarum* bv. *viciae* strains were lowered. Drouin et al. (2000) observed that low temperature conditions affected fatty acid composition of all rhizobial strains, regardless of their cold adaptation level. The proportion of unsaturated fatty acids also increased significantly with the decrease in the growth temperature from 25 to 5°C. A specific fatty acid (*cis*-12 octadecanoic acid) was detected in artic rhizobial strains during growth at 5°C.

4.2 Role of Cryoprotectants in Cold Tolerance

Cold-tolerant microorganisms are endowed with the ability to synthesize several cryoprotectants compounds, such as glycine betaine (a bacterial cryoprotectant), glycerol, trehalose, sorbitol, manintol, glucose, and fructose, to overcome the ill effects of cold temperature induced stress. Such cryoprotectants are thought to act as chemical chaperones at cold temperatures (Margesin and Schinner 1999; Russell 1998). Glycine betaine a cryoprotectant of bacterial origin was detected in the food borne pathogen *L. monocytogenes*, which survives at low temperatures and high osmolarity (Angelidis and Smith 2003). The exact mechanism of action of glycine betaine is not yet clear. However, it is thought to function as a chemical chaperone, which prevents the aggregation of cellular proteins during stress conditions. Chatopadhyay (2002) proposed that the possible function of glycine betaine is to regulate the fluidity of membrane at lower temperatures.

Trehalose is a nonreducing disaccharide (α-D-glucopyranosil-1, 1-α-D-glucopyranoside) found in many prokaryotic and eukaryotic organisms, known to be an important protectant against heat-shock and osmotic stress in microorganisms (Kandror et al. 2002). The main function of trehalose is the stabilization of the cell membrane and proteins by replacing water and preservation of intracellular water structure (Sano et al. 1999). Exogenous trehalose helps to protect a variety of organisms against freezing and the maximum protection happens when trehalose is present on both sides of the cell membrane (Herbraud and Potier 1999). The increased viability of *E. coli* under cold-shock conditions is attributed to the enhanced accumulation of trehalose (Kaasen et al. 1992; Mitta et al. 1997). Trehalose synthesis is regulated by the genes *otsA* and *otsB* that encode the enzymes, trehalose-6-phosphate synthases and trehalose-6-phosphatase, respectively (Kaasen et al. 1992).

4.3 Cold Acclimation Proteins

Cold-tolerant bacteria produce a set of ~20 permanent proteins called the cold acclimation proteins (Caps) during continuous growth at low temperatures. The Caps are fundamental to life in the cold and ensure improved protein synthesis at low temperature (Margesin et al. 2007). Some of the Caps identified in cold-adapted bacteria function as Csps in mesophiles, a typical example being the RNA chaperone CspA. It has been proposed that cold acclimatization proteins are essential for the maintenance of both growth and cell cycle at low temperatures, but their function is still poorly understood. A cold acclimation protein (Hsc 25) produced in the ice-nucleating bacterium, *Pantoea ananas* KUIN-3, was found to be capable of refolding enzymes, which were denatured by heat, cold, and guanidine hydrochloride, but it had high affinity for cold denatured enzymes than for heat-denatured enzymes (Kawahara et al. 2000).

4.4 Cold-Shock Proteins

A sudden decrease in temperature from the mesophilic range to cold temperatures (10–15°C) creates a stress situation. Microbial cells respond to such a situation by a specific adaptative mechanism, which allows their survival and subsequent growth at lower temperatures. Although such rapid down shifts are unlikely to occur in natural environments, it provides interesting laboratory conditions that have largely contributed to the elucidation of the molecular mechanisms by which cells responds to cold (Herbraud and Potier 1999). A sudden decrease in temperature initiates the cold-shock response (Jones et al. 1987; Graumann et al. 1996; Jones and Inouye 1996), which is evidently not confined to psychrophilic (cold loving) and psychrotrophic (cold-tolerant) microorganisms but constitutes the beginning of cold adaptation in all microbes. It involves the induction and synthesis of Csps for the regulation of protein synthesis and mRNA folding. Bacterial Csps consist of a single nucleic acid-binding domain, called the cold-shock domain (CSD). The CSD is considered to be an ancient molecule present even prior to the advent of single cell life and is the most evolutionarily conserved nucleic acid-binding domain within prokaryotes and eukaryotes (Graumann and Marahiel 1998). Owing to the design of prokaryotic transcriptional machinery, the cold-induced RNA secondary structure may impose premature transcription termination. The functional significance of bacterial Csps is therefore directly related to the formation of stable secondary RNA structures in response to low temperature stress (Polissi et al. 2003). The cold-shock proteins Csp A, Csp C, and Csp E were confirmed to possess in vivo and in vivo transcription antitermination activity (Bae et al. 2000). CspA has been proposed to function as an RNA chaperone at low temperature and has been implicated in transcriptional regulation of two cold-shock genes *hrs* and *gyr A*. The 5′ end of the Csp A mRNA contains a regulatory sequence (cold box), which may be responsible for the cold-shock induction (Jiang et al. 1997).

The number of Csps seems to increase with the severity of the cold shock. The major Csps accounts for more than 10% of total protein synthesized during the acclimation phase of *E. coli* (Goldstein et al. 1990). Radiolabelling of total cellular proteins of *Pseudomonas* spp. 30-3 revealed the elevated expression of an 8 kDa protein at 4°C, which suggests that the protein Cap B plays a pivotal role in survival and tolerance at cold and subzero temperatures (Panicker et al. 2002). Among the bacterial species, the Csps of *E. coli* are the most studied and a fair degree of homology has been observed with the Csps from other microorganisms such as *B. subtilis* and *Streptomyces clavuligerus*. Similarly, a 248 bp DNA fragment in *Pseudomonas* spp. 30-3 that was amplified using Cap B gene specific primers showed a 98% amino acid sequence homology with Cap B of *Pseudomonas fragi* and 62% homology with Csp A of *E. coli* (Michel et al. 1997; Panicker et al. 2002).

Drouin et al. (2000) isolated cold-adapted strains of *R. leguminosarum* bv. *viciae* from the legumes *Lathyrus japonicus* and *Lathyrus pratensis* in northern Quebec (Canada). When these strains were compared with a poorly adapted strain and a cold sensitive strain for freezing survival, protein induction, and fatty acid composition following a cold shock from 25°C to 10, 5, and 0°C, a common 6.1 kDa Csp was induced in all the strains, but the total number of Csps synthesized at 0°C was higher in the cold-adapted strains than in the cold sensitive strains. Csps have also been detected in several other agriculturally important bacterial species (Table 1).

The regulation of the expression of Csps and their homologues is complex involving autoregulation and is controlled at the level of transcription and translation as well as by the stability of mRNA and proteins. Response to cold shock might be controlled at the transcriptional or translational level, though the two possibilities are mutually exclusive (Horn et al. 2007).

Table 1 Major cold-shock proteins identified in some agriculturally important psychrotrophic bacterial species

Organism	Cold-shock protein	Biochemical features	References
Bacillus megaterium ATCC 14581	Unnamed	Responsible for the hyper induction of desaturase	Fujii and Fulco (1977)
Bacillus cereus WSBC 10201	Csp A	M.W. 7.5 kDa	Mayr et al. (1996)
Bacillus subtilis	Csp B Csp C Csp D Csp 7.4	M.W. 7.365 kDa M.W. 8.0 kDa M.W. 13.0 kDa	Schindelin et al. (1993) Fujii and Fulco (1977) Schnuchel et al. (1993)
Rhizobium sp. (Temperate strains)	Unnamed	M.W. 56.1,37.1,34.4, 17.3,11.1 kDa	Cloutier et al. (1992)
Rhizobium sp. (Artic strains)	Unnamed	M.W. 52.0,38.0,23.4,22.7 and 11.1 kDa	Cloutier et al. (1992)
Pseudomonas fragi	C 7.0 C 8.0	M.W. 7.0 kDa M.W. 8.0 kDa	Hebraud et al. (1994)
Listeria monocytogenes	Unnamed	M.W. 48 kDa M.W. 21.1 kDa M.W. 19.7 kDa M.W. 18.8 kDa	Bayles et al. (1996)
Streptococcus thermophilus	Unnamed	M.W. 7.5 kDa M.W. 21.5 kDa	Perin et al. (1999)
Lactococcus lactis	Unnamed	M.W. 7 kDa	Wouters et al. (1999)
Streptoccocus thermophilus	Clp L	M.W. 75 kDa	Varcamonti et al. (2006)

4.5 *Role of Ice Nucleators and Antifreeze Proteins in Cold Tolerance*

Ice nucleators are proteins which either limit super cooling or induce freezing at temperatures below 0°C by mimicking the structure of an ice crystal surface. They impose an ice crystal like arrangement on the water molecule with their surface and reduce the energy necessary for the initiation of ice formation. Ice-nucleating agents either facilitate cold-protection due to the released heat of fusion or establish protective extracellular freezing in place of lethal intracellular freezing (Zachariassen and Kristiansen 2000). The "ice plus" bacteria posses Ina protein (Ice nucleation-active protein) on the outer bacterial wall which acts as the nucleating center for ice crystals (Lee et al. 1995). This facilitates ice formation at high subzero temperatures, while "ice minus" bacteria do not posses Ina proteins and therefore lower the ice nucleation temperature. Very potent ice nucleators, active at high subfreezing temperature, are produced by bacteria such as *Erwinia herbicola* (Kozloff et al. 1983). Other bacterial genera viz., *Pseudomonas*, *Pantoea* (*Erwinia*), and *Xanthomonas* can nucleate the crystallization of ice from supercooled water (Lindow et al. 1978; Maki et al. 1974; Obata et al. 1990).

Another possible strategy used by microorganisms to survive freezing temperature is the production of antifreeze proteins (AFPs), a structurally diverse group of proteins that have the ability to modify ice crystal structure (Raymond and DeVries 1977) and inhibit recrystallization of ice (Knight et al. 1988). AFPs inhibit further binding of water molecules and affect the shape of ice crystal, even at very low concentrations. The Arctic plant growth promoting rhizobacteria *Pseudomonas putida* GR 12-2 secretes an AFP that enhances its survival at subzero temperature. Expression of *afp A* gene of *P. putida* in *E. coli* yielded an intracellular 72 kDa protein that exhibited lower levels of antifreeze and ice nucleation activities. The AfpA sequence was most similar to cell wall associated proteins and less similar to ice nucleation proteins (INPs). Hydropathy plots revealed that the amino acid sequence of AfpA was more hydrophobic than those of the INPs in the domain that forms the ice template, thereby suggesting that AFPs and INPs interact differently with ice (Muryoi et al. 2004).

4.6 *Cold-Adapted Enzymes*

The most important selective pressure of low temperatures is exerted toward chemical reaction rates, most of which exponentially drop with decreasing temperature. Despite this, psychrophiles produce cold-adapted enzymes that have high specific activities at low temperatures. The commonly accepted hypothesis for this cold adaptation is the activity–stability–flexibility relationship, which suggests that psychrophilic enzymes increase the flexibility of their structure to compensate for the "freezing effect" of cold habitats (Johns and Somero 2004). This increased flexibility might concern the entire protein or might be restricted to parts of the structure; especially those implicated in catalysis and are probably also responsible for the generally observed low stability of cold-adapted proteins (D'Amico et al. 2003).

4.7 Role of RNA Degradosomes

The degradosome, a protein-complex of several ribonucleases, is the major determinant factor for stability of cellular RNA. The degradosome of an antarctic bacterium *Pseudomonas syringae* has been found to contain an endoribonuclease RNAse E and a RNA helicase. But instead of polynucleotide phosphorylase, the exoribonuclease found in *E. coli*, the degradosome of the antarctic bacterium contains another exoribonuclease, called RNAse R. In *E. coli*, this enzyme is known to play an important role in ensuring the quality control of rRNA. The significance of the association of this enzyme with RNAse E in the Antarctic bacterium is not definitely known. But it is believed that RNAse R can degrade RNA molecules with extensive secondary structures. This eliminates the necessity of ATP, required by helicase, thereby helping the cell conserve energy at low temperatures (Purusharth et al. 2005).

4.8 Other Mechanisms of Cold Tolerance in Rhizobia

In *Rhizobium* strain DDSS69, it was observed by Sardesai and Babu (2001b) that the specific activities of key enzymes of the pentose phosphate pathway, viz., glucose-6-phosphate dehydrogenase and 6-phosphogluconate dehydrogenase, were enhanced under cold stress to rapidly generate energy to overcome the stress. They also reported diversity in the switching mechanisms of carbon metabolism among cold-acclimated and noncold-acclimated *Rhizobium* isolates. In another study, they detected a rapid induction of two high molecular weight membrane polypeptides of 135 and 119 kDa within 15 min of exposure to 5°C in the *Rhizobium* strain DDSS69. PAGE membrane protein profiles of stressed and nonstressed cells revealed differential regulation of genes (Sardesai and Babu 2001b).

5 Agricultural Importance of Cold-Tolerant Microorganisms

Microorganisms play a major role in sustaining the production and productivity of any agro-ecosystem through a myriad of roles that extend form nitrogen fixation, nutrient solubilization, nutrient mobilization, plant growth promotion, and the suppression of harmful pathogens and insects. A unique feature of temperate agro-ecosystems around the world is the short growing periods, which are interspersed by suboptimal temperatures. Under such a scenario most microbial processes are bound to slow down or worse even come to a standstill, thereby have an adverse effect on the productivity. This effect is most pronounced in the case of nutrient transformations, where microbes play an enormous role. In such a scenario where time and temperatures are crucial determinants of both crop growth and microbial growth, cold-tolerant microbes which retain their activity in suboptimal temperature conditions are indispensable. But unfortunately not many efforts have been undertaken in understanding the nature and properties of these microbes and the quantum of information on the tolerance mechanisms of other agriculturally important microorganisms is very scarce.

5.1 Plant Growth Promotion by Cold-Tolerant Microbes

The volume of soils surrounding roots is influenced chemically, physically, and biologically by the plant root and is commonly referred to as the rhizosphere. This is a highly favorable habitat for the proliferation of microorganisms which exert a potential impact on plant health and soil fertility. The plant growth promoting rhizobacteria (PGPR), which are an important component of the rhizosphere microbial community, were first defined by Kloepper and Schroth (1978). In recent years, the term has been modified as Plant Growth Promoting Bacteria (PGPB) to accommodate other strains that are nonrhizospheric in origin (Andrews and Harris 2003). In temperate climates, the growth and activity of such rhizospheric communities are highly dependent on the root zone temperature since most physiological processes that influence plant growth virtually come to a standstill at suboptimal temperatures. In such a scenario, it's important that the root colonizing bacteria retain their metabolic versatility at low temperatures, since plant growth promotion is achieved by the action of several metabolic intermediates and end products. One of the major mechanisms of plant growth promotion is the production of the stimulatory phytohormones, by PGPR/PGPB within the root zone; these hormones stimulate the density and length of root hairs resulting in the enhanced uptake of water and mineral nutrients from soil (Volkmar and Bremer 1998). Apart from phytohormone production, plant growth promotion is known to be mediated by a variety of mechanisms including siderophore production (Katiyar and Goel 2004), antagonism toward deleterious root microorganisms (Misaghi et al. 1982), deamination of the precursor molecule of the phytohormone ethylene whose accumulation in root tissue is known to be detrimental to root growth and development (Glick et al. 1998), and induction of systemic resistance to plant pathogenic microorganisms (Lavania et al. 2006).

The auxin, indole-3-acetic acid (IAA), is an important phytohormone produced by PGPR, and treatment with auxin-producing rhizobacteria has been shown to increase the plant growth (Patten and Glick 2002). The IAA producing capability of microorganisms is useful in their identification and provides a valuable marker when examining the physiological roles or ecological significance of IAA in the establishment and persistence of the organisms (Bric et al. 1991). Auxin production in bacteria is regulated though the proline-linked pentose phosphate pathway (McCue et al. 2000). Selvakumar et al. (2008a, b) reported the occurrence of cold-tolerant plant growth promoting bacterial strains *Pantoea dispersa* strain 1A and *Serratia marcescens* strain SRM from the North-Western Indian Himalayas. These strains retained their IAA producing ability at 4 and 15°C. Seed bacterization with these bacterial strains significantly enhanced plant biomass and nutrient uptake of wheat seedling grown at cold temperatures. The genus *Pseudomonas* is an important component of the rhizospheric microbial community and often plays an important role in plant growth promotion. Mishra et al. (2008, 2009a) described the cold tolerance and IAA production by *Pseudomonas* sp. strain PGERs17 and NARs9 at cold temperature. Seed inoculation with these strains enhanced the seed germination, root and shoot lengths of wheat seedlings grown at low temperatures. Considering the metabolic versatility of Pseudomonads, it is possible to unearth a whole of cold-tolerant plant growth promoting bacterial species in the future.

Another bacterial mechanism that positively influences plant growth is the production of the enzyme, 1-aminocyclopropane-1-carboxylate (ACC) deaminase. This enzyme plays a significant role in the regulation of the plant hormone, ethylene and thus, influences the growth and development of plants. Bacterial strains containing ACC deaminase can, in part, at least alleviate the stress-induced ethylene-mediated negative impact on plants. Like many other abiotic and biotic factors, accelerated ethylene production under high and chilling temperatures has widely been reported by researchers both in plant tissues and microbial species in the rhizosphere. Plants with ACC deaminase expression may cope with this unfavorable situation by lowering ethylene level like that under other environmental stresses (Saleem et al. 2007). A psychrotolerant ACC deaminase producing bacterium *P. putida* UW4 was found to promote canola plant growth at low temperature under salt stress (Cheng et al. 2007). Considering the role of ethylene in stress physiology, it can be rightly said that much more efforts are needed to decipher the role of ACC deaminase producing bacterial strains in plant growth promotion under cold temperature conditions.

Iron the fourth most abundant element in the earth's crust is required for growth of nearly all forms of life (Howard 1999). However, its availability to the organism is very limited due to the rapid oxidation of ferrous (Fe^{++}) to ferric (Fe^{+++}) state. Ferric ion is highly insoluble under physiological conditions and makes its acquisition by microorganisms a considerable challenge (Neilands 1995). Microorganisms have evolved specialized mechanisms for the assimilation of iron, including the production of low molecular weight iron chelating compounds, known as siderophores, which transport this element into their cells. Siderophores have been implicated for both direct and indirect enhancement of plant growth by rhizospheric microorganisms (Neilands 1981). Siderophores provide an advantage in survival of both plants and bacteria of because they mediate competition that results in exclusions of fungal pathogens and other microbial competitors in the rhizosphere by a reduction in the availability of iron for their survival (Masalha et al. 2000,

Wang et al. 2000). The role of siderophores in biocontrol of plant pathogens was first demonstrated with pseudobactin, the siderophore produced by plant growth promoting *Pseudomonas* strain B_{10} (Kloepper et al. 1980). A cold-tolerant mutant of *Pseudomonas fluorescens* with a 17-fold increase in siderophore production and increased rhizosphere colonization was developed by Katiyar and Goel (2004). This mutant strain promoted growth of *Vigna radiata* plants at 25 and 10°C. Studies on siderophore-mediated growth promotion by psychrotolerant bacteria still remain in its infancy and need to be probed further.

An important facet of the competitiveness of a biocontrol agent is its ability to persist and proliferate. However, it is often difficult to predict the behavior of a particular microbe in the soil since the soil persistence of a bacterium may be influenced by a number of different factors including soil temperature. Many fungal phytopathogens are most destructive when the soil temperature is low, hence it is reasonable to expect that the biocontrol agents are also cold-tolerant. McBeath (1995) reported the isolation of several strains of *Trichoderma* sp. that acted as biocontrol agents at low temperatures (i.e., 4–10°C) against a range of different pathogenic fungi. Negi et al. (2005) have characterized a group of cold-tolerant Pseudomonads from the Garhwal region of the Indian Himalaya. These strains produced siderophores and exhibited plant growth promotion activity at temperatures ranging from 4 to 25°C. Seed inoculation with these isolates resulted in the suppression of major root borne diseases of garden pea. A novel bacterium *Exiguobacterium acetylicum* strain 1P isolated from a high altitude soil in the N.W. Indian Himalaya, which has ability to produce siderophores at 4°C and inhibited the growth and development of *Rhizoctonia solani, Sclerotium rolfsii, Phythium*, and *Fusarium oxysporium* under in vitro and pot culture conditions was described by Selvakumar et al. (2009c). Recently, Malviya et al. (2009) have isolated antagonistic, chitinolytic, psychrotolerant strains of *Streptomyces* from glacial sites of the Indian Himalayas. These stains were found to inhibit the growth of several plant pathogenic fungi. In the present scenario where the demand for pesticide free food products is on the rise, much more research efforts are required for identifying cold-tolerant strains of biocontrol agents for use in the temperate growing regions.

Freezing injury in plants is particularly complex because of the nonuniform behavior of different plant parts, e.g., stem, leaf, bud, flowers, etc. Ice nucleation in plants is frequently not endogenous, but is induced by catalytic sites present in microbial parasites, which can be found on leaves, fruits, or stems (Lindow 1983). Ice nucleating strains of *P. syringae* increase the frost susceptibility of tomato and soybean when sprayed on leaves prior to low temperature stress in addition to being a pathogen of these plants (Anderson et al. 1982). Recognition of the gene associated with ice nucleation in *P. syringae* first led to the synthesis of an "ice-minus" mutant, which was found to be inactive in promoting ice nucleation in plants leaves (Xu et al. 1998). Reducing the numbers of ice nucleating bacteria by different approaches is an effective and environmentally safe method of controlling freezing injury in plants and is considered a classic example of displacement of a bacterial pathogen by a biocontrol agent.

5.2 Diazotrophy Under Cold Temperature Conditions

Nitrogen fixation by symbiotic and asymbiotic bacterial genera is one of the major means by which life is sustained in this planet. But this process is hugely affected by cold temperature stress. The effects of low temperature on the activities of rhizobia include depression of nodule competitiveness and nodule functioning. The production of Nod metabolites by *Rhizobium leguminosarum* bv. *trifolii* is reduced by lowering the temperature, which in turn may affect the nodulation and yield of host legumes (McKay and Djordjevic 1993). Many studies have shown that suboptimal temperatures affect the competitiveness of rhizobia for nodulation, delay root infection, and inhibit nodule function (Lynch and Smith 1994). It has been estimated that under temperate conditions, the establishment of an effective symbiosis a week earlier in the crop-growing season could double the amount of nitrogen fixed and thus increase legume crop productivity (Sprent 1979). Therefore, it is imperative to select cold-adapted strains of rhizobia to overcome the cold induced stress. In a major step in this direction, Prevost et al. (1999) selected cold-adapted rhizobia from Canadian soils with the aim of improving the productivity of legumes that are subjected to low temperatures during the growing season. For this purpose, they used rhizobia associated with legume species indigenous to arctic and subarctic regions. The candidate rhizobia were *Mesorhizobium* sp. isolated from *Astragalus/Oxytropis* spp. and *Rhizobium leguminosarum* from *Lathryrus* spp. These rhizobia are considered psychrotrophs due to their ability to grow at 0°C. The advantages of cold-adapted arctic *Mesorhizobium* in improving legume symbiosis were demonstrated with the temperate forage legume sainfoin (*Onobrychis viciifolia*). In laboratory and field studies, arctic rhizobia were found to be more efficient than temperate (commercial) rhizobia in improving growth of sainfoin and were more competitive in forming nodules. Biochemical studies on cold adaptation revealed higher synthesis of Csps in the cold-adapted rhizobia, than their mesophilic counterparts. Since the arctic *Mesorhizobium* could not nodulate agronomically important legumes, the nodulation genes and the bacterial signals (Nod factors) were characterized as a first step to modifying the host specificity of nodulation.

Another approach was to screen for cold-adapted rhizobia naturally associated with agronomic legumes cultivated in temperate areas. It has been shown that the environment from which rhizobia are isolated, relates to their ability to enter into symbiosis with legumes under specific environmental conditions. Rhizobia originating from the cooler climes of North America were able to positively influence the nodulation and nitrogen fixation of soybean, compared with their counterparts originating from the warmer southern climes (Zhang et al. 2003). A superior strain of *Sinorhizobium meliloti* adapted for nodulation of alfalfa at low temperatures was selected and was found efficient in improving growth of alfalfa in laboratory and field studies. This strain also performed well in improving growth of alfalfa after over wintering under cold and anaerobic (ice encasement) stresses, indicating a possible cross-adaptation of selected rhizobia for various abiotic stresses inherent to temperate climates (Prevost et al. 2003). Ideal candidate rhizobia for temperate legumes would, therefore, require a high degree of nodule competitiveness and nitrogen fixing abilities combined with cold-tolerant traits. Such rhizobia would retain their membrane fluidity at low temperatures, thereby enabling the synthesis and activity of membrane-associated Nod factors that play a major role in the nodulation and host specificity.

Azospirillum is an associative symbiotic plant growth promoting bacterium that is predominantly associated with the grasses and cereal crops of the tropics. Tripathi and Klingmuller (1992) proposed that growth, survival, and activity of the bacterium are highly dependent on temperature. Kaushik et al. (2001) postulated that a low or nonsignificant effect of *Azospirillum* inoculation in winter crops has discouraged the large-scale use of this bacterium. Kaushik et al. (2000) selected Tn5::lacZ mutants isogenic to wild type *Azospirillum brasilense* that were capable of growing at cold temperatures. In field studies, they observed that two strains of *A. brasilense* were able to influence wheat growth at suboptimal temperatures (Kaushik et al. 2002). Though the temperature regime at which the isolates were evaluated for their plant response was not strictly temperate, this is one of the few studies on field performance of *Azospirillum* under suboptimal temperatures. Considering its agronomic significance *Azospirillum* is a candidate bacterium for the potential for exploration and development of cold-tolerant isolates.

5.3 Phosphate Solubilization by Cold-Tolerant Bacteria

Phosphate solubilization by rhizospheric microflora is one of the most important means of achieving plant growth promotion. Bacterial mineral phosphate solubilization has been mainly attributed to the activity of glucose dehydrogenase; a membrane-bound enzyme that is involved in the direct oxidation of glucose to gluconic acid (Goldstein 1995). Subsequently, gluconic acid is enzymatically converted to 2-ketogluconic acid and 2,5-diketogluconic acid. The 2-ketogluconic acid is more effective than gluconic acid in solubilizing phosphate (Kim et al. 2002). Earlier studies on this phenomenon were restricted to mesophilic temperatures (Chung et al. 2005; Chen et al. 2006). The first report of P solubilization at low temperatures was made by Das et al. (2003) who studied cold-tolerant *Pseudomonas* mutants for their phosphate solubilization activity at low temperature (10°C). They found that all the cold-tolerant mutants were more efficient than their respective wild type counterparts for phosphate solubilization activity at 10°C as compared with 25°C. P solubilization by *Pseudomonas* mutant's at the psychrotolerant range has also been reported (Katiyar and Goel 2003; Trivedi and Sa 2008). But considering the environmental stability of mutant strains, for commercial inoculant production it would be prudent to scout pristine environments for naturally occurring psychrotolerant strains. Most progress has been made in this direction, mainly from the Indian Himalayan Region.

Pandey et al. (2006) isolated a cold-tolerant phosphate solubilizing and antagonistic strain of *P. putida*, from a subalpine location of Indian central Himalaya. This strain solubilized phosphate in the temperature range of 4–28°C. Phosphate solubilization by a cold-tolerant strain of *P. fragi* was reported by Selvakumar et al. (2009a). This is a novel discovery since *P. fragi* is generally associated with spoilage of dairy foods under refrigerated conditions. This strain solubilized phosphate at temperatures ranging from 4 to 30°C, besides significantly increasing the percent germination, rate of germination, plant biomass, and nutrient uptake of wheat seedlings under cold temperature conditions. A rhizosphere competent phosphate solubilizing strain of *Acinetobacter rhizosphaerae* was isolated from the cold deserts of the Indian Himalayan region by Gulati et al. (2009).Though phosphate solubilization at cold temperatures by this bacterium was not described, this is an early report on the occurrence of this bacterium in cold environments. Vyas et al. (2009) screened 19 efficient phosphate-solubilizing fluorescent *Pseudomonas* isolates from the cold deserts of the trans-Himalayas, for tolerance against temperature, alkalinity, salinity, calcium salts, and desiccation-induced stresses. Phylogenetic analysis based on 16S rRNA gene sequencing placed these bacteria under three groups with 14 strains in Group I including *Pseudomonas trivialis* and *P. poae*, two strains in Group II together with *Pseudomonas kilonensis* and *P. corrugata*, and three strains in Group III along with *Pseudomonas jessenii* and *P. moraviensis*. In a recent study, Selvakumar et al. (2009b) reported that the genetic clustering of cold-tolerant phosphate solubilizing Pseudomonads was affected by their geographical origin. Repetitive element PCR profiles revealed that isolates originating from the warmer southern slopes formed a distinct cluster, while their counterparts from the cooler north facing slopes formed the second cluster. The studies that have been mentioned above are mostly of exploratory nature, while the real need of the hour is the development of a commercially viable cold-tolerant PSB inoculant that can be profitably used in temperate agriculture.

5.4 Induction of Resistance to Cold Stress by PGPB

Cold temperature stress affects the metabolic activity of plants in multiple ways and causes significant yield reduction. To overcome this, several exploratory studies using microbial strains have been carried out. In vitro inoculation of *Vitis vinifera* cv. Chardonnay explants with a plant growth-promoting rhizobacterium, *Burkholderia phytofirmans* strain PsJN, increased grapevine growth and physiological activity at a low temperature. There was a relationship between endophytic bacterial colonization of the grapevine plantlets and their growth at both ambient (26°C) and low (4°C) temperatures and their sensitivities to chilling. The major benefits of bacterization were observed on root growth (11.8 and 10.7-fold increases at 26°C and 4°C, respectively) and plantlet biomass (6 and 2.2-fold increases at 26°C and 4°C, respectively). The inoculation with PsJN also significantly improved plantlet cold tolerance compared with that of the nonbacterized control. Moreover, relative to the noninoculated controls, bacterized plantlets had significantly increased levels of starch, proline, and phenolics. These increases correlated with the enhancement of cold tolerance of the grapevine plantlets (Barka et al. 2006).

Mishra et al. (2009b) examined the effect of seed inoculation with 12 cold-tolerant plant growth promoting *Pseudomonas* strains on wheat growth and physiological changes under green house conditions at 10 ± 2°C. It was observed that bacterization with Pseudomonads significantly improved root length (27.9–70.5%), shoot length (4.7–26.1%), dry root biomass (1.69–3.19-fold increases), dry shoot biomass (1.27–1.66-fold increase) compared with nonbacterized control. Bacterized wheat plants showed enhanced levels of total chlorophyll, anthocyanin, free proline, total phenolics, and starch contents, while a decrease was observed in the Na^+/K^+ ratio and electrolyte leakage vaues. These parameters are critical to the plant's ability to tolerate cold stress conditions. In another study, they observed that inoculation with cold-tolerant bacterium *Pseudomonas* spp. strain PPERs23 significantly improved root length (41%), shoot length (11.9%), dry root biomass (44.4%), dry shoot biomass (53.8%), total chlorophyll (3.1%), total phenolics (37.3%), and amino acid (39.4%) content of wheat seedlings. In this study also, increased levels of physiologically available iron, protein, anthocyanin, proline and relative water contents coupled with a decrease in Na^+/K^+ ratio and electrolyte leakage values were observed in bacterized wheat plants. These parameters indicate the ability of bacterium to alleviate cold induced stress in wheat seedlings (Mishra et al. 2009c).

6 Industrial Potential of Psychrotolerant Microorganisms

The unique properties of cold-tolerant microorganism make them potential candidates for exploitation in industry. Microbial cryoprotectants like trehalose have immense biotechnological potential and can be used in a wide range of applications (Lillford and Holt 2002). Similarly, the Antifreeze Proteins (Afps) from bacteria can be used in a wide variety of ways (Tange et al. 2003). Cold active proteases are used for the industrial peeling of leather by proteases at tap water temperatures instead of heating to 37°C. During cold storage β-galactosidases are used to remove lactose in milk, while cold active pectinases are used for the clarification of fruit juices. Another interesting application is the use of a heat labile alkaline phosphatase, which does not interfere with end labeling of polynucleotide kinase after heat treatment. Cold-tolerant microbes and enzymes can be used for the bioremediation of polluted soils and waste waters during winter in temperate regions, when the degradative capacities of the endogenous microflora is impaired by low temperatures (Feller and Gerday 2003).The development of transgenic plants with increased frost tolerance is another exciting application. The introduction of genes from microorganisms or even whole biosynthetic pathways in plants has already been shown to improve freeze tolerance.

Arabidopsis thaliana plants transformed with the *codA* gene encoding choline oxidase and accumulating glycine betaine in the chloroplast showed a significant improvement in freeze tolerance (Sakamoto et al. 2000).

7 Conclusion

Cold-tolerant microorganisms are widely distributed in the agro-ecosystem and play a variety of roles extending from nitrogen fixation, plant growth promotion, and alleviation of cold stress in plants. Though most research work conducted so far has largely focused on rhizobia, it is a welcome sign that many agriculturally important resourceful microbes are being described from various parts of the earth. But serious attempts are needed to study the activity of enzymes such as nitrogenases in cold-adapted microorganisms. Another interesting area where research needs to be focused is the identification of cold active decomposing microorganisms, since temperature is a major determinant of decomposition, and most decomposition processes come to a standstill at suboptimal temperatures. If research efforts succeed in identifying consortia of potential decomposers that retain their enzymatic potential at lower temperatures, it would be of immense use in agriculture the world over.

References

Anderson JA, Buchanan DW, Stall RE, Hall CB (1982) Frost injury of tender plants increased by *Pseudomonas syringae* van Hall. Am Soc Hort Sci 107:123–125

Andrews JH, Harris RF (2003) The ecology and biogeography of microorganisms on plant surfaces. Annu Rev Phytopathol 38:145–180

Angelidis AS, Smith GM (2003) Role of glycine betaine and carnitine transporters in adaptation of *Listeria monocytogenes* to chill stress in defined medium. Appl Environ Microbiol 69:7492–7498

Annous BA, Becker LA, Bayles DO, Labeda DP, Wilkinson BJ (1997) Critical role of anteiso-$C_{15:0}$ fatty acid in the growth of *Listeria monocytogenes* at low temperatures. Appl Environ Microbiol 63:3887–3894

Bae W, Xia B, Inouye M, Severinov K (2000) *E. coli*, Csp A – family RNA chaperone are transcription antiterminators. Proc Nat Acad Sci USA 97:7784–7789

Baleiras-Couto MM, Huis-In't-Veld JHJ (1995) Influence of ethanol and temperature on the cellular fatty acid composition of *Zygosaccharomyces bailii* spoilage yeasts. J Appl Bacteriol 78:327–333

Barka EA, Nowak J, Clement C (2006) Enhancement of chilling resistance of inoculated grapevine plantlets with a plant growth-promoting rhizobacterium *Burkholderia phytofirmans* strain PsJN. Appl Environ Microbiol 72:7246–7252

Bayles DO, Annous BA, Wilkinson BJ (1996) Cold stress proteins induced in *Listeria monocytogenes* is responsible for temperature downshock and growth at low temperatures. Appl Environ Microbiol 62:1116–1119

Bric JM, Bostock RM, Silverstone SR (1991) Rapid in situ assay for indole acetic acid production by bacteria immobilized on a nitrocellulose membrane. Appl Microbiol 57:535–538

Chattopadhyay MK (2002) Glycine betaine is a bacterial cryoprotectant and is believed to act by stabilizing cellular proteins and membranes at low temperature. Resonance 7:59–63

Chen YP, Rekha PD, Arun AB, Shen FT, Lai WA, Young CC (2006) Phosphate solubilizing bacteria from subtropical soil and their tricalcium phosphate solubilizing abilities. Appl Soil Ecol 34:33–41

Cheng Z, Park E, Glick BR (2007) 1-Aminocyclopropane-1-carboxylate (ACC) deaminase from *Pseudomonas putida* UW4 facilitates the growth of canola in the presence of salt. Can J Microbiol 53:912–918

Chung H, Par M, Madhaiyan M, Seshadri S, Song J, Cho H, Sa T (2005) Isolation and characterization of phosphate solubilizing bacteria from the rhizosphere of crop plant of Korea. Soil Biol Biochem 37:1970–1974

Cloutier J, Prevost D, Nadeau P, Antoun H (1992) Heat and cold shock protein synthesis in artic and temperate strains of rhizobia. Appl Environ Microbiol 58:2846–2853

D'Amico S, Marx JC, Gerday C, Feller G (2003) Activity–stability relationships in extremophilic enzymes. J Biol Chem 278:7891–7896

D'Amico S, Colins T, Marx JC, Feller G, Gerday C (2006) Psychrophilic microorganisms: a challenge for life. EMBO Rep 7:385–389

Das K, Katiyar V, Goel R (2003) P solubilization potential of plant growth promoting *Pseudomonas* mutants at low temperature. Microbiol Res 158:359–362

Deming JW (2002) Psychrophiles and Polar regions. Curr Opin Microbiol 5:301–309

Denarie J, Debelle F, Rosenberg C (1992) Signalling and host range variation in nodulation. Annu Rev Microbiol 46:497–531

Drouin P, Prevost D, Antoun H (2000) Physiological adaptation to low temperatures of strains of *Rhizobium leguminosarum* bv. viciae associated with *Lathyrus* spp. FEMS Microbiol Ecol 32:111–120

Dufrenne J, Delfgou E, Ritmeester W, Notermans S (1997) The effect of previous growth conditions on the lag phase time of some foodborne pathogenic microorganisms. Int J Food Microbiol 34:89–94

Escalana NG, Fey A, Hofle MG, Espejo RT, Guzman CA (2006) Quantitative reverse transcription polymerase chain reaction analysis of *Vibrio cholerae* cells entering the viable but non-culturable state and starvation in response to cold shock. Environ Microbiol 8:658–666

Evans RI, McClure PJ, Gould GW, Russel NJ (1998) The effect of growth temperature on the phospholipid and fatty acyl compositions of non-proteolytic *Clostridium botulinum*. Int J Food Microbiol 40:159–167

Feller G, Gerday C (2003) Psychrophilic enzymes: hot topic in cold adaptation. Nat Rev Microbiol 1:200–208

Forster J (1887) Ueber einige Eigenschaften leuchtender Bakterien Centr. Bakteriol Parasitenk 2:337–340

Fujii DK, Fulco A (1977) Biosynthesis of unsaturated fatty acids by Bacilli, hyper induction and modulation of desaturase synthesis. J Biol Chem 252:3660–3670

Geiger O, Spaink HP, Lugtenberg BJJ (1993) Biosynthesis of lipo-oligosaccharides: phospholipids of *Rhizobium* contain nod E-determined highly unsaturated fatty acid moieties. In: Palacios R, Mora J, Newton WE (eds) New horizons in nitrogen fixation. Kluwer, Dordrecht, p 233

Glick BR, Penrose DM, Jiping L (1998) A model for the lowering plant ethylene concentrations by plant growth promoting bacteria. J Theor Biol 190:63–68

Goldstein A (1995) Recent progress in understanding the molecular genetics and biochemistry of calcium phosphate solubilization by Gram negative bacteria. Biol Agric Hort 12:185–193

Goldstein J, Pollitt NS, Inouye M (1990) Major cold shock proteins of *Escherichia coli*. Proc Natl Acad Sci USA 87:283–287

Graumann P, Marahiel MA (1998) A superfamily of proteins containing the cold shock domain. Trends Biochem Sci 23:286–290

Graumann P, Schroder K, Schmid R, Marahiel MA (1996) Cold shock stress induced proteins in *Bacillus subtilis*. J Bacteriol 178:4611–4619

Greenland D, Losleben M (2001) Structure and function of an alpine ecosystem. In: Bowman WD, Seastedt TR (eds) Climate. Oxford University Press, Niwot Ridge, CO, pp 15–31

Gulati A, Vyas P, Rai P, Kasana RC (2009) Plant growth promoting and rhizosphere-competent *Acinetobacter rhizosphaerae* strain BIHB 723 from the cold deserts of the Himalayas. Curr Microbiol 58:371–377

Hebraud M, Dubois E, Potier P, Labadie J (1994) Effect of growth temperature on the protein levels in the psychrotrophic bacterium *Pseudomonas fragi*. J Bacteriol 176:4017–4024

Herbert RA (1986) The ecology and physiology of psychrotrophic micro-organism. In: Herbert RA, Codd GA (eds) Society for gen microbiology. Academic, London, pp 1–24

Herbraud M, Potier P (1999) Cold shock response and low temperature adaptation in psycrophilic bacteria. J Mol Microbiol Biotechnol 1:211–219

Hoang LC, Dumomt F, Marechal PA, Thanh ML, Gervais P (2007) Rates of chilling to 0°C: implication for the survival of microorganisms and relationship with membrane fluidity modification. Appl Microbiol Biotechnol 77:1379–1387

Horn G, Hofweber W, Kremer W, Kalbitzer HR (2007) Structure and function of bacterial cold shock proteins. Cell Mol Life Sci 64:1457–1470

Howard DH (1999) Acquisition, transport and storage of iron by pathogenic fungi. Clin Microbiol Rev 12:394–404

Jiang W, Hou Y, Inouye M (1997) The major cold shock proteins of *E. coli*, is an RNA chaperone. J Biol Chem 272:196–202

Johns GC, Somero GN (2004) Evolutionary convergence in adaptation of proteins to temperature: A4-lactate dehydrogenases of Pacific damselfishes (*Chromis* spp.). Mol Biol Evol 21:314–320

Jones PG, Inouye M (1996) RbfA, a 30S ribosomal binding factor, is a cold-shock protein whose absence triggers the cold-shock response. Mol Microbiol 21:1207–1218

Jones PG, VanBogelen RA, Neidhart (1987) Induction of proteins on response to low temperature in *E. coli*. J Bacteriol 169:2092–2095

Kaasen I, Falkenberg P, Styrvold OB, Stroem AR (1992) Molecular cloning and physical mapping of the *ots*BA genes, which encode the osmo-regulatory trehalose pathway of *Escherichia coli*: evidence that transcription is activated by Kat F(AppR). J Bacteriol 174:889–898

Kandror O, DeLeon A, Goldberg AL (2002) Trehalose synthesis is induced upon exposure of *Escherichia coli* to cold and is essential for viability at low temperatures. Proc Natl Acad Sci USA 99:9727–9732

Karner MB, Delong EF, Karl DM (2001) Archeal dominance in the mesophilic zone of Pacific Ocean. Nature 409:507–510

Katiyar V, Goel R (2003) Solubilization of inorganic phosphate and plant growth promotion by cold tolerant mutants of *Pseudomonas fluorescens*. Microbiol Res 158:163–168

Katiyar V, Goel R (2004) Siderophore mediated plant growth promotion at low temperature by a mutant fluorescent pseudomonad. Plant Growth Regul 42:239–244

Kaushik R, Saxena AK, Tilak KVBR (2000) Selection of Tn5::*lac Z* mutants isogenic to wild type *Azospirillum brasilense* strains capable of growing at sub-optimal temperature. World J Microbiol Biotechnol 16:567–570

Kaushik R, Saxena AK, Tilak KVBR (2001) Selection and evaluation of *Azospirillum brasilense* strains capable of growing at sub-optimal temperature in rhizocoenosis with wheat. Folia Microbiol 46:327–332

Kaushik R, Saxena AK, Tilak KVBR (2002) Can *Azospirillum* strains capable of growing at a sub-optimal temperature perform better in field-grown-wheat rhizosphere. Biol Fertil Soils 35:92–95

Kawahara H, Koda N, Oshio M, Obata H (2000) A cold acclimation protein with refolding activity on frozen denatured enzyme. Biosci Biotechnol Biochem 64:2668–2774

Kim KY, Hwangbo H, Kim YW, Kim HJ, Park KH, Kim YC, Seoung KY (2002) Organic acid production and phosphate solubilization by *Enterobacter intermedium* 60-2G. Korean J Soil Sci Fert 35:59–67

Klein W, Weber MHW, Marahiel MA (1999) Cold shock response of *Bacillus subtilis*: isoleucine dependent switch in the fatty acid branching pattern adaptation to low temperatures. J Bacteriol 181:5341–5349

Kloepper JW, Schroth MN (1978) Plant growth-promoting rhizobacteria on radishes. In: Proceedings of the 4th international conference on plant pathogenic bacteria, Gilbert-Clarey, Tours, France, pp 879–882

Kloepper JW, Leong J, Teintze M (1980) Enhanced plant growth by siderophores produced by plant growth promoting rhizobacteria. Nature (London) 286:885–886

Knight CA, Hallett J, Devries AL (1988) Solute effects on ice recrystallization: an assessment technique. Cryobiology 25:55–60

Kottmier ST, Sullivan CW (1990) Bacterial biomass and production in pack ice of Antarctica marginal ice age zones. Deep-Sea Res 37:1311–1330

Kozloff LM, Schofield MA, Lute M (1983) Ice nucleating activity of *Pseudomonas syringae* and *Erwinia herbicola*. J Bacteriol 153:222–231

Lavania M, Chauhan PS, Chauhan SVS, Singh HB, Nautiyal CS (2006) Induction of plant defense enzymes and phenolics by treatment with plant growth-promoting rhizobacteria *Serratia marcescens* NBRI1213. Curr Microbiol 52:363–368

Lee RE, Warren GJ, Gusta LV (1995) Biochemistry of bacterial ice nuclei. In: Ray F, Paul WK (eds) Biological ice nucleation and its application. APS Press, St Paul, MN, pp 63–83

Lillford PJ, Holt CB (2002) In vitro uses of biological cryoprotectants. Philos Trans R Soc Lond B Biol Sci 357:945–951

Lindow SE (1983) The role of bacterial ice nucleation in frost injury to plants. Annu Rev Phytopathol 21:363–384

Lindow SE, Arny DC, Upper CD (1978) *Erwinia herbicola*: a bacterial ice nucleus active in increasing frost injury to corn. Phytopathol 68:523–527

Lynch DH, Smith DL (1994) The effects of low temperature stress on two soybean (*Glycine max*) genotypes when combined with *Bradyrhizobium* strains of varying geographic origin. Physiol Plant 90:105–113

Maki IR, Galyon EL, Chang-Chien M, Cald WDR (1974) Ice nucleation induced by *Pseudomonas syringae*. Appl Microbiol 28:456–460

Malviya MK, Pandey A, Trivedi P, Gupta G, Kumar B (2009) Chitinolytic activity of cold tolerant antagonistic species of *Streptomyces* isolated from glacial sites of Indian Himalaya. Curr Microbiol 59:502–508

Margesin R, Schinner F (1999) Cold adapted organisms Ecology, Physiology, Enzymology, and Molecular biology. Springer, Berlin

Margesin R, Neuner G, Storey KB (2007) Cold-loving microbes, plants, and animals-fundamental and applied aspects. Naturewisenschaften 94:77–99

Masalha J, Kosegarten H, Elmaci O, Mengel K (2000) The central role of microbial activity for iron acquisition in maize and sunflower. Biol Fertil Soil 30:433–439

Mastronicolis SK, German JB, Megoulas N, Petrou E, Foka P, Smith GM (1998) Influence of cold shock on the fatty acid composition of different lipid classes of the food borne pathogen *Listeria monocytogenes*. Food Microbiol 15:299–306

Mayr B, Kaplan T, Lechner S, Scherer S (1996) Identification and purification of a family of dimeric major cold shock protein homologous from the psychrotrophic *Bacillus cereus* WSBC 10201. J Bacteriol 178:2916–2925

McBeath J (1995) Cold tolerant *Trichoderma*. US Patent #5,418,165

McCue P, Zheng Z, Pinkham JL, Shetty K (2000) A model for enhanced pea seedling vigour following low pH and salicylic acid treatments. Proc Biochem 35:603–613

McGibbon L, Russel NJ (1983) Fatty acid positional distribution in phospholipids of a psychrophilic bacterium during changes in growth temperature. Curr Microbiol 9:241–244

McKay IA, Djordjevic MA (1993) Production and excretion of nod metabolites by *Rhizobium leguminosarum* bv. *trifolii* are disrupted by the same environmental factors that reduce nodulation in the field. Appl Environ Microbiol 59:3385–3392

Michel V, lehoux I, Hebraud (1997) The cold shock response of the psychrotrophic bacterium *Pseudomonas fragi*. Curr Microbiol 33:16–25
Misaghi IJ, Stowell LJ, Grogan RG, Spearman LC (1982) Fungistatic activity of water-soluble fluorescent pigments of fluorescent pseudomonads. Phytopathology 72:33–36
Mishra PK, Mishra S, Selvakumar G, Bisht SC, Bisht JK, Gupta HS (2008) Characterization of a psychrotolerant plant growth promoting *Pseudomonas sp*. strain PGERs17 (MTCC 9000) isolated from North Western Indian Himalayas. Ann Microbiol 58:561–568
Mishra PK, Bisht SC, Pooja R, Joshi P, Suyal P, Bisht JK, Srivastva AK (2009c) Enhancement of chilling tolerance and productivity of inoculated wheat with cold tolerant plant growth promoting *Pseudomonas* spp. PPERs23. Abstract 4th USSTC. Nov 10–12, 2009
Mishra PK, Bisht SC, Ruwari P, Selvakumar G, Bisht JK (2009b) Enhancement of chilling tolerance of inoculated wheat seedlings with cold tolerant plant growth promoting Pseudomonads from N.W. Himalayas. Abstract Ist Asian PGPR conference. Jun 22–24, 2009
Mishra PK, Mishra S, Bisht SC, Selvakumar G, Kundu S, Bisht JK, Gupta HS (2009c) Isolation, molecular characterization and growth-promotion activities of a cold tolerant bacterium *Pseudomonas* sp. NARs9 (MTCC9002) from the Indian Himalayas. Biol Res 42:305–313
Mitta M, Fang L, Inouye M (1997) Deletion analysis of *cspA* of *Escherichia coli*: requirement of the AT-rich UP element for *cspA* transcription and the downstream box in the coding region for its cold shock induction. Mol Microbiol 26:321–335
Morita RY (1975) Psycrophilic bacteria. Bacteriol Rev 39:144–167
Muryoi N, Sato M, Kaneko S, Kawaahara H, Obata H, Yaish MWF, Griffth M, Glick BR (2004) Cloning and expression of *afp*A, a gene encoding an antifreeze protein from the Arctic plant growth-promoting rhizobacterium *Pseudomonas putida* GR12-2. J Bacteriol 186:5661–5671
Negi YK, Kumar J, Garg SK (2005) Cold-tolerant fluorescent *Pseudomonas* isolates from Garhwal Himalayas as potential plant growth promoting and biocontrol agents in pea. Curr Sci 89:2151–2156
Neilands JB (1981) Microbial iron compounds. Annu Rev Biochem 50:715–731
Neilands JD (1995) Siderophores: structure and function of microbial iron transport compounds. J Biol Chem 270:26723–26726
Obata H, Kakinami K, Tanishita J, Hasegawa Y (1990) Identification of new Ice-nucleating bacterium and its ice nucleation properties. Agric Biol Chem 54:725–730
Olson JC, Nottingham PM (1980) Temperature in microbial ecology of foods volume 1: factors affecting life and death of microorganisms. International Commission on Microbiological specifications for foods, Academic Press, London, pp 1–37
Pandey A, Trivedi P, Palni LMS (2006) Characterization of phosphate solubilizing and antagonistic strain of *Pseudomonas putida* (BO) Isolated from a sub-alpine location in the Indian Central Himalaya. Curr Microbiol 53:102–107
Panicker G, Aislabie SD, Bej AK (2002) Cold tolerance of *Pseudomonas* sp. 30-3 isolated from oil contaminated soil, Antarctica. Polar Biol 25:5–11
Panoff JM, Thammavongs B, Gueguen M, Boutibonnes P (1998) Cold stress responses in mesophilic bacteria. Cryobiology 36:75–83
Patten CL, Glick BR (2002) Role of *Pseudomonas putida* indoleacetic acid in development of the host plant root system. Appl Environ Microbiol 68:3795–3801
Perin C, Guimont C, Bracquart P, Gaillard J-L (1999) Expression of a new cold shock protein of 21.5 kDa and of the major cold shock protein by *Streptococcus thermophilus* after cold shock. Curr Microbiol 39:342–347
Polissi A, De Laurentis W, Zangrossi S, Briani F, Loghi V, Pesole G, Deho G (2003) Changes in *Escherichia coli* transcriptome during acclimatization at low temperature. Microbiol Res 154:573–580
Ponder MA, Gilmour SJ, Bergholz PW, Mindock CA, Hollingsworth R, Thomashow MW, Tiedje JM (2005) Characterization of potential stress response in ancient Siberian permafrost psychroactive bacteria. FEMS Microbiol Ecol 53:103–115

Prevost D, Drouin P, Antoun H (1999) The potential use of cold adapted rhizobia to improve nitrogen fixation in legumes cultivated in temperate regions. In: Margesin R, Schinner F (eds) Biotechnological application of cold-adapted organisms. Springer, Berlin, pp 161–176
Prevost D, Drouin P, Laberge S, Bertrand A, Cloutier J, Levesque G (2003) Cold-adapted rhizobia for nitrogen fixation in temperate regions. Can J Bot 81:1153–1161
Puettman M, Ade N, Hof H (1993) Dependence of fatty acid composition of *Listeria* spp. on growth temperature. Microbiol Res 144:279–283
Purusharth RI, Klein F, Sulthana S, Jager S, Jagannadham MV, Hackenberg EE, Ray MK, Klug G (2005) Exoribonuclease R interacts with endoribonuclease E and RNA helicase in the psychrotrophic bacterium *Pseudomonas syringae* Lz4W. J Biol Chem 280:14572–14578
Raymond JA, DeVries AL (1977) Adsorption inhibition as a mechanism of freezing resistance in polar fishes. Proc Natl Acad Sci USA 74:2589–2593
Russell NJ (1984) Mechanisms of thermal adaptation in bacteria: blueprints for survival. Trends Biochem Sci 9:108–112
Russell NJ (1990) Cold adaptation of micro-organism. Philos Trans Soc Lond 326:595–611
Russell NJ (1998) Molecular adaptation in psychrophilic bacteria: potential for biotechnological application. Adv Biochem Eng Biotechnol 61:1–2
Russell NJ, Evans RI, TerSteeg PF, Hellemons J, Verheul A, Abee T (1995) Membranes as a target for stress adaptation. Int J Food Microbiol 28:255–261
Sakamoto A, Valverde RA, Chen TH, Murata N (2000) Transformation of *Arabidopsis* with the *cod A* gene for choline oxidase enhances freezing tolerance of plants. Plant J 22:449–453
Saleem M, Arshad M, Hussain S, Bhatti AS (2007) Perspective of plant growth promoting rhizobacteria (PGPR) containing ACC deaminase in stress agriculture. J Ind Microbiol Biotechnol 34:635–648
Sano F, Asakawa N, Inouye Y, Sakurai M (1999) A dual role for intracellular trehalose in the resistance of yeast cells to water stress. Cryobiology 39:80–87
Sardesai N, Babu CR (2001a) Cold stress induced high molecular weight membrane polypeptides are responsible for cold tolerance in *Rhizobium* DDSS69. Microbiol Res 156:279–284
Sardesai N, Babu CR (2001b) Poly-β-hydroxybutyrate metabolism is affected by changes in respiratory enzymatic activities due to cold stress in two psychrotrophic strains of *Rhizobium*. Curr Microbiol 42:53–58
Schindelin H, Marahiel MA, Heinemann U (1993) Universal nucleic acid-binding domain revealed by crystal structure of the *Bacillus subtilis* major cold shock proteins. Nature 364:164–168
Schnuchel A, Wiltscheck R, Czisch M, Herrier M, Willimsky G, Graumann P, Marahiel MA, Holak TA (1993) Structure in solution of the major cold shock protein from *Bacillus subtilis*. Nature 364:169–171
Selvakumar G, Kundu S, Joshi P, Gupta AD, Nazim S, Mishra PK, Gupta HS (2008a) Characterization of a cold-tolerant plant growth-promoting bacterium *Pantoea dispersa* 1A isolated from a sub-alpine soil in the North Western Indian Himalayas. World J Microbiol Biotechnol 24:955–960
Selvakumar G, Mohan M, Kundu S, Gupta AD, Joshi P, Nazim S, Gupta HS (2008b) Cold tolerance and plant growth promotion potential of *Serratia marcescens* strain SRM (MTCC 8708) isolated from flowers of summer squash (*Cucurbita pepo*). Lett Appl Microbiol 46:171–175
Selvakumar G, Joshi P, Mishra PK, Bisht JK, Gupta HS (2009a) Mountain aspect influences the genetic clustering of psychrotolerant phosphate solubilizing Pseudomonads in the Uttarakhand Himalayas. Curr Microbiol 59:432–438
Selvakumar G, Joshi P, Nazim S, Mishra PK, Bisht JK, Gupta (2009b) Phosphate solubilization and growth promotion by *Pseudomonas fragi* CS11RH1 (MTCC 8984) a psychrotolerant bacterium isolated from a high altitude Himalayan rhizosphere. Biologia 64:239–245
Selvakumar G, Joshi P, Nazim S, Mishra PK, Kundu S, Gupta HS (2009c) *Exiguobacterium acetylicum* strain 1P (MTCC 8707) a novel bacterial antagonist from the North Western Indian Himalayas. World J Microbiol Biotechnol 25:131–137
Sprent JI (1979) The biology of nitrogen-fixing organisms. McGraw-Hill, NewYork, NY

Tange AN, Dijck PV, Thevelein JM (2003) Determinants of freeze tolerance in microorganisms, physiological importance, and biotechnological applications. Adv Appl Microbiol 53:129–167

Theberge MC, Prevost D, Chalifour FP (1996) The effect of different temperatures on the fatty acid composition of *Rhizobium leguminosarum* bv. *viciae* in the faba bean symbiosis. New Phytol 134:657–664

Tripathi AK, Klingmuller W (1992) Temperature sensitivity of nitrogen fixation in *Azospirillum* sp. Can J Microbiol 38:1238–1241

Trivedi P, Sa T (2008) *Pseudomonas corrugata* (NRRL B-30409) mutants increased phosphate solubilization, organic acid production, and plant growth at low temperatures. Curr Microbiol 56:140–144

Varcamonti M, Arsenijevic S, Martirani L, Fusco D, Naclerio G, Felice MD (2006) Expression of the heat shock gene *clpL* of *Streptococcus thermophilus* is induced by both heat and cold shock. Microb Cell Fact 5:6. doi:10.1186/1475-2859-5-6

Volkmar KM, Bremer E (1998) Effects of seed inoculation with strain of *Pseudomonas fluorescens* on root growth and activity of wheat in well-watered and drought stressed grass-fronted rhizotrons. Can J Plant Sci 78:545–551

Vyas P, Rahi P, Gulati A (2009) Stress tolerance and genetic variability of phosphate-solubilizing fluorescent *Pseudomonas* from the cold deserts of the Trans-Himalayas. Microb Ecol 58:425–434

Wang C, Knill E, Defago G (2000) Effect of transferring 1-aminocyclopropane-1-carboxylic acid (ACC) deaminase genes into *Pseudomonas fluorescens* strain CHAO and its *gacA* derivative CHA96 on their growth-promoting and disease-suppressive capacities. Can J Microbiol 46:898–907

Wilkins PO (1973) Psychrotrophic Gram-positive bacteria: temperature effects on growth and solute uptake. Can J Microbiol 19:909–915

Witter LD, Campbell MF, Azuma Y (1966) Formation of bacterial pigments at low temperature by psychrophillic pseudomonads. Dev Ind Microbiol 7:231–239

Wouters JA, Jeynov B, Rombouts FM, de Vos WM, Kuipers OP, Abee T (1999) Analysis of the role of 7 kDa cold-shock proteins of *Lactococcus lactis* MG1363 in cryoprotection. Microbiology 145:3185–3194

Xu H, Griffith M, Patten CL, Glick BR (1998) Isolation and characterization of an antifreeze protein with ice-nucleation activity from the plant growth promoting rhizobacterium *Pseudomonas putida* GR12-2. Can J Microbiol 44:64–73

Zachariassen KE, Kristiansen E (2000) Ice nucleation and anti-nucleation in nature. Cryobiology 41:257–279

Zhang H, Prithiviraj B, Charles TC, Driscoll BT, Smith DL (2003) Low temperature tolerant *Bradyrhizobium japonicum* strains allowing improved nodulation and nitrogen fixation of soybean in a short season (cool spring) area. Eur J Agron 19:205–213

The Role of Plant Growth Promoting Rhizobacteria in Sustainable and Low-Input Graminaceous Crop Production

Stephen P. Cummings and Caroline Orr

Contents

1 Introduction 298
2 The Role of PGPR in Low-Input and Sustainable Agriculture 299
3 Mechanisms of Action of PGPR 303
  3.1 Biological Nitrogen fixation 303
  3.2 Production of Phytohormones 305
  3.3 Enhanced Nutrient Availability 306
  3.4 Enhanced Stress Tolerance 307
  3.5 Indirect Effects 307
4 Application of PGPR to Soil 308
5 Future Work 309
  5.1 Engineering the Rhizosphere 309
  5.2 Bespoke Inocula 310
6 Conclusions 310
References 311

Abstract Plant growth stimulating rhizobacteria that improve the yield of graminaceous crops have been studied since the 1930s. Increases in crop yield have often been inconsistent, reflecting a lack of understanding by which PGPR exert their effects. Many PGPR are able to fix N_2, which was initially assumed to boost crops by supplementing soil N. Subsequently, it became clear that for most free-living PGPR other mechanisms affecting root development and nutrient uptake can explain the increased crop yields. Endophytic bacteria have demonstrated some potential to contribute to the N budget of certain graminaceous crops but require more robust assessment of their potential. Here, we review the current state of our understanding of PGPR in graminaceous crop cultivation, identifying their potential

S.P. Cummings (✉) and C. Orr
School of Applied Sciences University of Northumbria, Ellison Building, NE1 8ST, Newcastle upon Tyne, UK
e-mail: stephen.cummings@northumbria.ac.uk

D.K. Maheshwari (ed.), *Plant Growth and Health Promoting Bacteria*,
Microbiology Monographs 18, DOI 10.1007/978-3-642-13612-2_13,

contribution to more sustainable agricultural practices but also highlighting issues that need to be addressed before this technology can be appropriately assessed as a replacement for inorganic N addition.

1 Introduction

The latter part of the twentieth century saw a remarkable rise in the productivity of agricultural systems; today, the global production of food is 145% greater than that of 1960 (Pretty 2008). In parallel with increases in agricultural output, the world population has doubled from three to six billion over the same period; nevertheless, on a per capita basis there is 25% more food for each individual compared to 50 years ago. China has demonstrated the most significant developments, increasing overall food production fivefold and per capita production threefold since 1960. In contrast, Africa has observed a 10% decline in per capita food production over the same period (Pretty 2008), Currently, the global population could be fed by the present level of agricultural output; however, the regional inequalities reflect continuing political, economic, and social challenges preventing agriculture to feed the current population, 1.2 billion of whom live in crippling poverty (Hazell and Wood 2008).

The gravest issue for sustainable food security is the predicted increase in global population, which is projected to rise from the current 6.8 billion and surpass 9 billion people by 2050. The burden of feeding these additional people will be felt most keenly by developing countries, whose populations are projected to rise from 5.6 billion in 2009 to 7.9 billion in 2050 (https://www.unfpa.org/public/).

Most developing countries have environmental constraints that will impede the development of agricultural systems able to meet this challenge. These include lack of water, desertification and insufficient cultivable land. Moreover, the increase in urbanisation of the global population poses additional challenges. That will require a 500% increase in economic activity driven by a 300% increase in both energy consumption and manufacturing activity (Miller 2008). For many people in rapidly developing economies, an increased disposable income coincides with the adoption of a diet with a greater consumption of meat and processed cereal products. To meet this demand, livestock will need to be raised intensively on a diet of cereals and oils (Pretty 2008). This, in turn, will place an increased pressure on the available agricultural land and how it is farmed. As a result, it has been argued that current models of low-input agriculture relying on biological nitrogen fixation (BNF) and requiring large areas of land will be unlikely to provide the annual requirement of an extra 15 million tonnes of protein by 2050 to stave off widespread hunger (Jenkinson 2001; Smil 2001).

The technological advances that have enabled agriculture to intensify its production practices in order to keep pace with the increasing demand have relied on the application of huge amounts of inorganic N fertiliser that effectively meets the demand of crops throughout the growing season and removes the requirement for a fertility enhancing cycle in crop rotations. The exploitation of inorganic N fertiliser has contributed to a 4% increase in aggregate global cereal grain production in the 40 years since 1960, during which period fertiliser consumption increased from 10.8 to 85.6 Mt N year^{-1} (Crews and Peoples 2004). The significance of inorganic N fertiliser and the Haber–Bosch process that generates has been contextualised by Smil (2001) who asserts that by 2050 over half of the human population will owe its existence to synthetic N fertilisers.

While the application of inorganic N has had significant benefits for agricultural food productivity and global food security in the short term, there are increasing concerns around the sustainability of this technology to provide a long-term solution to ensure that food production keeps pace with the burgeoning population. The management of agricultural soil is fundamental to ensuring a sustainable agricultural system; however, it is becoming clear that intensive agricultural systems leads to the degradation of agricultural soils as a result of, among other factors, the loss of organic matter, compaction and increased salinity, and leaching of inorganic nitrate, along with associated costs such as fuel requirements and the loss of water resources (Kibblewhite et al. 2008; Peoples et al. 1995; Smil 2001).

Consequently, there is increasing interest in developing agricultural management systems that embrace the principles of sustainability. While such concepts are not novel, there is an increasing urgency in developing and implementing them because of increasing alarm that current conventional agricultural management systems cannot continue linearly increasing their reliance on fertilizer consumption, pesticide application, the expansion of agricultural land and machine usage indefinitely, without detriment to the environment (Kitzes et al. 2008).

Pretty (2008) articulated three factors that defined sustainable agricultural practices and technologies:

1. They have no adverse effects on the environment.
2. They are accessible and effective for farmers.
3. They lead to improvements in food productivity and have positive effects on environmental goods and services.

The aim of this chapter is to map the contribution of plant growth promoting rhizobacteria (PGPR) that are indigenous or inoculated into soils to the sustainable cultivation of graminaceous crops. We examine how this technology have and may continue to contribute to more sustainable approaches to the production of these key crops in the context of increasing population growth and other environmental pressures.

2 The Role of PGPR in Low-Input and Sustainable Agriculture

PGPR have been extensively studied and used as inoculants on graminaceous crops, to increase yield and simultaneously reduce the requirement for inorganic fertiliser application. There are numerous studies, utilising a range of bacterial taxa, which explore the effects of PGPR inoculation on such crops. In Table 1, a summary of the

Table 1 A summary of the effects of PGPR on graminaceous crops under laboratory and field conditions

PGPR	Crop	Effect of inoculation	Proposed mode of action	Reference
Azospirillum; *Flavobacterium*; *Pseudomonas*	Rice (*Oryzia sativa* L.)	Maximum rice yields were observed when PGPR inoculum was combined with inorganic fertiliser (30 kg N ha^{-1})	More efficient utilisation of fertilizer N	Ali et al. (1995)
Isolates from rice rhizosphere		Isolate inoculation resulted in a significant increase in germination rates, plant height, root length and dry matter production in rice seedlings.	Produce phytohormones and solubilise phosphate	Ashrafuzzaman et al. (2009)
Azospirillum lipoferum		Increased root length, root surface area and root volume	Phytohormone synthesis and siderophore production	Boyer et al. (2008)
Azospirillum		Grain yield and N content was improved	Increased N fixation	Pedraza et al. (2009)
Pseudomonas spp.		Increased IAA levels	Phytohormone synthesis	Karnwal (2009)
Azospirillum; Enterobacter; *Aeromonas veronii*		Increase in root area, plant biomass and N fixation.	Increased N fixation and phytohormone synthesis	Mehnaz et al. (2001)
Pseudomonas spp.; *Azospirillum*		Increased shoot biomass and grain yield	Increased N fixation and phytohormone production	Mirza et al. (2006)
Herbaspirillum sp. strain B501 gfp1		Increased dry and fresh weight	Increased N fixation and phytohormone synthesis	Zakria et al. (2007)
Burkholderia vietnamiensis		Increased shoot and root weight and leaf surface	Mechanism is not addressed but hypothesize that increased N fixation and phytohormone production are involved	Van et al. (2000)
Isolates from wheat rhizosphere	Wheat (*Triticum aestivum* L.)	Improved yield and higher N content in grain and straw when used in combination with recommended chemical fertilizers.	Improved N use efficiency	Akhtar et al. (2009)
Azospirillum brasilense		Increased the quantity of photosynthetic pigments resulting in greener plants.	Enhanced photosynthetic pigment production.	Bashan et al. (2006)
Isolates from wheat rhizosphere (reference strains *Azospirillum* sp.)		Increased shoot and root biomass	Siderophore or phytohormone production	Fischer et al. (2007)

Pseudomonas cepacia R55, R85, *P. aeruginosa* R80, *P. fluorescens* R92; *P. putida* R104		Increased root dry weight but results were very inconsistent	Interaction with AMF alters nutrient and water uptake but leads to inconsistent results	Germida and Walley (1996)
Pseudomonas fluorescens		Significant increase in yield	Regulate production of ethylene and elongate roots by hydrolyzing 1-aminocyclopropane-1-carboxylic acid	Naveed et al. (2008a)
Pseudomonas fluorescens		Plant protected from soil borne fungal pathogens	Suppress soil borne fungal pathogens	Okubara and Bonsall (2008)
Pseudomonas fluorescens SBW25; *Paenibacillus brasilensis* PB177		In some combinations increased plant dry weight	Protection from pathogenic fungus	Jaderlund et al. (2008)
30 Isolates taken from wheat rhizosphere		Increased growth and yield	Increased phytohormone biosynthesis	Khalid et al. (2004)
Pseudomonas spp.; *Burkholderia caryophylli*		Significant root elongation, root weight and grain and straw yields.	Increased ACC-deaminase activity, chitinase activity, phytohormone production and P solubilization	Shaharoona et al. (2008)
Pseudomonas fluorescens		Significantly improved growth when increased NPK are also added to the soil	ACC deaminase activity	Shaharoona et al. (2008)
Pseudomonas fluorescens	Rye (*Secale cereale*)	Significant increase in foliar dry mass	Siderophore production and suppression of fungal pathogens	Kurek and Jaroszuk-Scisel (2003)
Bacillus licheniformis RCO2; *Rhodobacter capsulatus* RCO4; *Paenibacillu polymyxa* RCO5; *Pseudomonas putida* RCO6	Barley (*Hordeum vulgare* L.)	Increased root and shoot weight and increased uptake of Fe, N, Mn and Zn	Increased N fixation and production of phytohormones	Cakmakci et al. (2007)
Commercially available Plant Growth Activator (PGA)	Maize (*Zea mays* L.)	Greater plant height	More efficient uptake of N and P	Adesemoye et al. (2008)

(*continued*)

Table 1 (continued)

PGPR	Crop	Effect of inoculation	Proposed mode of action	Reference
Azospirillum brasilense Az39; *Bradyrhizobium japonicum* E109	Maize (*Zea mays* L.)	Seed germination and nodule formation were promoted	Production of phytohormones	Cassan et al. (2009)
Bacillus sp.; *Ochrobactrum* sp.		Significantly increased shoot and root dry weight	Suppressed fungal pathogens	Principe et al. (2007)
Azospirillum lipoferum CRT1		Root growth was enhanced	No explanation given	El Zemrany et al. (2006)
Rhizobium spp.; *Sinorhizobium* spp.		Increased shoot and root dry biomass	Production of phytohormones and siderophores.	Hossain and Martensson (2008)
Bacillus megaterium; . *B subtilis*; *Pseudomonas corrugata*		Increase in grain yield	Increase in fixed nitrogen, production of phytohormones, phosphate solubilization, production of antibiotics and siderophores	Kumar et al. (2007)
Pseudomonas spp.		Increased grain yield and nutrient uptake	Hydrolyses ACC	Naveed et al. (2008a)

effects of PGPR inoculation on a range of graminaceous crops including rice, wheat and maize are presented, derived from studies that have been conducted at both a laboratory and field scale. What is striking is the number of PGPR that have N_2-fixing ability but also the additional array of mechanisms of action, proposed to account for the enhanced plant growth observed. Many of the studies report a response of the crop under study and link it, often without robust data sets, to a particular bacterial activity such as N_2 fixation or production of phytohormones. It is clear that the issues around the lack of reproducibility of response to PGPR inoculation require a far more systematic approach before the technology can be effectively deployed in the field. Such studies need, among other things, to be designed to robustly test the mechanism(s) by which the plant responds to PGPR inoculation and its reproducibility from one growing season to the next, a consistent measure of the plants response (Vessey 2003) and an appraisal of the persistence of the inoculated PGPR in the soil (Strigul and Kravchenko 2006).

3 Mechanisms of Action of PGPR

In much of the earliest work on the exploitation of PGPR in graminaceous crop production, the mechanism of action of the bacteria was presumed to be a result of the increased input of fixed nitrogen into the soil. Many of the taxa identified as being effective PGPR are capable of fixing atmospheric nitrogen (Table 1). Subsequent work has revealed that there are a variety of mechanisms through which plant growth can be facilitated, including hormone production, enhanced nutrient acquisition, pathogen suppression and N_2-fixation, often working in parallel to produce the observed response. These effects have been extensively studied and reviewed. Here, we summarise the key findings suggesting that PGPR frequently exert their effect through multiple mechanisms working simultaneously.

3.1 Biological Nitrogen fixation

Biological Nitrogen fixation (BNF) can occur in bulk or rhizospheric soil. Fixed nitrogen can then be acquired through root uptake and contribute to the nitrogen budget of the crop. The earliest large-scale experiments, exploiting PGPR potential to enhance crop productivity used N_2-fixing bacteria, with the implicit assumption that it was this activity that was producing the enhanced crop yields. For example, large-scale field trials in the 1950s used N_2-fixing bacteria, principally *Azotobacter chroococcum* as an inoculum on several million hectares of graminaceous crops including wheat (*Triticum aestivum* L.) and barley (*Hordeum vulgare* L.) (Cooper 1959); however, owing to inconsistent results the trials were abandoned in the 1960s (Andrews et al. 2003). Other bacterial taxa, including *Azospirillum* spp. and *Agrobacterium radiobacter*, were also extensively studied and trialled as potential substitutes for N fertiliser (Zavalin et al. 2001). One study in Russia to test the potential of a strain of *A. radiobacter*, isolated from the rhizosphere of rice (*Oryza sativa* L.), on winter wheat and spring barely appeared to give significant increases (5–30%) in yield in 2 out of 3 years. At the same time, it was estimated that the contribution of N_2 fixation to total N assimilation was between 23 and 32% (Bairamov et al. 2001). However, the lack of consistency in the results from one year to the next reflected that of the earlier studies (Andrews et al. 2003). More significantly, in this example, *A. radiobacter*, now reclassified as *Rhizobium radiobacter* (Young et al. 2001), was a taxa that had never demonstrated the ability to fix N_2. Subsequent studies on this strain demonstrated unequivocally that, as with all members of this taxon, *R. radiobacter* was not capable of fixing atmospheric N_2, nor did it form a physical association with the roots of barley. The plant growth promoting substances it produced were most probably responsible for the increase in yields of graminaceous crops (Humphry et al. 2007).

N_2-fixing activity has been confirmed in PGPR in many other cases. *Azospirillum* species have, for example, been implicated in the enhancement of rice (Pedraza et al. 2009), maize (Montanez et al. 2009) and wheat (Sala et al. 2007) through BNF mechanisms. As many of the PGPR exhibit N_2-fixing abilities, it will always remain a temptation to invoke this activity to explain some of the perceived enhancement of yields observed when such bacteria are used as inoculants on graminaceous crops (Andrews et al. 2003). However, it is apparent by careful analyses of the literature that their mechanisms of action in enhancing crop yields are often due to a range of other activities which ironically, can reduce soil N rather than supplement it. What is clear is that none of the PGPR effects, studied to date, can match N fertiliser application as a consistent replacement for soil N deficiency (Andrews et al. 2003).

In contrast to free-living PGPR diazotrophs in soil, there is evidence that endophytic bacteria, particularly *Azoarcus* spp. in Kallar grass (*Leptochloa fusca* (L.) kunth.) and *Gluconacetobacter diazotrophicus* and *Herbaspirillum* spp. in sugar cane (*Saccharum* spp.) may play a major role in N_2 fixation (Hurek et al. 2002; Baldani et al. 1997). Endophytic bacteria colonise plants without causing damage to the host and fix nitrogen in situ. In graminaceous crops, most interest has focussed on the potential of endophytic BNF to replace or reduce the requirement for inorganic N fertiliser largely as a result of intensive studies on the effects of acetic acid bacteria, particularly *G. diazotrophicus*. This bacterium has been shown to make a substantive contribution to the nitrogen requirement of sugar cane in several regions where this crop is cultivated, most notably in Brazil (Boddey et al. 1991). The contribution of this bacterium to the nitrogen balance of sugar cane has been convincingly demonstrated with ^{15}N isotope dilution and ^{15}N natural abundance studies (James 2000). It has also been isolated from other tropical grasses, including kallar grass (Malik et al. 1997), rice (Muthukumarasamy et al. 2005) and Cameroon grass (Reis et al. 1994). Moreover, the closely related taxa *G. azotocaptans* was isolated from maize, and *Swaminathania salitolerans* and *Acetobacter peroxydans* from wild and wetland rice, respectively (Loganathan and Nair 2004; Muthukumarasamy et al. 2005). The ecology and contribution of *G. diazotrophicus* and other taxa to BNF in graminaceous crops has been extensively reviewed (Pedraza 2008; Saravanan et al. 2008); however, it is worth summarising some of the key findings identified from such literature surveys. The mechanisms, by which colonisation of the host by *G. diazotrophicus* is achieved, remains to be thoroughly characterised; the bacterium evades the host defences and the plant may control the initial colonization event through specific signalling events (Nogueira et al. 2001; Vargas et al. 2003).

The interaction between *G. diazotrophicus* and sugar cane does not always result in demonstrable BNF activity; data from Australia, South Africa and Japan indicated no significant BNF from sugar cane colonised with *G. diazotrophicus* (Saravanan et al. 2008). It is also clear that as well as BNF other plant growth stimulating activities, including hormone synthesis, nutrient mobilisation and pathogen suppression, (Saravanan et al. 2007, 2008 and references therein) play a role in the enhancement of plant growth. The potential of endophytes on the cultivation of more economically significant graminaceous crops has yet to be thoroughly investigated. However, endophytic *Azoarcus* species have been shown to colonise the interior roots of rice and express nitrogen-fixing systems (Hurek and Reinhold-Hurek 2003).

3.2 Production of Phytohormones

The early large scale studies of PGPR using *Azotobacter* (in the 1930s–1950s) and *Azospirillum* spp. (between 1976 and late 1980s) demonstrated that, in field trials, it was possible to observe significant increases in yields with a number of graminaceous crops (Andrews et al. 2003 and references therein). However, results were inconsistent and as a result the technology was never adopted. The original hypothesis, that the increased crop yields were due to BNF by the PGPR increasing the soil N budget, could not be substantiated. However, subsequent work has demonstrated that there are a range of mechanisms at play, the most significant of which is the production of phytohormones that increase root weight, length and surface area (Vessey 2003).

There are a number of studies in which the inoculation of PGPR, along with the addition of inorganic N fertiliser, results in an increase in crop yields comparable or greater than that observed when conventional quantities of inorganic N are applied. A study on wheat demonstrated maximum increases in yields of grain and straw were observed in treatments where PGPR were used in combination with recommended dosages of inorganic fertiliser (Akhtar et al. 2009). A further study indicated that PGPR, which demonstrated ACC-deaminase activity, such as *Pseudomonas fluorescens* and *Pseudomonas putida*, could both improve wheat and maize yield and reduce the dependence on inorganic N by 25%, while giving an increase in wheat grain yield to 96% (Naveed et al. 2008a, b). Other workers have demonstrated positive responses on wheat yields with reductions in the requirement for inorganic fertiliser with strains of *P. fluorescens* and *Azospirillum brasilense*

(Shaharoona et al. 2008; Sala et al. 2007). Such work needs to be rigorously followed up for several seasons.

The mechanism of crop yield enhancement and the reduction on the requirement for inorganic N may reflect short-term enhancement of N uptake from the pool already present in the soil complementing that provided by the fertiliser. Over time, as the residual N in the soil is depleted such applications of PGPR, with reduced levels of inorganic N addition, may result in deficits for crop growth. A consequence could be a reduction or inconsistent response of yield to this protocol, typical of those that bedevilled earlier attempts to develop PGPR as a tool for enhancing sustainability in agriculture.

3.3 Enhanced Nutrient Availability

A number of studies have proposed that the addition of PGPR to crops can enhance yields by increasing uptake of nutrients, including nitrogen, phosphorus, potassium and iron. The uptake of nutrients by plants represents an interaction between the plant root, the physical and chemical environment of the soil and the rhizospheric microbial community. PGPR may increase the surface area or roots through the production of phytohormones, enabling greater uptake of key nutrients (Cakmakci et al. 2007). *G. diazotrophicus* has been shown to solubilise Zn, an essential micronutrient, a deficiency of which is common in sugar cane plants in which this bacterium is an endophyte (Saravanan et al. 2008) or may mobilise key nutrients by the production of siderophores (Fischer et al. 2007). Frequently, the mechanisms underlying the observed crop growth enhancement are not understood and, as a result, are attributed to a specific activity of the organism involved. In the case of free-living diazotrophs, the additional provision of N to the plant is assumed to be significant in observed increases in yields; however, such organisms do not seem able to directly release fixed N to the plant and this occurs only through the turnover of the microbial biomass (Richardson et al. 2009). In tandem with the N_2 fixation, many PGPR also produce phytohormones that have a significant effect on the crop root biomass and surface area, as seen in studies on rice (Mirza et al. 2006) and maize (Kumar et al. 2007). As a consequence, the increases in grain yield may reflect the indirect enhancement of plant nutrition through the increased root surface area, as opposed to a direct effect of increased fixed N being available to the plant form the diazotrophic bacteria. The effect of phytohormones on crop root growth probably explains the increased N use efficiency in rice (Van et al. 2000) and wheat (Akhtar et al. 2009) inoculated with PGPR.

Similarly, studies on rice, wheat and maize have all demonstrated that bacteria with P-solubilising activity can have a positive effect on plant growth (Bashan et al. 2006; Adesemoye et al. 2008). However, the mechanisms remain ambiguous and whether these organisms mobilise sufficient P to make a substantive contribution to plant nutrition has not been resolved and phytohormones may once again play a role in the positive increase in crop yields. Certainly, field studies have failed to consistently demonstrate such a response and few studies attempt to address the significance of P solubilisation by demonstrating a negation of the response when higher concentrations are applied (Richardson et al. 2009).

3.4 Enhanced Stress Tolerance

Large areas of agricultural land have been degraded by poor irrigation practice, resulting in damage such as salinization which affects 20% of total irrigated areas. Moreover, climate change appears to be a contributing factor to increased variability in rainfall (Hazell and Wood 2008). As a result, the impact of environmental stresses, such as drought and salinity, on crop yields is significant (Kibblewhite et al. 2008).

There is some evidence that the inoculation of crops with PGPR enhances the tolerance of crops to such environmental stress. *Pseudomonas* spp. inoculated on legumes were shown to ameliorate the effects of drought stress on the growth and yield of the crop (Arshad et al. 2008). However, effective inoculation of crops cultivated in soils subject to environmental stress requires that the bacteria deployed can tolerate these conditions and remain effective in promoting plant growth. Paul and Nair (2008) demonstrated that *P. fluorescens* MSP-393, used as biocontrol agent of soil pathogens, remained capable of effectively colonising plant roots even in high salinity soils. However, development of PGPR inocula for soils subjected to one or several environmental stresses need to validate that they remain effective under such conditions.

3.5 Indirect Effects

The application of PGPR to graminaceous crops may result in improved yields because of other indirect effects. The most widely studied is the ability of many such bacteria to suppress plant pathogens present in the rhizosphere. These effects have been extensively reviewed (Francis et al. 2010; Richardson et al. 2009; Vessey 2003). Here, we report some examples indicating that such effects are as applicable to graminaceous cultivation as they are to legume and vegetable crops. *Azotobacter* and *Azospirillum* strains have been shown to inhibit *Rhizoctonia solani* in the wheat rhizophere (Fatima et al. 2009) and *Pseudomonas* spp. have demonstrated similar activity in rice and maize against a range of fungal pathogens (Lawongsa et al. 2008). This response can be due to the production of antimicrobial compounds or competitive exclusion of the pathogen, as well as by inducing systemic resistance in plants, but typically due to multiple mechanisms (Francis et al. 2009).

PGPR have also been shown to promote the interaction of beneficial fungi with the plant host, for example, *A. brasilense* stimulated the root colonisation of maize by arbuscular mycorrhizae that enhance the uptake of various soil nutrients.

4 Application of PGPR to Soil

The use of bacterial inoculants to enhance crop production has been widely practised in the cultivation of legumes for many years. As a result, there are well-established technologies to add bacterial inoculants either as a liquid to coat the seeds or directly to the soil, typically using a carrier, such as peat or other materials like perlite, composted cork or bagasse (Albareda et al. 2008). In agricultural applications, peat carriers have been the most widely used on a commercial scale; they have a number of advantages, including, a long shelf life and better survival of the bacteria compared to liquid inoculants added directly to the seed. However, they have frequently resulted in inconsistent effects on crop yield, because of either the quality of the inoculant being low (Brockwell and Bottomley 1995) or the bacteria being unable to survive in the soils to which they are added as a result of either adverse environmental conditions, competition from native bacterial flora (Catroux et al. 2001) or a combination of these two factors.

The use of PGPR on graminaceous crops is a different issue to their use on legumes, the mechanisms of action may occur in the rhizosphere (phytohormone production, pathogen suppression, enhanced nutrient uptake) or be associated with the colonization of the plant roots (phytohormones, BNF). In the first case, the aim of the inoculation process is to engineer the rhizosphere to accommodate the bacteria. The competitiveness of the introduced bacteria will reflect how well it adapts to soil conditions and competes with the indigenous flora. Studies utilizing genetically engineered *P. putida* strains in the wheat rhizosphere, inoculated by broth culture application to the seed coat, have shown a rapid decrease in the numbers of introduced bacteria by five orders of magnitude between sowing and harvesting (Viebahn et al. 2003). The experiment was conducted over two growing seasons, in the first some perturbation of the indigenous microbial flora was observed but not in the second. Moreover, the effect of the genetically modified PGPR on increased plant growth was no greater than that observed after a conventional crop rotation event.

A recent study on the impact of inoculation of rice seeds with *A. brasilense* on the diversity of bacteria in the phyllosphere showed no significant impact (Pedraza et al. 2009). In another study, *Azospirillum lipoferum* was shown to significantly shift the rhizosphere population of field grown maize up to 35 days after sowing (Baudoin et al. 2009).

The influence of the plant genotype on the microbial community of the rhizosphere has been understood for almost 40 years, following studies using several wheat lines (Neal et al. 1970). This reflects the differential rhizodeposition of different plant species and varieties. Ryan et al. (2008) have recently reviewed data from a number of studies indicating the differential population of *P. fluorescens* found in the rhizospheres of both different wheat varieties (Mazzola et al. 2004) and plant hosts (Bergsma-Vlami et al. 2005).

The application and fate of inoculants on field-grown crops needs to be carefully validated to ensure that they can produce some demonstrable benefit to yields.

Recently, attempts have been made to mathematically model PGPR inoculation into the rhizosphere (Strigul and Kravchenko 2006). Such approaches are welcome as they enable the impacts of the different abiotic and biotic factors on PGPR survival to be considered. Strigul and Kravchenko (2006) demonstrated, through mathematical simulations, that the most significant factor affecting PGPR survival was the competition for limiting resources with indigenous flora, followed by the compatibility between the rhizodeposition of compounds by the plant host and the ability of the inoculated bacteria to utilise them. Such work is useful in framing ongoing studies in the use of PGPR, enabling a prediction of the success of a PGPR inoculation in a particular soil with a specific variety of crop to be made.

5 Future Work

Most of the PGPR inhibit the deleterious phytopathogens by involving proteins, peptides, etc. Their gene manipulation may help in engineering proteins, etc., which ultimately diffuse out in the rhizosphere. PGPR or bacterial inoculums adapted to a specific soil and crop varieties are in the form of 'bespoke inocula' proved beneficial in increasing yield.

5.1 Engineering the Rhizosphere

Engineering the rhizosphere of crops to improve productivity and plant health has been studied through a number of mechanisms, including manipulating the plants to modify their rhizosphere to promote nutrient availability, suppress pathogens or encourage PGPR bacterial growth (Ryan et al. 2008). Similarly, the inoculation of soil with a PGPR leading to enhancement of crop yields implies that the bacteria have become established in the rhizosphere of the plant and are exerting a stimulatory effect via one or several mechanisms described above. As a result, there is an implicit assumption that the rhizosphere has been manipulated or engineered by the inoculation process. Such a response can be demonstrated in the field; for example, *A. lipoferum* inoculated onto the seed of field grown maize produced a statistically significant shift in the composition of the indigenous rhizobial community (Baudoin et al. 2009). However, several studies including a field-based study on wheat have indicated that such inoculation effects are transient as a result of a rapid decline in inoculant numbers after the bacteria are added (Viebahn et al. 2003). Advances in our understanding of the ecological effects of inoculation will also be significant in enabling more effective modelling of the inoculation. Recent studies indicate invading bacteria might release anti-competitor toxins or parasitic phage to overcome the barrier presented by the resident flora in the rhizosphere (Brown et al. 2006). More explicit manipulation has been demonstrated by engineering PGPR strains to enhance their ability to suppress pathogens or inhibit the production of stress hormones by the plant (Ryan et al. (2008)). It is unlikely that genetically engineered strains offer a realistic mechanism to exploit PGPR effectively in the short and medium term, as they would have to satisfy stringent regulatory criteria, demonstrate a reproducible positive impact on crop yield and in some areas significant public antipathy to such technology.

5.2 *Bespoke Inocula*

The effective utilisation of PGPR in the future will demand that there is a much more rational approach to the choice and delivery of the particular bacterium into the field. This will depend on a range of variables that require consideration (Trivedi et al. 2005). The development of 'bespoke inocula' that are adapted to specific soil and crop varieties is essential if the full benefit of PGPR increase in crop yields is to be realised (Cummings and Andrews 2003). However, a consequence of such parochial inoculants is that the cost of development and production may outweigh the benefits in terms of increased yields and reduce the size of the potential market for such products such that they are not economically viable.

6 Conclusions

At the outset of this discussion, we aimed to map the potential contribution of PGPR to the sustainable cultivation of graminaceous crops. We described the three factors that defined sustainable agricultural practices and technologies (Pretty 2008):

1. They have no adverse effects on the environment.
   PGPR represent a less significant threat to the environment than the use of inorganic N or pesticide application. However, in the longer term, the consequence of inoculation of soils with PGPR on microbial soil diversity is unknown. Most studies indicate that such bacteria rapidly reduce in competition with the indigenous flora. Genetically engineered strains are possible but remain an expensive and potentially more controversial approach to the technology. However, until it has been demonstrated to be a robust and reproducible method of crop yield enhancement this approach does not appear to be viable.
2. They are accessible and effective for farmers.
   The technology has had a long and chequered history, while the production of inoculants is relatively cheap, until they can be proven to produce a return for the additional cost it is unlikely to be widely taken up by farmers. Inoculant technology has developed significantly in recent years, in terms of scale and quality, particularly for legumes. The mechanisms by which PGPR seem to exert their most significant effect on crop growth is by enhanced nutrient uptake.

However, they do not offer significant reproducible gains in graminaceous crop yield year on year. More systematic approaches to research questions should be adopted to determine how PGPR can be most effectively deployed to improve agricultural productivity

3. They lead to improvements in food productivity and have positive effects on environmental goods and services.
   Questions remain whether PGPR inoculation of graminaceous crops could offer a long-term increase in productivity – the evidence does suggest that the key response of graminaceous crops to PGPR inoculation is an improvement in nutrient uptake from the soil and a number of studies have shown (Table 1) that as a result more efficient utilisation of inorganic fertiliser can be observed resulting in the requirement for the application of lower amounts to the soil without compromising yield – and whether this is sufficient to offset the additional cost of the inoculation itself requires systematic study, but this would give a positive effect on sustainability.

References

Adesemoye AO, Torbert HA, Kloepper JW (2008) Enhanced plant nutrient use efficiency with PGPR and AMF in an integrated nutrient management system. Can J Microbiol 54:876–886

Akhtar MJ, Asghar HN, Shahzad K, Arshad M (2009) Role of plant growth promoting rhizobacteria applied in combination with compost and mineral fertilizers to improve growth and yield of wheat (*Triticum aestivum* L.). Pak J Bot 41:381–390

Albareda M, Rodriguez-Navarro DN, Camacho M, Temprano FJ (2008) Alternatives to peat as a carrier for rhizobia inoculants: solid and liquid formulations. Soil Biol Biochem 40:2771–2779

Ali S, Hamid N, Rasul G, Malik KA (1995) Use of biofertilizers to enhance rice yield, nitrogen uptake and fertilizer-N use efficiency in saline soils. Pak J Bot 27:275–281

Andrews M, James EK, Cummings SP, Zavalin AA, Vinogradova LV, McKenzie BA (2003) Use of nitrogen fixing bacteria inoculants as a substitute for nitrogen fertiliser for dryland graminaceous crops: progress made, mechanisms of action and future potential. Symbiosis 35:209–229

Arshad M, Shaharoona B, Mahmood T (2008) Inoculation with *Pseudomonas* spp. containing ACC-deaminase partially eliminates the effects of drought stress on growth, yield, and ripening of pea (*Pisum sativum* L.). Pedosphere 18:611–620

Ashrafuzzaman M, Hossen FA, Ismail MR, Hoque MA, Islam MZ, Shahidullah SM, Meon S (2009) Efficiency of plant growth-promoting rhizobacteria (PGPR) for the enhancement of rice growth. Afr J Biotechnol 8:1247–1252

Bairamov LE, Vinogradova LV, Zavalin AA (2001) Nitrogen nutrition and productivity of barley as conditioned by the application of associative diazotrophs. Asp Appl Biol 63:135–139

Baldani JI, Caruso L, Baldani VLD, Goi SR, Dobereiner J (1997) Recent advances in BNF with non-legume plants. Soil Biol Biochem 29:911–922

Bashan Y, Bustillos JJ, Leyva LA, Hernandez JP, Bacilio M (2006) Increase in auxiliary photoprotective photosynthetic pigments in wheat seedlings induced by *Azospirillum brasilense*. Biol Fertil Soil 42:279–285

Baudoin E, Nazaret S, Mougel C, Ranjard L, Moenne-Loccoz Y (2009) Impact of inoculation with the phytostimulatory PGPR *Azospirillum lipoferum* CRT1 on the genetic structure of the rhizobacterial community of field-grown maize. Soil Biol Biochem 41:409–413

Bergsma-Vlami M, Prins ME, Raaijmakers JM (2005) Influence of plant species on population dynamics, genotypic diversity and antibiotic production in the rhizosphere by indigenous *Pseudomonas* spp. FEMS Microbiol Ecol 52:59–69
Boddey RM, Urquiaga S, Reis V, Dobereiner J (1991) Biological nitrogen-fixation associated with sugar cane. Plant Soil 137:111–117
Boyer M, Bally R, Perrotto S, Chaintreuil C, Wisniewski-Dye F (2008) A quorum-quenching approach to identify quorum-sensing-regulated functions in *Azospirillum lipoferum*. Res Microbiol 159:699–708
Brockwell J, Bottomley PJ (1995) Recent advances in inoculant technology and prospects for the future. Soil Biol Biochem 27:683–697
Brown SP, Le Chat L, De Paepe M, Taddei F (2006) Ecology of microbial invasions: amplification allows virus carriers to invade more rapidly when rare. Curr Biol 16:2048–2052
Cakmakci R, Donmez MF, Erdogan U (2007) The effect of plant growth promoting rhizobacteria on barley seedling growth, nutrient uptake, some soil properties, and bacterial counts. Turk J Agric For 31:189–199
Cassan F, Perrig D, Sgroy V, Masciarelli O, Penna C, Luna V (2009) *Azospirillum brasilense* Az39 and *Bradyrhizobium japonicum* E109, inoculated singly or in combination, promote seed germination and early seedling growth in corn (*Zea mays* L.) and soybean (*Glycine max* L.). Eur J Soil Biol 45:28–35
Catroux G, Hartmann A, Revellin C (2001) Trends in rhizobial inoculant production and use. Plant Soil 230:21–30
Cooper R (1959) Bacterial fertilisers in the Soviet Union. Soil Fertil 22:327–333
Crews TE, Peoples MB (2004) Legume versus fertilizer sources of nitrogen: ecological tradeoffs and human needs. Agri Ecosys Environ 102:279–297
Cummings S, Andrews M (2003) Use of specific N_2 fixing genotypes as crop inoculants: progress made and potential for stressful soil environments. In: Tiezzi E, Brebbia CA, Usó JL (eds) Ecosystems and sustainable development, vol 2. WIT press, Southampton, pp 755–765
El Zemrany H, Cortet J, Lutz MP, Chabert A, Baudoin E, Haurat J, Maughan N, Felix D, Defago G, Bally R, Moenne-Loccoz Y (2006) Field survival of the phytostimulator *Azospirillum lipoferum* CRT1 and functional impact on maize crop, biodegradation of crop residues, and soil faunal indicators in a context of decreasing nitrogen fertilisation. Soil Biol Biochem 38:1712–1726
Fatima Z, Saleemi M, Zia M, Sultan T, Aslam M, Riaz Ur R, Chaudhary MF (2009) Antifungal activity of plant growth-promoting rhizobacteria isolates against *Rhizoctonia solani* in wheat. Afr J Biotechnol 8:219–225
Fischer SE, Fischer SI, Magris S, Mori GB (2007) Isolation and characterization of bacteria from the rhizosphere of wheat. World J Microbiol Biotechol 23:895–903
Francis I, Holsters M, Vereecke D (2010) The Gram-positive side of plant-microbe interactions. Environ Microbiol 12(1):1–12
Germida JJ, Walley FL (1996) Plant growth-promoting rhizobacteria alter rooting patterns and arbuscular mycorrhizal fungi colonization of field-grown spring wheat. Biol Fertil Soil 23:113–120
Hazell P, Wood S (2008) Drivers of change in global agriculture. Philos Trans R Soc B-Biol Sci 363:495–515
Hossain MS, Martensson A (2008) Potential use of *Rhizobium* spp. to improve fitness of non-nitrogen-fixing plants. Acta Agric Scand B-Soil Plant Sci 58:352–358
Humphry DR, Andrews M, Santos SR, James EK, Vinogradova LV, Perin L, Reis VM, Cummings SP (2007) Phylogenetic assignment and mechanism of action of a crop growth promoting *Rhizobium radiobacter* strain used as a biofertiliser on graminaceous crops in Russia. Antonie Van Leeuwenhoek Int J Gen Mol Microbiol 91:105–113
Hurek T, Reinhold-Hurek B (2003) Azoarcus sp strain BH72 as a model for nitrogen-fixing grass endophytes. J Biotechnol 106:169–178
Hurek T, Handley LL, Reinhold-Hurek B, Piche Y (2002) *Azoarcus* grass endophytes contribute fixed nitrogen to the plant in an unculturable state. Mol Plant-Microbe Interact 15:233–242

Jaderlund L, Arthurson V, Granhall U, Jansson JK (2008) Specific interactions between arbuscular mycorrhizal fungi and plant growth-promoting bacteria: as revealed by different combinations. FEMS Microbiol Lett 287:174–180

James EK (2000) Nitrogen fixation in endophytic and associative symbiosis. Field Crops Res 65:197–209

Jenkinson DS (2001) The impact of humans on the nitrogen cycle, with focus on temperate arable agriculture. Plant Soil 228:3–15

Karnwal A (2009) Production of indole acetic acid by fluorescent *Pseudomonas* in the presence of L-tryptophan and rice root exudates. J Plant Pathol 91:61–63

Khalid A, Arshad M, Zahir ZA (2004) Screening plant growth-promoting rhizobacteria for improving growth and yield of wheat. J Appl Microbiol 96:473–480

Kibblewhite MG, Ritz K, Swift MJ (2008) Soil health in agricultural systems. Philos Trans R Soc B-Biol Sci 363:685–701

Kitzes J, Wackernagel M, Loh J, Peller A, Goldfinger S, Cheng D, Tea K (2008) Shrink and share: humanity's present and future Ecological Footprint. Philos Trans R Soc B-Biol Sci 363:467–475

Kumar B, Trivedi P, Pandey A (2007) Pseudomonas corrugata: a suitable bacterial inoculant for maize grown under rainfed conditions of Himalayan region. Soil Biol Biochem 39:3093–3100

Kurek E, Jaroszuk-Scisel J (2003) Rye (*Secale cereale*) growth promotion by *Pseudomonas fluorescens* strains and their interactions with *Fusarium culmorum* under various soil conditions. Biol Control 26:48–56

Lawongsa P, Boonkerd N, Wongkaew S, O'Gara F, Teaumroong N (2008) Molecular and phenotypic characterization of potential plant growth-promoting *Pseudomonas* from rice and maize rhizospheres. World J Microbiol Biotechnol 24:1877–1884

Loganathan P, Nair S (2004) *Swaminathania salitolerans* gen. nov., sp nov., a salt-tolerant, nitrogen-fixing and phosphate-solubilizing bacterium from wild rice (*Porteresia coarctata* Tateoka). Int J Syst Evol Microbiol 54:1185–1190

Malik KA, Bilal R, Mehnaz S, Rasul G, Mirza MS, Ali S (1997) Association of nitrogen-fixing, plant-growth-promoting rhizobacteria (PGPR) with kallar grass and rice. Plant Soil 194:37–44

Mazzola M, Funnell DL, Raaijmakers JM (2004) Wheat cultivar-specific selection of 2, 4-diacetylphloroglucinol-producing fluorescent Pseudomonas species from resident soil populations. Microb Ecol 48:338–348

Mehnaz S, Mirza MS, Haurat J, Bally R, Normand P, Bano A, Malik KA (2001) Isolation and 16S rRNA sequence analysis of the beneficial bacteria from the rhizosphere of rice. Can J Microbiol 47:110–117

Miller FP (2008) After 10,000 years of agriculture, whither agronomy? Agron J 100:22–34

Mirza MS, Mehnaz S, Normand P, Prigent-Combaret C, Moenne-Loccoz Y, Bally R, Malik KA (2006) Molecular characterization and PCR detection of a nitrogen-fixing *Pseudomonas* strain promoting rice growth. Biol Fertil Soil 43:163–170

Montanez A, Abreu C, Gill PR, Hardarson G, Sicardi M (2009) Biological nitrogen fixation in maize (*Zea mays* L.) by N-15 isotope-dilution and identification of associated culturable diazotrophs. Biol Fertil Soil 45:253–263

Muthukumarasamy R, Cleenwerck I, Revathi G, Vadivelu M, Janssens D, Hoste B, Gum KU, Park KD, Son CY, Sa T, Caballero-Mellado JC (2005) Natural association of *Gluconacetobacter diazotrophicus* and diazotrophic *Acetobacter peroxydans* with wetland rice. Syst Appl Microbiol 28:277–286

Naveed M, Khalid M, Jones DL, Ahmad R, Zahir ZA (2008a) Relative efficacy of *Pseudomonas* spp., containing ACC-Deaminase for improving growth and yield of maize (*Zea mays* L.) in the presence of organic fertilizer. Pak J Bot 40:1243–1251

Naveed M, Zahir ZA, Khalid M, Asghar HN, Akhtar MJ, Arshad M (2008b) Rhizobacteria containing ACC-Deaminase for improving growth and yield of wheat under fertilized conditions. Pak J Bot 40:1231–1241

Neal JL, Atkinson TG, Larson RI (1970) Changes in rhizosphere microflora of spring wheat induced by disomic substitution of a chromosome. Can J Microbiol 16:153–158

Nogueira ED, Vinagre F, Masuda HP, Vargas C, de Padua VLM, da Silva FR, dos Santos RV, Baldani JI, Cavalcanti P, Ferreira G, Hemerly AS (2001) Expression of sugarcane genes induced by inoculation with *Gluconacetobacter diazotrophicus* and *Herbaspirillum rubrisubalbicans*. Genet Mol Biol 24:199–206

Okubara PA, Bonsall RF (2008) Accumulation of Pseudomonas-derived 2, 4-diacetylphloroglucinol on wheat seedling roots is influenced by host cultivar. Biol Control 46:322–331

Paul D, Nair S (2008) Stress adaptations in a Plant Growth Promoting Rhizobacterium (PGPR) with increasing salinity in the coastal agricultural soils. J Basic Microbiol 48:378–384

Pedraza RO (2008) Recent advances in nitrogen-fixing acetic acid bacteria. Int J Food Microbiol 125:25–35

Pedraza RO, Bellone CH, de Bellone S, Sorte PMB, Teixeira KRD (2009) Azospirillum inoculation and nitrogen fertilization effect on grain yield and on the diversity of endophytic bacteria in the phyllosphere of rice rainfed crop. Eur J Soil Biol 45:36–43

Peoples MB, Ladha JK, Herridge DF (1995) Enhancing legume N_2 fixation through plant and soil management. Plant Soil 174:83–101

Pretty J (2008) Agricultural sustainability: concepts, principles and evidence. Philos Trans R Soc B-Biol Sci 363:447–465

Principe A, Alvarez F, Castro MG, Zachi L, Fischer SE, Mori GB, Jofre E (2007) Biocontrol and PGPR features in native strains isolated from saline soils of Argentina. Curr Microbiol 55:314–322

Reis VM, Olivares FL, Dobereiner J (1994) Improved methodology for isolation of *Acetobacter diazotrophicus* and confirmation of its endophytic habitat. World J Microbiol Biotechnol 10:401–405

Richardson AE, Barea JM, McNeill AM, Prigent-Combaret C (2009) Acquisition of phosphorus and nitrogen in the rhizosphere and plant growth promotion by microorganisms. Plant Soil 321:305–339

Ryan RP, Germaine K, Franks A, Ryan DJ, Dowling DN (2008) Bacterial endophytes: recent developments and applications. FEMS Microbiol Lett 278:1–9

Sala VMR, Cardoso E, de Freitas JG, da Silveira APD (2007) Wheat genotypes response to inoculation of diazotrophic bacteria in field conditions. Pesq Agropec Bras 42:833–842

Saravanan VS, Madhaiyan M, Thangaraju M (2007) Solubilization of zinc compounds by the diazotrophic, plant growth promoting bacterium *Gluconacetobacter diazotrophicus*. Chemosphere 66:1794–1798

Saravanan VS, Madhaiyan M, Osborne J, Thangaraju M, Sa TM (2008) Ecological occurrence of *Gluconacetobacter diazotrophicus* and nitrogen-fixing Acetobacteraceae members: their possible role in plant growth promotion. Microb Ecol 55:130–140

Shaharoona B, Naveed M, Arshad M, Zahir ZA (2008) Fertilizer-dependent efficiency of Pseudomonads for improving growth, yield, and nutrient use efficiency of wheat (*Triticum aestivum* L.). Appl Microbiol Biotechnol 79:147–155

Smil V (2001) Enriching the earth. MIT Press, Cambridge, MA

Strigul NS, Kravchenko LV (2006) Mathematical modeling of PGPR inoculation into the rhizosphere. Environ Model Soft 21:1158–1171

Trivedi P, Pandey A, Palni LMS (2005) Carrier-based preparations of plant growth-promoting bacterial inoculants suitable for use in cooler regions. World J Microbiol Biotechol 21:941–945

Van VT, Berge O, Ke SN, Balandreau J, Heulin T (2000) Repeated beneficial effects of rice inoculation with a strain of *Burkholderia vietnamiensis* on early and late yield components in low fertility sulphate acid soils of Vietnam. Plant Soil 218:273–284

Vargas C, De Padua VLM, Nogueira ED, Vinagre F, Masuda HP, Da Silva FR, Baldani JI, Ferreira PCG, Hemerly AS (2003) Signaling pathways mediating the association between sugarcane and endophytic diazotrophic bacteria: a genomic approach. Symbiosis 35:159–180

Vessey JK (2003) Plant growth promoting rhizobacteria as biofertilizers. Plant Soil 255:571–586
Viebahn M, Glandorf DCM, Ouwens TWM, Smit E, Leeflang P, Wernars K, Thomashow LS, van Loon LC, Bakker P (2003) Repeated introduction of genetically modified *Pseudomonas putida* WCS358r without intensified effects on the indigenous microflora of field-grown wheat. Appl Environ Microbiol 69:3110–3118
Young JM, Kuykendall LD, Martinez-Romero E, Kerr A, Sawada H (2001) A revision of *Rhizobium* Frank 1889, with an emended description of the genus, and the inclusion of all species of *Agrobacterium* Conn 1942 and *Allorhizobium undicola* de Lajudie et al. 1998 as new combinations: *Rhizobium radiobacter*, *R. rhizogenes*, *R. rubi*, *R. undicola* and *R. vitis*. Int J Syst Evol Microbiol 51:89–103
Zakria M, Njoloma J, Saeki Y, Akao S (2007) Colonization and nitrogen-fixing ability of *Herbaspirillum* sp strain B501 gfp1 and assessment of its growth-promoting ability in cultivated rice. Microbes Environ 22:197–206
Zavalin AA, Vinogradova LV, Dukhanina TM, Vaulin AV, Christotin MV, Sologub DB, Gabibov M, Lekomtsev PV, Pasynkov AV (2001) Geographical regularities of effect of inoculation with associative diazotrophs on the productivity of cereals. Aspects Appl Biol 63:123–127

Rice Endophytic Diazotrophic Bacteria

Janpen Prakamhang, Nantakorn Boonkerd, and Neung Teaumroong

Contents

1 Introduction 318
2 The Diversity of Endophytic Diazotrophic Bacteria 319
3 Colonization Sites and Infection Pathways 320
4 Complexity of Endophytic Diazotrophic Bacteria Community Structure 323
5 Rice Growth and Endophytic Bacteria 325
6 Conclusion 329
References 329

Abstract The extension of nitrogen-fixing symbioses to cultivated rice has been a long-standing goal in the field of biological nitrogen fixation. Endophytic bacteria have been found in virtually every plant studied, where they colonize the internal tissues of their host plant and do not cause any harmful effect to their host plant. Therefore, there is a need to use endophytic diazotrophic bacteria that can make biologically fixed nitrogen available for the growth of rice plants. However, prior to introducing any selected endophytic diazotrophic strain into rice plant, the port of entry of the endophytic bacteria, the interaction via this bacteria and their host plant should be clarified. Furthermore, the complexity of bacterial community such as the behavior of native species inside the rice tissue and their interaction with inoculated endophytic strain should be clearly demonstrated. Moreover, the mechanism of plant response to those of bacteria should also be revealed. Consequently, the diversity of endophytic diazotrophic bacteria, the colonization sites and infection pathways, the effect of diazotrophic bacteria on rice growth, as well as the complexity of endophytic diazotrophic bacteria community structure will be reviewed and discussed in this chapter.

J. Prakamhang, N. Boonkerd, and N. Teaumroong (✉)
School of Biotechnology, Institute of Agricultural Technology, Suranaree University of Technology, Nakhon Ratchasima 30000, Thailand

D.K. Maheshwari (ed.), *Plant Growth and Health Promoting Bacteria*, Microbiology Monographs 18, DOI 10.1007/978-3-642-13612-2_14,

1 Introduction

Rice (*Oryza sativa*) is the most important staple crop in the developing world. In the next three decades, the world will need to produce about 60% more rice than today's global production to feed the extra billion people (Ladha and Reddy 2003). Nitrogen is the most important input required for rice production. In order to make rice cultivation sustainable and less dependent on chemical nitrogen fertilizer, it has been shown that the proportion of plant growth promoting bacteria, which is bacterial endophytes, is higher than the case of bacteria found on the rhizoplane or in the rhizosphere (Hallmann et al. 1997). Therefore, an endophytic diazotrophic bacterium is the high potential group of biofertilizers that can be used for rice cultivation.

The term "endophyte" is defined as an organism, inhabiting plant organs that at some time in its life can colonize internal plant tissue without causing apparent harm to the host (Petrini 1991). Endophytes have been discovered in high numbers within different tissues of various plants. Various endophytic nitrogen fixing bacteria, named "endophytic diazotrophs" have been detected most frequently in the nonsymbiotic roots and vascular tissues of several nonleguminous plants (Hallmann et al. 1997). Endophytic diazotrophs have been proposed to be responsible for the supply of biologically fixed nitrogen to their host plant such as *Pantoea* sp. and *Ochrobactrum* sp. to deep-water rice (Verma et al. 2004), *Herbaspirillum* sp. B501 to wild rice (You et al. 2005), *Pantoea agglomerans* YS19 (Feng et al. 2006) and *Azoarcus* sp. strain BH72 (Reinhold-Hurek et al. 2006) to rice. These endophytes do not cause damage to the host organism but they promote plant growth by the production and secretion of plant growth regulators (Verma et al. 2001), the antagonistic activity against phytopathogens (Downing and Thomson 2000) and the supply of biologically nitrogen fixation (Ladha et al. 1997). Therefore, cultivated rice fields are considered to be ideal niches for biological nitrogen fixation (BNF), especially for endophytic diazotroph bacteria.

It is well known that a remarkable diversity of N_2-fixing bacteria is naturally associated with field-grown rice (Balandreau 1986). However, in the case of wetland rice, even when specific varieties have been shown to fix N_2 (Ladha et al. 1997), it will be extremely difficult to isolate the organisms responsible, because approximately 90% of the bacteria isolated from surface-sterilized rice plants (several species and varieties, plus some related genera) using N-deficient media are nondiazotrophs (Barraquio et al. 1997). In addition, the culturable diazotrophic population is extremely varied, and so far virtually uncharacterized (Stoltzfus et al. 1997). Also, the behavior of native species inside rice tissues, and the natural association and endophytic interaction of diazotrophs with rice are considered very promising. The microbial community in rice is inherently complex, and assessments performed with such a complex population do not always reveal its specific components. Moreover, the community structure of the bacterial population, both culturable and unculturable strains, inside the rice should be considered especially in relation to the actual rice field soil.

It is widely recognized that endophytic diazotroph inoculum is capable of fixing N more efficiently than diazotrophs that remain in the rhizosphere or on the rhizoplane. This may be due to the fact that the plants directly provide the endophytic diazotroph bacteria with their nutrient requirement. Therefore, they do not need to compete with other soil microbes for scarce resources. In return for providing this niche, the bacteria provide fixed N and/or plant growth-promoting compounds to the host plant.

In the present chapter, we focus on the diversity of endophytic diazotrophic bacteria, and the colonization sites and infection pathways will be discussed.

2 The Diversity of Endophytic Diazotrophic Bacteria

Diverse endophytic diazotrophic bacteria have been isolated from rice plants. The endophytic niche offers protection from the environment for those bacteria that can colonize and establish in planta. These bacteria generally colonize the intercellular spaces, and they have been isolated from all plant compartments including seeds (Kaga et al. 2009).

Bacteria belonging to the genera *Azospirillum*, *Herbaspirillum* and *Azoarcus* are found as endophytes of rice mostly from the tropical regions. *Azospirillum* has been found in the elongation and root hair zones of roots, and some strains of both *A. lipoferum* and *A. brasilense* are either facultatively or obligately endophytic (Baldani et al. 1997). Strains of *A. brasilense* can colonize plant tissues differently; some strains live only on root surfaces, whereas others colonize cortical intercellular spaces or even the vascular tissue (James and Olivares 1998). Besides, the ability of fixing nitrogen, both *A. brasilense* and *A. lipoferum* can produce plant growth hormone auxin (Costacurta and Vanderleyden 1995). Nitrogen-fixing bacteria belonging to the genus *Azoarcus* has been found mainly in roots of Kallar grass (*Leptochloa fusca*) and rice in the intercellular spaces, xylem vessels, and dead root cells. *Azoarcus* has been demonstrated to spread systemically within the plant via the xylem vessels (Hurek et al. 1994). In addition to the plant roots, this bacterium has been discovered in close interaction with a rhizosphere fungus (Hurek et al. 1997). The genus *Herbaspirillum* contains an unusual group of endophytes in the respect that these bacteria may become pathogenic to their host under certain conditions. *H. seropedicae* strain Z67 colonized mainly subepidermal regions of rice roots (Roncato-Maccari et al. 2003). There are several bacterial species, in addition to the most well studied root endophytes, which have been isolated from gramineous plants, whereas certain species are less studied, but connected by their ability to fix nitrogen (James and Olivares 1998). Genera *Burkholderia* and *Klebsiella* are preferably regarded as endophytes (Baldani et al. 2000; Palus et al. 1996). The isolation of presumptive endophytic diazotroph bacteria from rice has also been reported. For example, *K. oxytoca* and *Enterobacter cloacae* have been isolated from the rhizosphere of wetland rice (Fujii et al. 1987). *Serratia marcescens* IRBG500 have been observed within rice root, stem, and leaves and could also increase the root length and root dry weight of the inoculated plants (Gyaneshwar et al. 2001). Teaumroong et al. (2001) found that five endophytic bacteria isolates from Thai rice showed a high N_2-fixation potential and three strains were able to produce the plant growth promoting substance IAA. *P. agglomerans* (Remus et al. 2000), *Alcaligenes faecalis* (You and Zhou 1989), and a few other bacteria belonging to the genera *Pseudomonas*, *Enterobacter* and *Bacillus* (Lindberg et al. 1985; Persello-Cartieaux et al. 2003; Watanabe and Lin 1984) have also been considered as endophytic bacteria.

A recent study published by Minamisawa et al. (2004) reported the existence of anaerobic N_2-fixing consortia (ANFICOs) in many gramineous plants consisting of N_2-fixing clostridia and diverse nondiazotrophic accompanying bacteria which were phylogenetically dispersed in the β- and *γ-Proteobacteria* and the high C+G content and low G+C content Gram-positive lineages (Fig. 1a). The phylogenetic analyses of 40 anaerobic N_2-fixing isolates from various origins categorized them exclusively into clusters I and XIVa among the 17 clusters of *Clostridium* spp. on the basis of their 16S rRNA gene sequences (Fig. 1b). The clostridial isolates were further subdivided into groups I and II in cluster XIVa and groups III, IV, and V in cluster I. These clusters and groups were not clearly correlated with the plant species, plant tissue, or location of isolation. Their work indicated that clostridia should be candidates for diazotrophic endophytes in grasses, and also demonstrated a new principle in environmental microbiology, i.e., consortium of bacteria, rather than monocultures, may be responsible for a particular activity within a very complex environment. Recently, Prakamhang et al. (2009) reported the population of viable endophytic bacterial communities within each plant part and growth stage of rice under different soil conditions in cultivated rice (*O. sativa* L. cultivar KDML-105), single isolates from each diazotrophic consortium were shown to be capable of both the inhibition and promotion of N-fixation and found closely related to *E. dissolvens, Brevundimonas aurantiaca, P. agglomerans, Pseudomonas* spp., *Rheinheimera* sp. and Enterobacteriaceae. This is the first report of diazotrophic nature of *Rheinheimera* strain, although it has been reported that this bacteria is associated with spores of the arbuscular mycorrhizal fungi (Roesti et al. 2005) and have association to the root of the tomato plant (Kim et al. 2006).

3 Colonization Sites and Infection Pathways

According to Dobereiner's report (1997), endophytic diazotrophs, by inhabiting the interior of the plants, can avoid the competition with rhizospheric bacteria and derive nutrients directly from the host plants. In return, as the plant interior may provide an environment conducive to N-fixation by being low in O_2 and relatively high in carbon, the bacteria can fix N more efficiently to the host (James and Olivares 1998). The stele of plants has been considered to be colonized by pathogens only (Campbell and Greaves 1990) or by saprophytes (Gagné et al. 1987). Vessels of nondiseased plants were thought to be sterile. This is not true for

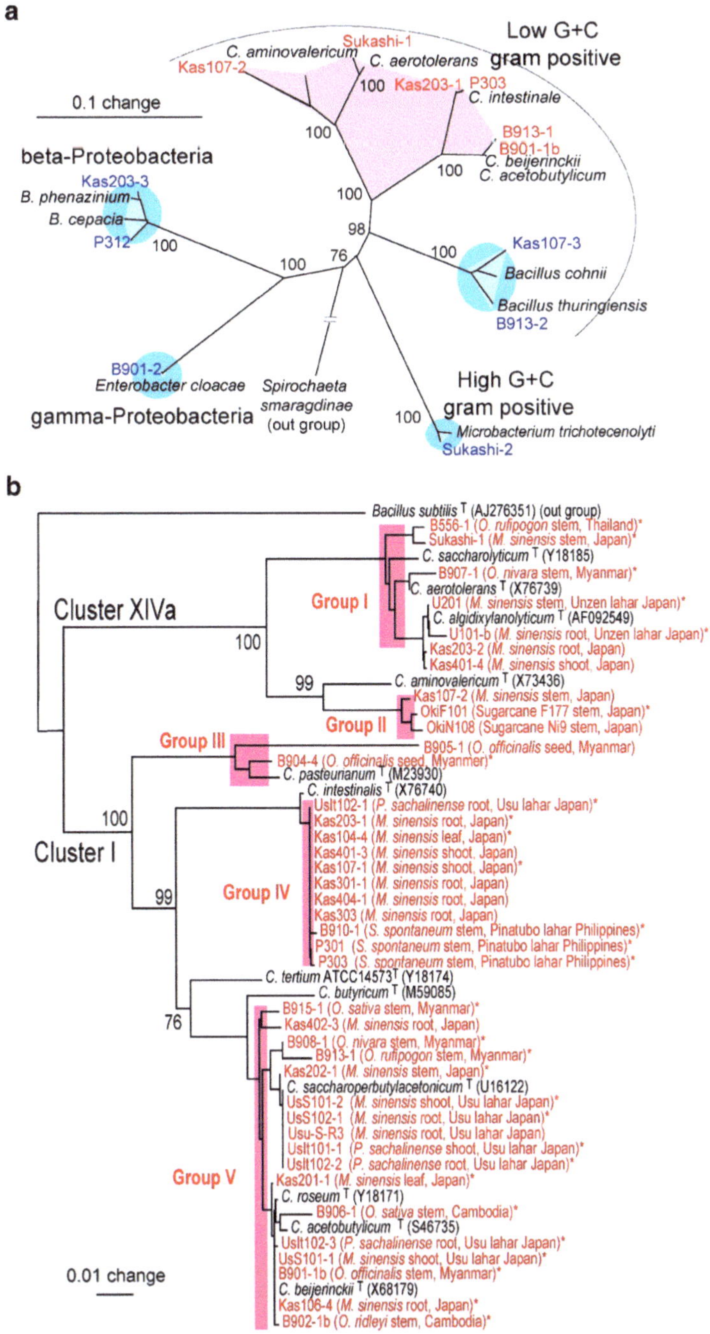

Fig. 1 Phylogenetic tree of anaerobic nitrogen-fixing bacteria and accompanying bacteria from various origins and representative close relatives by 16S rRNA gene sequences. (**a**) Representative members of ANFICOs; *grey area* indicated nitrogen-fixing bacteria, *grey oval area* indicated

endophytic diazotrophs, as first found in *Azoarcus* sp. BH72 in Kallar grass and rice. The bacteria were present in vessels of roots in gnotobiotic cultures demonstrated by immunogold-labeling using genus-specific antibodies (Hurek et al. 1991, 1994). Microscopical studies using immunological approaches and reporter genes have clearly shown similar colonization patterns for several nitrogen-fixing grass endophytes, such as *Azoarcus* sp. BH72 (Hurek et al. 1994), *H. seropedicae* (James and Olivares 1998), *Gluconacetobacter diazotrophicus* (Cavalcante and Dobereiner 1988) and certain strains of *Azospirillum* spp. (Schloter and Hartmann 1998).

In plants showing no symptoms of disease, *Azoarcus* sp. BH72 colonizes the original host Kallar grass and also rice seedlings in a similar way. Outer cell layers, epidermis and the root cortex are colonized inter- and intracellularly within 2–3 weeks, the aerenchyma which occurs in waterlogged plants being the main site for large microcolonies (Egener et al. 1999; Hurek et al. 1994). The main intercellular colonization pattern raises questions on the delivery of nutrients, especially carbon sources for the bacteria (Hurek et al. 1994). Rarely, the bacteria penetrate deeply into plant roots into the stele where they may present in the parenchyma and in xylem vessels. The detection of *Azoarcus* sp. in stelar parenchymatic cells of the culm and in vessels of Kallar grass and rice (Hurek et al. 1991, 1994) suggested that systemic spreading into shoots may be mediated through the transport in vessels. However, shoot colonization of *Gramineae* appears to be more obvious in *G. diazotrophicus* (James and Olivares 1998) and *H. seropedicae* (Gyaneshwar et al. 2002). Kaga et al. (2009) hypothesized that endophytic bacteria are considered to originate from the external environment. To examine this hypothesis, endophytic bacteria were isolated from the rice (*O. sativa*, cultivar Kinuhikari) seeds, the shoots, remains of the seeds, and roots of rice seedlings that were aseptically cultivated in vitro from surface-disinfected seeds. Of the various bacterial strains isolated, the closest relatives, identified by 16S rRNA gene sequencing, were; *Bacillus firmus*, *B. fusiformis*, *B. pumilus*, *Caulobacter crescentus*, *Kocuria palustris*, *Micrococcus luteus*, *Methylobacterium fujisawaense*, *Me. radiotolerans*, and *P. ananatis*. The latter three species have been detected frequently inside both rice seedlings and mature rice plants. These results indicate that rice seeds are an important source of endophytic bacteria. The bacteria that colonize the seed interior appear to infect the subsequent generation via seeds and become the dominant endophytic species in the mature plant. The presence of diazotrophic bacteria was also detected in roots, stems and leaves (Prakamhang et al. 2009). The location of

---

Fig. 1 (continued) *accompanying bacteria. (**b**) Tree of 40 isolates of anaerobic nitrogen-fixing bacteria from various pioneer plants and wild rice species, including M. sinensis, S. spontaneum* (wild sugarcane), *Polygonum sachalinense, Saccharum* hybrid sp. (sugarcane), *Oryza sativa* (cultivated rice), and *O. rufipogon, O. nivara, O. officinalis,* and *O. redleyi* (wild rice species). Grey box indicated the nitrogen-fixing bacteria. The trees are based on >1.2 kb of DNA sequences and were constructed by the neighbor-joining method. Bootstrap values (percentages from 1,000 replications) are indicated. Bold letter indicated the reference strain, regular letter represent isolated strain. The utilization of carbon sources was tested for 30 isolates, which are indicated with *asterisks* (Adapted from Minamisawa et al. 2004)

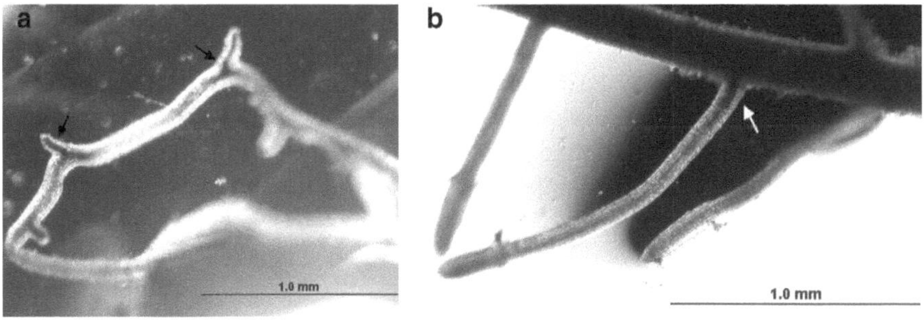

Fig. 2 Light micrographs of GUS-stained roots, stems, and leaves of rice at 5 days after inoculation with strain VFR5-3 marked with GUS. *Arrows* indicate the most intense GUS activity on the lateral roots (**a**), root junction (**b**)

the *Rheinheimera* sp. J3-AN42 in root of rice on the fifth day after inoculation (DAI) was observed with the most intense color development on the lateral roots (Fig. 2a) and root junctions (Fig. 2b).

Endophytic microorganisms differ remarkably from highly developed root nodule symbioses, in which rhizobia enter the plants through root hairs via infection threads. The infection of grasses by endophytes is similar to the crack entry. One site of primary colonization is the points of emergence of lateral roots, where bacterial microcolonies can readily be detected, and bacterial cells have been found between the cell layers of the lateral root and the cortex of the main root. Another route of entry is the root tip at the zone of elongation and differentiation. The bacteria can invade intercellular and intracellular and may penetrate into the central tissues (Hurek et al. 1991, 1994), with the exception of *Azospirillum* spp. which is mainly regarded as a rhizoplane colonizer (Steenhoudt and Vanderleyden 2000). The entry of bacteria into the root is most likely an active process, which might be mediated by enzymes degrading plant cell wall polymers. Two types of cellulolytic enzymes, cellobiohydrolase and β-glucosidase have been detected in *Azoarcus* sp. BH72 (Reinhold-Hurek and Hurek 1998), pectinase and cellulase production were also detected (Prakamhang et al. 2009). Further insights into the cellular machinery for plant invasion, and a comparison with pathogens and symbionts, will be fostered by the genome analysis.

4 Complexity of Endophytic Diazotrophic Bacteria Community Structure

While the search for a natural association and endophytic interaction of diazotrophs with rice is considered very promising, the microbial community in rice is also inherently complex, and assessments performed with such a complex population do not always reveal its specific components. The PCR–DGGE analysis conducted directly on rice tissue samples obtained in Thailand using 16S rDNA primers was used to elucidate the structure of the endophytic bacterial communities (Fig. 3). Almost all of the samples contained two major bands of DGGE–PCR products except that from reproductive stage of rice. Each sequence retrieved from the bands a–d showed similarity to different strains. While band a has a high similarity to *E. dissolvens,* band b showed similarity to *B. aurantiaca,* band c to *P. agglomerans*, and band d to *Pseudomonas* sp. (Prakamhang et al. 2009).

Recently, the endophytic–endophytic consortium interaction within rice plant has been reported. Since the N-fixing activities of other single culture occurred more than those of the original combinations (Prakamhang et al. 2009), this suggests the presence of the accompanying bacteria that produced specific metabolites of consortium that induced/reduced the N-fixation as well as association of nondiazotrophic endophytes in culture. The surface-sterilized rice plant materials were mechanically macerated and then cultured in N-free semisolid medium for determinations of the N-fixing bacteria as a consortia or original mixed culture. Each single isolate was tested for inhibition/promotion to the other isolate in the same consortium. This suggested that one single isolate affected other single isolates in the same consortium by producing agents that can kill the bacteria (bactericidal effect) as shown by the clear zone in the bacterial layer (Fig. 4). However, some consortia do not show any clear zone around the spotted culture filtrate of each single isolate. Perhaps one single isolate of this consortium produced agents that can inhibit only N-fixing activity (bacteriostatic effect). Similarly, a major feature of ANFICOs is that N- fixation by the anaerobic clostridia is supported by the consumption of oxygen by the accompanying bacteria in the culture and the presence of unknown metabolites (Minamisawa et al. 2004).

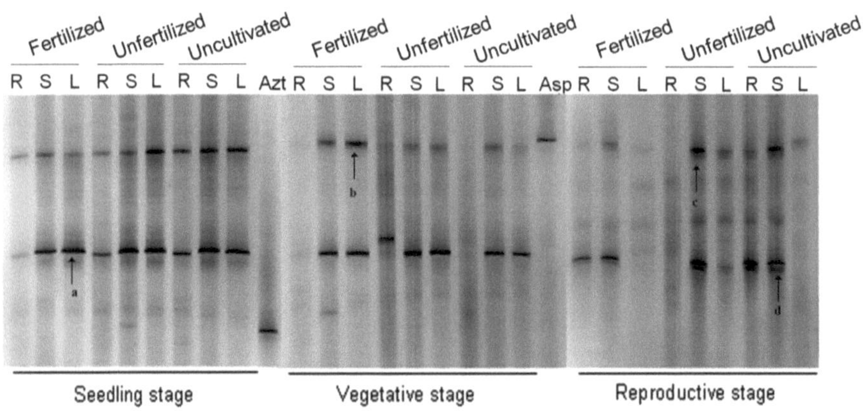

Fig. 3 PCR–DGGE analysis of 16S rRNA gene banding patterns from rice endophytic bacteria in samples from plants grown in different types of soil (fertilized paddy soil, unfertilized paddy soil and uncultivated forest soil). *Arrows* show excised and sequenced bands (a–d). Azt., *Azotobacter* sp.; Asp., *Azospirillum* sp.; R, root; S, stem; L, leaf. (Adapted from Prakamhang et al. 2009)

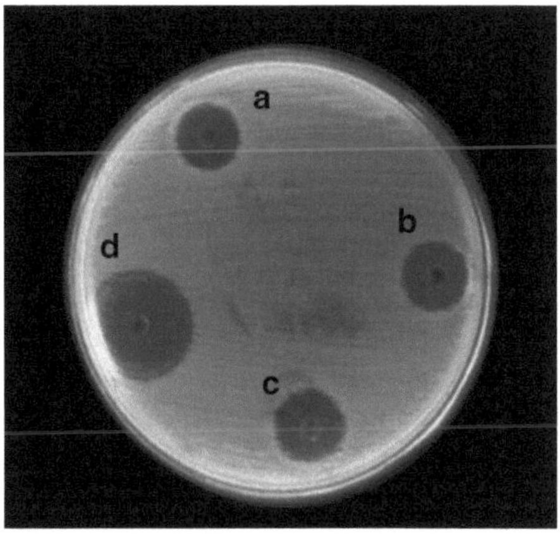

Fig. 4 Antimicrobial activities of culture filtrate of strain VFR3-1 corresponding to bacterial growth as show by the inhibition zones in the bacterial layer. The N-free plate was cultured with VFR3-1-1 inoculated with 1 μl (**a**), 2 μl (**b**), 3 μl (**c**), and 4 μl (**d**) of single culture VFR3-1-2

5 Rice Growth and Endophytic Bacteria

Microbial promotion of plant growth may be the outcome of several additional factors besides the nitrogen fixation. For example, an indirect plant growth promotion is the production of phytohormones, which has been considered to be the main function in the symbioses. Almost all root endophytes also fix nitrogen (Baldani et al. 1997). However, the benefit of their nitrogen fixing ability for the plant has not been demonstrated indisputably (James 2000). *P. agglomerans* can infect and colonize in the rice roots and produce IAA and have been shown to be a potent biological control agent against fungal disease (Verma et al. 2001). Production of auxins and gibberellins is also typical for many root associated endophytic bacteria such as *Azospirillum* sp., *G. diazotrophicus*, and *H. seropedicae* (Bastián et al. 1998). The flavonoids, quercetin and diadzein, significantly increased the endophytic colonization ability of *Serratia* sp. than growth hormones. The induced colonization of *Serratia* sp. due to quercetin proportionally increased the *in planta* nitrogenase activity which is reflected in the increased plant height, protein and chlorophyll contents of rice seedlings (Sandhiya et al. 2005). However, apart from the roots, the importance of the microbial production of phytohormones has been evaluated to be low, and the significance of these products for the plant has remained ambiguous (Zinniel et al. 2002). Therefore, endophytic function which is considered distinctively beneficial for the plant appears to be the protection of the host against pathogens. As, not all endophytes are responsible for producing antagonistic substances, their role is yet to be discovered. Nevertheless, it appears that the function of an endophyte may be composed of several diverse factors that may together have a positive influence on the plant.

Nitrogen fixation is catalyzed by the enzyme nitrogenase complex. More than, 20 genes have been identified as controlling the structure and function of the nitrogenase system, and much functional detail has been defined. A substantial molecular diversity of N fixing bacteria has been detected in field grown rice based on retrieval of *nifH* gene fragments from root DNA (Ueda et al. 1995). The diazotrophic endophyte of rice, *Serratia* sp., was marked with enhanced green fluorescence protein (egfp)-Km marker gene by biparental mating, and was used for colonization studies in rice. The conjugants established themselves endophytically in rice root, stem and leaves, with the stem being most colonized (Sandhiya et al. 2005). The expression of *nif* genes of *H. seropedicae* LR15 strain occurred in roots, stems and leaves as detected by the GUS reporter system, and the colonization of plant tissue by *H. seropedicae* did not depend on the nitrogen-fixing ability (Roncato-Maccari et al. 2003). To detect N-fixing bacteria in a plant without using culture methods, *nifH* gene segments were amplified with degenerate primers from DNA extracted from stems and leaves of rice plant. Furthermore, the study of Ueda et al. (1995) demonstrated the extent of phylogenetic diversity of diazotrophic bacteria associated with rice roots by characterizing 23 *nifH* gene sequences derived directly from rice roots without culturing the organisms. This study also showed a variety of significant components of the diazotrophic community dominated mainly by proteobacteria. Similar results were obtained by Prakamhang et al. (2009), in which nested PCR–DGGE analysis with *nifH* primer demonstrated less diazotrophic bacterial diversity in the roots of rice cultivated in paddy soil amended with nitrogen fertilizer than in unfertilized and previously uncultivated soil, and plant tissue type was found to dictate the endophytic diazotrophic community structure rather than the type of soil or fertilizer amendment (Fig. 5). Furthermore, most isolates were detected both by culturable approach and by DGGE, suggesting that the molecular approach directly reported culturable endophyte bacteria. However, some isolates such as those from leaves with no-fertilizer soil of all stages of growing were not detected by ARA procedure, suggesting that these endophytic bacteria are nonculturable (or not yet cultured) endophytic bacteria from rice plants. The dendrogram constructed from the PCR–DGGE of *nifH* gene band pattern of endophytic diazotrophic bacteria within each parts (root, stem, and leaf) of rice

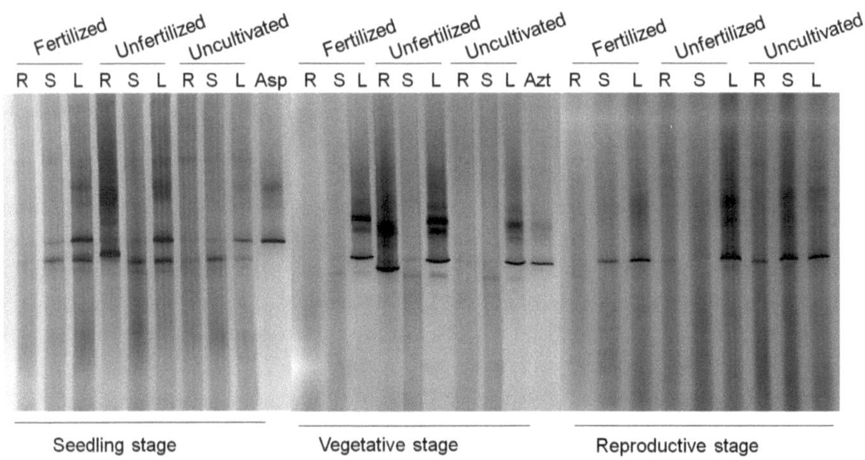

Fig. 5 Nested PCR–DGGE of *nifH* banding patterns from rice endophytic bacteria. R, root; S, stem; L, leaf; types of soil as in Fig. 3

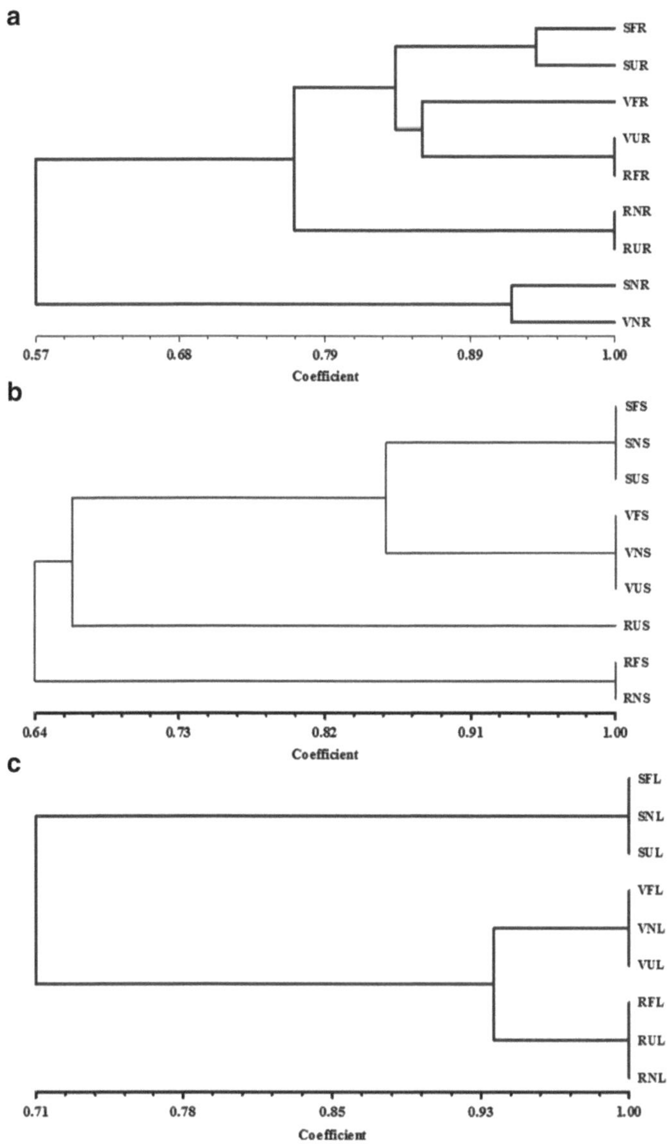

Fig. 6 Dendrogram obtained from UPGMA cluster analysis of PCR–DGGE of *nifH* gene band pattern of rice endophytic diazotrophic bacteria within stage of growth and part of rice plant; seedling stage (**a**), vegetative stage (**b**), and reproductive stage (**c**). *Abbreviations*, First letter is growth stage of rice (S; seedling stage, V; vegetative stage, R; reproductive stage), second letter is type of soil (F; fertilized, N; unfertilized, U; uncultivated soil), and third letter is rice part (R; root, S; stem, L; leaf)

plant represented high complexity of community in the rice root (Fig. 6a). On the other hand, in stem and leaf, stages of growth seem to dictate endophytic diazotrophic bacteria community. For example, in stem and leaf, bacteria of each stages of growth were grouped together (Fig. 6b, c). This result suggested that plant tissue types may dictate the endophytic diazotrophic community structure rather than the type of soil or fertilizer amendment.

However, only the presence of nitrogenase gene does not indicate that bacteria are actively fixing nitrogen (James 2000). Culturing techniques have been used to determine the type of individual species present, but these techniques yield biased results and a misrepresentation of the types of bacterial species that are active in the environment. The reverse transcriptase PCR (RT-PCR) makes it possible to assay for cells that are actively expressing specific gene at the time of sampling and it has been used recently to detect *nifH* expression of *Azoarcus sp* BH72 in Kallar grass and rice (Hurek et al. 2002), and *Herbaspirillum* sp. B501 associated in the shoot (leaf and stem) of wild rice (You et al. 2005). Recent study of Prakamhang et al. (2009) showed that the *nifH* gene expression could be differently detected in each part and growth stage of rice plants as well as could be influenced by soil nitrogen status (Fig. 7). The expression level of the *nifH* gene in all roots from plants grown in N-fertilized soil was the lowest among the treatments studied. The results confirm the complexity of the endophytic diazotrophic bacterial community, and indicate that the type of plant tissue seems to influence the community structure.

In addition to fertilization with nitrogen, variations in the growth stage and part of rice plant and the environmental conditions caused large differences in the population structure of endophytic diazotrophs, as demonstrated in a culturable approach. Nitrogen fertilization has been reported as a leading repression factor of nitrogenase genes and inactivation of nitrogenase activity in most diazotrophic bacteria (Egener et al. 1999; Fuentes-Ramirez et al. 1999; Martin and Reinhold-Hurek 2002). Nevertheless, the diazotrophs abundant in rice plants may be either rapidly decaying or overgrown by others after fertilizer application. Colonization of maize plants by diazotrophic bacteria was inhibited by high N-fertilization (18.5 mg kg^{-1}) during the early stages of growth but not during subsequent stages (Roesch et al. 2006), whereas the inhibitory effect of high N fertilization (148 mg kg^{-1}) on diazotrophic bacterial numbers could be reduced by the application of

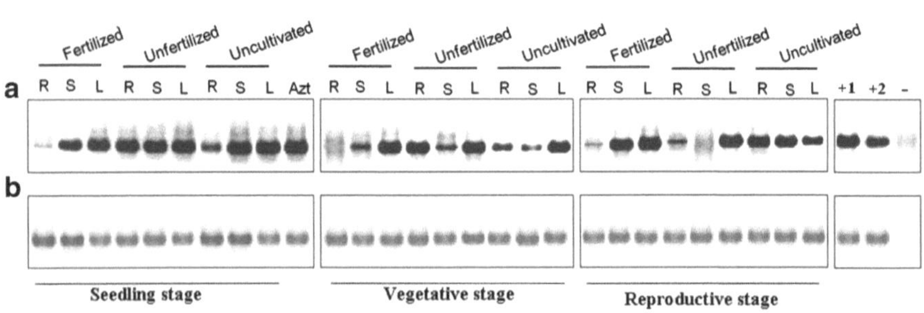

Fig. 7 Gel electrophoresis analysis of RT-PCR for *nifH* gene expression. Reverse transcription and nested PCR amplification was performed using *nifH* primers (**a**). The 16S rRNA gene was used as a standard to calibrate the amount of RNA (**b**). R, root; S, stem; L, leaf; types of soil as in Fig.3 Azt., *Azotobacter* sp.; +1, rice inoculated with *Azotobacter* sp. as positive control; +2, Positive Control; –, negative control (no-template)

compost (Muthukumarasamy et al. 2007). In the case of cultivated rice in Thailand, the *nifH* pattern indicated lower diversity in plants grown in soil fertilized according to local custom (35.9 mg kg^{-1}) and in uncultivated soil than that in plants from unfertilized soil (Prakamhang et al. 2009). Since the N-fertilization did not affect the diazotrophic bacterial population in all stages of growth, the observed effect does not seem to be a direct negative effect of the fertilizer on the bacteria. *Herbaspirillum* spp. was also found to occur both in N-fertilized and unfertilized samples (Muthukumarasamy et al. 1999). Nitrogen alters the physiological state of the plant, and this subsequently affects its association with the diazotrophic bacterial population (Muthukumarasamy et al. 1999; Reis et al. 2000). This suggests that the original diazotrophic community consists mainly of autochthonic bacteria on which N-depletion conferred a selective advantage. These results also demonstrate that rice represents a highly dynamic microenvironment for bacteria.

6 Conclusion

In the present chapter, we have shown that there is a high diversity in endophytic diazotrophic bacteria community. The presence of diazotrophic bacteria was detected in roots, stems, and leaves in the colonization sites. The complexity of the endophytic diazotrophic bacterial community revealed that the type of plant tissue seems to influence the community structure. The understanding about how and when endophytic communities form in rice plants and about their interaction is essential for an investigation of the ability of endophytic diazotrophic bacteria, especially the culturable strains, to compete with other endophyte strains, and to contribute nitrogen to the host plants prior to applying endophytic diazotroph bacteria as rice biofertilizer for rice farming. Thus, it is strongly proposed that endophytic diazotrophic bacteria provide an agricultural benefit which is of definite ecological and economic significance.

References

Balandreau J (1986) Ecological factors and adaptive processes in N_2-fixing bacterial populations of the plant environment. Plant Soil 90:73–92

Baldani J, Caruso L, Baldani VLD, Goi SR, Dobereiner J (1997) Recent advances in BNF with non-legume plants. Soil Biol Biochem 29:911–922

Baldani VLD, Baldani JI, Döbereiner J (2000) Inoculation of rice plants with the endophytic diazotrophs *Herbaspirillum seropedicae* and *Burkholderia spp*. Biol Fertil Soils 30:485–491

Barraquio WL, Revilla L, Ladha JK (1997) Isolation of endophytic diazotrophic bacteria from wetland rice. Plant Soil 194:15–24

Bastián F, Cohen A, Piccoli P, Luna V, Bottini R, Baraldi R, Bottini R (1998) Production of indole-3-acetic acid and gibberellins A 1 and A 3 by *Acetobacter diazotrophicus* and *Herbaspirillum seropedicae* in chemically-defined culture media. J Plant Growth Regul 24:7–11

Campbell R, Greaves MP (1990) The rhizosphere. Wiley, Chichester, England

Cavalcante VA, Dobereiner J (1988) A new acid-tolerant nitrogen-fixing bacterium associated with sugarcane. Plant Soil 108:23–31

Costacurta A, Vanderleyden J (1995) Synthesis of phytohormones by plant-associated bacteria. Crit Rev Microbiol 21:1–18

Dobereiner J (1997) Biological nitrogen fixation in the tropics: social and economic contributions. Soil Biol Biochem 29:771–774

Downing KJ, Thomson JA (2000) Introduction of the *Serratia marcescens chiA* gene into an endophytic *Pseudomonas fluorescens* for the biocontrol of phytopathogenic fungi. Can J Microbiol 46:363–369

Egener T, Hurek T, Reinhold-Hurek B (1999) Endophytic expression of *nif* genes of *Azoarcus* sp. strain BH72 in rice roots. Mol Plant Microbe Interact 12:813–819

Feng Y, Shen D, Song W (2006) Rice endophyte *Pantoea agglomerans* YS19 promotes host plant growth and affects allocations of host photosynthates. J Appl Microbiol 100:938–945

Fuentes-Ramirez LE, Caballero-Mellado J, Sepulveda J, Martinez-Romero E (1999) Colonization of sugarcane by *Acetobacter diazotrophicus* is inhibited by high N-fertilization. FEMS Microbiol Ecol 29:117–128

Fujii T, Huang YD, Higashitani A, Nishimura Y, Iyama S, Hirota Y, Yoneyama T, Dixon RA (1987) Effect of inoculation with *Klebsiella oxytoca* and *Enterobacter cloacae* on dinitrogen fixation by rice-bacteria associations. Plant Soil 103:221–226

Gagné S, Richard C, Rousseau H, Antoun H (1987) Xylem-residing bacteria in alfalfa roots. Can J Microbiol 33:996–1000

Gyaneshwar P, James EK, Mathan N, Reddy PM, Reinhold-Hurek B, Ladha JK (2001) Endophytic colonization of rice by a diazotrophic strain of *Serratia marcescens*. J Biotechnol 183: 2634–2645

Gyaneshwar P, James EK, Reddy PM, Ladha JK (2002) *Herbaspirillum* colonization increases growth and nitrogen accumulation in aluminium-tolerant rice varieties. New Phytol 154:131–145

Hallmann J, Quadt-Hallmann A, Mahaffee WF, Kloepper JW (1997) Bacterial endophytes in agricultural crops. Can J Microbiol 43:895–914

Hurek T, Reinhold-Hurek B, van Montagu M, Kellenberger E (1991) Infection of intact roots of Kallar grass and rice seedlings by *Azoarcus*. In: Polsinelli M, Materassi R, Vincenzini M (eds) Nitrogen fixation. Kluwer, Dordrecht, pp 235–242

Hurek T, Reinhold-Hurek B, Van Montagu M, Kellenberger E (1994) Root colonization and systemic spreading of *Azoarcus* sp. strain BH72 in grasses. Appl Environ Microbiol 176:1913–1923

Hurek T, Wagner B, Reinhold-Hurek B (1997) Identification of N_2-fixing plant- and fungus-associated *Azoarcus* species by PCR-based genomic fingerprints. Appl Environ Microbiol 63: 4331–4339

Hurek T, Handley LL, Reinhold-Hurek B, Piche Y (2002) *Azoarcus* grass endophytes contribute fixed nitrogen to the plant in an unculturable state. Mol Plant Microbe Interact 15: 233–242

James EK (2000) Nitrogen fixation in endophytic and associative symbiosis. Field Crops Res 65:197–209

James EK, Olivares FL (1998) Infection and colonization of sugarcane and other Graminaceous plants by endophytic diazotrophs. Crit Rev Plant Sci 17:77–119

Kaga H, Mano H, Tanaka F, Watanabe A, Kaneko S, Morisaki H (2009) Rice seeds as sources of endophytic bacteria. Microbes Environ 24:154–162

Kim JS, Dungan RS, Kwon SW, Weon HY (2006) The community composition of root-associated bacteria of the tomato plant. World J Microbiol Biotechnol 22:1267–1273

Ladha JK, Reddy PM (2003) Nitrogen fixation in rice systems: state of knowledge and future prospects. Plant Soil 252:151–167

Ladha JK, de Bruijn FJ, Malik KA (1997) Introduction: assessing opportunities for nitrogen fixation in rice – a frontier project. Plant Soil 194:1–10

Lindberg T, Granhall U, Tomenius K (1985) Infectivity and acetylene reduction of diazotrophic rhizosphere bacteria in wheat (*Triticum aestivum*) seedlings under gnotobiotic conditions. Biol Fertil Soils 1:123–129

Martin DE, Reinhold-Hurek B (2002) Distinct roles of PII-like signal transmitter proteins and *amtB* in regulation of *nif* gene expression, nitrogenase activity, and posttranslational modification of NifH in *Azoarcus* sp. strain BH72. J Bacteriol 184:2251–2259

Minamisawa K, Nishioka K, Miyaki T, Ye B, Miyamoto T, You M, Saito A, Saito M, Barraquio WL, Teaumroong N, Sein T, Sato T (2004) Anaerobic nitrogen-fixing consortia consisting of Clostridia isolated from gramineous plants. Appl Environ Microbiol 70:3096–3102

Muthukumarasamy R, Revathi G, Lakshminarasimhan C (1999) Influence of N fertilisation on the isolation of *Acetobacter diazotrophicus* and *Herbaspirillum* spp. from Indian sugarcane varieties. Biol Fertil Soils 29:157–164

Muthukumarasamy R, Kang UG, Park KD, Jeon WT, Park CY, Cho YS, Kwon SW, Song J, Roh DH, Revathi G (2007) Enumeration, isolation and identification of diazotrophs from Korean wetland rice varieties grown with long-term application of N and compost and their short-term inoculation effect on rice plants. J Appl Microbiol 102:981–991

Palus JA, Borneman J, Ludden PW, Triplett EW (1996) A diazotrophic bacterial endophyte isolated from stems of *Zea mays* L. and *Zea luxurians* Iltis and Doebley. Plant Soil 186: 135–142

Persello-Cartieaux F, Nussaume L, Robaglia C (2003) Tales from the underground: molecular. plant-rhizobacteria interactions. Plant Cell Environ 26:189–199

Petrini O (1991) Fungal endophytes of tree leaves. In: Andrews JH, Hirano SS (eds) Microbial ecology of leaves. Springer, New York, pp 179–197

Prakamhang J, Minamisawa K, Teamtaisong K, Boonkerd N, Teaumroong N (2009) The communities of endophytic diazotrophic bacteria in cultivated rice (*Oryza sativa* L.). Appl Soil Ecol 42:141–149

Reinhold-Hurek B, Hurek T (1998) Interactions of gramineous plants with *Azoarcus* spp. and other diazotrophs: identification, localization, and perspectives to study their function. Crit Rev Plant Sci 17:29–54

Reinhold-Hurek B, Maes T, Gemmer S, Van Montagu M, Hurek T (2006) An endoglucanase is involved in infection of rice roots by the not-cellulose-metabolizing endophyte *Azoarcus* sp. strain BH72. Mol Plant Microbe Interact 19:181–188

Reis VM, Baldani JI, Baldani VLD, Dobereiner J (2000) Biological dinitrogen fixation in *Gramineae* and palm trees. Crit Rev Plant Sci 19:227–247

Remus R, Ruppel S, Jacob HJ, Hecht-Buchholz C, Merbach W (2000) Colonization behaviour of two enterobacterial strains on cereals. Biol Fertil Soils 30:550–557

Roesch LFW, Olivares FL, Pereira Passaglia LM, Selbach PA, de Sa ELS, de Camargo FAO (2006) Characterization of diazotrophic bacteria associated with maize: effect of plant genotype, ontogeny and nitrogen-supply. World J Microbiol Biotechnol 22:967–974

Roesti D, Ineichen K, Braissant O, Redecker D, Wiemken A, Aragno M (2005) Bacteria associated with spores of the arbuscular mycorrhizal fungi *Glomus geosporum* and *Glomus constrictum*. Appl Environ Microbiol 71:6673–6679

Roncato-Maccari LDB, Ramos HJO, Pedrosa FO, Alquini Y, Chubatsu LS, Yates MG, Rigo LU, Steffens MBR, Souza EM (2003) Endophytic *Herbaspirillum seropedicae* expresses *nif* genes in gramineous plants. FEMS Microbiol Ecol 45:39–47

Sandhiya GS, Sugitha TC, Balachandar D, Kumar K (2005) Endophytic colonization and *in planta* nitrogen fixation by a diazotrophic *Serratia* sp. in rice. Indian J Exp Biol 43:802–807

Schloter M, Hartmann A (1998) Endophytic and surface colonization of wheat roots (*Triticum aestivum*) by different *Azospirillum brasilense* strains studied with strain-specific monoclonal antibodies. Symbiosis 25:159–179

Steenhoudt O, Vanderleyden J (2000) *Azospirillum*, a free-living nitrogen-fixing bacterium closely associated with grasses: genetic, biochemical and ecological aspects. FEMS Microbiol Rev 24:487–506

Stoltzfus JR, So R, Malarvithi PP, Ladha JK, de Bruijn FJ (1997) Isolation of endophytic bacteria from rice and assessment of their potential for supplying rice with biologically fixed nitrogen. Plant Soil 194:25–36

Teaumroong N, Teamtaisong K, Sooksa-ngun T, Boonkerd N (2001) The diazotrophic endophytic bacteria in Thai rice. In: Suriyaphan O, Hansakdi E, Jongruaysup S, Simons R (eds) Proceeding of the 5th ESAFS international conference on rice environments and rice products. Krabi, Thailand, pp 147–160

Ueda T, Suga Y, Yahiro N, Matsuguchi T (1995) Remarkable N_2-fixing bacterial diversity detected in rice roots by molecular evolutionary analysis of *nifH* gene sequences. J Biotechnol 177:1414–1417

Verma SC, Ladha JK, Tripathi AK (2001) Evaluation of plant growth promoting and colonization ability of endophytic diazotrophs from deep water rice. J Biotechnol 91:127–141

Verma SC, Singh A, Chowdhury SP, Tripathi AK (2004) Endophytic colonization ability of two deep-water rice endophytes. *Pantoea* sp. and *Ochrobacterium* sp. using green fluorescent protein reporter. Biotechnol Lett 26:425–429

Watanabe I, Lin C (1984) Response of wetland rice to inoculation with *Azospirillum lipoferum* and *Pseudomonas* sp. Soil Sci Plant Nutr (Tokyo) 30:117–124

You C, Zhou F (1989) Non-nodular endorhizospheric nitrogen fixation in wetland rice. Can J Microbiol 35:403–408

You M, Nishiguchi T, Saito A, Isawa T, Mitsui H, Minamisawa K (2005) Expression of the *nifH* gene of a *Herbaspirillum* endophyte in wild rice species: daily rhythm during the light-dark cycle. Appl Environ Microbiol 71:8183–8190

Zinniel DK, Lambrecht P, Harris NB, Feng Z, Kuczmarski D, Higley P, Ishimaru CA, Arunakumari A, Barletta RG, Vidaver AK (2002) Isolation and characterization of endophytic colonizing bacteria from agronomic crops and Prairie plants. Appl Environ Microbiol 68:2198–2208

Bacillus and *Paenibacillus* spp.: Potential PGPR for Sustainable Agriculture

Venkadasamy Govindasamy, Murugesan Senthilkumar, Vellaichamy Magheshwaran, Upendra Kumar, Pranita Bose, Vikas Sharma, and Kannepalli Annapurna

Contents

1 Introduction 334
2 Taxonomy and Phylogeny of Genus *Bacillus* and *Paenibacillus* 335
3 Ecology and Distribution of *Bacillus* and *Paenibacillus* spp. 338
4 PGPR Potentials of *Bacillus* and *Paenibacillus* spp. 339
4.1 Promotion of Host Plant Nutrition and Growth 340
4.2 Antagonism Against Plant Fungal and Bacterial Pathogens 345
4.3 Antagonism Against Insect Pests and Nematode 347
4.4 Stimulation of Plant Host Defense Mechanisms Through Induced Systemic Resistance 348
5 Production of Peptide Antibiotics by *Bacillus* and *Paenibacillus* spp. 350
6 Endophytic Colonization and Biofilm Formation by *Bacillus* and *Paenibacillus* spp. 352
7 Conclusion 356
References 356

Abstract The Gram-positive aerobic endospore-forming bacteria (AEFB) belonging to the genus *Bacillus* and *Paenibacillus* are essentially ubiquitous and occur abundantly in most rhizospheric soils. In the rhizosphere, species of these two genera are involved in atmospheric nitrogen fixation, solubilization of soil phosphorus and uptake of micronutrients, and production of phytohormones and antimicrobial metabolites. Multiple species of *Bacillus* and *Paenibacillus* affect the crop growth and its health by three different ecological mechanisms viz, promotion of host plant nutrition and growth, antagonism against fungal, bacterial, nematode pathogens and insect pests, and stimulation of host defence mechanisms. Specific strains of both *Bacillus* and *Paenibacillus* spp. are known to elicit induced systemic

V. Govindasamy, M. Senthilkumar, V. Magheshwaran, U. Kumar, P. Bose, V. Sharma, and K. Annapurna (✉)
Division of Microbiology, Indian Agricultural Research Institute, New Delhi 110 012, India
e-mail: annapurna93@yahoo.co.in

D.K. Maheshwari (ed.), *Plant Growth and Health Promoting Bacteria*,
Microbiology Monographs 18, DOI 10.1007/978-3-642-13612-2_15,

resistance (ISR) similar to that of *Pseudomonas* spp. which leads to the stimulation of host defence mechanisms against multiple pathogens on diverse crop plants. Several species of *Bacillus* and *Paenibacillus* are the major source of broad spectrum peptide antibiotics that are active against various microbial and nematode pathogens. Endophytic colonization and biofilm formation by these two genera are also reported. These plant growth promoting abilities of *Bacillus* and *Paenibacillus* can make them suitable plant growth promoting rhizobacteria for their application in sustainable agriculture.

1 Introduction

The microbial world is the largest unexplored reservoir of biodiversity and act as a major resource for agricultural, industrial, and medicinal applications (Handelsman et al. 1998; Daniel 2005; Lorenz and Eck 2005). Bacteria are the most dominant group of this diversity which exists in diverse ecological niches, including extreme environments present in both lithosphere and hydrosphere, where their metabolic abilities play a critical role in geochemical nutrient cycling (Daniel 2005). Rhizosphere, as coined by Hiltner (1904), is one such well-characterized ecological niche consisting of layer of soil with highest bacterial population and their activities are much influenced by the surrounding plant roots. The bacterial populations in the rhizosphere is 100–1,000 times higher than in bulk soil, and up to 15% of root surface is covered by microcolonies of a variety of bacterial species (Gray and Smith 2005). The rhizosphere effect is due to the fact that a substantial amount of carbon fixed by the plant as photosynthates (5–21%) is secreted in to rhizosphere mainly as root exudates that can be utilized as nutrients by bacterial populations. In return, the metabolic activities of these bacteria in the rhizosphere stimulate mineral nutrient delivery and uptake by plant roots (Glick 1995). These beneficial bacterial populations of rhizosphere are commonly called as plant growth promoting rhizobacteria or PGPR (Kloepper and Schroth 1978; Glick 1995). They promise plant growth promotion by secreting a variety of metabolites and employing various mechanisms (Glick et al. 1999).

A number of different bacterial groups being considered as plant growth promoting rhizobacteria include *Acinetobacter*, *Agrobacterium*, *Arthobacter*, *Azotobacter*, *Azospirillum*, *Burkholderia*, *Bradyrhizobium*, *Rhizobium*, *Frankia*, *Serratia*, *Thiobacillus*, Pseudomonads, and Bacilli (Glick 1995; Vessey 2003). Among them, *Bacillus* and *Paenibacillus* of aerobic endospore-forming bacteria (AEFB) are essentially ubiquitous in agricultural systems. The native populations of these two genera occur abundantly in most rhizosphere soils and plant tissues are differently colonized by distinct subpopulations (Mahaffee and Kloepper 1997; Seldin et al. 1998). Multiple species of *Bacillus* and *Paenibacillus* can promote plant growth and health in a variety of ways. Some species can promote plant growth directly by synthesizing plant hormones or increasing mineral nutrient uptake by fixing atmospheric nitrogen, solubilization of soil phosphorus, and other methods. Some populations suppresses plant pathogens and insect pests by producing antibiotic metabolites, while others stimulate plant host defenses prior to pathogen infection (Glick et al. 1999; Van loon 2007; Govindasamy et al. 2008), which indirectly contributes to increased crop productivity. Published reports on endophytic colonization and biofilm formation by *Bacillus* and *Paenibacillus* spp. have suggested that the endophytic colonization and biofilm formation improves the bacterium ability to act as a biocontrol agent against plant pathogens (Hallman et al. 1997; Davey and O'Toole 2000; Timmusk et al. 2005). In recent years, *Bacillus* and *Paenibacillus* spp. attracted considerable attention because of their advantages over other PGPR strains in inoculant formulations, stable maintenance in rhizosphere soil, and greater potentials in sustainable agriculture.

2 Taxonomy and Phylogeny of Genus *Bacillus* and *Paenibacillus*

Taxonomically, the genus *Bacillus* and *Paenibacillus* is coming under gram-positive, aerobic, or facultative endospore-forming bacteria. The genus *Bacillus* has undergone considerable taxonomic changes. Early attempts at classification of *Bacillus* species were based on two characteristics: aerobic growth and endospore formation. Starting off with two prominent and truly endospore-forming species, *Bacillus anthracis* and *B. subtilis*, the number of species allocated to this genus increased to an incredible 146 in the fifth edition of *Bergey's Manual of Determinative Bacteriology* (Bergey et al. 1939). Meticulous comparative studies on 1,114 strains of AEFB helped to reduce this number to 22 well-defined species in the eighth edition of *Bergey's Manual of Determinative Bacteriology* (Buchanan and Gibbons 1974). In 1980, with the publication of the *Approved Lists of Bacterial Names* 38 species of AEFB were listed, of which 31 were allocated to the genus *Bacillus* and 7 to other aerobic endospore-forming genera (Skerman et al. 1980). In Bergey's Manual of Systematic Bacteriology (1st edn., 1986), the G+C content of known species of *Bacillus* ranges from 32 to 69%. This observation, as well as DNA hybridization tests, revealed the genetic heterogeneity of the genus. Not only was there variation from species to species, but there were sometimes profound differences in G+C content within strains of a species. For example, the G+C content of the *Bacillus megaterium* group ranged from 36 to 45% (Claus and Berkeley 1986).

In Bergey's Manual of Systematic Bacteriology (2nd edn., 2004), phylogenetic classification schemes landed the two most prominent types of endospore-forming bacteria, clostridia, and bacilli, in two different Classes of Firmicutes, Clostridia, and Bacilli. Clostridia includes the Order *Clostridiales* and Family *Clostridiaceae* with 11 genera including, *Clostridium*. Bacilli included in the Order *Bacillales* and the Family *Bacillaceae*. In this family, there are 37 new genera on the level with *Bacillus*. This explains the heterogeneity in G+C content observed in the 1986 genus *Bacillus*. Their taxonomic hierarchy (Table 1) is Kingdom: Bacteria;

Table 1 Systematic position of the gram-positive aerobic endospore-forming bacteria (AEFB) based on 16S rRNA/DNA sequences as per Bergey's Manual of Systematic Bacteriology (2nd edn., 2004)

Systematic position/taxonomic hierarchy	No. of genus	No. of species/subsp.
Domain: Bacteria		
PhylumBXII: Firmicutes phy nov		
Class III: Bacilli		
OrderI: Bacillales		
Family I: Bacillaceae	17	
Key genus: *Bacillus*		88/2
Family II: Alicyclobacillaceae	3	
Key genus: *Alicyclobacillus*		8/2
Family III: Caryophanaceae	1	
Key genus: *Caryophanaon*		1
Family IV: Listeraceae	2	
Key genus: *Listera*		1
Family V: Paenibacillaceae	7	
Key genus: *Paenibacillus*		45/2
Family VI: Planococcaceae	5	
Key genus: *Planococcus*		1
Family VII: Sporolactobacillaceae	2	
Key genus: *Sporolactobacillus*		3
Family VIII: Staphylococcaceae	5	
Key genus: *Staphylococcus*		2
Family IX: Thermoactinomycetaceae	1	
Key genus: *Thermoactinomyces*		6
Family X: Turicibacteraceae	1	
Key genus: *Turicibacter*		1

Phylum: Firmicutes; Class: Bacilli; Order: Bacillales; Family: Acyclobacillaceae (genus: *Acyclobacillus*); Family: Bacillaceae (genus: Bacillus, *Geobacillus*); Family: Paenibacillaceae (genus: *Paenibacillus*, *Brevibacillus*); Family: Planococcaceae (genus: *Sporosarcina*). The phylogenetic approach to *Bacillus* taxonomy has been accomplished largely by the analysis of 16S rRNA molecules by oligonucleotide sequencing. This technique, of course, also reveals phylogenetic relationships. Surprisingly, *Bacillus* species showed a kinship with certain non-spore-forming species, including *Enterococcus*, *Lactobacillus*, and *Streptococcus* at the Order level, and *Listeria* and *Staphylococcus* at the Family level. Otherwise, some former members of the genus *Bacillus* were gathered into new families, including *Acyclobacillaceae*, *Paenibacillaceae,* and *Planococcaceae*, now on the level with *Bacillaceae*. All in all, today (2004) over 200 species of AEFB allocated to about 25 genera have been validly published. Notable former members of the genus *Bacillus* that have been moved to new families and/or genera are given in Table 2.

Taxonomy of the genus *Bacillus* consists of two groups of organisms vernacularly called the *B. subtilis* group and the *B. cereus* group. Species of the *B. subtilis* group are closely related and thus not easily distinguishable which included the two subspecies of *B. subtilis* (*B. subtilis* subsp. *subtilis* and *B. subtilis* subsp. *spizizenii*), *B. pumilus*, *B. licheniformis*, *B. amyloliquefaciens*, *B. mojavensis*, *B. sorensis* and

Table 2 Important species reassignments in the Genus *Bacillus* as per the recent approved lists of bacterial names (1986–2004)

Sl. no.	Bergey's manual of systematic bacteriology (1st edn., 1986)	Bergey's manual of systematic bacteriology (2nd edn., 2004)
1	*Bacillus acidocalderius*	*Acyclobacillus acidocalderius*
2	*Bacillus agri*	*Brevibacillus agri*
3	*Bacillus alginolyticus*	*Paenibacillus alginolyticus*
4	*Bacillus amylolyticus*	*Paenibacillus amylolyticus*
5	*Bacillus alvei*	*Paenibacillus alvei*
6	*Bacillus azotofixans*	*Paenibacillus azotofixans*
7	*Bacillus brevis*	*Brevibacillus brevis*
8	*Bacillus globisporus*	*Sporosarcina globisporus*
9	*Bacillus larvae*	*Paenibacillus larvae*
10	*Bacillus laterosporus*	*Brevibacillus laterosporus*
11	*Bacillus lentimorbus*	*Paenibacillus lentimorbus*
12	*Bacillus macerans*	*Paenibacillus macerans*
13	*Bacillus pasteurii*	*Sporosarcina pasteurii*
14	*Bacillus polymyxa*	*Paenibacillus polymyxa*
15	*Bacillus popilliae*	*Paenibacillus popilliae*
16	*Bacillus psychrophilus*	*Sporosarcina psychrophilia*
17	*Bacillus stearothermophilus*	*Geobacillus stearothermophilus*
18	*Bacillus thermodenitrificans*	*Geobacillus thermodenitrificans*

B. vallismortis. Even more loosely attached to this group are the species *B. firmus*, *B. lentus*, and *B. sporothermodurans*, which are clearly distinguishable from the other species of this group (Claus and Berkeley 1986). The species of *B. cereus* group includes closely related species such as *B. cereus*, *B. thuringiensis* (both motile), *B. mycoides,* and *B. pseudomycoides.* The species *B. weihenstephanensis* seems to consist of strains of *B. mycoides* and *B. cereus* (Jackson et al. 1999).

The genus *Paenibacillus* was created by Ash et al. (1993) to accommodate the former "group 3" of the genus *Bacillus.* It comprises over 30 species of facultative anaerobes and endospore-forming, neutrophilic, periflagellated heterotrophic, low G+C gram-positive bacilli. The name reflects this fact, in Latin *paene* means *almost*, and therefore the *Paenibacillus* is almost a *Bacillus.* Comparative 16S rRNA sequence analyses revealed that rRNA "group 3" bacilli represents a phylogenetically distinct group and exhibit high intragroup sequence relatedness and is only remotely related to *B. subtilis,* the type species of the genus *Bacillus.* The taxon contains various species such as *B. alvei*, *B. amylolyticus*, *B. azotofixans*, *B. gordonae*, *B. larvae*, *B. macerans*, *B. macquariensis*, *B. pabuli*, *B. polymyxa*, *B. pulvifaciens*, and *B. validus* (Ash et al. 1993). Phenotypically, species of this group react weakly with gram's stain and even young cultures appear gram-negative. They differentiate into ellipsoidal spores which distinctly swell the mother cell. The combination of morphology and physiology is sufficient to distinguish rRNA "group 3" bacilli from all other mesophilic species of *Bacillus* with the exception of *B. circulans*, *B. lautus*, *B. lentimorbus,* and *B. popilliae.* The latter four species are, however, phylogenetically only remotely related to *B. polymyxa* and its relatives and the described rRNA "group 3" specific gene probe provides an unequivocal method for distinguishing these taxa (Ash et al. 1991). The genus *Bacillus* of which, *B. subtilis* is the type is and an established model organism for research on gram positive bacteria. Recently, the genome of *B. subtilis* was sequenced completely and it represents the first published genome for a soil-living bacterium (Kunst et al. 1997; Wipat and Harwood 1999). Among the 51,713 Firmicutes sequences listed in Ribosomal Database Project (RDP) II, *Paenibacillaceae* comprises 1,057 16S rRNA sequences with 74 as *P. polymyxa* (as on January 2008). Complete sequencing of the genome of the plant growth promoting strain *P. polymyxa* E681, isolated from winter barley roots, is in progress.

3 Ecology and Distribution of *Bacillus* and *Paenibacillus* spp.

The species of *Bacillus* and *Paenibacillus* are metabolically diverse; the primery habitat of genus *Bacillus* is the soil and associated plants, rivers, and estuarine waters, although some species are pathogenic for mammals (e.g., *B. anthracis*) and insects (e.g., *B. sphaericus*, *B. thuringiensis*). The species of *Paenibacillus* inhabits different niches such as soils, roots, and rhizosphere of various crop plants including wheat, maize, sorghum, sugarcane and barley, and forest trees such as lodgepole pine, douglas fir, and marine sediments etc (Holl and Chanway 1992; von Der Weid et al. 2000). Multiple *Bacillus* and *Paenibacillus* spp. can be readily cultured from both bulk and rhizosphere soils. Culturable counts of these bacteria generally range from log 3 to log 6 cells per gram fresh weight, with soil counts typically exceeding those obtained from the rhizosphere (Mahaffee and Kloepper 1997; Seldin et al. 1998). Standard isolations on complex media typically yield multiple isolates of phylogenetically and phenotypically similar species related to *B. subtilis* and *B. cereus*. Most distinctive among these morphologically is *B. mycoides*, which often confound attempts to accurately enumerate cultured populations by virtue of their rapid mycelial-like growth patterns on agar media. *B. megaterium* has been reported to be one of the most abundant in some soils (Liu and Sinclair 1992), but it seems unlikely that a single species will dominate numerically in most soils. While multiple species of *Paenibacillus* can be detected in the soils and rhizosphere (Seldin et al. 1998), less work has been done to indicate which might be the most commonly isolated species.

Some species were initially defined based on the extreme physical or chemical conditions under which they were first isolated (e.g., *B. psychrophilus*), but few examples of obligate extremophiles exist (e.g., *B. stearothermophilus*, which are typically isolated from thermophilic composts) (Priest 1993). Instead, niche specificity and important ecological activities in *Bacillus* and *Paenibacillus* spp. appear to span phylogenetic boundaries. Most species can survive as saprophytes in soils, which are considered the primary reservoirs of these bacteria; however, most viable cells probably occur as inactive spores at any given time (Nicholson 2002). Culture-independent analyses of soil DNA have confirmed the presence of the easily cultured species and revealed additional, uncultured diversity in both the *Bacillus* and *Paenibacillus* rRNA lineages (Felske et al. 1999; Smalla et al. 2001; Garbeva et al. 2003). However, contradictory evidence exists on the relative abundance of cultured and uncultured representatives of these genera in different soils. Some reports indicated that the large majority of *Bacillus*-like sequences cloned from soils were highly similar to known species. But, others report that the dominant *Bacillus* sequences present in a different soil are not the same as those present in easily cultured isolates (Smalla et al. 2001; Garbeva et al. 2003). Interestingly, the substantial effort leading to the isolation of this previously uncultured lineage (referred to as DA001) also led to the isolation of even more microdiversity that had not been previously directly detected in DNA clone banks of sequences obtained from the same soil (Felske et al. 1999).

At the species level, most *Bacillus* and *Paenibacillus* are globally distributed and such widespread occurrence of more defined subspecies of *B. subtilis* and *B. cereus* with the capacity to suppress plant pathogens has also been reported (Priest 1993; Stabb et al. 1994; Pinchuk et al. 2002). Other studies have reported only a limited degree of geographic endemicity in *B. thuringiensis* (Chak et al. 1994; Bravo et al. 1998) and *Paenibacillus azotofixans* (Seldin et al. 1998) over spatial scales. Recently, ribosomal sequences amplified from environmental samples have been used to characterize the relative distribution of *Bacillus* and *Paenibacillus* spp. between soils and plant tissues. Overall, the structure of soil bacterial communities is known to vary with soil type more than with management regime (Garbeva et al. 2003); however, the magnitude of such variation may be relatively small for *Bacillus* and *Paenibacillus* spp.

4 PGPR Potentials of *Bacillus* and *Paenibacillus* spp.

Bacillus and *Paenibacillus* spp. are known to have wide PGPR potentials; in the rhizosphere, they are involved in nitrogen fixation, soil phosphorus solubilisation, the production of antibiotics, chitinase, hydrolytic enzymes, and exopolysaccharides and in the enhancement of soil porosity. Numerous *Bacillus* and *Paenibacillus* strains express these activities which promote plant growth and suppress soilborne plant pathogens. A number of these strains already have been developed commercially as plant growth promoters and biocontrol agents (Table 3) and their use in agriculture has recently been reviewed (Lacey et al. 2001; Paulitz and Belanger 2001). Similarly, many strains of *Paenibacillus* were isolated and characterized functionally (Table 4) for their potential use in agriculture as plant growth promoters (Timmusk and Wagner 1999; Timmusk et al. 2003; Senthilkumar et al. 2007a). Improvements in plant growth and productivity by the applications of *Bacillus* and *Paenibacillus* spp. are mediated by three different ecological mechanisms: promotion of host plant nutrition and growth, antagonism against plant pathogens and insect pests, and stimulation of plant host defense mechanisms.

Table 3 Commercially available *Bacillus* spp. based plant growth promoters and biocontrol products

Bacillus species/strains	Activity/function	Product name
Bacillus polymyxa and other species	Atmospheric nitrogen fixation	Wide variety of products
Bacillus megaterium and *B. coagluans*	Mineral phosphate solubilization	Phosphobacter
B. subtilis QST 713	Fungi and bacteria on vegetables and fruit	Serenade
B. licheniformis	Fungi on turf	Ecoguard
B subtilis GB03	Fungi on cotton and soybeans	Kodiak
B. pumilis GB34	Fungi on soybeans	Yield Shield
B. amyloliquefasciens and *B. subtilis* GB122	Fungi on bedding plants	BioYield
B. subtilis MBI600	Fungi on cotton and soybeans	Subtilex
B. subtilis MBI600 and *Rhizobium*	Fungi on soybeans	HiStick

4.1 Promotion of Host Plant Nutrition and Growth

Bacillus and *Paenibacillus* spp. promote plant growth directly by providing nitrogen to the host plant. They also solubilize insoluble phosphates in soil by various mechanisms and secrete phytohormones. Such activities lead to induced plant growth and development.

4.1.1 Biological N_2 Fixation

Biological nitrogen fixation by soil prokaryotic microorganisms is considered one of the major mechanisms by which plant benefit from the association of microbial partners. One of the benefits that diazotrophic bacteria provide to plants is fixed nitrogen in exchange of fixed carbon released as root exudates. Isolates of nitrogen-fixing bacilli from plant rhizospheres were determined by an acetylene reduction assay (ARA) for nitrogenase activity and by amplifying and sequencing part of *nifH* gene. Xie et al. (1998) reported that the following species were nitrogen-fixing bacteria based on nitrogenase activity: *Bacillus megaterium*, *Bacillus cereus*, *Bacillus pumilus*, *Bacillus circulans*, *Bacillus licheniformis*, *B. subtilis*, *Bacillus brevis,* and *Bacillus firmus*. The three former Bacillus species, *Paenibacillus azotofixans*, *Paenibacillus macerans,* and *P. polymyxa*, were nitrogen fixers, based on nitrogenase activity (Seldin et al. 1984). Recently, *Paenibacillus odorifer*, *Paenibacillus graminis*, *Paenibacillus peoriae,* and *Paenibacillus brasilensis* have been described as nitrogen fixers (Berge et al. 2002; von der Weid et al. 2002). However, *nifH* gene was only detected in the following *Paenibacillus* species: *P. azotofixans*, *P. macerans*, *P. polymyxa*, *P. graminis,* and *P. odorifer* (Berge et al. 2002). Ding et al. (2005) isolated and identified nitrogen-fixing bacilli from plant

Table 4 Functional characteristics of isolated *Paenibicillus polymyxa*

Sl. no.	Isolates/strains	Source	Activity/functions	References
1	*P. polymyxa* strain B1 and B2	Wheat rhizosphere	Nitrogen fixation and formation of biofilm	Lindberg et al. (1985), Timmusk et al. (2005)
2	*P. polymyxa* strain B2	Wheat rhizosphere	Cytokinin production	Timmusk et al. (1999)
3	*P. polymyxa* strains B2, B3 and B4	Wheat rhizosphere	Increased resistance to plant pathogens (biotic stress) and drought resistance(abiotic stress)	Timmusk and Wagner (1999)
4	*P. polymyxa* strains B5 and B6	Wheat rhizosphere	Biocontrol of the oomycete plant pathogens, *Phytophora palmivora* and *Phytim aphanidermatum*	Timmusk et al. (2003)
		Soil around peanut roots	Production of exopolysaccharides, biocontrol against *Aspergillus niger* in roots and seeds of peanut plants	Haggag and Timmusk (2008)
5	*P. polymyxa* PMD66, PMD112, PMD128, PMD216 and 230	Wheat rhizoplane, rhizosphere, soil	Production of auxin and other indolic and phenolic compounds	Labuhn et al. (1997)
	P. polymyxa		Production of chitinase	Mavingui and Heulin (1994)
6	*P. polymyxa* SCE2	Soil(Brazil)	Proteases production, production of antimicrobial compounds active against human pathogenic microorganisms	Seldin et al. (1999)
	P. polymyxa T129	Soil	Biocontrol against *Fusarium oxysporum*	Dijksterhuis et al. (1999)
7	*P. polymyxa* JB115, *P. polymyxa* 1460 and *P. polymyxa* BY-28	Soil	Production of β-glucan, lectin, and flocculants production	Jung et al. (2007), Karpunina et al. (2003), Gong et al. (2003)
8	*P. polymyxa* strains CM5-5 and CM5-6	Barley rhizosphere	Production of hydrolytic enzymes, multitarget and medium-independent type of fungal antagonism	Nielsen and Sorensen (1997)
	P. polymyxa E681	Winter barley roots	Fusaricidin biosynthesis, biocontrol of fungal pathogens on sesame plants	Choi et al. (2007), Ryu et al. (2006)
9	*P. polymyxa* HKA-15	Root and nodule tissues of soybean	Biocontrol against charcoal rot pathogen *R. bataticola* and production of cyclic and depsipeptides	Senthilkumar et al. (2007a, b)
10	*P. polymyxa* OSY-DF	Fermented foods	Coproduction of polymyxin E1 and lantibiotic	He et al. (2007)
	P. polymyxa P13	Fermented sausages	Polyxin production and biosorption of heavy metal	Piuri et al. (1998)

rhizospheres in Beijing region and reported that *nifH* gene exists in both genera *Bacillus* and *Paenibacillus*. Nitrogen-fixing ability by *P. polymyxa* was demonstrated by Guemori-Athmani et al. (2000). These authors measured nitrogenase activity of some representative isolates of *P. polymyxa* recovered from Algerian soil by ARA. Results showed that only 14 of the 23 strains tested were able to reduce acetylene. Some of them were very active: strain SGH1 reduced C_2H_2 at a similar rate to *P. azotofixans* ATCC 35681T, which is a very efficient nitrogen-fixing bacterium (Seldin and Penido 1986). In India, numerous reports are available on the application of free-living diazotrophs, including *Bacillus* spp., for increased yield of various crops.

4.1.2 Solubilization of Phosphorus and Uptake of Minor Nutrients

Phosphorus (P) is a major growth-limiting nutrient, and unlike the case for nitrogen, there is no large atmospheric source that can be made biologically available (Ezawa et al. 2002). Microorganisms enhance the P availability to plants by mineralizing organic P in soil and by solubilizing precipitated phosphates (Fig. 1) (Kucey et al. 1989; Pradhan and Sukla 2005; Chen et al. 2006). Inorganic forms of P are solubilized by a group of heterotrophic bacteria excreting organic acids that dissolve phosphatic minerals and/or chelate cationic partners of the P ions, i.e., PO_4^{3-} directly, releasing P into solution. Bacterial biomass assimilates soluble P, and prevents it from adsorption or fixation. These bacteria in the presence of labile carbon serve as a sink for P by rapidly immobilizing it even in low P soils (Bünemann et al. 2004; Khan and Joergensen 2009). Subsequently, phosphate-solubilizing bacteria (PSB) become a source of P to plants upon its release from their cells. PSB are being used as biofertilizer since 1950s and release of P by PSB from insoluble and fixed/adsorbed forms is an important aspect regarding P availability in soils (Igual et al. 2001).

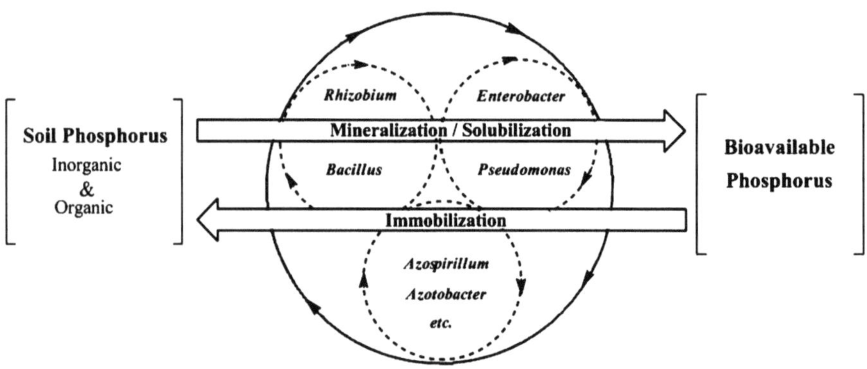

Fig. 1 Schematic diagram of soil phosphorus mineralization, solubilization and immobilization by rhizobacteria

Phosphate solubilization takes place through various microbial processes/ mechanisms including organic acid production and proton extrusion. Inorganic P is solubilized by the action of organic and inorganic acids secreted by PSB in which hydroxyl and carboxyl groups of acids chelate cations (Al, Fe, and Ca) and decrease the pH in basic soils. The PSB dissolve the soil P through the production of low molecular weight organic acids, mainly gluconic and keto gluconic acids (Goldstein 1995; Deubel et al. 2000), in addition to lowering the pH of rhizosphere. Among the soil bacterial communities, ectorhizospheric strains from *Pseudomonas* and Bacilli, and endosymbiotic rhizobia have been described as effective phosphate solubilizers. *Bacillus megaterium*, *B. circulans*, *B. coagulans*, *B. subtilis*, *B. polymyxa*, *B. sircalmous*, and *Pseudomonas striata* could be referred as the most important strains (Subbarao 1988; Kucey et al. 1989). Gluconic acid and 2-ketogluconic acid seems to be the most frequent agent of mineral phosphate solubilization. Other organic acids, such as glycolic, oxalic, malonic, and succinic acid, have also been identified among phosphate solubilizers. Strains of *Bacillus* were found to produce mixtures of lactic, isovaleric, isobutyric, and acetic acids in addition to the major organic acids. Phophorus-solubilizing activity of *B. megaterium* enhanced the number of nodules, dry weight of nodules, yield components, grain yield, nutrient availability, and uptake in soybean crop and enhanced the seedling length of *Cicer arietinum* (Son et al. 2006; Sharma et al. 2007), while coinoculation of *Bacillus* spp. along with other PGPR strains reduced P application by 50% without affecting corn yield (Yazdani et al. 2009). Inoculation with phosphate-solubilizing *B. megaterium* increased sugarcane yield by 12.6% (Sundara et al. 2002). Phosphate-solubilizing *B. subtilis* strains have been reported to synergistically increase plant nitrogen and phosphate-accumulation when coinoculated with *Glomus intraradices*. Toro et al. (1997) evaluated the interactive effect of Phosphate-solubilizing bacteria (*Bacillus subtilis*) and arbuscular mycrorrhizal (AM) fungi (*Glomus intraradices*) on onion with a soil of low P content. Coinoculation of these both significantly increased the vegetative biomass and N, P accumulation in onion tissues. Combined inoculation of arbuscular mycorrhiza and Phosphate-solubilizing *Bacillus* and *Paenibacillus* spp. give better uptake of both native P from the soil and P coming from the phosphatic rock (Goenadi et al. 2000; Cabello et al. 2005).

Almost half of the microorganisms in soil and plant roots possess P mineralization potential under the action of phosphatases. The largest portion of extracellular soil phosphatases is derived from the microbial population (Dodor and Tabatabai 2003). Alkaline and acid phosphatases use organic phosphate as a substrate to convert it into inorganic form. Principal mechanism for mineralization of soil organic P is the production of acid phosphatases by rhizobacteria (Hilda and Fraga 2000). Three out of the four strains investigated were identified as *B. amyloliquefaciens* and were able to degrade extracellular phytate (*myo*-inositol hexakisphosphate). The highest extracellular phytase activity was detected in strain FZB45, and diluted culture filtrates of this strain stimulated growth of maize seedlings under phosphate limitation in the presence of phytate (Idriss et al. 2002). Mixed cultures of phosphate-solubilizing rhizobacteria including *Bacillus* and *Paenibacillus* spp. are most effective in mineralizing organic phosphate.

Most studies are indicating that PGPR isolates may increase the mobility and availability of micronutrients including iron (Fe) by the formation of high-affinity siderophores. Chemically, siderophores are low molecular weight compounds of either catecholate or hydroxamate types that complex with Fe^{2+} and render it available to crop plants (Leong 1986). The widespread production of siderophores by diverse rhizobacterial genera included as *Bacillus*, *Rhizobium*, *Pseudomonas,* and *Agrobacterium* at low iron levels are reported by Neilands (1986). Numerous plants are capable of using rhizobacterial Fe siderophore complexes as a means of obtaining Fe from soil (Wang et al. 1993). This view is supported by the findings of Hughes et al. (1992) who reported enhanced Fe uptake in oats because of siderophore production. *Bacillus* and *Paenibacillus* spp. produces both types of siderophores; the bacterium *B. megaterium* ATCC 19213 is known to produce two hydroxamate siderophores, schizokinen and N-deoxyschizokinen, under iron-limited conditions. In addition to their high affinity for ferric ions, these siderophores also chelate aluminum (Hu and Boyer 1996). Wilson Melissa et al. (2006) reported that three *B. anthracis* strains (USAMRIID, 7702, and 34F2) and *B. cereus* ATCC 14579 excrete two catecholate siderophores, petrobactin (which contains 3,4-dihydroxybenzoyl moieties) and bacillibactin (which contains 2,3-dihydroxybenzoyl moieties). However, the insecticidal organism *B. thuringiensis* ATCC 33679 makes only bacillibactin. More details on the production of siderophores by *Bacillus* and *Paenibacillus* spp. and their role in enhancing Fe uptake have been reported by different researchers in variety of crop plants.

4.1.3 Production of Phytohormones and Growth Stimulants

Plant hosts may also be affected by hormones known to be produced by various microbial species, including *Bacillus and Paenibacillus.* There are five classes of well-known phytohormones, namely, auxins, gibberellins, cytokinins, ethylene, and abcisic acid and soil microorganisms, particularly the rhizosphere bacteria, are potential sources of these hormones (Patten and Glick 1996; Arshad and Frankenberger 1998). These phytohormones are known to mediate processes such as plant cell enlargement, division, and extension in symbiotic as well as nonsymbiotic roots. Among these hormones, most attention has focused on auxins in which the most common and well characterized is indole-3-acetic acid (IAA), which is known to stimulate both rapid (e.g., increase in cell elongation) and long-term (e.g., cell division and differentiation) responses in crop plants. Gutierrez-Manero et al. (2001) isolated *B. pumilus* and *B. licheniformis* from the rhizosphere of *Alnus glutinosa* shown to produce physiologically active gibberellins which had strong growth promoting activity on alder. The hormone ethylene production and its effect on plant growth by *Bacillus licheniformis*, *Bacillus subtilis,* and *Bacillus mycoides* also had been reported (Fukuda et al. 1989).

The production of plant growth promoting compounds by *P. polymyxa,* similar in activity to indole-3-acetic acid (IAA), has been suggested to stimulate growth in crested wheatgrass (Holl et al. 1988). It also releases iso-pentenyladenine and one unknown cytokinin-like compound during its stationary phase of growth which promotes seed germination, *de novo* bud formation, release of buds from apical dominance, stimulation of leaf expansion and reproductive development and retardation of senescence in wheat (Mok 1994). Some strains of *B. subtilis*, and *B. amyloliquefaciens*, promote plant growth by releasing volatiles such as 2,3-butanediol and acetoin. The highest level of growth promotion was observed with mutants of *B. amyloliquefaciens* IN937a and *B. subtilis* GB03, blocked in the biosynthesis of these compounds, were inactive in plant growth promotion (Ryu et al. 2003). More recently, Zhang et al. (2008) found that *B. subtilis* GB03 increases the photosynthetic efficiency and chlorophyll content of *A. thaliana* through the modulation of endogenous signalling of glucose and abscisic acid sensing. Enzyme cellulase (CMCase) activities were also shown in *Bacillus pumilus*, *Bacillus sphaericus*, and *Bacillus circulans*, which showed that most plant-associated microorganisms might have cellulase activity for adoption or establishment of a plant microbe interaction (Emitizi et al. 2007). The effect of inoculation with *P. polymyxa* on growth parameters of wheat and spinach plants and the activities of enzymes present in the leaves of these plants such as glucose-6-phosphate dehydrogenase, 6-phosphogluconate dehydrogenase, glutathione reductase, and glutathione S-transferase were also observed (Cakmakci et al. 2007).

4.2 Antagonism Against Plant Fungal and Bacterial Pathogens

Bacillus and *Paenibacillus* spp. suppress phytopathogens by producing various antifungal metabolites and impart induced systemic resistance (ISR) against insects and nematodes.

4.2.1 Control of Fungal Pathogens

Direct antagonism against plant fungal pathogens by *Bacillus* spp. has been well exploited in agriculture as biocontrol agents. The most thoroughly studied of these include *B. subtilis* (Leifert et al. 1995; Asaka and Shoda 1996; Pinchuk et al. 2002). Additionally, a number of studies have reported direct antagonism by other species including *B. amyloliquefaciens*, *B. cereus*, *B. licheniformis*, *B. megaterium*, *B. mycoides*, and *B. pumilus* as well as isolates of unidentified species from the genus (Handelsman et al. 1990; Leifert et al. 1995; Liu and Sinclair 1992). Although less frequently reported in the literature, some isolates of *P. macerans* and *P. polymyxa* may also be antagonistic to plant pathogens (Timmusk and Wagner 1999). Most of these studies focused on control of fungal and oomycete pathogens. The culture filtrate of *B. amyloliquefaciens* RC-2 showed activity against *Colletotrichum demantium*, *Rosellina necatrix*, *Pyricualria oryzae*, *Agrobacterium tumefaciens*, and *Xanthomonas campestris* pv. campestris (Yoshida et al. 2001). A soil bacterium *Bacillus* sp. strain BC121 isolated from the

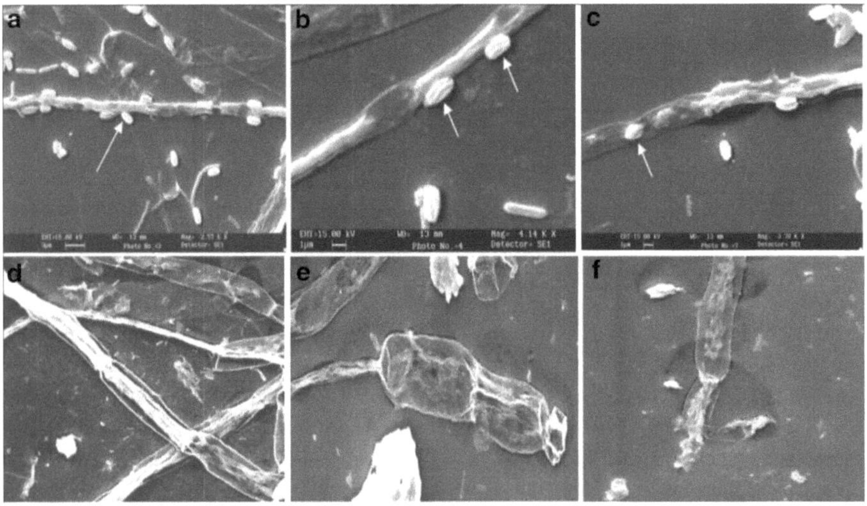

Fig. 2 Scanning electron microphotographs (SEM) showing the antagonistic interaction of *Paenibacillus polymyxa* HKA-15 against charcoal rot pathogen *Rhizoctonia bataticola* (Senthilkumar et al. 2007b)

rhizosphere of sorghum showed high antagonistic activity against *Curvularia lunata*. The strain produces chitinase protein which showed clear hyphal lysis in in vitro observations (Basha and Ulaganathan 2002). *Bacillus subtilis* strain PRBS-1 and AP-3 inhibited five soybean seed pathogenic fungi viz., *Rhizoctonia solani*, *Colletotrichum truncatum*, *Sclerotinia sclerotium*, *Macrophomina phaseolina*, and *Phomopsis* spp. under in vitro conditions (Araujo et al. 2005). *B. amyloliquefaciens* strains conferred protection of oil seed rape (*Brassica napus*) toward all fungal pathogens such as *Alternaria brassicae*, *Botrytis cinerea*, *Leptosphaeria maculans*, and *Verticillum longisporum* (Danielsson et al. 2006).

The in vitro antagonistic activity of *P. polymyxa* against the fungus *Gaeumannomyces graminis* var. *tritici* that causes take-all of wheat and the plant pathogenic fungus *Fusarium oxysporum* that causes *Fusarium* wilt disease has been reported by Heulin et al. (1994). Ryu et al. (2006) demonstrated that *P. polymyxa* strain E681 effectively controlled preemergence and postemergence damping-off diseases on sesame plants. *Paenibacillus polymyxa* HKA-15 was active against *R. bataticola* causing charcoal rot in soybean (Senthilkumar et al. 2007a). Many workers have carried out light microscopic studies on the effect of biocontrol isolates on fungal hyphal morphology. The necrotrophic effect and sequential lysis of *R. bataticola* fungal hyphae by *Paenibacillus polymyxa* HKA-15 cells under light and scanning electron microscope was demonstrated (Fig. 2) (Senthilkumar et al. 2007a). Microscopic analysis on the effect of antagonist on *Magnoporthe grisea* revealed the inhibition of spore germination under light microscope (Tendulkar et al. 2007). Recently, Zhou et al. (2008) isolated *Paenibacillus* strain HT16 from locusts, which showed strong inhibition to *Penicillium expansum* and produced antifungal protein with the molecular weight of 4,517 Da.

4.2.2 Control of Bacterial Pathogens

The role of lipopeptides produced by *Bacillus* sp. against *Xanthomonas campestris*, has been widely studied (Monteiro et al. 2005; Salerno and Sagardoy 2003). However, the effect of metabolites on bacterial cell morphology has also been reported (Hashizume et al. 1996; Nakao et al. 1981). Antagonistic activities of epiphytic bacteria from soybean leaves against *Pseudomonas syringae* pv. glycineae in vitro and in planta was tested. In in planta assay, *Pseudomonas syringae* pv. glycineae and each isolate were simultaneously inoculated in to wounds of pin-pricked leaves of greenhouse-grown plants. Out of 82 isolates, 19 isolates were able to suppress the pathogen. The mixtures of isolate and pathogen were inoculated at ratios >1 (May et al. 1996). Under green-house conditions, inoculation of the isolate *Bacillus subtilis* 210, 72 h before the inoculation of the pathogenic bacteria, significantly reduced the number of lesions caused by *X. campestris* (Salerno and Sagardoy 2003). The species of *Bacillus* and *Paenibacillus* also showed effective antagonism against other bacterial plant pathogens of economically important crops.

4.3 Antagonism Against Insect Pests and Nematode

Antagonism of insect pests and pathogen populations by *Bacillus* sp. and closely related AEFB takes many forms. Some species are pathogens of insects or nematodes (Siddiqui and Mahmood 1999; Lacey et al. 2001). Perhaps, the most studied of the insect pathogens are those classified as *B. thuringiensis*. This species is distinguished from the common saprophytic species *B. cereus* by the occurrence of plasmids that encode pathogenicity factors that make the strains pathogenic to various invertebrates. The production of the crystalline inclusion bodies (Cry proteins) within their spores allow for opportunistic growth when consumed by soil invertebrates. While the crystalline proteins are widely known to be disruptive to the digestive tracts of numerous Lepidopertera and Diptera larvae, evidence also exists for their toxicity to nematodes (Wei et al. 2003). The wide variation in *cry* gene structure and the known occurrence of tolerance to the protein toxins produced by various isolates indicates that a range of virulence exists in nature. *B. sphaericus* are pathogenic to various Diptera species, but the species appears to be more effective at controlling insects that bite animals and humans rather than those that damage crops. *B. sphearicus* also produce protein toxins, but these are deposited outside the spore coat by the mother cell. *P. popilliae* and *P. lentimorbus* cause milky disease in the larvae of some beetles (Order: Coleoptera) including those that can damage crops. Antagonistic activity of *P. polymyxa* was also demonstrated against the root-knot nematode, *Meloidogyne javanica*. The inoculation of *P. polymyxa* alone or together with *Rhizobium* increased lentil plant growth both in *M. javanica*-inoculated and uninoculated plants (Siddiqui et al. 2007).

4.4 Stimulation of Plant Host Defense Mechanisms Through Induced Systemic Resistance

Recently, "induced resistance" to diseases, or plant "immunization," has received increasing attention (Uknes et al. 1992). This refers to a process in which plants exhibit an increased level of resistance to infection by a pathogen after appropriate stimulation. Induced resistance can be triggered by, e.g., infection with a necrotizing pathogen, or by treatment with certain chemicals, e.g., salicylic acid (SA). This response is referred to as systemic acquired resistance (SAR). Induced resistance can also be a result of root colonization by PGPR (Alström 1991; Wei et al. 1991). The latter response is called induced systemic resistance (ISR), and has been shown to protect against disease in several plant species (Thomashow and Weller 1995; van Wees et al. 1997). Elicitation of ISR by plant-associated bacteria was initially demonstrated using *Pseudomonas* spp. and other gram-negative bacteria. Fewer published accounts of ISR by *Bacillus* spp. are also available. The specific strains of the species *B. amyloliquefaciens*, *B. subtilis*, *B. pasteurii*, *B. cereus*, *B. pumilus*, *B. mycoides*, and *B. sphaericus* elicit significant reductions in the incidence or severity of various diseases on a diversity of hosts (Kloepper et al. 2004). Elicitation of ISR by these strains has been demonstrated in greenhouse or field trials on tomato, bell pepper, muskmelon, watermelon, sugar beet, tobacco, *Arabidopsis* sp., cucumber, loblolly pine, and two tropical crops (long cayenne pepper and green kuang futsoi) (Van Loon 2007). Protection resulting from ISR elicited by *Bacillus* spp. has been reported against leaf-spotting fungal and bacterial pathogens, systemic viruses, a crown-rotting fungal pathogen, root-knot nematodes, and a stem-blight fungal pathogen as well as damping-off, blue mold, and late blight diseases (Van Loon et al. 1998). Reductions in populations of three insect vectors have also been noted in the field: striped and spotted cucumber beetles that transmit cucurbit wilt disease and the silverleaf whitefly that transmits *Tomato mottle virus*. In most cases, *Bacillus* spp. that elicits ISR also promotes plant growth (Zehnder et al. 1997, 2000).

Many individual bacterial components induce ISR, such as LPS, flagella, salicylic acid, and siderophores (Van Loon 2007). More recently, cyclic lipopeptides, the antifungal factor Phl, the signal molecule acyl homoseine lactone (AHL), and volatile blends produced by *B. subtilis* GB03 and, to a lesser extent, the individual volatiles acetoin and 2,3-butanediol have been added to the list (Ryu et al. 2004). Studies on mechanisms indicate that elicitation of ISR by *Bacillus* spp. is associated with ultrastructural changes in plants during pathogen attack and with cytochemical alterations. Investigations into the signal transduction pathways of elicited plants suggest that *Bacillus* spp. activate some of the same pathways as *Pseudomonas* spp. and some additional pathways (Fig. 3). For example, ISR elicited by several strains of *Bacillus* spp. is independent of salicylic acid but dependent on jasmonic acid, ethylene, and the regulatory gene *NPR1* – results that are in agreement with the model for ISR elicited by *Pseudomonas* spp. (Van Loon 2007). However, in other cases, ISR elicited by *Bacillus* spp. is dependent on salicylic acid and independent of jasmonic acid and *NPR1*. In addition, while ISR by *Pseudomonas* spp.

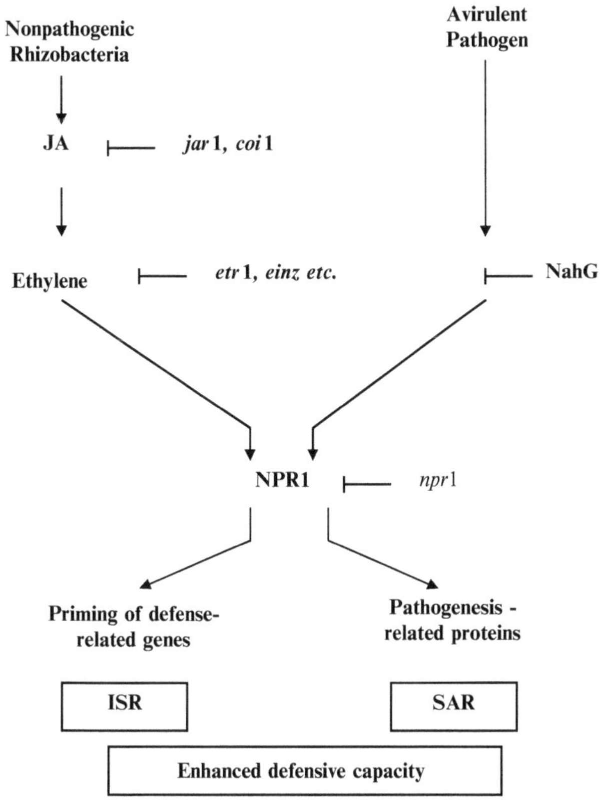

Fig. 3 Current model of signal-transduction pathways leading to pathogen-induced systemic acquired resistance (SAR) and rhizobacteria-induced systemic resistance (ISR). Some species of *Bacillus* spp. may trigger a SA-dependent signaling pathway that leads to a state of induced resistance resembling SAR (Van Loon 2007)

does not lead to accumulation of the defense gene *PR1* in plants, in some cases, ISR by *Bacillus* spp. does. For example, elicitation of ISR in sugar beet by *B. mycoides* strain Bac J and *B. pumilus* strains 203-6 and 203-7 was associated with enhanced peroxidase activity and increased production of one chitinase isozyme and two isozymes of β-1,3-glucanase (Bargabus et al. 2002, 2004). In the tobacco blue mold system, Zhang et al. (2002) reported that plants treated with *B. pumilus* strain SE34 had greatly increased levels of salicylic acid, compared with that of nontreated plants or plants treated with two gram-negative bacteria, 1 day after challenge-inoculation with the pathogen. In a recent study, Timmusk and Wagner (1999) reported that natural isolates of *P. polymyxa* B_2 induces changes in *Arabidopsis thaliana* gene expression and confers significant resistance to plant pathogen *Erwinia caratovora* upon challenge inoculation. This isolate also induces drought tolerance and these effects were observed in both gnotobiotic and soil systems. Similarly, *P. polymyxa* isolates B_2, B_3, and B_4 induces ISR toward Oomycete plant pathogens *Phytophthora palmivora* and

Pythium aphanidermatum causing damping-off in *Arabidopsis thaliana* (Timmusk et al. 2003).

5 Production of Peptide Antibiotics by *Bacillus* and *Paenibacillus* spp.

Several species of *Bacillus* and *Paenibacillus* are known to produce toxins that are inhibitory to the growth and/or activities of fungal, bacterial, and nematode pathogens of plants. Catabolic enzymes (e.g., proteases, chitinases, and glucanases), peptide antibiotics, and small molecules can be secreted by various species (Priest 1993) and may all contribute to pathogen suppression. *Bacillus* spp. and its related genera have been identified as potential biocontrol agent as they produce wide range of peptide antibiotics (Fig. 4) active against various microorganisms

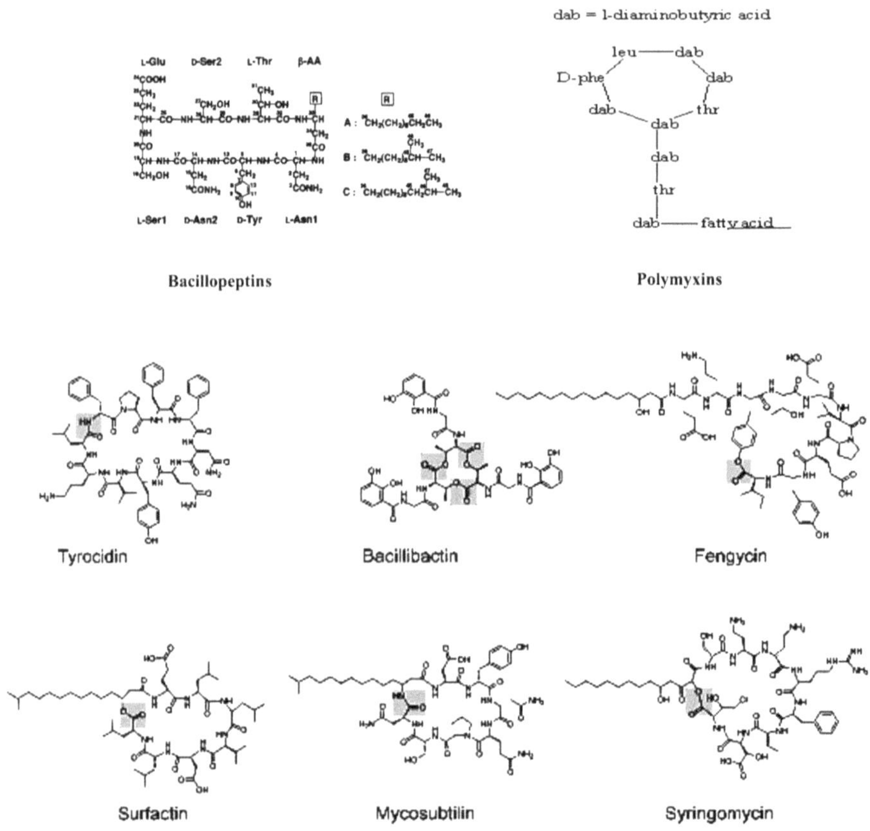

Fig. 4 Chemical structures of important peptide antibiotics produced by *Bacillus* and *Paenibacillus* spp. (Selim et al. 2005)

(Kim et al. 2003). On the basis of the chemical nature, these peptides may be classified into two broad groups viz., cyclic and linear peptides. The identification and characterization of these peptides from different strains showed that they are Bacillopeptins (Kajimura et al. 1995), fusaridicin group of peptides (Beatty and Jensen 2002), matacin (Polymyxin M) (Martin et al. 2003), Gavaserin and Saltavalin (Pichard et al. 1995), and Polymyxin B (Selim et al. 2005).

Iturin A2, a cyclic peptide was identified by NMR and FAB-MS analysis from the culture filtrate of *B. amyloliquefaciens* RC-2 (Yoshida et al. 2001). Crude antibiotic extracts of *B. cereus* were purified and the active fraction $X_{16s}I$ was stable at a wide range of temperature, pH, and polar organic solvents (Safdi et al. 2002). Bacillomycin F, a new family of iturin group antibiotics was isolated from *Bacillus subtilis* (Mhammedai et al. 1982). Acid hydrolysis of the antibiotic gave a peptide moiety which contains 7 mol of amino acids and a lipid moiety which is a mixture of two main long-chain β-aminoacids. Tendulkar et al. (2007) reported biologically active fractions isolated from the culture filtrate of *B. licheniformis* BC 98. These were further fractionated by RP-HPLC and characterized by 500 MHz ^1H NMR analysis and identified as surfactin with the molecular mass of 1,035 Da. Two active methanol fractions viz., KB-8A and KB-8B were extracted from the culture filtrate of *Bacillus polymyxa*. The purified fraction KB 8A had minimum inhibitory concentration (MICs) of 12.8 μg/ml for *Fusarium oxysporum* and *Alternaria mali* (Hyun et al. 1999).

Two new antibacterial substances viz., Gavaserin and Saltavalin were isolated from *Bacillus polymyxa* with the molecular mass of 911 and 903 Da, respectively (Pichard et al. 1995). Bacillopeptins, a new iturin group antifungal antibiotic was isolated from *B. subtilis* FR-2 from the rhizosphere of garlic suffering from basal rot caused by *F. oxysporum*. Their structures were elucidated to be cyclic lipopeptides similar to bacillomycin L (Kajimura et al. 1995). Bacillomycin F produced by *Bacillus subtilis*, isolated from honey, showed antagonism against *Byssochlamys fulva* H25 and its structure had varying lengths of the fatty acid chain moiety from C14 to C16. Tamehiro et al. (2002) reported a novel phospholipid antibiotic (bacilysocin) produced by *Bacillus subtilis* 168 and the structure of Bacilysocin elucidated was 1-(12-methyltetradecanoyl)-3-phosphoglyceroglycerol using NMR and Mass Spectrometry analysis. The antimicrobial peptide cerein 8A was isolated from *B. cereus* and its purified substance corresponded to 26 κDa peptide band (Bizani et al. 2005). The importance of antibiotic production to plant disease suppression by *Bacillus* spp. has been demonstrated. *B. subtilis* strains that produce the lipopeptide antibiotics iturin A and surfactin could suppress damping-off in tomato while mutants could not (Asaka and Shoda 1996). And, in *B. cereus*, production and resistance to zwittermicin A have been correlated to suppression of damping-off in alfalfa (Raffel et al. 1996). Pueyo et al. (2009) showed a large group of lipopeptides produced by soil bacterium *B. megaterium* and their antagonistic activity similar to surfactins, lichenysins, itrurin A, and fengycins.

Physicochemical characterization of antimicrobial metabolite produced by *Paenibacillus peoriae* strain NRRL BD-62 showed that the compound retained the activity after autoclaving at 121°C for 10 min. The compound was stable after treatment with organic solvents and hydrolytic enzymes, and its activity was preserved at a wide range of pH (Weid et al. 2003). *Paenibacillus* sp. strain B2, isolated from the mycorrhizosphere of sorghum colonized by *Glomus mosseae*, produced three active antagonistic compounds. The first peptide compound had the same retention time as polymyxin B1 with the molecular mass of 1,184.7 and contains a 2,3-didehydrobutyrine residue with a molecular mass of 101 Da replacing a threonine at the A_2 position of polymyxin side chain and this could explain the broader range of antagonistic activity of this peptide compared to that of polymyxin B (Selim et al. 2005). Most studies on the biocontrol activity of *P. polymyxa* have been concentrated on the production of different antibiotic substances. Fusaricidin, a peptide antibiotic consisting of six amino acids, has been identified as a potential antifungal agent from *P. polymyxa* E681 (Choi et al. 2007). Various analogs of fusaricidins were isolated and characterized from *P. polymyxa*; these included LI-F03, LIF04, LI-F05, LI-F06, LI-F07, and LI-F08 as well as fusaricidins A–D (Kajimura and Kaneda 1996, 1997).

Fusaricidins have an excellent antifungal activity against plant pathogenic fungi such as *Fusarium oxysporum*, *Aspergillus niger*, *Aspergillus oryzae*, *Penicillium thomii,* and fusaricidin B has particularly antagonistic activity against *Candida albicans* and *Saccharomyces cerevisiae*. Fusaricidins also have an excellent germicidal activity to gram-positive bacteria such as *Staphylococcus aureus* (Kajimura and Kaneda 1996, 1997). *Paenibacillus polymyxa* PKB1 produces fusaridicin peptides with molecular masses of 883, 897, 948, and 960 Da. The characterization of 897 Da component was determined to be cyclic depsipeptide and has antifungal activity against *Leptosphaeria maculans*, which causes black root rot of canola (Beatty and Jensen 2002). The antifungal metabolite produced by *Paenibacillus polymyxa* strain HKA-15 showed strong antagonism against *Rhizoctonia bataticola* causing charcoal rot disease in soybean. Two bioactive fractions collected from the culture filtrate of *Paenibacillus polymyxa* strain HKA-15 by preparative HPLC were characterized as cyclic peptide and depsipeptide (Senthilkumar et al. 2007b). *Paenibacillus lentimorbus* strain WJ5, a soil isolate, produced antifungal metabolite which was extracted with *n*-butanol. The FT-IR spectrum of the antifungal metabolite confirmed the presence of the peptide and glycosidic bonds. (Lee et al. 2008).

6 Endophytic Colonization and Biofilm Formation by *Bacillus* and *Paenibacillus* spp.

Some bacteria and fungi present in the rhizosphere are capable of entering the plant as endophytes that do not cause harm and could establish a mutualistic association. Plants constitute vast and diverse niches for endophytic organisms. Endophytic bacteria have been isolated from a large diversity of plants and most likely, there is not a single plant species devoid of endophytes (Hallman et al. 1997). In general, endophytic bacteria occur at lower population densities than rhizospheric bacteria or bacterial pathogens. Endophytic populations, like rhizospheric populations, are conditioned by biotic and abiotic factors but endophytic bacteria could be better protected from biotic and abiotic stresses than rhizospheric bacteria (Hallman et al. 1997). The population density of endophytes is highly variable, depending mainly on the bacterial species and host genotypes but also on the host developmental stage, inoculum density, and environmental conditions. Endophytic bacteria in a single plant host are not restricted to a single species but comprise several genera and species (Araujo et al. 2002). The presence of different endophytic species in soybean depended on the plant genotype, the plant age, the tissue sampled, and also on the season of isolation. It seems that the bacteria best adapted for living inside plants are naturally selected. Mavingui et al. (1992) found that there are different populations of *Bacillus polymyxa* in rhizosphere soil and rhizoplane, and that wheat roots select specific populations. The analysis by genomic fingerprinting of the diversity of *B. pumilus* isolated from surface-disinfected leaves showed that populations inside citrus do not seem to be clones derived from a single genotype (Araujo et al. 2002).

Similar to rhizosphere bacteria plant growth stimulation mechanisms by endophytic bacteria is also a consequence of nitrogen fixation or by enhancing availability of minerals or the production of phytohormones, biocontrol of phytopathogens in the root zone through production of antifungal or antibacterial agents, siderophore production, nutrient competition, and induction of systematic acquired host resistance (Sessitsch et al. 2004; Rosenblueth and Martínez-Romero 2006). Endophytic N_2-fixing bacteria seem to constitute only a small proportion of total endophytic bacteria and increasing N_2-fixing populations in plants has been considered as a possibility to increase nitrogen fixation. Nitrogen-fixing bacteria were identified in sweet potato in N-poor soils with an analysis that consisted of amplifying nitrogenase (*nifH*) genes by polymerase chain reaction and the resulting sequences, presumably derived from endophytic *Paenibacillus odorifer* (Reiter et al. 2003). The endophytic *B. megaterium* isolated from maize, sweet corn, carrot, and citrus plants is known to solubilize insoluble phosphates (McInroy and Kloepper 1995; Surette et al. 2003). Bacterial endophytes were also isolated from root, nodule, and stem tissues of wild (*Glycine soja*) and cultivated (*Glycine max*) soybean varieties. Many were phytohormone indole acetic acid (IAA) producers and 33% of them secreted extra cellular enzymes cellulase and pectinase (Hung et al. 2007). Bacterial endophytes are capable of suppressing the proliferation of fungal and nematode pathogens, and this may benefit other crops in rotation with the host plants. Nine of the soybean bacterial endophytes belong to *Bacillus* spp., reported to have antifungal activity against major soilborne plant pathogens like *Rhizoctonia bataticola*, *Macrophomina phaseolina*, *Fusarium udam*, and *Sclerotium rolfsii*. These endophytes suppressed the pathogens under in vitro plate assay shown to possess biocontrol traits such as production of hydrogen cyanide (HCN), siderophores, hydrolytic enzymes, and antibiotics. The endophytic *Bacillus* sp. HKA-121 was reported as effective suppressor of charcoal rot disease as well as plant growth promotion in soybean (Senthilkumar et al. 2009).

Fig. 5 Schematic diagram showing protection of plant roots against pathogen infection by bacterial biofilm formation on roots

The elucidation of these mechanisms promoting plant growth by bacterial endophytes will help favor species and conditions that lead to greater plant benefits.

It is now commonly known that bacteria persist in their natural environments by forming biofilms. Biofilms are highly structured, surface-attached communities of cells encased in a self-produced extracellular matrix. Bacteria seems to initiate biofilm formation in response to specific environmental cues, such as nutrient and oxygen availability, biofilms undergo dynamic changes during their transition from free-living to sessile biofilm cells, including the specific production of secondary metabolites and a significant increase in the resistivity to biological, chemical, and physical assaults (Davey and O'Toole 2000). Recent studies have suggested that the biofilm formation is important for the bacteriums' ability to act as a biocontrol agent against plant pathogens (Fig. 5). Bacterial biofilms established on plant roots could protect the colonization sites and act as a sink for the nutrients in the rhizosphere, hence reducing the availability of root exudate nutritional elements for pathogen stimulation or subsequent colonization on the root (Weller and Thomashow 1994). In addition, these biofilm-forming bacterial species can produce a variety of antimicrobial metabolites which include broad spectrum lipopeptides of *Bacillus* and *Paenibacillus*, such as surfactins that are potent biosurfactants and important for maintaining the aerial structure of biofilms (Bais et al. 2004). The presence of surfactin-producing *B. subtilis* 6051 biofilms is expected to prevent the planktonic cells of other microbes colonizing biological surfaces including plant roots. Bais et al. (2004) have reported that the biocontrol of *Pseudomonas syringae* by biofilm-forming *B. subtilis* 6051 is related to surfactin production on the surface of the root. Upon root colonization, *B. subtilis* 6051 forms a stable, extensive biofilm and secretes surfactin, which acts together to protect plants against infection by other pathogenic bacteria.

The biofilm-forming strains of *B. thuringiensis* suppress the quorum-sensing-dependent virulence of the plant pathogen *Erwinia carotovora* through a new form of microbial antagonism called signal interference (Dong and Zhang 2004). *E. carotovora* produces and responds to AHL quorum sensing signals to regulate antibiotic production and the expression of virulence genes, whereas *B. thuringiensis* strains posses AHL-lactonase, which is a potent AHL-degrading enzyme. *B. thuringiensis* does not seem to interfere with the normal growth of *E. carotovora*; however, it abolishes the accumulation of the AHL-signal when they are cocultured (Zhang and Dong 2004). In plants, *B. thuringiensis* significantly decreases the incidence of *E. carotovora* infection and symptom development of potato soft rot caused by the pathogen (Dong and Zhang 2004). In the recent studies, Timmusk et al. (2005) reported that the natural isolates of plant growth promoting rhizobacterium *P. polymyxa* B_1 and B_2 forms biofilms in *Arabidopsis thaliana*. They studied intracellular colonization of these isolates by tagging with plasmid-borne green fluorescent protein (GFP). Fluorescence microscopy and scanning electron microscopy indicated that the bacteria colonized predominantly the root tip, intercellular spaces outside the vascular cylinder where they formed biofilms (Fig. 6). Similarly, the intracellular colonization and biofilm formation in root tips and nodules of soybean by a biocontrol bacterium *P. polymyxa* HKA-15 tagged with GFP was also

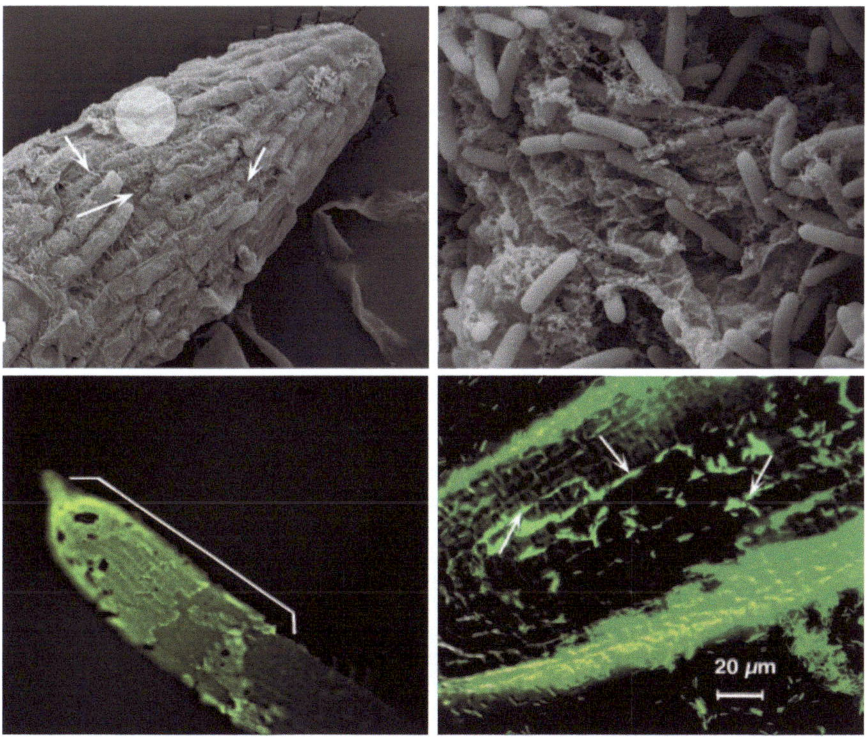

Fig. 6 Scanning electron microphotographs (*top row*) and fluorescence micrographs (*bottom row*) showing the biofilm formation by Gfp tagged isolates of *Paenibacillus polymyxa* B1 on roots (Timmusk et al. 2005)

observed under fluorescence microscopy (data not published). The biofilm-forming *P. polymyxa* on plant roots protected the plant from pathogen infection. Two *P. polymyxa* isolates B_1 and B_2 characterized to form biofilms prevented root colonization and infection by *Aspergillus niger* causing crown rot disease in peanuts (Haggag and Timmusk 2008).

7 Conclusion

The various plant growth promoting properties, together with endospore-forming ability of *Bacillus* and *Paenibacillus*, enable the strain formulations to resist a wide range of environmental stresses. Strains of *Bacillus subtilis* and *Paenibacillus polymyxa* are well-established model organisms for research on molecular plant–microbe interactions. The complete genome of *Bacillus subtilis* has been published and genome sequencing of root colonizing bacterium *Paenibacillus polymyxa* is underway. This genomic information will be available to investigate molecular responses of *Bacillus* and *Paenibacillus* in soil and the crop rhizosphere in particular. Further, biotechnology can be applied to create transgenic strains with multiple mechanisms of action and strains with specific formulation qualities (stability of inoculants and better root colonization). Continued research with endophytic colonization and biofilm formation by these bacterial genera also holds potential for developing biofertilizer and biocontrol agents that may be self-perpetuating within the colonizing host plants. Focusing research in these areas will make *Bacillus* and *Paenibacillus* spp. as promising/potential PGPR. Their applications will significantly reduce the use chemical fertilizers and pesticides which will be essential for achieving sustainable crop yield in agriculture.

References

Alström S (1991) Induction of disease resistance in common bean susceptible to halo blight bacterial pathogen after seed bacterisation with rhizosphere pseudomonads. J Gen Appl Microbiol 37:495–501

Araujo WL, Marcon J, Maccheroni W Jr, Van Elsas JD, Van Vuurde JWL, Azevedo JL (2002) Diversity of endophytic bacterial populations and their interaction with *Xylella fastidiosa* in citrus plants. Appl Environ Microbiol 68:4906–4914

Araujo FF, Henning AA, Hungria M (2005) Phytohormones and antibiotics produced by *Bacillus subtilis* and their effects on seed pathogenic fungi and on soybean root development. World J Microbiol Biotechnol 21:1639–1645

Arshad M, Frankenberger WT Jr (1998) Plant growth regulating substances in the rhizosphere: microbial production and functions. Adv Agron 62:46–151

Asaka O, Shoda M (1996) Biocontrol of *Rhizoctonia solani* damping-off of tomato with *Bacillus subtilis* RB14. Appl Environ Microbiol 62:4081–4085

Ash C, Farrow JAE, Wallbanks S, Collins MD (1991) Phylogenetic heterogeneity of the genus *Bacillus* revealed by comparative analysis of small subunit – ribosomal RNA sequences. Lett Appl Microbiol 13:202–206

Ash C, Priest FG, Collins MD (1993) Molecular identification of rRNA group 3 bacilli (Ash, Farrow, Wallbanks and Collins) using a PCR probe test: proposal for the creation of a new genus *Paenibacillus*. Antonie Leeuwenhoek 64:253–260

Bais HP, Fall R, Vivanco JM (2004) Biocontrol of *Bacillus subtilis* against infection of arbidopsis roots by *Pseudomonas syringae* is facilitated by biofilm formation and surfactin production. Plant Physiol 134:307–319

Bargabus RL, Zidack NK, Sherwood JW, Jacobsen BJ (2002) Characterization of systemic resistance in sugar beet elicited by a non-pathogenic, phyllosphere-colonizing *Bacillus mycoides*, biological control agent. Physiol Mol Plant Pathol 61:289–298

Bargabus RL, Zidack NK, Sherwood JW, Jacobsen BJ (2004) Screening for the identification of potential biological control agents that induce systemic acquired resistance in sugar beet. Biol Control 30:342–350

Basha S, Ulaganathan K (2002) Antagonism of *Bacillus* species (strain BC121) towards *Curvularia lunata*. Curr Sci 82:1457–1463

Beatty PH, Jensen SE (2002) *Paenibacillus polymyxa* produces fusaricidin-type antifungal antibiotics active against *Leptosphaeria maculans*, the causative agent of blackleg disease of canola. Can J Microbiol 48:159–169

Berge OMH, Guinebretièr W, Achouak P, Normand T, Heulin P (2002) *Paenibacillus graminis* sp. nov. and *Paenibacillus odorifer* sp. nov., isolated from plant roots, soil and food. Int J Syst Evol Microbiol 52:607–616

Bergey DH, Breed RS, Murray EGD, Hitchens AP (1939) Bergey's manual of determinative bacteriology, 5th edn. Williams and Wilkins, Baltimore

Bizani D, Dominguez APM, Brandelli A (2005) Purification and partial chemical characterization of the antimicrobial peptide cerein 8A. Lett Appl Microbiol 41:269–273

Bravo A, Sarabia S, Lopez L, Ontiveros H, Abarca C, Ortiz A, Ortiz M, Lina L, Villalobos FJ, Pena G, Nunez-Valdez ME, Soberon M, Quintero R (1998) Characterization of *cry* genes in a Mexican *Bacillus thuringiensis* strain collection. Appl Environ Microbiol 64:4965–4972

Buchanan RE, Gibbons NE (1974) Bergey's manual of determinative bacteriology, 8th edn. Williams and Wilkins, Baltimore

Bünemann EK, Bossio DA, Smithson PC, Frossard E, Oberson A (2004) Microbial community composition and substrate use in a highly weathered soil as affected by crop rotation and P fertilization. Soil Biol Biochem 36:889–901

Cabello M, Irrazabal G, Bucsinszky AM, Saparrat M, Schalamuck S (2005) Effect of an arbuscular mycorrhizal fungus, *G. mosseae* and a rock-phosphate-solubilizing fungus, *P. thomii* in *Mentha piperita* growth in a soilless medium. J Basic Microbiol 45:182–189

Cakmakci R, Erat M, Erdogan U, Domez MF (2007) The influence of plant growth–promoting rhizobacteria on growth and enzyme activities in wheat and spinach plants. J Plant Nutr Soil Sci 170:288–295

Chak K, Chao D, Tseng M, Kao S, Tuan S, Feng T (1994) Determination and distribution of *cry*-type genes of *Bacillus thuringiensis* isolates from Taiwan. Appl Environ Microbiol 60:2415–2420

Chen YP, Rekha PD, Arunshen AB, Lai WA, Young CC (2006) Phosphate solubilizing bacteria from subtropical soil and their tricalcium phosphate solubilizing abilities. Appl Soil Ecol 34:33–41

Choi SK, Park SY, Kim R, Lee CH, Kim JF, Park SH (2007) Identification and functional analysis of the fusaricidin biosynthetic gene of *Paenibacillus polymyxa* E681. Biochem Biophys Res Commun 365:89–95

Claus D, Berkeley RCW (1986) Bergey's manual of systematic bacteriology. Williams and Wilkins, Baltimore, pp 1105–1139

Daniel R (2005) The metagenomics of soil. Nat Rev Microbiol 3:470–478

Danielsson J, Reva O, Meijer J (2006) Protection of oilseed rape (*Brassica napus*) toward fungal pathogens by strains of plant-associated *Bacillus amyloliquefaciens*. Microb Ecol 54:134–140

Davey ME, O'Toole AG (2000) Microbial biofilms: from ecology to molecular genetics. Microbiol Mol Biol Rev 64:847–867
Deubel A, Gransee A, Merbach W (2000) Transformation of organic rhizodeposits by rhizoplane bacteria and its influence on the availability of tertiary calcium phosphate. J Plant Nutr Soil Sci 163:387–392
Dijksterhuis J, Sanders M, Gorris LGM, Smid EJ (1999) Antibiosis plays a role in the context of direct interaction during antagonism of *Paenibacillus polymyxa* towards *Fusarium oxysporum*. J Appl Microbiol 86:13–21
Ding Y, Wang J, Liu Y, Chen S (2005) Isolation and identification of nitrogen-fixing bacilli from plant rhizospheres in Beijing region. J Appl Microbiol 99:1271–1281
Dodor DE, Tabatabai AM (2003) Effect of cropping systems on phosphatases in soils. J Plant Nutr Soil Sci 166:7–13
Dong YH, Zhang LH (2004) Insecticidal *Bacillus thuringiensis* silences *Erwinia caratovora* virulence by a new form of microbial antagonism, signal interference. Appl Environ Microbiol 70:954–960
Ezawa T, Smith SE, Smith FA (2002) P metabolism and transport in AM fungi. Plant Soil 244:221–230
Emtiazi G, Pooyan M, Shamalnasab M (2007) Cellulase activities in Nitrogen Fixing *Paenibacillus* isolated from soil in N-free media. World J Agril Sci 3:602–608
Felske A, Wolterink A, Van Lis R, de Vos WM, Akkermans ADL (1999) Searching for predominant soil bacteria: 16S rDNA cloning versus strain cultivation. FEMS Microbiol Ecol 30:137–145
Fukuda H, Takahashi M, Fujii T, Ogawa T (1989) Ethylene production from L-methionine by *Cryptococcus albidus*. J Ferment Bioeng 67:173–175
Garbeva P, van Veen JA, van Elsas JD (2003) Predominant *Bacillus* spp. in agricultural soil under different management regimes detected via PCR-DGGE. Microbiol Ecol 45:302–316
Glick BR (1995) The enhancement of plant growth promotion by free living bacteria. Can J Microbiol 41:109–117
Glick BR, Patten CL, Holguin G, Penrose DM (1999) Biochemical and genetic mechanisms used by plant growth promoting bacteria. Imperial College Press, London
Goenadi DH, Siswanto A, Sugiarto Y (2000) Bioactivation of poorly soluble phosphate rocks with a phosphorus-solubilizing fungus. Soil Sci Soc Am J 64:927–932
Goldstein AH (1995) Recent progress in understanding the molecular genetics and biochemistry of calcium phosphate solubilization by Gram-negative bacteria. Biol Agric Horttic 12:185–193
Govindasamy V, Senthilkumar M, UpendraKumar AK (2008) PGPR-biotechnology for management of abiotic and biotic stresses in crop plants. In: Maheshwari DK, Dubey RC (eds) Potential microorganisms for sustainable agriculture. IK International Publishing, New Delhi, pp 26–48
Gray EJ, Smith DL (2005) Intracellular and extracellular PGPR: Commonalities and distinctions in the plant bacterium signaling process. Soil Biol Biochem 37:395–412
Guemouri-Athmani S, Berge O, Bourrain M, Mavingui P, Thiy JM, Bhatnagar T, Heulin T (2000) Diversity of *Paenibacillus polymyxa* in the rhizosphere of wheat (*Triticum durum*) in Algerian soils. Eur J Soil Biol 36:149–159
Gong XY, Luan ZK, Pei YS, Wang SG (2003) Culture conditions for floculant production by *Paenibacillus polymyxa* BY-28. J Environ Sci Health Part A 38:657–669
Gutierrez-Manero FG, Ramos-Solano B, Probanza A, Mehouachi J, Tadeo FR, Talon M (2001) The plant-growth-promoting rhizobacteria bacillus pumilus and bacillus licheniformis produce high amounts of physiologically active gibberellins. Physiol Plant 111:206–211
Haggag WM, Timmusk S (2008) Colonization of peanut roots by biofilm-forming *Paenibacillus polymyxa* initiates biocontrol against crown rot disease. J Appl Microbiol 104:961–969
Hallman J, Quadt-Hallman A, Mahafee WF, Kloepper JW (1997) Bacterial endophytes in agricultural crops. Can J Microbiol 43:895–914

Handelsman J, Raffel S, Mester E, Wunderlich L, Grau C (1990) Biological control of damping-off of alfalfa seedlings with *Bacillus cereus* UW85. Appl Environ Microbiol 56:713–718

Handelsman J, Rondon MR, Brady SF, Clardy J, Goodman RM (1998) Molecular biological access to the chemistry of unknown soil microbes: a new frontier for natural products. Chem Biol 5:245–249

Hashizume T, Nakamura K, Nakagawa S (1996) Affinities of BO- 2727 for bacterial penicillin-binding proteins and morphological change of gram negative rods. J Antibiot 50:139–142

He Z, Kisla D, Zhang L, Yuan C, Green-Church KB, Yousef AE (2007) Isolation and identification of a *Paenibacillus polymyxa* strain that coproduces a novel lantibiotic and polymyxin. Appl Environ Microbiol 73:168–178

Heulin T, Berge O, Mavingui P, Gouzou L, Hebbar KP, Balandreau J (1994) *Bacillus polymyxa* and *Rahnella aquatilis*, the dominant N_2-fixing bacteria associated with wheat rhizosphere in French soils. Eur J Soil Biol 30:35–42

Hilda R, Fraga R (2000) Phosphate solubilizing bacteria and their role in plant growth promotion. Biotechnol Adv 17:319–359

Hiltner L (1904) U¨ ber neuere Erfahrungen und Probleme auf dem Gebiete der Bodenbakteriologie unterbessonderer Berucksichtigung der Grundung und Brache. Arb Dtsch Landwirtsch Ges Berl 98:59–78

Holl FB, Chanway CP (1992) Rhizosphere colonization and seedling growth promotion of lodgepole pine by *Bacillus polymyxa*. Can J Microbiol 38:303–308

Holl FB, Chanway CP, Turkington R, Radley RA (1988) Response of crested wheatgrass (*Agropyron cristatum* L.), perennial ryegrass (*Lolium perenne* L.) and white clover (*Trifolium repens* L.) to inoculation with *Bacillus polymyxa*. Soil Biol Biochem 20:19–24

Hu X, Boyer GL (1996) Siderophore-mediated aluminum uptake by *Bacillus megaterium* ATCC 19213. Appl Environ Microbiol 62:4044–4048

Hughes DF, Jolley VD, Brown JC (1992) Role of potassium in iron-stress response mechanisms of strategy I and strategy II plants. J Plant Nutr 15:1821–1839

Hung PQ, Senthilkumar M, Govindasamy V, Annapurna K (2007) Isolation and characterization of endophytic bacteria from wild and cultivated soybean varieties. Biol Fertil Soils 44:155–162

Hyun JW, Kim YH, Lee YS, Park WM (1999) Isolation and evaluation of protective effect against *Fusarium* wilt of sesame plants of antibiotic substance from *Bacillus polymyxa* KB-8. Plant Pathol J 15:152–157

Idriss EE, Makarewicz O, Farouk A, Rosner K, Greiner R, Bochow H, Richter T, Borriss R (2002) Extracellular phytase activity of *Bacillus amyloliquifaciens* FZB45 contributes to its plant-growth-promoting effect. Microbiology 148:2097–2109

Igual JM, Valverde A, Cervantes E, Velázquez E (2001) Phosphate-solubilizing bacteria as inoculants for agriculture: use of updated molecular techniques in their study. Agronomie 21:561–568

Jackson PJ, Hill KK, Laker MT, Ticknor LO, Keim P (1999) Genetic comparison of *B. anthracis* and its close relatives using RFLP and PCR analysis. J Appl Microbiol 87:263–269

Jung HK, Hong JH, Park SC, Park BK, Nam DH, Kim SD (2007) Production and physico-chemical characterization of β-glucan produced by *Paenibacillus polymyxa* JB115. Biotechnol Bioprocess Eng 12:713–719

Kajimura Y, Kaneda M (1996) Fusaricidin A, a new depsipeptide antibiotic produced by *Bacillus polymyxa* KT-8: taxonomy, fermentation, isolation, structure elucidation, and biological activity. J Antibiot 49:129–135

Kajimura Y, Kaneda M (1997) Fusaricidins B, C and D, new depsipeptide antibiotics produced by *Bacillus polymyxa* KT-8: isolation, structure elucidation and biological activity. J Antibiot 50:220–228

Kajimura Y, Sugiyama M, Kaneda M (1995) Bacillopeptins, new cyclic lipopeptide antibiotics from *Bacillus subtilis* FR-2. J Antibiot 48:1095–1103

Karpunina LV, Mel'nikova UY, Konnova SA (2003) Biological role of lectins from the nitrogen-fixing *Paenibacillus polymyxa* strain 1460 during bacterial-plantroot interactions. Curr Microbiol 47:376–378

Khan KS, Joergensen RG (2009) Changes in microbial biomass and P fractions in biogenic household waste compost amended with inorganic P fertilizers. Bioresour Technol 100: 303–309

Kim HS, Park J, Choi SW, Choi KH, Lee GP, Ban SJ, Lee CH, Kim CS (2003) Isolation and characterization of *Bacillus* strains for biological control. J Microbiol 41:196–201

Kloepper JW, Schroth MN (1978) Plant growth promoting rhizobacteria on radishes. In: Proceedings of VIth international conference on plant pathogenic bacteria, Angres, France, vol 2, pp 879–882

Kloepper JW, Ryu CM, Zhang S (2004) Induced systemic resistance and promotion of plant growth by *Bacillus* spp. Phytopathology 94:1259–1266

Kucey RMN, Janzen HH, Legget ME (1989) Microbial mediated increases in plant available phosphorus. Adv Agron 42:199–228

Kunst F, Ogasawara N, Moszer I, Albertini AM, Alloni GV, Azevedo M, Bertero G, Bessières P, Bolotin A, Borchert S (1997) The complete genome sequence of the Gram-positive bacterium Bacillus subtilis. Nature 390:249–256

Lacey LA, Frutos R, Kaya HK, Vail P (2001) Insect pathogens as biological control agents: Do they have a future? Biol Control 21:230–248

Lee H, Churey JJ, Worobo RW (2008) Purification and structural characterization of bacillomycin F produced by a bacterial honey isolate active against *Byssochlamys fulva* H25. J Appl Microbiol 105:663–673

Leifert C, Li H, Chidburee S, Hampson S, Workman S, Sigee D, Epton HAS, Harbour A (1995) Antibiotic production and biocontrol activity by *Bacillus subtilis* CL27 and *Bacillus pumilus* CL45. J Appl Bacteriol 78:97–108

Leong J (1986) Siderophores: their biochemistry, and possible role in the biocontrol of plant pathogens. Ann Rev Phytopathol 24:187–209

Lebuhn M, Heulin T, Hartmann A (1997) Production of auxin and other indolic and phenolic compounds by *Paenibacillus polymyxa* strains isolated from different proximity to plant roots. FEMS Microbiol Ecol 22: 325–334

Lindberg T, Granhall U, Tomenius K (1985) Infectivity and acetylene reduction of diazotrophic rhizosphere bacteria in wheat (*Triticum aestivum*) seedlings under gnotobiotic conditions. Biol Fertil Soils 1:123–129

Liu ZL, Sinclair JB (1992) Population dynamics of *Bacillus megaterium* strain B153-2-2 in the rhizosphere of soybean. Phytopathology 82:1297–1301

Lorenz P, Eck J (2005) Metagenomics and industrial applications. Nat Rev Microbiol 3:510–516

Mahaffee WF, Kloepper JW (1997) Temporal changes in the bacterial communities of soil, rhizosphere, and endorhiza associated with field grown cucumber (*Cucumis sativus* L.). Microbiol Ecol 34:210–223

Martin NL, Hu H, Moake M, Churey JJ, Whittal R, Worobo RW, Vederas JC (2003) Isolation, structural characterization, and properties of mattacin (Polymyxin M), a cyclic peptide antibiotic produced by *Paenibacillus kobensis*. J Bio Chem 278:13124–13132

Mavingui P, Heulin T (1994) In vitro chitinase and antifungal activity of a soil, rhizosphere and rhizoplane population of *Bacillus polymyxa*. Soil Biol Biochem 26:801–803

Mavingui P, Laguerre G, Berge O, Heulin T (1992) Genetic and phenotypic diversity of *Bacillus polymyxa* in soil and in the wheat rhizosphere. Appl Environ Microbiol 58:1894–1903

May R, Volksh B, Kampmann G (1996) Antagonistic activities of epiphytic bacteria from soybean leaves against *Pseudomonas syringae* pv. glycinea in vitro and in planta. Microbiol Ecol 34:118–124

McInroy JA, Kloepper JW (1995) Survey of indigenous bacterial endophytes from cotton and sweet corn. Plant Soil 173:337–342

Mhammedai J, Peypoux F, Besson F, Michel G (1982) Bacillomycin F, a new antibiotic of iturin group: isolation and characterization. J Antibiot 35:306–311

Mok MC (1994) Cytokinins and plant development – an overview. In: Mok DWS, Mok MC (eds) Cytokinins: chemistry, activity and function. CRC, New York, pp 115–166

Monteiro L, Mariano R, de Lima R, Souto-Maior AM (2005) Antagonism of *Bacillus* spp. against *Xanthomonas campestris* pv. *campestris*. Braz Arch Biol Technol 48:23–29
Nakao M, Kondo M, Tsuchiya K (1981) Light and Electron Microscopy of the morphological response of *Eshcherichia coli* and *Serratia marescens* to Cefmenoxime (SCE-1365): a new broad spectrum Cephalosprorin. J Antibiot 34:1046–1054
Neilands JB (1986) A saga of siderophores. In: Swinburne TR (ed) Siderophores and plant diseases. Plenum, New York, pp 289–298
Nicholson WL (2002) Roles of *Bacillus* endospores in the environment. Cell Mol Life Sci 59:410–416
Nielsen P, Sorensen J (1997) Multi-target and medium independent fungal antagonism by hydrolytic enzymes in *Paenibacillus polymyxa* and *Bacillus pumilus* strains from barley rhizosphere. FEMS Microbiol Ecol 22:183–192
Patten CL, Glick BR (1996) Bacterial biosynthesis of indole-3-acetic acid. Can J Microbiol 42:207–220
Paulitz TC, Belanger RR (2001) Biological control in greenhouse systems. Annu Rev Phytopathol 39:103–133
Pichard B, Larue JP, Thouvenot D (1995) Gavesrin and saltavalin, new peptide antibiotics produced by *Bacillus polymyxa*. FEMS Microbiol Lett 133:215–218
Pinchuk IV, Bressollier P, Sorokulova IB, Verneuil B, Urdaci MC (2002) Amicoumacin antibiotic production and genetic diversity of *Bacillus subtilis* strains isolated from different habitats. Res Microbiol 153:269–276
Piuri M, Sanchez-Rivas C, Ruzal SM (1998) A novel antimicrobial activity of a *Paenibacillus polymyxa* strain isolated from regional fermented sausages. Lett Appl Microbiol 27:9–13
Pradhan N, Sukla LB (2005) Solubilization of inorganic phosphate by fungi isolated from agriculture soil. Afr J Biotechnol 5:850–854
Priest F (1993) Systematics and ecology of *Bacillus*: *Bacillus subtilis* and other gram-positive bacteria, biochemistry, physiology, and molecular genetics. American Society for Microbiology, Washington DC, pp 3–16
Pueyo MT, Bloch CJ, Carmona RAM, Masico P (2009) Lipopeptides produced by a soil *Bacillus megaterium* strain. Microbiol Ecol 57:367–378
Raffel SJ, Stabb EV, Milner JL, Handelsman J (1996) Genotypic and phenotypic analysis of zwittermicin A-producing strains of *Bacillus cereus*. Microbiol 142:3425–3436
Reiter B, Bürgmann H, Burg K, Sessitsch A (2003) Endophytic *nifH* gene diversity in African sweet potato. Can J Microbiol 49:549–555
Rosenblueth M, Martínez-Romero E (2006) Bacterial endophytes and their interactions with hosts. Mol Plant Microbe Interact 19:827–837
Ryu CM, Farag MA, Hu CH, Reddy MS, Wie HX (2003) Bacterial volatiles promote growth of *Arabidopsis*. Proc Natl Acad Sci USA 100:4927–4932
Ryu CM, Farag MA, Hu CH, Reddy MS, Kloepper JW, Pare PW (2004) Bacterial volatiles induce systemic resistance in Arabidopsis. Plant Physiol 134:1017–1026
Ryu CM, Kima J, Choi O, Kima SH, Park CS (2006) Improvement of biological control capacity of *Paenibacillus polymyxa* E681 by seed pelleting on sesame. Biol Control 39:282–289
Safdi N, Cherif M, Hajlaui MR, Boudabbous A, Belanger R (2002) Isolation and partial purification of antifungal metabolites produced by *Bacillus cereus*. Ann Microbiol 52:323–337
Salerno CM, Sagardoy MA (2003) Antagonistic activity by Bacillus subtilis against Xanthomonas campestris pv. glycines under controlled conditions. Span J Agri Res 1:55–58
Seldin L, Penido EGC (1986) Identification of *Paenibacillus azotofixans* using API tests. Antonie Leeuwenhoek 52:403–409
Seldin L, van Elsas JD, Penido EGC (1984) *Buciffusazotofixans* sp. nov., a nitrogen-fixing species from Brazilian soils and grass roots. Int J Syst Bacteriol 34:451–456
Seldin L, Soares Rosado A, da Cruz DW, Nobrega A, van Elsas JD, Paiva E (1998) Comparison of *Paenibacillus azotofixans* strains isolated from rhizoplane, and non-root-associated

soil from maize planted in two different Brazilian soils. Appl Environ Microbiol 64: 3860–3868

Seldin L, de Azevedo FS, Alviano DS, Alviano CS, de Freire Bastos MC (1999) Inhibitory activity of *Paenibacillus polymyxa* SCE2 against human pathogenic micro-organisms. Lett Appl Microbiol 28:423–427

Selim S, Negrel J, Govaerts C, Gianinazzi S, Tuinen DV (2005) Isolation and partial characterization of antagonistic peptides produced by *Paenibacillus* sp. Strain B2 isolated from the sorghum mycorhizosphere. Appl Environ Microbiol 71:6501–6507

Senthilkumar M, Govindasamy V, Annapurna K (2007a) Role of antibiosis in suppression of charcoal rot disease by soybean endophyte *Paenibacillus* sp. HKA-15. Curr Microbiol 55:25–29

Senthilkumar M, Govindasamy V, Dureja P, Annapurna K (2007b) Purification and partial characterization of antifungal peptides from soybean endophyte-*Paenibacillus* sp strain HKA-15. J Plant Biochem Biotechnol 16:131–134

Senthilkumar M, Swarnalakshmi K, Govindasamy V, Lee YK, Annapurna K (2009) Biocontrol potential of soybean bacterial endophytes against charcoal rot fungus, *Rhizoctonia bataticola*. Curr Microbiol 58:288–293

Sessitsch A, Reiter B, Berg G (2004) Endophytic bacterial communities of field-grown potato plants and their plant-growth-promoting and antagonistic abilities. Can J Microbiol 50:239–249

Sharma K, Dak G, Agrawal A, Bhatnagar M, Sharma R (2007) Effect of phosphate solubilizing bacteria on the germination of *Cicer arietinum* seeds and seedling growth. J Herb Med Toxicol 1:61–63

Siddiqui ZA, Mahmood I (1999) Role of bacteria in the management of plant parasitic nematodes: a review. Bioresour Technol 69:167–179

Siddiqui ZA, Baghel G, Akhtar MS (2007) Biocontrol of *Meloidogyne javanica* by *Rhizobium* and plant growth-promoting rhizobacteria on lentil. World J Microbiol Biotechnol 23:435–441

Skerman VBD, McGowan V, Sneath PHA (1980) Approved lists of bacterial names. American Society for Microbiology, Washington DC

Smalla K, Wieland G, Buchner A, Zock A, Parzy J, Kaiser S, Roskot N, Heuer H, Berg G (2001) Bulk and rhizosphere soil bacterial communities studied by denaturing gradient gel electrophoresis: plant dependent enrichment and seasonal shifts revealed. Appl Environ Microbiol 67:4742–4751

Son TTN, Diep CN, Giang TTM (2006) Effect of bradyrhizobia and phosphate solubilizing bacteria application on Soybean in rotational system in the Mekong delta. Omonrice 14:48–57

Stabb EV, Jacobson LM, Handelsman J (1994) Zwittermicin A producing strains of *Bacillus cereus* from diverse soils. Appl Environ Microbiol 60:4404–4412

Subbarao NS (1988) Phosphate solubilizing micro-organism. In: Biofertilizer in agriculture and forestry. Regional Biofertilizer Development Centre, Hissar, India, pp 133–142

Sundara B, Natarajan V, Hari K (2002) Influence of phosphorus solubilizing bacteria on the changes in soil available phosphorus and sugarcane yields. Field Crops Res 77:43–49

Surette MA, Sturz AV, Lada RR, Nowak J (2003) Bacterial endophytes in processing carrots (*Daucus carota* L. var. sativus): their localization, population density, biodiversity and their effects on plant growth. Plant Soil 253:381–390

Tamehiro N, Okamoto-Hosoya Y, Okamoto S, Ubukata M, Hamada M, Naganawa H, Ochi K (2002) Bacilysocin, a Novel Phospholipid antibiotic produced by *Bacillus subtilis* 168. J Antibiot 46:315–320

Tendulkar SR, Saikumar YK, Patel V, Raghotama Munshi TK, Balaram P, Chatoo BB (2007) Isolation, purification and characterization of an antifungal molecule produced by *Bacillus licheniformis* BC98, and its effect on phytopathogen *Magnoporthe grisea*. J Appl Microbiol 103:2331–2339

Thomashow LS, Weller DM (1995) Current concepts in the use of introduced bacteria for biological control: mechanisms and antifungal metabolites. In: Stacey G, Keen N (eds) Plant-Microbe Interactions, vol 1. Chapman and Hall, New York, pp 187–235

Timmusk S, Wagner EGH (1999) The plant-growth-promoting rhizobacterium *Paenibacillus polymyxa* induces changes in *Arabidopsis thaliana* gene expression: a possible connection between biotic and abiotic stress responses. Mol Plant Microbe Interact 12:951–959

Timmusk S, Nicander B, Granhall U, Tillberg E (1999) Cytokinin production by *Paenibacillus polymyxa*. Soil Biol Biochem 31:1847–1852

Timmusk S, van West P, Gow NAR, Wagner EGH (2003) Antagonistic effects of *Paenibacillus polymyxa* towards the oomycete plant pathogens Phytophthora palmivora and Pythium aphanidermatum. In: Mechanism of action of the plant growth promoting bacterium *Paenibacillus polymyxa*. Uppsala University, Uppsala Sweden, pp 1–28

Timmusk S, Grantcharova N, Wagner EGH (2005) *Paenibacillus polymyxa* invades plant roots and forms biofilms. Appl Environ Microbiol 71:7292–7300

Toro M, Azcón R, Barea JM (1997) Improvement of arbuscular mycorrhizal development by inoculation with phosphate-solubilizing rhizobacteria to improve rock phosphate bioavailability (32P) and nutrient cycling. Appl Environ Microbiol 63:4408–4412

Uknes S, Mauch-Mani B, Moyer M, Potter S, Williams S, Dincher S, Chandler D, Slusarenko A, Ward E, Ryals J (1992) Acquired resistance in *Arabidopsis*. Plant Cell 4:645–656

Van Loon LC (2007) Plant responses to plant growth-promoting bacteria. Eur J Plant Pathol 119:243–254

Van Loon LC, Bakker PAHM, Pieterse CMJ (1998) Systemic resistance induced by rhizosphere bacteria. Annu Rev Phytopathol 36:453–483

van Wees S, Pieterse C, Trijssenaar A, Van't Westende Y, Hartog F, van Loon LC (1997) Differential induction of systemic resistance in *Arabidopsis* by biocontrol bacteria. Mol Plant Microbe Interact 10:716–724

Vessey JK (2003) Plant growth promoting rhizobacteria as biofertilizers. Plant Soil 255:571–586

von der Weid IA, Paiva E, Norega A, van Elsas JD, Seldin L (2000) Diversity of *Paenibacillus polymyxa* strains isolated from the rhizosphere of maize planted in Cerrado soil. Res Microbiol 151:369–381

von der Weid IA, Duarte GF, van Elsas JD (2002) *Paenibacillus brasilensis* sp. nov., a novel nitrogen-fixing species isolated from the maize rhizosphere in Brazil. Int J Syst Evol Microbiol 52:2147–2153

Wang Y, Brown HN, Crowley DE, Szaniszlo PJ (1993) Evidence for direct utilization of a siderophore ferrioxaminae B in axenically grown cucumber. Plant Cell Environ 16: 579–585

Wei G, Kloepper JW, Tuzun S (1991) Induction of systemic resistance of cucumber to *Colletotrichum orbiculare* by selected strains of plant growth-promoting rhizobacteria. Phytopathology 81:1508–1512

Wei JZ, Hale K, Carta L, Platzer E, Wong C, Fang SC, Aroian RV (2003) *Bacillus thuringiensis* crystal proteins that target nematodes. Proc Nat Acad Sci USA 100:2760–2765

Weid VD, Alviano DS, Santos ALS, Soares RMA, Alviano CX, Seldin L (2003) Antimicrobial activity of *Paenibacillus peoriae* strain NRRL BD-62 against a broad spectrum of phytopathogenic bacteria and fungi. J Appl Microbiol 95:1143–1151

Weller DM, Thomashow LS (1994) Current challenges in introducing beneficial microorganisms into the rhizosphere. In: O'Gara F, Dowling DN, Boesten B (eds) Molecular ecology of rhizosphere microorganisms. VCH Weinheim, Germany, pp 1–18

Wilson Melissa K, Abergel Rebecca J, Raymond Kenneth N, Arceneaux Jean EL, Byers BR (2006) Siderophores of *Bacillus anthracis*, *Bacillus cereus*, and *Bacillus thuringiensis*. Biochem Biophys Res Commun 348:320–325

Wipat A, Harwood CR (1999) The *Bacillus subtilis* genome sequence: the molecular blueprint of a soil bacterium. FEMS Microbiol Ecol 28:1–9

Xie GH, Su BL, Cui ZJ (1998) Isolation and identification of N^2-fixing strains of Bacillus in rice rhizosphere of the Yangtze River valley. Acta Microbiol Sin 38:480–483

Yazdani M, Bahmanyar MA, Pirdashti H, Esmaili MA (2009) Effect of Phosphate solubilization microorganisms (PSM) and plant growth promoting rhizobacteria (PGPR) on yield and yield components of Corn (*Zea mays L.)*. Proc World Acad Science. Eng Technol 37:90–92

Yoshida S, Hiradate S, Tsukamot T, Hatakeda K, Shirata A (2001) Antimicrobial activity of culture filtrate of *Bacillus amyloliquefaciens* RC-2 isolated from mulberry leaves. Biol Control 91:181–187

Zehnder G, Kloepper J, Tuzun S, Yao C, Wei G, Chambliss O, Shelby R (1997) Insect feeding on cucumber mediated by rhizobacteria induced plant resistance. Entomol Exp Appl 83:81–85

Zehnder GW, Yao C, Murphy JF, Sikora EJ, Kloepper JW (2000) Induction of resistance in tomato against cucumber mosaic cucumovirus by plant growth-promoting rhizobacteria. Biol Control 45:127–137

Zhang LH, Dong YH (2004) Quorum sensing and signal interference: diverse implications. Mol Microbiol 53:1563–1571

Zhang S, Moyne AL, Reddy MS, Kloepper JW (2002) The role of salicylic acid in induced systemic resistance elicited by plant growth-promoting rhizobacteria against blue mold of tobacco. Biol Control 25:288–296

Zhang H, Xie X, Kim MS, Kornyeyev DA, Holaday S, Paré PW (2008) Soil bacteria augment Arabidopsis photosynthesis by decreasing glucose sensing and abscisic acid levels in planta. Plant J 56:264–273

Zhou WW, Huang JX, Niu TG (2008) Isolation of an antifungal *Paenibacillus* strain HT 16 from locusts and purification of its medium-dependent antagonistic compound. J Appl Microbiol 105:912–919

The Role of ACC Deaminase Producing PGPR in Sustainable Agriculture

Meenu Saraf, Chaitanya Kumar Jha, and Dhara Patel

Contents

1 Introduction ... 366
2 Ethylene Biosynthesis in Higher Plants ... 368
3 Characteristics of ACC Deaminase Enzyme ... 369
4 Crystal Structure of 1-Aminocyclopropane-1-Carboxylate Deaminase ... 370
5 Mechanism of ACC Deaminase Action ... 370
6 Role of Bacterial ACC Deaminase Under Stress Agricultural Conditions ... 372
6.1 Pathogenicity Stress ... 372
6.2 Remediation of High/Heavy Metal Concentration ... 374
6.3 Drought Stress ... 374
6.4 Organic Contaminants Stress ... 375
6.5 Waterlogging Stress ... 375
6.6 Temperature Stress ... 375
6.7 Flower Senescence ... 376
6.8 Salinity Stress ... 376
6.9 Ethylene–IAA Cross-talk ... 377
6.10 Air Pollutants Stress ... 378
6.11 Rhizobial Infection ... 378
7 Microbe–Microbe Interactions Benefiting Sustainable Agro-Ecosystem Development ... 378
8 ACC Deaminase Gene-Containing Transgenic Plants ... 379
9 Conclusions and Future Trends ... 379
References ... 380

Abstract The plant rhizosphere is a multidimensional and dynamic ecological environment of complicated microbe–plant interactions for harnessing essential macro and micronutrients from a limited nutrient pool. Certain plant growth

M. Saraf (✉), C.K. Jha, and D. Patel
Department of Microbiology, School of Sciences, Gujarat University, Ahmedabad 380 009, Gujarat, India
e-mail: sarafmeenu@gmail.com

D.K. Maheshwari (ed.), *Plant Growth and Health Promoting Bacteria*, Microbiology Monographs 18, DOI 10.1007/978-3-642-13612-2_16,

promoting rhizobacteria (PGPR) contain a vital enzyme, 1-aminocyclopropane-1-carboxylic acid (ACC) deaminase (EC 4.1.99.4), which regulates ethylene production by metabolizing ACC (an intermediate precursor of ethylene biosynthesis in higher plants) into α-ketobutyrate and ammonia. The microbial enzyme 1-aminocyclopropane-1-carboxylate deaminase cleaves ACC irreversibly, this being the immediate precursor of ethylene in plants. ACC deaminase-expressing PGPR protect plants against the growth inhibition that might otherwise result following flooding, extremes of temperature, the presence of organic and inorganic toxicants, phytopathogens, drought or high salt concentrations. Organisms containing ACC deaminase genes have been reported to be useful in promotion of early root development from either seeds or cuttings, increasing the life of horticultural flowers, protecting plants against a wide range of environmental stresses, facilitating the production of volatile organic compounds responsible for aroma formation and phytoremediation of contaminated soils.

1 Introduction

Certain plant growth promoting rhizobacteria (PGPR) contain a vital enzyme, 1-aminocyclopropane-1-carboxylic acid (ACC) deaminase (EC 4.1.99.4), which regulates ethylene production by metabolizing ACC (an intermediate precursor of ethylene biosynthesis in higher plants) into α-ketobutyrate and ammonia. This pyridoxal phosphate (PLP) enzyme was first isolated in 1978 from *Pseudomonas* sp. strain ACP and from the yeast *Hansenula satrunus* (Honma and Shimomura 1978) since then, it has been detected in fungi and in a number of other bacteria. When ACC deaminase-containing plant growth-promoting bacteria (PGPB) are bound to a plant, they act as a sink for ACC ensuring that plant ethylene levels do not become elevated to the point.

Conceptually, PGPR can have an impact on plant growth and development in two different ways: indirectly or directly. The indirect promotion of plant growth occurs when bacteria decrease or prevent some of the deleterious effects of a phytopathogenic organism by one or more mechanisms. On the other hand, the direct promotion of plant growth by PGPR generally entails providing the plant with a compound that is synthesized by the bacterium or facilitating the uptake of nutrients from the environment (Glick 1995; Glick et al. 1999). Rhizosphere bacteria multiply to high densities on plant root surfaces where root exudates and root cell lysates provide ample nutrients. Sometimes, they exceed 100 times to those densities found in the bulk soil (Campbell and Greaves 1990). Certain strains of these plant-associated bacteria stimulate plant growth in multiple ways: (1) they may fix atmospheric nitrogen, (2) reduce toxic compounds, (3) synthesize phytohormones and Siderophores, or (4) suppress pathogenic organisms (Bloemberg and Lugtenberg 2001). Research on the "biocontrol" activity of rhizobacteria has seen considerable progress in recent years. Disease suppression of soilborne pathogens includes competition for nutrients and production of antimicrobial compounds or lytic enzymes for fungal cell walls or nematode structures (Persello-Cartieaux 2003). By contrast, systemic resistance can also be induced by rhizosphere-colonizing *Pseudomonas* and *Bacillus* species where the inducing bacteria and the challenging pathogen remained spatially separated excluding direct interactions (Van Loon et al. 1998; Ryu et al. 2004).

Etiolated pea seedlings are very sensitive to ethylene. The most widely renowned example of the effect of ethylene on plant growth is the classical "triple" response in etiolated dicot seedlings in the presence of ethylene. This effect consists of three distinct morphological changes in the shape of seedlings, inhibition of stem elongation, increase in stem diameter and horizontal growth (Akhtar et al. 2005; Khalid et al. 2006). This "triple" response reaction of etiolated seedlings has been a reliable bioassay for ethylene action (Guzman and Ecker 1990). Shaharoona et al. (2007) observed the effect of inoculation with ACC utilizing and ethylene-producing rhizobacteria and compared through highly ethylene specific classical "triple" response bioassay. In this study, the effect of inoculation with rhizobacteria having different ACC-deaminase activities on extenuating the classical "triple" response in etiolated pea seedlings was investigated.

ACC deaminase-containing PGPB up-regulate genes involved with plant growth and protein production while down-regulating plant genes involved with ethylene stress and defence signaling pathways (Hontzeas et al. 2004a). The ACC deaminase-containing PGPB, in part, alleviate the need for the plant to actively defend itself against various environmental stresses (Hontzeas et al. 2004b; Van Loon and Glick 2004). The crystal structure has been determined for the yeast (Minami et al. 1998), and recently for the bacteria (Karthikeyan et al. 2004) ACC deaminase enzymes; the biochemical and thermodynamic properties of the ACC deaminase from *Pseudomonas putida* UW4 have been measured (Hontzeas et al. 2004b).

ACC deaminase from *Pseudomonas* sp. ACP, *P. putida, P. fluorescens* (Glick 1995), *Enterobacter cloacae* CAL2 and UW4 (Shah et al. 1998), *Kluyvera ascorbata* SUD165 (Burd et al. 1998), *Hansenula saturnus* (Honma and Shimomura 1978), and *Penicillium citrinum* (Jia et al. 2006) have been reported.

This enzyme facilitates plant growth as a consequence of the fact that it sequesters and cleaves plant produced ACC, thereby lowering the level of ethylene in the plant. In turn, decreased ethylene levels allow the plant to be more resistant to a wide variety of environmental stresses, all of which induce the plant to increase its endogenous level of ethylene; stress ethylene exacerbates the effects of various environmental stresses. The ACC deaminase-containing soil bacteria decrease a significant portion of the physiological damage to plants following environmental stresses including phytopathogen infection, exposure to extremes of temperature, high salt, flooding, drought, exposure to metals and organic contaminants, and insect predation. For many plants a burst of ethylene is required to break seed dormancy but, following germination, a sustained high level of ethylene can be inhibitory to root elongation. PGPB that contain the enzyme ACC deaminase, when bound to a plant root or to the seed coat of a developing seedling, may act as a mechanism for insuring that the ethylene level within the plant's tissues does not become elevated to the point where root (or shoot) growth is impaired. By facilitating the formation of longer roots and shoots, these bacteria may enhance the survival of some seedlings, especially during the first few days after the seeds are planted.

2 Ethylene Biosynthesis in Higher Plants

Ethylene, which is produced in almost all plants, mediates a range of plant responses and developmental step. Ethylene is involved in seed germination, tissue differentiation, formation of root and shoots primordial, root elongation, lateral bud formation, flowering initiation, anthocyanin synthesis, flower opening and senescence, fruit ripening and degreening, production of aroma, leaf and fruit abscission and response of plant to biotic and abiotic stresses. (Saraf and Tank 2005). Ethylene is a potent plant growth regulator that affects diverse developmental processes, including fruit ripening, senescence, and stress responses (McKeon and Yang 1987; Reid 1987). Chemical inhibitors of ethylene synthesis or action completely block ripening in fruits and senescence in flowers of many plant species.

At a molecular level, ethylene is known to induce expression of a number of genes involved in ripening (Lincoln and Fischer 1988) and pathogen response (Ecker and Davis 1987). In some instances, ethylene is stimulatory while in others it is inhibitory.

When plants are exposed to conditions that threaten their ability to survive, the same mechanism that produces ethylene for normal development instead produces "stress ethylene" which may be defined as an acceleration of ethylene biosynthesis associated with biological and environmental stresses, and pathogen attack (Abeles et al. 1992; Hyodo 1991; VanLoon 1984). Ethylene is synthesized from S-adenosyl L-methionine (AdoMet) by way of the intermediate ACC (McKeon and Yang 1987).

While working on the ethylene biosynthesis pathway, Adams and Yang (1979) found that when ACC was applied to various plant organs, an increase in ethylene production was obtained. From their observations, ACC, as a key intermediate that linked the methionine cycle and ethylene biosynthesis, was deemed to be the direct precursor of ethylene production with its level directly controlling ethylene synthesis in plants (Fig. 1).

Ethylene biosynthesis consists of three steps (1) L-methionine is converted to AdoMet, a reaction catalyzed by methionine S-adenosyl transferase. AdoMet is also utilized in other cellular reactions such as ethylation and polyamine synthesis, (2) The conversion of AdoMet to ACC which is catalyzed by ACC synthase. The ACC synthase step is considered to be the rate-limiting step in the pathway (3) ACC is further metabolized to ethylene, carbon dioxide and cyanide by ACC oxidase.

Since all plants respond differently to stress, it has been difficult to detail the functioning of stress ethylene. Increased ethylene levels in plants exposed to various types of stress including chilling, heat, wounding, pathogen infection, salt, metals and nutritional stress, with increased damage as the result has been documented. Stress ethylene, though its role is unclear, is deleterious to plants in many instances (Saravanakumar and Samiyappan 2007).

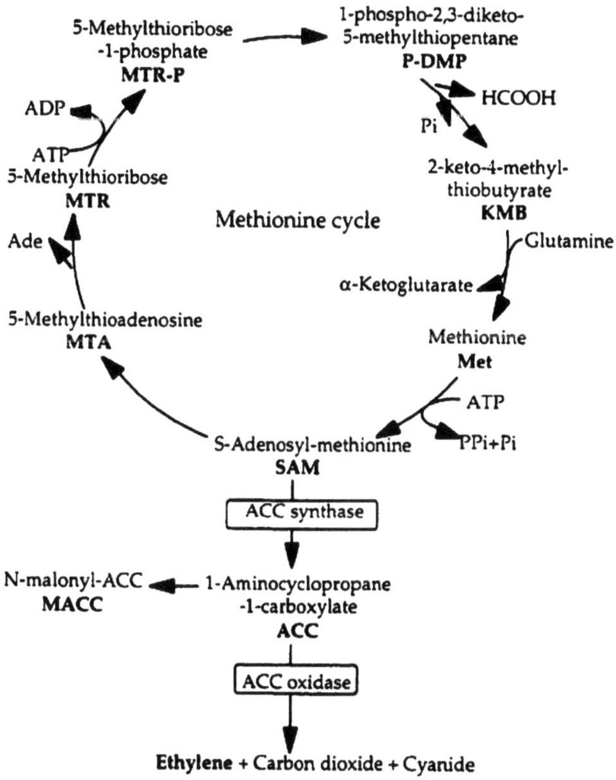

Fig. 1 Pathway of ethylene biosynthesis from the methionine cycle in higher plants. Modified figure adapted from the source reference Li (1999)

3 Characteristics of ACC Deaminase Enzyme

Enzymatic activity of ACC deaminase is assayed by monitoring the production of either ammonia or α-ketobutyrate, the products of ACC hydrolysis. ACC deaminase has been found only in microorganisms, and there are no microorganisms that synthesize ethylene via ACC (Fukuda et al. 1993). ACC Deaminase is a multimeric enzyme (homodimeric or homotrimeric) with a subunit molecular mass of approximately 35-42 kDa. It is a sulfhydral enzyme in which one molecule of the essential cofactor PLP is tightly bound to each subunit. Interestingly, this enzyme is cytoplasmically localized so that the substrate ACC must be exuded by plant tissues and subsequently taken up by an ACC deaminase-containing microorganism before it is cleaved (Glick et al. 1998).

The enzyme–substrate relationship demonstrates Km values of ACC deaminase for ACC estimated at pH 8.5, in all instances examined, to be approximately 1.5–17.4 mM indicating that the enzyme does not have a particularly high affinity for ACC (Honma and Shimomura 1978). Moreover ACC levels in plants are typically in μM range, therefore in most plant tissues the ACC concentration will be dramatically below the Km of ACC deaminase for this substrate so that based on the Michaelis–Menton rate equation for enzyme catalyzed reaction a small increase in the ACC concentration will result in a parallel increase in the rate of ACC cleavage.

4 Crystal Structure of 1-Aminocyclopropane-1-Carboxylate Deaminase

PLP-dependent enzymes catalyze many important reactions that act upon amino acids, including transamination, decarboxylation, β,γ-replacement/elimination, and racemization. In all of these reactions (except in the case of the glycogen phosphorylase family), the two basic chemical properties of the PLP are conserved; it forms an external aldimine between its aldehyde group and the α-amino group of the substrates and withdraws electrons from the substrate by serving as an electron sink. As a PLP-dependent enzyme, the ACCD's ring opening reaction starts with a transformation reaction from an internal aldimine between the PLP and the enzyme to an external aldimine. These enzymes have been classified based on their three dimensional structure, into four folding types: (1) tryptophan synthase, (2) aspartate aminotransferase, (3) D-amino acid aminotransferase and (4) alanine racemase. In most of the PLP-dependent enzymes, the next step is the nucleophilic abstraction of the α-substituent, either an α-proton or a carboxylate group, to form an α-carbanionic intermediate. This reaction mechanism cannot be applied to ACCD because the substrate (ACC) does not contain α-hydrogen and the carboxyl group is retained in the product. Therefore, the ring-opening reaction of ACC must be initiated without obvious accessibility to an α-carbanionic intermediate, which is, for PLP-dependent enzymes, the common entry for catalysis. One proposed reaction mechanism is the nucleophilic addition to Cγ followed by the cleavage of the Cα–Cγ bond and β-proton abstraction. As PLP, acts as an electron sink, external aldimine is fairly electrophilic, and the nucleophilic addition to Cγ to rupture the cyclopropane ring of ACC is mechanistically feasible (Yao et al. 2000) (Fig. 2).

5 Mechanism of ACC Deaminase Action

A model is proposed to explain how ACC deaminase-containing PGPB can lower plant ethylene levels and in turn stimulate plant growth (Glick et al. 1998), especially under stress conditions. PGPB bind to the surface of either the seed or root of a developing plant in response to tryptophan and other small molecules in the seed or root exudates the PGPB synthesize and secrete the auxin, Indoleacetic acid (IAA), some of which is taken up by the plant. This IAA together with endogenous plant IAA can stimulate plant cell proliferation and elongation, or it can induce the activity

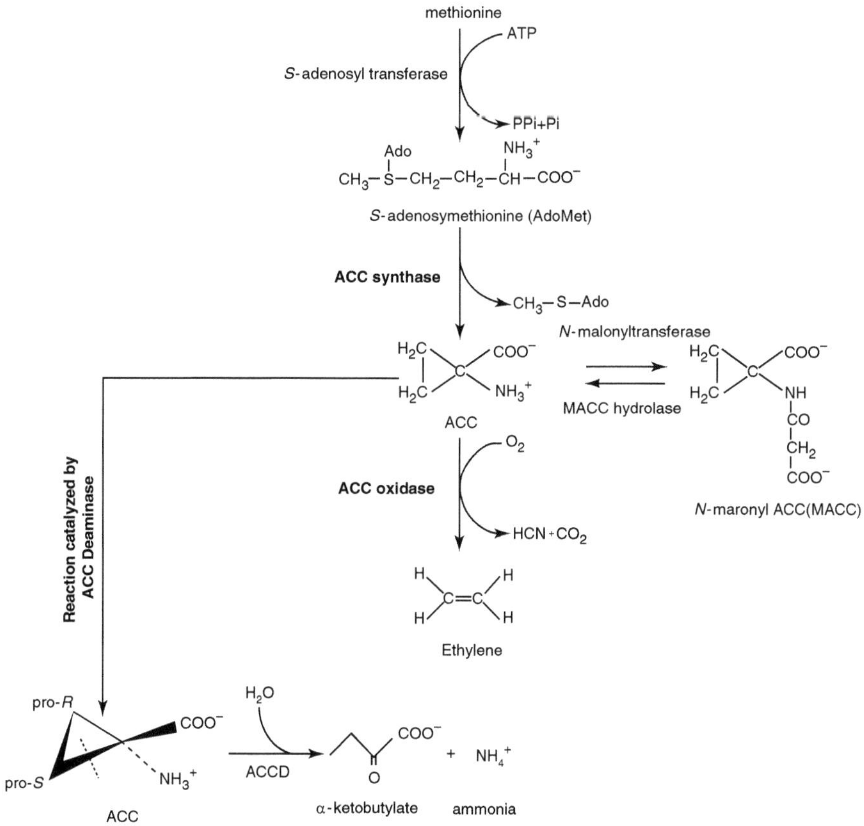

Fig. 2 The enzymatic reaction catalyzed by ACCD. Modified figure adapted from the source reference Ose et al. (2003)

of ACC synthase to produce ACC (Penrose and Glick 2001). Some of the plant's ACC will be exuded along with other small molecules such as sugars, organic acids and amino acids. The exudates may be taken up by the bacteria and utilized as a food source of the rhizosphere bacteria. ACC may be exuded together with the other components of the root or seed exudates. ACC may be cleaved by ACC deaminase to form ammonia and α-ketobutyrate, compounds that are readily further metabolized by the bacteria (Holguin and Glick 2001). The presence of the bacteria induces the plant to synthesize more ACC than it would otherwise need and also, stimulates the exudation of ACC from the plant (some of which may occur as a consequence of plant cell wall loosening caused by bacterial IAA). Thus, PGPB are supplied with a unique source of nitrogen in the form of ACC that enables them to proliferate/survive under conditions in which other soil bacteria may not readily flourish (Hontzeas et al. 2006). As a result of acting as a sink for ACC and lowering its level within the plant, the amount of ethylene that is produced by the plant is also

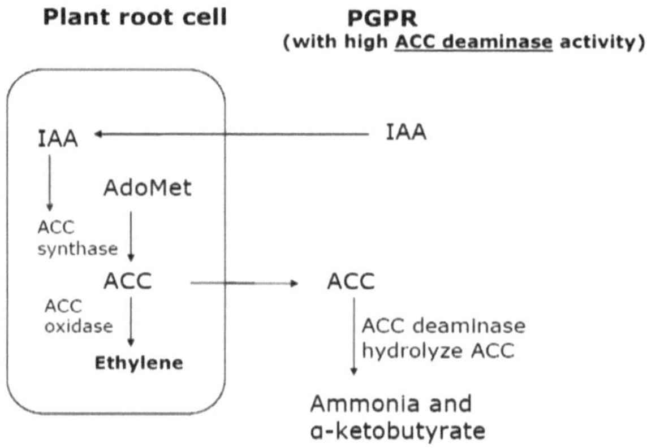

Fig. 3 The ACC deaminase in PGPR degrades the ethylene precursor ACC. The ACC deaminase in PGPR lowers ethylene level in plants by degrading ACC to ammonia and α-Ketobutyrate. Lowering ethylene in plants can alleviate stress and thereby improve plant growth. Some PGPR can also produce plant regulator IAA and further stimulate plant growth. Modified figure adapted from the source reference Glick and Pasternak (2003)

reduced. Thus, the inhibition of plant growth by ethylene (especially during periods of stress) is decreased and these plants generally have longer roots and shoots and greater biomass (Fig. 3).

6 Role of Bacterial ACC Deaminase Under Stress Agricultural Conditions

PGPR containing ACC Deaminase activity eliminates heavy metal toxicity, imparts resistant to drought, other abiotic stresses such as salinity, extremes of temperature and pH in soil apart from antagonism against phytopathogens. Ethylene regulation in plants due to PGPR is now well established (Table 1).

6.1 Pathogenicity Stress

Pathogenic microorganisms are a major and serious threat to food production and ecosystem stability worldwide. PGPR mediated biocontrol in terms of competition for an ecological niche or a substrate and producing allelo-chemicals and inducing systemic resistance (ISR) in host plants to a broad spectrum of pathogens (Compant et al. 2005).

ACC deaminase bacteria, apart from directly antagonizing pathogens, support the plant resistance against pathogen attack. Beneficial rhizobacteria do not obviously damage their host/cause localized necrosis, therefore, the eliciting factors

Table 1 List of ACC deaminase producing bacteria

Strain	ACC deaminase activity (nM αKB $mg^{-1}h^{-1}$)	Reference(s) or Sources
Achromobacter xylosoxidans A551	400 ± 4	Belimov et al. (2001, 2005)
A. xylosoxidans Bm1	90 ± 4	Belimov et al. (2001, 2005)
Achromobacter sp. strain CM1	130 ± 3	Belimov et al. (2001, 2005)
Acidovorax facilis 4p-6	3,080 ± 120	Belimov et al. (2001, 2005)
Azospirillium brasilense Cd1843	–	Holguin and Glick (2003)
Enterobacter aerogenes CAL3	16 ± 12	Shah et al. (1998)
Pseudomonas putida UW4	3,030 ± 60	Hontzeas et al. (2006)
P. syringae GR12-2	3,470 ± 30	Belimov et al. (2001, 2005)
P. brassicacearum Am3	5,660 ± 12	Belimov et al. (2001, 2005)
P. putida BM3	3,780 ± 32	Belimov et al. (2001, 2005)
P. marginalis DP3	4,054 ± 27	Belimov et al. (2001, 2005)
Rhizobium leguminosarum128C53K	5 ± 1	Belimov et al. (2001, 2005)
R. hedysari ATCC 43676	20 ± 0.1	Ma et al. (2003)
R. leguminosarum 99A1	8 ± 3	Ma et al. (2003)
Rhodococcus sp. strain Fp2	7,320 ± 400	Belimov et al. (2001, 2005)
Rhodococcus sp. strain 4N-4	12,970 ± 440	Belimov et al. (2001, 2005)
Serratia quinivirans SUD165	12 ± 15	Belimov et al. (2001, 2005)
Variovorax paradoxus 3P-3	3,700 ± 90	Belimov et al. (2001, 2005)
V. paradoxus 5C-2	4,322 ± 100	Belimov et al. (2001, 2005)
V. paradoxus 2C-1	3,588 ± 26	Belimov et al. (2001, 2005)
P. putida ATCC17399	–	Shah et al. (1998)
Schizosaccharomyces pombe	–	Wood et al. (2002)
Hansenula saturnus	–	Honma and Shimomura (1978), Minami et al. (1998)
Penicillium citrinum	–	Jia et al. (2006)
Yersinia pestis	–	Parkhill et al. (2001)
Caulobacter crescentus	–	Nierman et al. (2001)
Bacillus anthracis	–	Read et al. (2002)
Mesorhizobium loti	–	Sullivan et al. (2002)
Burkholderia fungorum	–	NCBI microbial genome annotation project

produced by ISR-triggering rhizobacteria must be different from elicitors of pathogens. Expression of ISR is similar to systemic acquired resistance (SAR) upon challenge inoculation with pathogen wherein disease severity is reduced; the number of diseased plants also diminishes. This reduction is associated with decreased growth of the pathogen and reduced colonization of induced tissues which reflects upon the ability of plant to resist the pathogen (Dobbelaere et al. 2003). Salicylic acid is an important signaling molecule in both locally and systemically induced resistance responses; however, research on rhizobacteria mediated ISR signaling has demonstrated that jasmonic acid and ethylene play the key roles. Thus, expression of ISR is phenotypically quite similar to SAR, and relies not only on a different type of biological induction but occurs also through different defense-related activities (Domenech et al. 2006). It is emphasized that ISR-inducing PGPR is a useful tool to reduce diseases caused by pathogens that are sensitive to jasmonic acid and ethylene-dependent defenses. Rasche et al. 2006 reported that ACC deaminase bacteria were capable of antagonizing at least one of the two potato pathogens *Ralstonia solanacearum* and *Rhizoctonia solani.*

6.2 Remediation of High/Heavy Metal Concentration

High metal concentrations in soil have also been shown to cause increased ethylene production and inhibition of root development, to reduce CO_2 fixation and limit sugar translocation. ACC deaminase and siderophore producing PGPB can help plants to overcome many of the effects of high levels of metal (Burd et al. 1998, 2000). Phytoremediation of metals poses a significant challenge because most metal contaminants are tightly bound by soil particles and are not readily bioavailable to plants. Moreover, although many plants can tolerate the presence of excess metals in the soil, most will experience a decrease in plant growth and viability due to either the synthesis of stress ethylene and/or iron depletion. PGPR can alleviate some of the effects of metal toxicity in plants via several different mechanisms. For example bacterial siderophore bind iron with extremely high affinity and plants are able to take up and utilize the iron from these complexes. Thus PGPR are able to protect plants against the inhibitory effects of high concentration of metals by providing the plants with sufficient iron. Belimov et al. (2005) reported 11 cadmium-tolerant strain of PGPR isolated from the rhizosphere of *Brassica juncea* grown in cadmium-containing soils. *Variovorax paradoxus, Rhodococcus* sp. and *flavobacterium sp.* all stimulated root elongation in untreated and Cd-treated soils.

6.3 Drought Stress

Drought is one of the major environmental stresses that limit the growth of plants and the production of crops. The inhibitory effects of ethylene induced by drought stress might have been eliminated through the ACC deaminase activity of the PGPR. Inoculation of plants with PGPR containing ACC deaminase partially or completely eliminated the “drought stress imposed effects” on root and shoot growth, fresh and dry weights, and number of leaves per plant of peas. This might be due to suppression of the stress-induced accelerated synthesis of ethylene by the ACC deaminase activity of these PGPR in the inoculated roots. Sharp increases in ACC levels and, consequently, ethylene synthesis in plants under drought stress conditions has been frequently reported. (Apelbaum and Yang 1981). The rhizobacteria having ACC deaminase activity are effective in promoting plant growth and water use efficiency under drought conditions, by lowering the ethylene or ACC accumulation whose higher levels have inhibitory effects on root and shoot growth. It is highly likely that rhizobacteria containing ACC deaminase might have decreased the drought-stress induced ethylene in inoculated plants, which resulted in better growth of plants even at low moisture levels. Therefore, inoculation with rhizobacteria containing ACC deaminase could be helpful in eliminating the inhibitory effects of drought stress on the growth of plants. Dodd et al. (2005) investigated the physiological responses of pea (*pisum sativum* L.) to inoculation with ACC deaminase bacteria *V. paradoxus* 5C-2 under moisture stress and watering condition. The bacterial effects were more pronounced and more consistent under controlled soil drying (moisture stress conditions).

6.4 Organic Contaminants Stress

Many organic contaminants are recalcitrant and highly persistent in the environment, making them particularly difficult to remediate. Many of these compounds are hydrophobic and are bound tightly to soil particles. A few studies have revealed an accelerated production of ethylene in soil and plants treated with organic contaminants (Coupland and Jackson 1991). Reed and Glick (2005) have studied the growth of canola (*Brassica napus*) seeds treated with PGPR in copper-contaminated and creosote-contaminated soil. In creosote-contaminated soils, the native bacterium was the least effective, and the transformed encapsulated ACC deaminase bacterium was the most effective in growth promotion.

6.5 Waterlogging Stress

Waterlogging enhances the biosynthesis of ethylene in roots and stem of plants. In flooding, ACC, which is synthesized in roots, is transported to plant shoots where it is converted to ethylene by ACC oxidase (Bradford and Yang 1980). The molecular basis for the increase in ethylene production observed in shoots of flooded tomato plants is due to an increase in the activity of both ACC synthase in the submerged roots and ACC oxidase in the shoots (Chao et al. 1997). The accelerated production of ethylene in the shoots of flooded tomato plants is responsible for the phenotype to demonstrate abnormal growth under flooding conditions (Jackson 1997).

6.6 Temperature Stress

The heat stress in terms of so-called global warming is a serious threat to world agriculture (Mendelsohn and Rosenberg 1994). A fluctuation in temperature leads to hormonal imbalances in plants and thus their growth is significantly affected

(Cheikh and Jones 1994). It has been reported that PGPR containing ACC deaminase activity performs better when subjected to diurnal temperature regime. *Bacillus globiosporus* was inoculated to analyze the effect of diurnal temperature regime (i.e., 25°C days and 5°C night) on root and shoot length, fresh and dry weight were significantly increased in comparison to *B. subtilis* and magnesium sulphate controls (Ghosh et al. 2003).

6.7 Flower Senescence

Ethylene is a key signal in the initiation of willing in most plants. Typically flowers produce minute amount of ethylene until an endogenous rise of the phytohormone, which is responsible for flower senescence to occur (Mol et al. 1995). However, the senescence symptoms that are covered by ethylene differ from plant to plant. The use of ACC deaminase containing PGPR to lower ACC levels in cut flowers might be an environmentally friendly alternative to the available use of silver thiosulphate. An important characteristic of PGPR containing ACC deaminase activity has been shown to be the enhancement of shelf life of flowers incubated in suspension form (Nayani et al. 1998). On a commercial scale, shelf life of flowers could be increased to manifold by treating them with suspensions of PGPR containing ACC deaminase activity, which portends great prospects for the application of this biotechnological approach to commercial floriculture.

6.8 Salinity Stress

Salinity is one of the most severe environmental stresses on plants (White and Broadley 2001; Tester and Davenport 2003; Munns and Tester 2008). Salt primarily limits plant growth in three ways: (1) osmotic effects that lower the ability of plants to take up water from the soil, (2) ion-specific damage of excess Na^+ and Cl^-, and (3) nutrient deficiencies because elevated levels of Na^+ compete with the uptake of other nutrients by interfering with ion transporters (Tester and Davenport 2003). Symptoms of damage to plants include: growth inhibition, leaf discoloration, anatomical and morphological changes such as changes in cell wall structure (Tester and Davenport 2003). Highly saline soil (ECe $>$ 16 dS/m) can severely interfere with seed germination and growth of plants. As water and nutrients move from areas of low salt concentration to areas of high salt concentration, soil salinity prevents plant roots from taking up water and other nutrients, resulting in osmotic and nutrient imbalances that impair proper plant growth. A sudden increase in soil salinity will cause plant cells to shrink due to water loss and immediate changes in expansion rates resulted from the osmotic effects of salt around the roots (Cramer and Bowman 1991; Munns 2002; Neumann 1993). After several hours, plant cells can restore their original shape; however, a decrease in cell elongation rates is observed in both leaves and roots (Hsiao and Xu 2000; Munns 2002). Continued exposure for a few days results in a decrease in plant growth (i.e., slower cell division and impaired cell elongation). In this case, leaves are often more sensitive to salinity than roots (Hsiao and Xu 2000; Munns 2002). Changes in plant cell dimension are observed more for an area than depth, therefore, leaves appear to be smaller and thicker (Munns and Tester 2008). The effects of salinity become more apparent after a few weeks of exposure (Munns and Tester 2008). Yellowing or death of older leaves may be visible in salt-sensitive plants, where salt levels are high, due to increase uptake or inability to store salt in vacuoles (Karley et al. 2000; Munns and Tester 2008; Tester and Davenport 2003). Only the salt-tolerant plants are able to grow for several months under moderate salinity; but showed early flowering or decreased production of florets (Munns 2002).

Salinity stress boosts endogenous ethylene production in plants, which in most cases serves as a stress hormone (Blumwald 2000). It is very likely that reducing salinity-induced ethylene by any mechanism could decrease the negative impact of salinity on to plant growth. Recent studies have revealed that plants inoculated with PGPR containing ACC deaminase were able to thrive better through the salinity stress while demonstrating a normal growth pattern. Tank and Saraf (2010) have reported that increase in the salinity is directly proportional to the ACC deaminase activity which increases survival rate in saline soils. As the uptake and hydrolysis of ACC by the PGPR decreases the ACC level in plants, the biosynthesis of the "stress ethylene" is impeded, facilitating plant growth under stress conditions (Glick et al. 1998). It has been shown that PGPR promotes plant growth under saline conditions. The presence of PGPR with ACC deaminase may lower the levels of ethylene in developing or stressed plants, enhance the survival of some seedlings and facilitate the formation of longer roots.

6.9 *Ethylene–IAA Cross-talk*

It is well known that IAA can activate the transcription of ACC synthase (Kende 1993; Kim et al. 1992) but it is less well known that ethylene may inhibit IAA transport and signal transduction (Pratiyon et al. 2006). This feedback loop of ethylene inhibition of IAA synthesis and/or functioning limits the amount of ACC synthase, ACC and ultimately, ethylene following every stressful event in the life of the plant. When an ACC deaminase containing PGPB lowers the ethylene concentration in plant roots, these relieve the ethylene repression of auxin response factor synthesis, and indirectly increase plant growth. Thus ACC deaminase containing PGPR facilitate plant growth by decreasing ethylene inhibition and permitting IAA stimulation without the negative effects of increasing ACC synthase and plant ethylene levels.

6.10 Air Pollutants Stress

It is very likely that PGPR can be utilized as a gene source for genetic modification of plants expressing the enzyme ACC deaminase against plant damage by air pollutants. Air pollution, in addition to damaging plants, inhibits many enzyme systems and metabolic processes of plants (McCune 1975). Increased ethylene evolution by plants exposed to various environmental stresses i.e., air contaminants has been well documented (Wang et al. 2002) and this hormone is now considered a major regulator of plant defense reactions, including cell death, in response to pathogen attack and air contaminant stresses, i.e., O_3 exposure. Many researchers reported that the inhibition of ethylene biosynthesis resulted in a significant reduction of O_3-induced leaf lesion formation (Moeder et al. 2002). In this direction, the role of ACC deaminase in alleviation of air contaminants stresses has not been studied.

6.11 Rhizobial Infection

Considerable evidence suggests that the ethylene that is produced following infection of legumes with *rhizobia* is inhibitory to the process of nodulation. The latest evidence has demonstrated that PGPR containing ACC deaminase activity promotes nodulation in legumes through inhibition of ethylene biosynthesis and consequently, they enhance symbiosis and nitrogen fixation in plants (Okazaki et al. 2004). Uchiumi et al. (2004) reported that an up regulated gene in bacteroids, mlr5932, and encoding ACC deaminase activity was involved in enhanced nodulation in *Lotus japonicus*. Pandey et al. (2005) isolated an endophytic ACC deaminase bacterium capable of modulating nodulation in *Mimosa pudica*. Coinoculation with *Bradyrhizobium* plus ACC deaminase rhizobacteria increased nodulation in mung bean compared to inoculation with *Bradyrhizobium* spp. alone (Shaharoona et al. 2006).

7 Microbe–Microbe Interactions Benefiting Sustainable Agro-Ecosystem Development

Direct interactions occurring between members of different microbial types often result in the promotion of key processes benefiting plant growth and health. It is obvious that all interactions taking place in the rhizosphere are, at least indirectly, plant-mediated (Azcon-Aguilar and Barea 1992). However, this section will deal with direct microbe–microbe interactions themselves, with the plant as a "supporting actor" in the rhizosphere. Three types of interactions have a major role to play in bacteria–plant health development because of their relevance to the development of sustainable agro-ecosystems. These are (1) the cooperation between ACC deaminase producing PGPR and *Rhizobium* for improving N-fixation, (2) microbial antagonism for the biocontrol of plant pathogens, and (3) interactions between rhizosphere microbes and AM fungi to establish a functional mycorrhizosphere (Barea et al. 2005).

8 ACC Deaminase Gene-Containing Transgenic Plants

Transgenic plants express a bacterial ACC deaminase under the control of either the *35S* (constitutive) or *rolD* (root-specific) promoter as a treatment with ACC deaminase containing bacteria, although ethylene levels have been reported to be decreased by more than 95% in some ripen transgenic tomato fruit. Transgenic plants that express ACC deaminase are also significantly protected against the potentially deleterious effects of a variety of stresses including drought, flooding (Grichko and Glick 2001), high salt (Sergeeva et al. 2006), phytopathogens (Robison et al. 2001), arsenic (Nie et al. 2002), and several different metals (Grichko et al 2001). In all instances, transgenic plants, in which ACC deaminase was under the control of the *rolD* promoter, performed significantly better than the nontransformed plants (regardless of whether the plant was tomato, canola or tobacco) and the transgenic lines in which the ACC deaminase gene was under the control of the *rolD* promoter, yielded significantly more root and shoot biomass than either the nontransformed plants or transgenic plants in which the ACC deaminase gene was under the control of the *35S* or *prb-1b* (stress-specific) promoter. Transgenic plants in which ACC deaminase is under the control of the *rolD* promoter appear to mimic the behavior of nontransgenic plants treated with ACC deaminase-containing PGPB. However, the performance of plants treated with ACC deaminase-containing PGPB is almost always superior to the performance of transgenic plants expressing ACC deaminase under the control of the *rolD* promoter. This likely reflects the fact that the bacteria do more than merely lower plant ethylene levels. They also provide the plants with other "benefits" such as plant hormones and siderophores.

9 Conclusions and Future Trends

There is considerable experimental evidence that certain microorganisms are able to colonize the root–soil environments where they carry out a variety of interactive activities known to benefit plant growth and health, and also soil quality. Given the current reluctance of many consumers worldwide to embrace the use as foods of genetically modified plants, it may be advantageous to use PGPB as a means to promote growth by lowering plant ethylene levels or reduce disease through induction of resistance, rather than genetically modifying the plant itself to the same end.

Rhizobacteria having ACC deaminase activity are effective in promoting plant growth and water use efficiency under drought conditions, by lowering the ethylene or ACC accumulation whose higher levels have inhibitory effects on root and shoot growth. From the previous demonstrations, it is established that the microorganisms that possess ACC deaminase activity have the selective advantage over other bacteria during biotic and abiotic stress conditions. Besides the activity of ACC deaminase in alleviating ethylene-mediated abiotic and biotic stresses, the ecology of bacterium and physiology of the plant may also interact with plant system to increase resistance to stress. However, the defined mechanisms involved in the use of plant growth-promoting rhizobacteria which decrease the damage to plants that occurs under stress conditions is a potentially important adjuvant to agricultural practice in locales where stress is a major constraint.

From the agricultural and ecological viewpoints, the aims will be to increase food quality, and to improve sustainable plant productivity, while maintaining environmental quality. However, to achieve this, basic and strategic studies must be undertaken to improve our understanding of microbial interactions in the rhizosphere. Only then can the corresponding agro-biotechnology be applied successfully. Hence, future investigation in the field of microbial cooperation in the rhizosphere will include: (1) advances in visualization technology; (2) analysis of the molecular basis of root colonization; (3) signaling in the rhizosphere; (4) functional genomics; (5) mechanisms involved in beneficial cooperative microbial activities; (6) engineering of microorganisms for beneficial purposes; and (7) biotechnological developments for integrated management.

References

NCBI Microbial Genome Annotation project Residues 1 to 95851 of *Burkholderia fungorum*. Submitted (18-SEP-2002) National Center for Biotechnology Information, NIH, Bethesda, MD 20894, USA.

Abeles FB, Morgan PW, Saltveit ME (eds) (1992) Regulation of ethylene production by internal, environmental and stress factors. In: Ethylene in plant biology, 2nd edn. Academic Press, San Diego, pp 56–119

Adams DO, Yang SF (1979) Ethylene biosynthesis: identification of 1-aminocyclopropane-1-carboxylic acid as an intermediate in the conversion of methionine to ethylene. Proc Natl Acad Sci USA 76:170–174

Akhtar MJ, Arshad M, Khalid A, Mahmood HM (2005) Substrate-dependent biosynthesis of ethylene by rhizosphere soil fungi and its influence on etiolated pea seedlings. Pedobiologia 49:211–219

Apelbaum A, Yang SF (1981) Biosynthesis of stress ethylene induced by water deficit. Plant Physiol 68:594–596

Azcon-Aguilar C, Barea JM (1992) Interactions between mycorrhizal fungi and other rhizosphere micro-organisms. In: Allen MJ (ed) Mycorrhizal functioning: an integrative plant–fungal process. Chapman and Hall, New York, pp 163–198

Barea JM, Pozo MJ, Azcon R, Azcon-Aguilar C (2005) Microbial co-operation in the rhizosphere. J Exp Bot 56(417):1761–1778

Belimov AA, Safronova VI, Sergeyeva TA, Egorova TN, Matveyeva VA, Tsyganov VE, Borisov AY, Tikhonovich IA, Kluge C, Preisfeld A, Dietz KJ, Stepanok VV (2001) Characterization of plant growth promoting rhizobacteria isolated from polluted soils and containing 1-aminocyclopropane-1-carboxylate deaminase. Can J Microbiol 47:642–652
Belimov AA, Hontzeas N, Safronova VI, Demchinskaya SV, Piluzza G, Bullitta S, Glick BR (2005) Cadmium-tolerant plant growth-promoting bacteria associated with the roots of Indian mustard (*Brassica juncea* L. Czern.). Soil Biol Biochem 37:241–250
Bloemberg GV, Lugtenberg BJJ (2001) Molecular basis of plant growth promotion and biocontrol by rhizobacteria. Curr Opin Plant Biol 4:343–350
Blumwald E (2000) Sodium transport and salt tolerance in plants. Curr Opin Cell Biol 12:431–434
Bradford KJ, Yang SF (1980) Xylem transport of 1-aminocyclopropane- 1-carboxylic acid, an ethylene precursor, in waterlogged tomato plants. Plant Physiol 65:322–326
Burd GI, Dixon DG, Glick BR (1998) A plant growth-promoting bacterium that decreases nickel toxicity in seedlings. Appl Environ Microbiol 64:3663–3668
Burd GI, Dixon DG, Glick BR (2000) Plant growth-promoting bacteria that decrease heavy metal toxicity in plants. Can J Microbiol 46:237–245
Campbell R, Greaves MP (1990) Anatomy and community structure of the rhizosphere. In: Lynch JM (ed) The rhizosphere. Wiley, Chichester, England, pp 11–34
Chao Q, Rothenberg M, Solano R, Roman G, Terzaghi W, Ecker JR (1997) Activation of the ethylene gas response pathway in *Arabidopsis* by the nuclear protein and related proteins. Cell 89:1133–1144
Cheikh N, Jones RJ (1994) Disruption of maize kernel growth and development by heat stress (role of cytokinin/abscisic acid balance). Plant Physiol 106:45–51
Compant S, Duffy B, Nowak J, Clement C, Barka EA (2005) Use of plant growth-promoting bacteria for biocontrol of plant diseases: principles, mechanisms of action, and future prospects. Appl Environ Microbiol 71:4951–4959
Coupland D, Jackson MB (1991) Effects of mecoprop (an auxin analogue) on ethylene evolution and epinasty in two biotypes of *stellaria media*. Ann Bot 68:167–172
Cramer GR, Bowman DC (1991) Kinetics of maize leaf elongation.1. Increased yield threshold limits short-term, steady-state elongation rates after exposure to salinity. J Exp Bot 42(244): 1417–1426
Dobbelaere S, Vanderleyden J, Okon Y (2003) Plant growth-promoting effects of diazotrophs in the rhizosphere. Crit Rev Plant Sci 22:107–149
Dodd IC, Belimov AA, Sobeih WY, Safronova VI, Grierson D, Davies WJ (2005) Will modifying plant ethylene status improve plant productivity in water-limited environments? In: 4th International Crop Science Congress
Domenech J, Reddy MS, Kloepper JW, Ramos B, Gutierrez-Mañero J (2006) Combined application of the biological product LS213 with *Bacillus*, *Pseudomonas* or *Chryseobacterium* for growth promotion and biological control of soil-borne diseases in pepper and tomato. Biocontrol 51:245–258
Ecker J, Davis RW (1987) Plant defense genes are regulated by ethylene. Proc Natl Acad Sci USA 84:5202–5206
Fukuda H, Ogawa T, Tanase S (1993) Ethylene production by microorganisms. Adv Microb Physiol 35:275–306
Ghosh S, Penterman JN, Little RD, Chavez R, Glick BR (2003) Three newly isolated plant growth-promoting bacilli facilitate the seedling growth of canola, *Brassica campestris*. Plant Physiol Biochem 41:277–281
Glick BR (1995) The enhancement of plant growth by free-living bacteria. Can J Microbiol 41:109–117
Glick BR, Pasternak JJ (2003) Molecular biotechnology: principles and applications of recombinant DNA, 3rd edn. ASM, Washington
Glick BR, Penrose DM, Li J (1998) A model for the lowering of plant ethylene concentrations by plant growth-promoting bacteria. J Theor Biol 190:63–68

Glick BR, Patten CL, Holguin G, Penrose DM (1999) Biochemical and genetic mechanisms used by plant growth promoting bacteria. Imperial College Press, London

Grichko VP, Glick BR (2001) Amelioration of flooding stress by ACC deaminase containing plant growth-promoting bacteria. Plant Physiol Biochem 39:11–17

Grichko VP, Filby B, Glick BR (2000) Increased ability of transgenic plants expressing the bacterial enzyme ACC deaminase to accumulate Cd, Co, Cu, Ni, Pb and Zn. J Biotechnol 81:45–53

Guzman P, Ecker JR (1990) Exploiting the triple response of *Arabidopsis* to identify ethylene-related mutants. Am Soc Plant Physiol 2:513–523

Holguin G, Glick BR (2001) Expression of the ACC Deaminase Gene from *Enterobacter cloacae* UW4 in *Azospirillum brasilense*. Microb Ecol 41:281–288

Holguin G, Glick BR (2003) Transformation of *Azospirillum brasilense* Cd with an ACC deaminase gene from *Enterobacter cloacae* UW4 fused to the Tetr gene promoter improves its fitness and plant growth promoting ability. Microb Ecol 46:122–133

Honma M, Shimomura T (1978) Metabolism of l aminocyclopropane- 1-carboxylic acid. Agric Biol Chem 42:1825–1831

Hontzeas N, Saleh S, Glick BR (2004a) Changes in gene expression in canola roots induced by ACC-deaminase-containing plant-growth-promoting bacteria. Mol Plant Microbe Interact 12:951–959

Hontzeas N, Zoidakis J, Glick BR, Abu-Omar MM (2004b) Expression and characterization of 1-aminocyclopropane-1-carboxylate deaminase from the rhizobacterium *Pseudomonas putida* UW4: a key enzyme in bacterial plant growth promotion. Biochem Biophys Acta 1703:11–19

Hontzeas N, Hontzeas CE, Glick BR (2006) Reaction mechanisms of bacterial enzyme 1-aminocyclopropane-1-carboxylate deaminase. Biotechnol Adv 24:420–426

Hsiao TC, Xu LK (2000) Sensitivity of growth of roots versus leaves to water stress: biophysical analysis and relation to water transport. J Exp Bot 51(350):1595–1616

Hyodo H (1991) Stress/wound ethylene. In: Mattoo AK, Suttle JC (eds) The plant hormone ethylene. CRC, Boca Raton, pp 65–80

Jackson MB (1997) Hormones from roots as signal for the shoots of stressed plants. Trends Plant Sci 2:22–28

Jia YJ, Ito H, Matsui H, Honma M (2006) 1-aminocyclopropane-1-carboxylate (ACC) deaminase induced by ACC synthesized and accumulated in *Penicillium citrinum* intracellular spaces. Biosci Biotechnol Biochem 64:299–305

Karley AJ, Leigh RA, Sanders D (2000) Where do all the ions go? The cellular basis of differential ion accumulation in leaf cells. Trends Plant Sci 5(11):465–470

Karthikeyan S, Zhou Q, Zhao Z, Kao C, Tao Z, Robinson H (2004) Structural analysis of Pseudomonas 1-aminocyclopropane-1-carboxylate Deaminase complexes: insight into the mechanism of a unique pyridoxal-5-phosphate dependent cyclopropane ring opening reaction. Biochemistry 43:13328–13339

Kende H (1993) Ethylene biosynthesis. Annu Rev Plant Physiol Plant Mol Biol 44:283–307

Khalid A, Akhtar MJ, Mahmood MH, Arshad M (2006) Effect of substrate-dependent microbial produced ethylene on plant growth. Microbiology 75:231–236

Kim WT, Siverstone A, Yip WK, Dong JG, Yang SF (1992) Induction of 1-aminocyclopropane-1-carboxylate synthase mRNA by auxin in mung bean hypocotyls and cultured apple shoots. Plant Physiol 98:465–471

Li J (1999) Isolation, characterization and regulation of 1-aminocyclopropane-1-carboxylate deaminase genes from plant growth promoting rhizobacteria. Ph.D thesis, University of Waterloo, ON, Canada

Lincoln JE, Fischer RL (1988) Diverse mechanisms for the regulation of ethylene-inducible gene expression. Mol Gen Genet 212:71–75

Ma W, Guinel FC, Glick BR (2003) *Rhizobium leguminosarum* biovar viciae 1-aminocyclopropane-1-carboxylate deaminase promotes nodulation of pea plants. Appl Environ Microbiol 69:4396–4402

McCune JM (1975) Definition of invisible injury in plants. In: Treshow M (ed) Interaction of air pollutants and plant diseases, vol 122. Academic, New York, pp 307–334

McKeon T, Yang SF (1987) Biosynthesis and metabolism of ethylene. In: Davies PJ (ed) Plant hormones and their role in plant growth and development. Martinus Nijhoff, Boston, pp 94–112

Mendelsohn R, Rosenberg NJ (1994) Framework for integrated assessments of global warming impacts. Clim Change 28:15–44

Minami R, Uchiyama K, Murakami T, Kawai J, Mikami K, Yamada T (1998) Properties, sequence, and synthesis in Escherichia coli of 1-aminocyclopropane-1-carboxylate deaminase from *Hansenula saturnus*. J Biochem (Tokyo) 123:1112–1118

Moeder W, Barry CS, Tauriainen AA, Betz C, Tuomainen J, Utriainen M, Grierson D, Sandermann H, Langebartels C, Kangasjärvi J (2002) Ethylene synthesis regulated by biphasic induction of 1-aminocyclopropane-1-carboxylic acid synthase and 1-aminocyclopropane-1-carboxylic acid oxidase genes is required for hydrogen peroxide accumulation and cell death in ozone-exposed tomato. Plant Physiol 130:1918–1926

Mol JNM, Holton TA, Koes RE (1995) Floriculture: genetic engineering of commercial traits. Trends Biotechnol 13:350–355

Munns R (2002) Comparative physiology of salt and water stress. Plant Cell Environ 25(2): 239–250

Munns R, Tester M (2008) Mechanisms of salinity tolerance. Annu Rev Plant Biol 59:651–681

Nayani S, Mayak S, Glick BR (1998) The effect of plant growth promoting rhizobacteria on the senescence of flower petals. Ind J Exp Biol 36:836–839

Neumann PM (1993) Rapid and reversible modifications of extension capacity of cell walls in elongating maize leaf tissues responding to root addition and removal of NaCl. Plant Cell Environ 16(9):1107–1114

Nie L, Shah S, Burd GI, Dixon DG, Glick BR (2002) Phytoremediation of arsenate contaminated soil by transgenic canola and the plant growth-promoting bacterium *Enterobacter cloacae* CAL2. Plant Physiol Biochem 40:355–361

Nierman WC, Feldblyum TV, Laub MT, Paulsen IT, Nelson KE, Eisen JA, Heidelberg JF, Alley MR, Ohta N, Maddock JR, Potocka I, Nelson WC, Newton A, Stephens C, Phadke ND, Ely B, Deboy RT, Dodson RJ, Durkin AS, Gwinn ML, Haft DH, Kolonay JF, Sumit J, Craven MB, Khouri H, Shetty J, Berry K, Utterback T, Tran K, Wolf A, Vamathevan J, Ermolaeva M, White O, Salzberg SL, Venta JC, Shapiro L, Fraser CM, Eisen J (2001) Complete genome sequence of *Caulobacter crescentus*. Proc Natl Acad Sci USA 98: 4136–4141

Okazaki S, Nukui N, Sugawara M, Minamisawa K (2004) Rhizobial strategies to enhance symbiotic interactions: rhizobitoxine and 1-aminocyclopropane-1-carboxylate deaminase. Microbes Environ 19:99–111

Ose T, Fujino A, Yao M, Watanbe N, Honma M, Tanak I (2003) Reaction intermediate structure of 1-aminocyclopropane-1-carboxylate deaminase. J Biol Chem 278(4):41069–41076

Pandey P, Kang SC, Maheshwari DK (2005) Isolation of endophytic plant growth promoting *Burkholderia* sp. MSSP from root nodules of *Mimosa pudica*. Curr Sci 89:170–180

Parkhill J, Wren BW, Thomson NR, Titball RW, Holden MT, Prentice MB, Sebaihia M, James KD, Churcher C, Mungall KL, Baker S, Bashan D, Bentley SD, Brooks K, Cerdeno-Tarrage AM, Chillingworth T, Cronin A, Davies RM, Davis P, Dougan G, Feltwell T, Hamlin N, Holroyd S, Jagels K, Karlshev AV, Leather S, Moule S, Oyston PC, Quail M, Rutherford K, Simmonds M, Skelton J, Stevens K, Whitehead S, Barrell BG (2001) Genome sequence of *Yersinia pestis*, the causative agent of plague. Nature 413:523–527

Penrose DM, Glick BR (2001) Levels of 1-aminocyclopropane-1-carboxylic acid (ACC) in exudates and extracts of canola seeds treated with plant growth-promoting bacteria. Can J Microbial 47:368–372
Persello-Cartieaux F (2003) Tales from the underground: molecular plant–rhizobia interactions. Plant Cell Environ 26:189–199
Pratiyon J, Rolfe BG, Mathesius U (2006) The Ethylene-insensitive sickle mutant of Medicago truncatula shows altered auxin transport regulation during nodulation. Plant Physiol 142:168–180
Rasche F, Velvis H, Zachow C, Berg G, Van Elsas JD, Sessitsch A (2006) Impact of transgenic potatoes expressing anti-bacterial agents on bacterial endophytes is comparable with the effects of plant genotype, soil type and pathogen infection. J Appl Ecol 43:555–566
Read TD, Salzberg SL, Pop M, Shumway M, Umayam L, Jiang L, Holtzapple E, Busch JD, Smith KL, Schupp JM, Solomon D, Keim P, Fraser CM (2002) Comparative genome sequencing for discovery of novel polymorphisms in *Bacillus anthracis*. Science 296:2028–2033
Reed MLE, Glick BR (2005) Growth of canola (*Brassica napus*) in the presence of plant growth-promoting bacteria and either copper or polycyclic aromatic hydrocarbons. Can J Microbiol 51:1061–1069
Reid M (1987) Ethylene in plant growth, development and senescence. In: Davies PJ (ed) Plant hormones and their role in plant growth and development. Martinus Nijhoff, Boston, pp 257–279
Robison MM, Shah S, Tamot B, Pauls KP, Moffatt BA, Glick BR (2001) Reduced symptoms of Verticillium wilt in transgenic tomato expressing a bacterial ACC deaminase. Mol Plant Pathol 2:135–145
Ryu CM, Farag MA, Hu CH, Reddy MS, Kloepper JW, Pare PW (2004) Bacterial volatiles induce systemic resistance in Arabidopsis. Plant Physiol 134:1017–1026
Saraf M, Tank N (2005) Increased plant fitness by ACC deaminase containing bacteria. Agrobios Newsl 4(5):20–21
Saravanakumar D, Samiyappan R (2007) ACC deaminase from *Pseudomonas fluorescens* mediated saline resistance in groundnut (*Arachis hypogea*) plants. J Appl Microbiol 102:1283–1292
Sergeeva E, Shah S, Glick BR (2006) Tolerance of transgenic canola expressing a bacterial ACC deaminase gene to high concentrations of salt. World J Microbiol Biotechnol 22:277–282
Shah S, Li J, Moffatt BA, Glick BR (1998) Isolation and characterization of ACC deaminase genes from two different plant growth-promoting rhizobacteria. Can J Microbiol 44:833–843
Shaharoona B, Arshad M, Zahir ZA (2006) Effect of plant growth promoting rhizobacteria containing ACC-deaminase on maize (*Zea mays* L.) growth under axenic conditions and on nodulation in mung bean (*Vigna radiata* L.). Lett Appl Microbiol 42:155–159
Shaharoona B, Arshad M, Khalid A (2007) Differential response of etiolated pea seedling to 1-aminocyclopropane-1-carboxylate and/or L-methionine utilizing rhizobacteria. J Micrbiol 45(1):15–20
Sullivan JT, Trzebiatowski JR, Cruickshank RW, Gouzy J, Brown SD, Elliot RM, Fleetwood DJ, Mc Callum NG, Rossbach U, Stuart GS, Weaver JE, Webby RJ, De Bruijn FJ, Ronson CW (2002) Complete sequence analysis of the symbiosis island of *Mesorhizobium loti* strain R7A. J Bacteriol 184:3086–3095
Tank N, Saraf M (2010) Salinity resistant PGPR ameliorates NaCl stress on tomato plants. J Plant Interact 5(1):51–58
Tester M, Davenport R (2003) Na+ tolerance and Na+ transport in higher plants. Ann Bot 91(5):503–527
Uchiumi T, Oowada T, Itakura M, Mitsui H, Nukui N, Dawadi P, Kaneko T, Tabata S, Yokoyama T, Tejima T, Saeki K, Oomori H, Hayashi M, Maekawa T, Sriprang R, Murooka Y, Tajima S, Simomura K, Nomura M, Suzuki A, Shimoda S, Sioya K, Abe M, Minamisawa K (2004) Expression islands clustered on symbiosis island of *Mesorhizobium loti* genome. J Bacteriol 186:2439–2448

Van Loon LC, Glick BR (2004) Increased plant fitness by rhizobacteria. In: Sandermann H (ed) Molecular ecotoxicology of plants. Springer, Berlin, pp 177–205

Van Loon LC, Bakker PAHM, Pieterse CMJ (1998) Systemic resistance induced by rhizosphere bacteria. Annu Rev Phytopathol 36:453–483

VanLoon LC (1984) Regulation of pathogenesis and symptom expression in diseased plants by ethylene. In: Fuchs Y, Chalutz E (eds) Ethylene: biochemical, physiological and applied aspects. Martinus Nijhoff/Dr W. Junk, The Hague, pp 171–180

Wang KL, Li H, Ecker JR (2002) Ethylene biosynthesis and signaling networks. Plant Cell 14:131–151

White PJ, Broadley MR (2001) Chloride in soils and its uptake and movement within the plant: a review. Ann Bot 88:967–988

Wood V, Gwilliam R, Rajandream MA, Lyne M, Lyne R, Stewart A, Sgouros J (2002) The genome sequence of *Schizosaccharomyces pombe*. Nature 15:871–880

Yao M, Ose T, Sugimoto H, Horiuchi A, Nakagawa A, Wakatsuki S, Yokoi D, Murakami T, Honma M, Tanaka I (2000) Crystal structure of 1-aminocyclopropane-1-carboxylate deaminase from *Hansenula saturnus*. J Biol Chem 44(3):34557–34565

The Role of the C:N:P Stoichiometry in the Carbon Balance Dynamics of the Legume–AMF–Rhizobium Tripartite Symbiotic Association

Vincent M. Gray

Contents

1 Introduction 388
2 The Legume Tripartite Symbiotic Association 390
3 AMF Carbon Economy 392
4 Rhizobial Carbon Economy 393
5 N and P Control of Legume Photosynthesis 393
6 Microbial Symbiont Effect on Legumes 396
7 Microbial Effects on Legume Maintenance Respiration 396
8 Mycorrhizal C:P Exchange Dynamics 397
9 Rhizobial C:N Exchange Dynamics 400
10 Rhizobium Nodule Phosphorus Requirements 402
11 Interrelations Between the C:P and C:N Exchange Dynamics 403
12 Nitrogen Utilization in Legume Biomass Production 405
13 Phosphorus Utilization in Legume Biomass Production 407
14 Conclusion 409
References 410

Abstract Synergistic or additive interactions among the partners of the legume tripartite symbiotic association (*Rhizobium*–Arbuscular mycorrhizal fungi–legume) have been shown in most instances to increase legume productivity. Arbuscular mycorrhizal fungi (AMF) promote increased legume biomass production and photosynthetic rates by increasing the ratio of P to N accumulation. An increase in the P content in legume tissue due to the AMF symbiotic association has been consistently associated with an increase in N accumulation and N productivity in legumes with or without a *Rhizobium* association. Photosynthetic N use efficiency, irrespective of the inorganic source of N is usually enhanced by increased P supply because of the AMF association. Both light-saturated photosynthetic rates and

V.M. Gray
School of Molecular and Cell Biology, University of the Witwatersrand, Johannesburg, South Africa
e-mail: Vincent.Gray@wits.ac.za

D.K. Maheshwari (ed.), *Plant Growth and Health Promoting Bacteria*, Microbiology Monographs 18, DOI 10.1007/978-3-642-13612-2_17,

quantum yields increase in legumes in response to increasing N supply due to the *Rhizobium* symbiotic association. However, the maximum levels achieved for both light-saturated photosynthesis and quantum yield as a function of N supply concentration depend on both P and CO_2 supply rates. The N:P supply ratio controls the legume's growth and photosynthetic response to elevated atmospheric CO_2 concentrations. These findings indicate that the N:P:C supply ratio as influenced by the tripartite symbiotic associations plays a fundamental role in controlling the legume's photosynthetic rate and biomass productivity.

1 Introduction

Resource acquisition and allocation in legumes are dependent on a complex set of exchanges between the three members of the legume–*Rhizobium*–mycorrhizal tripartite symbiotic association (Fig. 1). The biological basis for the superiority

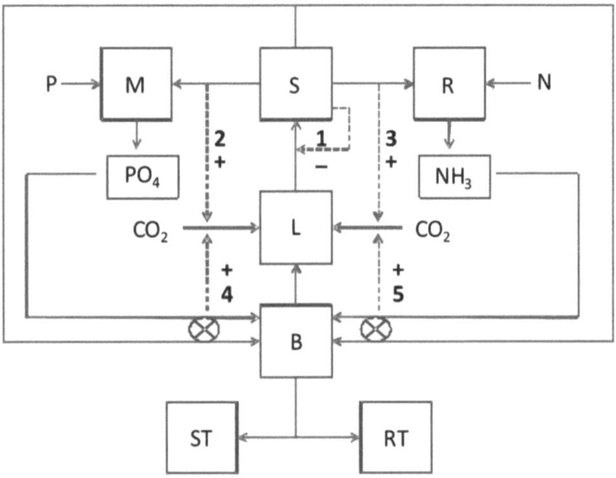

Fig. 1 Model of the three-way resource exchange system between the three symbiotic partners of the legume tripartite association. The *arrows* represent the fluxes or exchanges of C, N and P between partners of the tripartite symbiotic association. M, AM fungi; R, rhizobial nodule system; S, substrate carbon (C) storage pool; L, leaves which function as source of the carbon substrate; B, plants metabolic pool in which the processes of catabolism and anabolism take place; ST, represents the plant stems and RT represents the plant root system. The *broken arrows* with *plus* sign represent stimulation of photosynthesis by (a) C fluxes from the carbon storage pool to the AM fungi (2) and to the rhizobial nodules (3) both promote photosynthetic CO_2 assimilation; (b) fluxes of P (4) and N (5) into the metabolic pool promote the flux of substrate carbon (C) from the carbon storage pool into the metabolic pool which in turn promotes photosynthetic CO_2 assimilation. Build up of C in the carbon storage pool represses (*broken arrow* 1 with *minus* sign) photosynthetic CO_2 assimilation. The *cross* in the *circle symbol* indicates that the flux of C into the metabolic pool is controlled by the supply of N and P. Biosynthesis of the biomass polymers from C takes place in the metabolic pool

of legume crops derives from the three-way resource exchanges among members of the tripartite symbiotic association (Bethlenfalvay and Newton 1991; Barea et al. 1992). AMF and *Rhizobium* play an important role as microbial endosymbionts in the supply of P and N, respectively, to legumes growing in nutrient deficient soils (Azcón et al. 1979). In exchange for P or N, the two microbial symbionts receive C from the legume host. Thus, the formation of the tripartite symbiotic association (legume–AMF–*Rhizobium*) is codependent on a complex three-way source–sink relation involving C exchanges for P and C exchanges for N (Brown and Bethlenfalvay 1988). In most reported instances, these exchanges have had a positive influence on legume growth (Azcón et al. 1979; Paul and Kucey 1981; Harris et al. 1985; Brown and Bethlenfalvay 1988; Jia et al. 2004).

In general, the legume has been shown to have the capacity to fully compensate for any internal carbon deficit resulting from photosynthate transfers to the microbial endosymbionts (Jia et al. 2004; Kaschuk et al. 2009). Furthermore, in general, the evidence indicates that the C:P and C:N exchanges between the host and the two microbial symbionts under P and N limiting conditions do not diminish legume productivity relative to the productivity of plants that are not nutrient limited (Azcón et al. 1979; Paul and Kucey 1981; Harris et al. 1985; Brown and Bethlenfalvay 1988; Gray 1996). Plant growth is usually co-limited by both N and P supply (Jia and Gray 2004a, b). This observation is consistent with recent studies that have investigated the relationship between N:P stoichiometries and yield maximization in various crops (Ågren 2004; Sadras 2006). In previous studies, we have shown that the ratio of P to N was a major factor in determining the level of productivity in *Vicia faba* L (Jia et al. 2004 Jia and Gray 2004a, b). Recent studies have shown that legumes that were partners in the tripartite symbiotic association (*V. faba*–AMF–*Rhizobium*) had significantly higher elemental P to N ratio compared to plants with no symbiotic association (Jia et al. 2004). These results also confirmed the original observations of Brown and Bethlenfalvay (1988) that plants colonized by both AMF and *Rhizobium* had significantly higher photosynthetic nitrogen-use efficiencies and photosynthetic phosphorus-use efficiencies. Both P and N use efficiency has been shown to be strongly dependent on the P to N supply ratio (Jia and Gray 2004a, b, 2007). In N and P supply studies involving *V. faba* L for plants without any microbial symbiotic associations, it was found that the optimal values for the different photosynthetic parameters such as photon saturated net photosynthetic rates (P_{max}), quantum efficiency (α), intercellular CO_2 concentrations (C_i) and carboxylation efficiency (CE) were dependent on both N supply rate and leaf nitrogen content (Jia and Gray 2004a). It was also found that the level of N accumulation and the optimal values for the above photosynthetic parameters were positively influenced by the level of P supply (Jia and Gray 2004b; Jia and Gray 2007). These results indicate that the P or N exchanges from the microbial symbiont for host C also have a stimulatory effect on leaf photosynthetic capacity (Jia et al. 2004).

2 The Legume Tripartite Symbiotic Association

Growth of legumes under limiting nitrogen and phosphorus regimes is facilitated by the resource acquisitions efficiencies and capacities of the two microsymbionts, *Rhizobium* and AMF. AMF symbiotic associations with plant roots generally improve plant growth by enhancing the uptake of inorganic phosphorus (Jayachandran et al. 1992). Each of the symbiotic partners in the legume tripartite system has a specific source and sink function with regard to C, N and P exchanges. In response to C, N and P demand or supply, each of the tripartite symbiotic partners plays a specific role in this three-way source–sink exchange system (Fig. 1). It has been proposed that this three-way resource exchange may be subject to intersymbiont competition (Bayne et al. 1984; Bethlenfalvay 1992). The two microsymbionts may be regarded as the primary sources of P and N for legumes growing in soils deficient in plant-available forms of these two nutrients (Azcón et al. 1979, 1988; Barea and Azcón-Aguilar 1983; Piccini et al. 1988; Cihacek 1993). From the legume side of the symbiotic association, the exchange of resources involves the allocation of carbon to nodules in exchange for reduced N and to the AM fungi in exchange for P (Fig. 1). Phosphorus, the major plant growth limiting factor apart from N, is required for photosynthesis in the leaves of the legume and also for nitrogen fixation in the root nodules (Israel 1987; Haaker 1988; Bethlenfalvay and Newton 1991). The contribution of AM fungi to the tripartite symbiotic association is particularly significant for nodulated legumes growing under a soil regime where available inorganic nitrogen in the form of ammonium or nitrate is limiting, the reason being the high P requirement for nodulation (Daft 1978; Betlenfalvay and Yoder 1981) and N_2 fixation (Bergersen 1971).

With respect to the above ground and below ground source–sink interactions in legume systems, the roots, nitrogen-fixing nodules and mycorrhizal fungi all compete for a share of the below ground carbon allocation. An appreciation of this potential three-way below ground competition for carbon in legumes brings a new perspective to the conceptualization of source–sink dynamics in legumes. It has been reported that 42% or more of daily net photosynthate can be allocated to the belowground legume–*Rhizobium*–mycorrhizal association (Paul and Clark 1989). Paul and Kucey (1981) reported that 60% of the photosynthetic carbon flux was partitioned into the below ground root-nodule–mycorrhizal association. This below ground fraction of daily carbon allocation is nearly evenly distributed (12, 13 and 17%) to nodules, the root, and the mycorrhizal fungi (Paul and Clark 1989). In one report regarding the photosynthate allocation schedule in growing alfalfa plants, the partitioning of the carbon in the following proportions to the major organ systems was observed: 26.2% to the main stem; 12.7% to shoot apex unexpanded leaves on the main stem; 0.8% to the fully expanded leaves on the main stem; 27.1% to the auxiliary bud shoots on the main stem; 6.5% to the crown shoots; 3.8% to the crown; 19.1% the roots and 3.5% to the nodules (Cralle et al. 1987). If the crown is included as part of the taproot, then 73.3% of photosynthate is allocated to the shoot, and the remaining 26.7% of fixed carbon is allocated to the roots and nodules (Cralle et al. 1987). The carbon allocation to mycorrhizal fungi, which is an obligate symbiont, can constitute between 4 and 20% of host photosynthate as indicated in single host–fungus combinations (Azcón and Ocampo 1984; Douds et al. 1988; Pearson and Jakobsen 1993). Observations on the supply of photosynthate in legume systems confirm that photosynthate production is in excess of carbon demand by the nodules (Gordon et al. 1985; Kouchi et al. 1985; Hostak et al. 1987; Vance and Heichel 1991; Kaschuk et al. 2009).

While there is excess photosynthate supply capacity in legumes such as alfalfa as evidenced in the accumulation of starch in the taproots in these legumes (Vance and Heichel 1991), the above examples do demonstrate that legume root microsymbionts represent substantial carbon sinks. Taken together, a series of observations (Hostak et al. 1987; Walsh et al. 1987) indicate that the growth of legumes such as alfalfa or soybean was not limited by source photosynthetic capacity but rather by microsymbiont sink strength for carbon. This becomes especially significant if the flux of carbon to the microsymbionts is regulated by the plant in exchange for P or N. This idea of microsymbiont carbon sink strength defined in terms of carbon demand being coupled to the microsymbionts capacity to supply N or P to the legume needs to be more fully developed. Which of the possible two resource exchanges (C:P or C:N) is most limiting or constraining with respect to legume growth? Is it possible that the carbon demand by the microsymbiont sinks could limit or constrain legume growth? Is it possible that AMF microsymbiont's capacity to accumulate, transport and mobilize P in exchange for C could limit legume growth? Or alternatively is it possible that the rhizobial microsymbiont' capacity to fix and mobilize N in exchange for C and P could limit legume growth? Given that the supply of P or N or both are factors that in general limit plant growth, then all factors influence that effect and the C, N, and P exchange dynamics among symbionts of the tripartite association will have an impact on legume productivity. These questions will be explored in this chapter. There is evidence that the relative amounts of carbon allocated for storage or growth depend on the supply of N and P to the legume (Greenwood et al. 1991).

In Fig. 1, photosynthetic CO_2 assimilation is controlled by the flux of substrate carbon out of the carbon storage pool (Gray 2000). This flux is promoted by the C:P and C:N exchanges between the legume host and the two microsymbionts, AMF and *Rhizobium* respectively. Various consequences of the relationships and dynamics depicted in Fig. 1 will be investigated in this chapter. The focus of this investigation is the status of the hypothesis that the photosynthetic capacity of the legume host and its growth rate are constrained by the supply of N and P. In addition, it will be argued that the sink demand for C or the flux of C to the two microsymbionts does not limit legume growth as its photosynthetic capacity can be increased to compensate for C losses to the two microsymbionts.

3 AMF Carbon Economy

In an analysis of the carbon economy of the tripartite soybean–*Glomus–Rhizobium* symbiotic association, Harris et al. (1985) found that carbon was allocated in the following proportions: 30.49% to leaves, 20.52% to stems and petioles, 6.3% to shoot respiration, 7.8% to roots, 2.0% to nodules, 2.7% to AMF, 5.2% to root and soil respiration, 13.7% to AMF respiration and 9.4% to nodule respiration. In that study, it was reported that approximately 42.6% of photosynthate was allocated to below ground sinks. This below ground carbon allocation was distributed between the various sinks as follows: 38.6% to the AMF, 26.6% to nodules and 30.6% to roots. When compared to the case of nonsymbiotic plants with nonlimiting supplies of P and N, it appears that the complex C, N and P three-way source–sink relations between the members of the tripartite symbiotic association do not limit plant productivity (Brown and Bethlenfalvay 1988). This supports the hypothesis that the carbon demand by the microbial symbionts does not limit plant growth.

In the case of AMF, the sink demand created by fungal colonization could account for an extra 4–26% drain of photosynthate from the AM-infected host plant (Kucey and Paul 1982a, b; Snellgrove et al. 1982; Koch and Johnson 1984; Harris et al. 1985; Douds et al. 1988; Wang et al. 1989; Jakobsen and Rosendahl 1990; Black et al. 2000). The maximum hypothetical photosynthetic allocation to the AMF association may well be as high as 40–50% of the total photosynthate production (Stribley et al. 1980). Hence, it has been suggested that the carbon demand by the AMF symbiont has the potential to limit plant growth, and thereby bring about a decline in the plant growth efficiency (Buwalda and Goh 1982, Jia et al. 2004; Jia and Gray 2008; Kaschuk et al. 2009). It has been proposed that increases in the photosynthetic rate of the AM-infected host plant fully compensates for carbon losses from the plant because of increases in AM carbon demand (Brown and Bethlenfalvay 1988; Fitter 1991; Wright et al. 1998a, b; Kaschuk et al. 2009). If N and P status of AM-infected plant and nonmycorrhizal plants were similar, it would be found that mycorrhizal plants had higher photosynthetic rates but similar biomass to the nonmycorrhizal, indicating that the additional photosynthate had been allocated to the fungal symbiont (Wright et al. 1998a, b) as supported by the fact that increases in the photosynthetic rate of AM-infected plants growing under low P conditions may be mainly due to mycorrhizal-dependent increases in the plant P status (Allen et al. 1981; Fredeen and Terry 1988; Azcón et al. 1992; Black et al. 2000). Further increase in photosynthetic rates in AM-infected plants may be due to the combined effects of enhanced P status and the AM-dependent carbon sink (Black et al. 2000). An increase in the photosynthetic rate in the leaves of AM-infected cucumber was found to be due to an increase in the leaf P status and not because of compensatory increases in photosynthesis in response to increase in the mycorrhizal sink demand for assimilates (Black et al. 2000). In barley where there was no difference in P status between AM-infected plants and nonmycorrhizal plants, the AM-infected plants had enhanced photosynthetic rates, indicating a compensatory response to mycorrhizal colonization

(Fay et al. 1996). Also, at equivalent P:N ratios, AM-infected *Andropogon gerardii* had higher overall photosynthetic rates compared to the nonmycorrhizal plants (Miller et al. 2002). Enhancement of photosynthetic rates in mycorrhizal plants was found not to be indirectly related to the plant P or N status but directly related to compensatory responses to fungal colonization (Miller et al. 2002) as also evidenced by Wright et al. (1998a, b).

A number of attempts have been made to quantify the stoichiometric exchange of mycorrhizal acquired phosphorus for photosynthate (Douds et al. 1988; Eissenstat et al. 1993; Pearson and Jakobsen 1993). Schwab et al. (1991) have proposed a model involving the exchange of one triose-phosphate for one inorganic phosphate molecule. This gives a carbon/phosphorus exchange ratio of 3. A positive correlation has been observed to exist between the capacity of AMF to stimulate the growth of the host plant and the radiant flux density incident on the plant (Same et al. 1983; Son and Smith 1988). Thus, it could be argued that any enhancement of additional photosynthetic capacity above the levels of P unlimited nonmycorrhizal plants would be strongly dependent on the P supply from AMF.

4 Rhizobial Carbon Economy

In the case of the rhizobial endosymbiont, dinitrogen (N_2) fixation and nodule growth also like AMF create a sink demand for photosynthate (Bethlenfalvay and Brown 1985; Brown and Bethlenfalvay 1986). A number of estimates evaluating the carbon energetic cost of N_2-fixation have been attempted (Salsac et al. 1984; Twary and Heichel 1991; Vance and Heichel 1991). Total energy costs including energy utilized for the growth of nodulated roots, maintenance respiration of nodules, N_2-fixation and ammonia assimilation range from 0.4 to 19.4 g C g^{-1} N (Salsac et al. 1984). If the energetic costs associated with the growth and maintenance of nodules are omitted, then the estimates of the energy costs for N_2-fixation range from 2.5 to 8.0 g C g^{-1} N (Atkins 1984; Salsac et al. 1984, Minchin and Witty 2005). Gross and net carbon requirements for N_2-fixation in alfalfa range from 3.5 to 11.9 g C g^{-1} N and 2.5 to 8.8 g C g^{-1} N, respectively (Twary and Heichel 1991). The average value from over 35 estimates for the carbon cost of N_2-fixation was approximately 6.0 g C g^{-1} N (Vance and Heichel 1991).

5 N and P Control of Legume Photosynthesis

It is likely that the microbial symbionts exert a positive influence on the legumes photosynthetic capacity via control of the N and P supply to the host (Fig. 1). For example, inorganic nitrogen supply rates exert a direct influence on photon saturated net photosynthetic rates, quantum efficiency, intercellular CO_2 concentrations and carboxylation efficiency (Jia and Gray 2004a, b). In addition, the optimal values attained for the above photosynthetic parameters in response to N supply are in turn strongly modulated by P supply (Jia and Gray 2004a, b, 2007). As leaf nitrogen concentration increased, the α converged onto a maximum asymptotic value of 0.0664 $\pm$ 0.0049 μmol (CO_2) μmol (quantum)$^{-1}$. Also, as leaf nitrogen concentration increased, the C_i value fell to an asymptotic minimum of 115.80 $\pm$ 1.59 μmol mol^{-1}, and CE converged onto a maximum asymptotic value of 1.645 $\pm$ 0.054 μmol (CO_2) m^{-2} s^{-1} Pa^{-1} and declined to zero at a leaf nitrogen content equal to 0.596 $\pm$ 0.096 g(N) m^{-2}. Quantum efficiency fell to zero for a leaf nitrogen content of 0.660 $\pm$ 0.052 g(N) m^{-2}. In *V. faba* L as leaf nitrogen content increases, the value of photon saturated photosynthesis converged onto a maximum asymptotic value of 33.400 $\pm$ 2.563 μmol (CO_2) m^{-2} s^{-1}. Photon saturated photosynthesis fell to zero for a leaf nitrogen content equal to 0.710 $\pm$ 0.035 g(N) m^{-2}. Increased P supply increased the photosynthetic N use efficiency in terms of P_{max} and α. Increased P supply was also associated with an increase in CE and a decrease in $C_{i.}$.

It has been observed that the responses of plant communities to global warming and elevated CO_2 were influenced by leaf N:P ratios (Hedin 2004), which are in turn dependent on soil N and P supply. This proposal has received support from experiments which exhibited increases in biomass production in plants acclimatized to elevated CO_2 (440 and 600 CO_2 μL L^{-1}) relative to that in control plants (280 CO_2 μL L^{-1}) depending on the level of NPK supply (Grünweig and Korner 2003). These results are consistent with the observations that photosynthesis is co-limited by both N and P supply (Jia and Gray 2004a, b, 2007; Jia et al. 2004). In *Pinus pinaster* the magnitude of growth, photosynthetic rates and N partitioning into ribulose-1, 5-bisphosphate carboxylase oxygenase (RubisCO) in response to increasing N supply were also positively modulated by P supply (Warren and Adams 2001). This modulation of the photosynthetic response to N supply by P may take place either directly or indirectly. Direct modulation of photosynthetic activity by P may be facilitated through the influence of P on RubisCO activation (Marcus and Gurevitz 2000). Alternatively, indirect control of photosynthetic rates by P supply could be exerted through the chloroplast phosphate shuttle. Phosphate recycling between the chloroplast and cytoplasm has been observed to modulate the photosynthetic rate by influencing the rate of export of photosynthate from the chloroplast (Cockburn et al. 1967a, b; Usuda and Edwards 1982; Mächler et al. 1984; Rao and Terry 1989; Rao et al. 1989a, b; Usuda and Shimogawara 1991; Rao and Terry 1995). P supply may also indirectly modulate photosynthetic rate by influencing sink demand for photosynthate (Pieters et al. 2001). In general, the proposal that photosynthesis is usually co-limited by both N and P supply is consistent with the observations of recent studies that yield maximization in various crops was influenced by N:P supply stoichiometries (Ågren 2004; Sadras 2006).

Photosynthesis and plant growth are co-limited by N and P supply (Sterner and Elser 2002; Ågren 2004). Maintenance of high photosynthetic rates and plant growth depends on the capacity to mobilize accumulated substrate carbon (C) from the storage pools into anabolic and catabolic metabolic pathways. Mobilization of C from the substrate storage pools into the catabolic and anabolic metabolic pathways is tightly coupled to N and P supply (Fig. 1). When N and P supply becomes limiting, photosynthetic generated substrate carbon accumulates. Accumulation of storage carbon results in the repression of photosynthesis. Under these conditions, plant growth is not limited by carbon supply. However, the growth of the two heterotrophic microbial symbionts is always limited by C supply. In an idealized legume tripartite symbiotic system, the N and P supply rates from the microbial symbionts to the legume will be dependent on C:N and C:P exchange stoichiometries between the legume and the two microsymbionts. What would be the optimal N:P supply ratio for the achievement of maximum photosynthetic and plant growth rates? Also would the supply of CO_2 become limiting when N and P supply is nonlimiting with respect photosynthesis and growth? While an extensive literature now exists on the optimal N:P stoichiometries for crop growth (Sadras 2006), not much is known about the optimum C:N:P ratios for photosynthesis or plant growth. For the legume, the C in the C:N:P ratio would represent CO_2, the inorganic C substrate for plant growth. In the case of the microsymbionts, C represents organic carbon from the plant's substrate storage pool. The study of Grünweig and Korner (2003) does suggest that with long-term exposure to elevated CO_2, plant growth becomes co-limited by both N and P supply. Plants exposed to elevated CO_2 when adapted to ambient CO_2 levels show responses consistent with photosynthesis being co-limited by CO_2, N and P supply (Jia and Gray 2007). In addition, if the photosynthetic catalytic machinery in terms of leaf N concentration determines the source capacity of the plant canopy and P concentration influences the energetic efficiency of CO_2 assimilation into plant biomass, then the stoichiometric ratio of N:P (as a percentage of dry biomass) will determine plant productivity levels in response to CO_2 supply. Therefore, in general, it could also be argued that with increasing CO_2 supply, the sink demand of actively growing tissues for additional reduced carbon and source capacity for assimilating additional CO_2 would also be controlled by the N:P supply ratio. In addition, it can be further argued that with increasing CO_2 supply, sink demand of the legume's microbial symbiotic associations may be controlled by the N:P supply ratio from the microsymbionts.

In general, the evidence indicates that the C:P and C:N exchanges between the host and the two microbial symbionts under P and N limiting conditions do not diminish legume productivity relative to that of plants that are not nutrient limited (Azcón et al. 1979; Paul and Kucey 1981; Harris et al. 1985; Brown and Bethlenfalvay 1988; Gray 1996). In fact, plant growth is usually co-limited by both N and P supply (Jia and Gray 2004b). This observation is consistent with recent studies that have investigated the relationship between N:P stoichiometries and yield maximization in various crops (Ågren 2004; Sadras 2006). Earlier, it was observed that the ratio of P to N was a major factor in determining the level of productivity in *V. faba L* (Jia et al. 2004). Plants that were partners in a tripartite symbiotic association (*V. faba*–AMF–*Rhizobium*) had a significantly higher elemental P to N ratio compared to plants with no symbiotic association. These results also confirmed the original observations of Brown and Bethlenfalvay (1988) that plants colonized by both AMF and *Rhizobium* had significantly higher photosynthetic nitrogen and phosphorus-use efficiencies. Both photosynthetic P and N use efficiencies have been shown to be strongly dependent on the P to N supply ratio (Jia and Gray 2004b).

6 Microbial Symbiont Effect on Legumes

In the case of legumes, there have been several studies on maintenance respiration and growth yield (McCree and Silsbury 1978; Irving and Silsbury 1987); however, these studies did not include the effects of microbial symbiotic partners. Also, while many other studies have focused on the apparent carbon costs induced by the microsymbiotic partners in the tripartite symbiotic association, they did not directly investigate the impact of these costs on the growth yield (Azcón et al. 1979; Paul and Kucey 1981; Harris et al. 1985; Douds et al. 1988). What possible effects do microbial symbiotic associations have on the host growth yield (Y_g) and maintenance respiration under P and N limiting conditions? Increases in legume productivity correspond to increases in the growth yield. For various legumes the average values reported for Y_g are 0.75 for *Trifolium. subterraneum* L (McCree and Silsbury 1978), 0.7 for field bean, 0.72 for Lucerne and 0.68 for chick pea (Irving and Silsbury 1987). In the study undertaken with *V. faba* L, plants with both rhizobial and AMF associations under both low N and P supply conditions achieved values for $Y_g > 0.7$ (Jia and Gray 2008). Hence, the legume partner in the tripartite symbiotic complex can overcome the constraints on growth efficiency that arise as a consequence of low N and P concentrations in the soil, even with the concomitant C losses to the microsymbiotic partners. In broad bean, depending on the presence of microbial symbiotic associations, the Y_g values ranged from 0.44 to 0.78 (Jia and Gray 2008). Growth yield were found to be significantly higher in plants that had one or more microbial symbiotic association (Jia and Gray 2008). It was observed that all the legumes with one microbial symbiotic had similar Y_g values irrespective of the N supply level. Plants with two microbial associations had significantly higher Y_g values than plants with all N and P treatments but without any microbial association.

7 Microbial Effects on Legume Maintenance Respiration

Maintenance respiration (*m*) estimates obtained for field bean, lucerne, chick pea, pea and kidney bean have been as follows: 21.35, 24.11, 28.65, 26.53 and 16.51 mg CO_2 $(gDM)^{-1}$ day^{-1}, respectively (Irving and Silsbury 1987). For *Trifolium subterraneum*, maintenance respiration rates ranged from 14 mg CO_2 $(gDM)^{-1}$ day^{-1} grown at 10°C to 64 mg CO_2 $(gDM)^{-1}$ day^{-1} at 35°C (McCree and Silsbury 1978).

No significant differences were found in the maintenance respiration rates between legumes without any microbial association and those with one or two microbial associations (AMF or rhizobial) (Jia and Gray 2008). Depending on nitrogen supply, the *m* values fell between 12 and 36 $CO_2(gDM)^{-1}$ day^{-1}. Maintenance respiration rates were highest in plants with rhizobial and AMF associations grown under low N supply conditions. As with Y_g, the *m* rates also remained constant over the two harvest intervals for individual treatments. The above growth yield and maintenance respiration results are consistent with the hypothesis that in the legumes potential photosynthetic capacity exceeds the carbon demand of the *Rhizobium*–AMF symbiotic complex.

8 Mycorrhizal C:P Exchange Dynamics

In order to evaluate the impacts of mycorrhizal C demand and P supply on legume growth, two major considerations are (1) the C requirements of the mycorrhizal fungi and (2), the P acquisition efficiency of the extra radicle mycorrhiza hyphal system. Both the carbon requirement of the mycorrhizal symbiont and its efficiency in phosphate acquisition have an impact on the nitrogen and carbon economy of the tripartite legume symbiont system. Mycotrophic growth of plants is most frequently attributed to enhanced P uptake; however, there appears to be an optimal level of mycorrhizal colonization above which the plant receives no additional enhancement in P uptake or growth, yet the plant continues to support mycorrhizal metabolism (Koch and Johnson 1984; Douds et al. 1988). The conditions necessary for the maintenance of mycotrophic growth under steady-state levels of available soil phosphorus can be summed up in the following proposition: The maximum balanced-exponential growth rate under the conditions corresponding to a given steady-state soil phosphorus regime is only achievable for certain values of the following ratios – (a) the plant biomass:mycorrhizal biomass ratio; (b) the mycorrhizal C utilization rate:mycorrhizal P supply rate and (c) the plant specific photosynthetic rate (mg CO_2 g^{-1} dry leave mass s^{-1}):leaf phosphate concentration (% P of dry mass). The values of these specific ratios would represent the optimal values for a given soil phosphorus regime. To get some idea of the order of magnitude with respect to the above ratios, the following values have been reported (they are not necessarily the optimal values): A ratio of 140C/1P (g atom/g atom) corresponding to mycorrhizal C utilization and mycorrhizal P supply for soybean during mycotrophic growth has been reported (Harris et al. 1985). The standard critical leaf concentration for P-deficiency is 0.35% P on a dry mass basis (Scaife et al. 1983); 0.250–0.7 mg P g^{-1} leaf is the leaf P concentration range showing P-deficiency symptoms; 1.0–8.0 mg P g^{-1} leaf is P-sufficiency range for tissue phosphate concentrations (Bould et al. 1983). Leaf phosphate status affects the level of light saturated photosynthesis and plant growth rate (Rao and Terry 1995; Jia and Gray 2004b; Jia et al. 2004).

The regulatory control of carbon allocation to the various below ground sinks which includes the storage, and resource acquisition structural–functional components of the root require some elaboration. The extent of the fungal carbon requirements has not been well investigated; however, it appears that the carbohydrate flux is regulated by the host plant species, and is also dependent on the mycorrhizal fungal species (Sieverding 1991). It is estimated that AM fungi remove for their development and functional activity 1–17% of the carbohydrates allocated by the plant to root biomass production (Sieverding 1991). The maximum hypothetical photosynthate requirement and corresponding potential loss of plant dry matter production to the AM fungal association may well be as high as 40–50% (Stribley et al. 1980). This means that carbon demand by the fungus seems to have the potential to limit plant growth (Buwalda and Goh 1982).

Intraradical hyphae of VA mycorrhiza exhibit four categories of hyphae: (1) intracellular hyphae which exist as coils, often found in the outerlayers of the root cortex; (2) intercellular hyphae; (3) intracellular hyphae with highly ramified and invaginating membrane structures, known as arbuscules and (4), intercellular or intracellular hypertrophic hyphae, called vesicles (Scannnerini and Bonfante-Fasolo 1983). The arbuscules which are the sites of nutrient exchanges are essential for the functioning of VA mycorrhizal associations. The other three intraradical structures are involved in growth and storage functions. Arbuscules are found in the inner root cortex and are formed from a penetrating hypha that invaginates the host plasmalemma and repeatedly bifurcates to form a bushlike structure with progressively thinner branches (Wilcox 1991).

Potential carbon fluxes to mycorrhizal fungi via the arbuscules have been estimated to approach values up to 100 mg C g^{-1} root day^{-1} (Schwab et al. 1983). If the rate of C transfer per unit interfacial area in arbuscules is similar to that in mildew haustoria and C uptake by mycorrhizae occurs only in the arbuscules, then transfer of 5 mg C g^{-1} root day^{-1} to the fungus would require an interfacial surface area of 7×10^3 mm^2 g^{-1} root day^{-1} (Harris and Paul 1987). The magnitude of carbon or phosphate fluxes across the host–fungus interface will be influenced by the area of the arbuscule membrane surface within root cells. It has been calculated that the membrane surface area presented by arbuscules to colonized root cells ranges from 40 to 300 mm^2 of plasma membrane cm^{-1} length of root of onion and maize respectively (Toth and Miller 1984; Toth et al. 1990). Using the data of Cox et al. (1975), Harris and Paul (1987) estimate that 1.6×10^5 arbuscules g^{-1} root dry mass with an arbuscule membrane surface area in the region of 4×10^4 μm^2 $cell^{-1}$ would be necessary to facilitate the uptake of 5 mg C g^{-1} root day^{-1}. If there are approximately 1.4×10^8 cells g^{-1} root (Brown and Broadbent 1950), about 0.1% of cells must contain active arbuscules to facilitate the above daily transfer of C to the mycorrhizae.

Mycorrhizal carbon demand will depend on the carbon requirements necessary to support (1) fungal growth; (2) growth respiration and (3) maintenance respiration. Carbon demand by the fungus will depend on the specific growth rate of the mycorrhizal system. With respect to *Glomus fasciculatum*, Harris et al. (1985) have reported mycorrhizal fungal growth rates of 3.7 mg C day^{-1} when associated with the soybean–*Rhizobium–Glomus* tripartite symbiotic system. Bethlenfalvay et al. (1982) gave a specific growth rate of 0.064 h^{-1} with a doubling time of 11 days for mycorrhizal fungi associated with soybean. Instantaneous specific growth rates ranging from 0.03 to 0.04 day^{-1} have been reported for *G. fasciculatum* in *Glycine max–Rhizobium–G. fasciculatum* associations (Harris et al. 1985). Growth analysis studies of mycorrhizae have demonstrated that fungal biomass production has an exponential phase (Bethlenfalvay et al. 1982). During exponential growth of the mycorrhizal symbiont, substrate C is partitioned into growth of new fungal biomass; growth respiration and maintenance respiration. A maximum growth yield (Y_{max}) of 0.6 has been found for many aerobic heterotrophs growing on a variety of substrates (Payne 1970). The maintenance respiration coefficient, m, has been defined in terms of Y_{max} and the specific maintenance rate, ($time^{-1}$), by $m = b/Y_{max}$. Accurate estimations of mycorrhizal specific maintenance rate (b) have been difficult to obtain; however, the following respiration rates have been reported: 11.0 mg CO_2 g^{-1} h^{-1} for *Glomus mossease* in leeks (Snellgrove et al. 1982) and 11.7 mg CO_2 g^{-1} h^{-1} for *G. fasciculatum* in soybeans (Harris et al. 1985).

Carbon fluxes into the mycorrhizal association and the growth of AMF, or phosphate fluxes from the mycorrhizal symbiont into the legume root, can be meaningfully analysed as a function of three parameters: (1) the area of the intracellular arbuscule membranal exchange interface; (2) percentage of root cells with active arbuscules and (3) the stoichiometric exchange of mycorrhizal acquired phosphate for photosynthate. A number of attempts have been made to quantify the exchange of mycorrhizal acquired phosphorus for photosynthate (Douds et al. 1988; Eissenstat et al. 1993; Pearson and Jakobsen 1993). Schwab et al. (1991) have proposed a model involving the exchange of one triose-phosphate (TP) for one inorganic orthophosphate (Pi) molecule. The root plasma membrane contains a triose-phosphate:inorganic phosphate translocator similar to the triose-phosphate: inorganic phosphate shuttle system corresponding to the phosphate-translocator of chloroplast membranes. In the case of host–mycorrhizal carbon–phosphate exchange, the TPs would be dephosphorylated in the interfacial matrix with the Pi returned to the host via a proton motive force driven Pi pump. The unchanged trioses in turn will be taken up by the fungus down a gradient of chemical potential. The source of exchangeable Pi in the interfacial matrix would probably be derived from the hydrolysis of polyphosphate granules located in the arbuscular hyphae. Phosphate will be transferred down a concentration gradient from the arbuscular hyphae into the interfacial matrix and then be taken up the root cells in exchange for TPs.

Within a given plant species the level of AM fungal infection is positively correlated with (1) the root content of soluble carbohydrates (Same et al. 1983) and (2) the level of sugar exudation (Azcón and Ocampo 1981, 1984). As the host plant supplies photosynthates for AM fungal growth in exchange for phosphate, there should exist a positive correlation between the efficiency of AM fungi as a supplier of phosphate and the photosynthetic rate of the host. Phosphate supply efficiency depends on radiant flux density incident on the plant (Diedrichs 1983a; Tester et al. 1986; Son and Smith 1988) and the day length to which the plant is exposed (Diedrichs 1983b). Shading or defoliation depresses mycorrhizal growth (Same et al. 1983). Phosphorus deficiency is often associated with increased exudation of sugars and enhanced VA mycorrhizal infection of roots (Graham et al. 1981).

9 Rhizobial C:N Exchange Dynamics

It remains unresolved whether assimilation of nitrogen via NO_3^- or N_2 reduction is more favourable to growth and dry matter production in legumes (Atkins 1984). Carbon costs of N_2 fixation vary with host species, rhizobial strain, stage of plant development and method of measuring (Rainbird et al. 1984; Salsac et al. 1984; Skot et al. 1986). Reported values range from 1.3 to 22.8 mol C mol^{-1} N (Twary and Heichel 1991). Under steady-state growth conditions, it can be reasonably assumed that the rate of N acquisition by the rhizobium microsymbionts equals the rate of N supply to the legume host, which in turn is a function of host demand for N. With a dynamic functional equilibrium between plant and N_2 fixing symbiont, biomass production in alfalfa was found to be correlated with nodule mass per plant (Cralle et al. 1987). However, an increase in nodulation was associated with an overall increase in biomass, but the actual proportion of photosynthate partitioning to individual components of the plant (stem, leaves, crown, roots, nodules) remains unaltered. Reduction of N_2 by nitrogenase is an energy-intensive process that results in the consumption of 60–80% of nodule ATP (Heytler and Hardy 1984). Dinitrogen (N_2) fixation by legume–rhizobial symbioses is driven by shoot photosynthesis and therefore can theoretically decrease the amount of photosynthate partitioned to economic yield. If there is no compensatory increase in the photosynthetic rate in response to C demand associated with N_2 fixation, then the photosynthate consumed by the processes of N_2 fixation will be unavailable for plant dry matter production. Twary and Heichel (1991) found that dry matter accumulation in alfalfa was unrelated to the C cost of N_2 fixation.

The C consumed in nodule N_2 fixation is partitioned among the component processes of nodule growth and maintenance, reduction of N_2 by nitrogenase and assimilation plus transport from the nodules (Mahon 1983). Net CO_2 evolution by nodule respiration is usually measured to calculate the C costs of N_2 fixation (Schubert and Wolk 1982). A certain proportion of this CO_2 is refixed by phosphoenolpyruvate carboxylase in the nodule (Twary and Heichel 1991). It therefore follows that gross and net C costs of N_2 fixation need to be distinguished when evaluating the C energetic costs of N_2 fixation. Because of refixation of respiratory CO_2 in the nodules, the C costs of N_2 fixation can be decreased by 10–14% (Ryle et al. 1984).

A number of estimates evaluating the energetic cost of nitrogen fixation have been attempted (Phillips 1980; Rawsthorne et al. 1980; Schubert and Ryle 1980; Minchin et al. 1981; Schubert and Wolk 1982; Skot et al. 1986) in terms of the amount of carbon utilized for nitrogen fixation. In general, these estimates sum up the total costs involved in nitrogen acquisition from N_2 as the nitrogen source. Therefore, the total energetic cost includes the energy utilized in nodulated root growth, maintenance respiration, N_2 fixation and ammonia assimilation. The values range from 0.4 to 19.4 g C g^{-1} N (Salsac et al. 1984). If the energetic costs associated with the growth and maintenance of nodulated roots are omitted, then estimates of the energetic requirements for N_2 fixation range from 2.5 to 8.0 g C g^{-1} N (Atkins 1984; Salsac et al. 1984). Twary and Heichel (1991) observed that the specific N_2 fixation rates in alfalfa depended on the strain of *Rhizobium meliloti*. They reported specific N_2 fixation rates ranging from 3.39 to 16.49 nmol N min^{-1} g^{-1}dry wt nodule. The rhizobium microsymbionts rate of N acquisition would be dependent on the total nodule mass. Selection of alfalfa for greater nodule mass resulted in a proportional increase in the mass of all plant organs (Cralle and Heichel 1986; Cralle et al. 1987). Nodule N_2 fixation occurs unabated throughout both day and night (Vance and Heichel 1991). The carbon energy input that drives N_2 fixation in the nodules appears to be derived entirely from the shoot (Vance and Heichel 1991). During the day, translocation of photosynthate into the nodules drives N_2 fixation and at night mobilization and translocation of shoot carbon storage reserves support N_2 fixation. It would thus appear that the tap root storage reserves are not available for driving N_2 fixation in the nodules (Vance and Heichel 1991). Leaf nitrogen status determines growth potential and if the N input for plant growth is dependent on nodule N_2 fixation then the rate of N_2 fixation becomes the process that limits growth. Under these conditions, the legume growth depends on the plant's nitrogen utilization efficiency and the rate at which fixed nitrogen is supplied to the host. Achievement of maximum growth rates of the nodule system and the plant may be constrained by the exchange rates associated with C export from the plant to the rhizobial nodules and N export from the rhizobial nodules to the plant system.

Under conditions of balance exponential growth, a functional equilibrium relationship with respect to the C:N exchange ratio should exist between the legume and the rhizobium microsymbiont. Under these conditions a system of regulatory control would maintain a dynamic functional equilibrium between the activity of the N_2 fixing nodule system and the photosynthesizing shoot such that the legume's tissue C:N ratio would be kept poised at a constant value during steady-state exponential growth. Given this state of affairs, the legume's specific growth rate would be a function of the nodule:legume mass ratio. Thus, if the nodule's specific N_2 fixation rate and the plant's nitrogen concentration are known, then theoretically the legume's specific growth rate can be determined as a function the legume's nodule:plant mass ratio. Measured mean relative growth rates range from 0.120 g g^{-1} day^{-1} for soybean (Sa and Israel 1995) to 0.237 g g^{-1} day^{-1} for alfalfa shoots (Philippot et al. 1991). On the basis of these relationships, it should be possible to compare predictions for legume specific growth rates, on the basis of nodule N_2 fixation rates, plant nitrogen concentrations and nodule:plant dry mass ratios, with measured mean relative growth rates (RGRs).

Values for nodule:plant mass ratios range from 0.01 to 0.03 for alfalfa (Cralle et al. 1987) and 0.017 to 0.08 for soybean (Ribet and Drevon 1995; Sa and Israel

1995). Assuming the carbon content of alfalfa to be 45%, the C:N ratios vary from 14.5 to 19 (Twary and Heichel 1991). With a plant carbon content of 40%, the C:N values will vary from 8 to 13.4. The *critical nutrient range* (CNR) for tissue nitrogen concentration in alfalfa is about 3% of dry mass and the plant sufficiency levels for tissue nitrogen in alfalfa range from 3.76 to 5.5% of dry mass (Miller and Donahue 1990). Assuming a plant carbon content of approximately 40%, the average C:N ratio for a typical legume such as alfalfa growing on arable soils is 13 (Miller and Donahue 1990, pp 188–189), which is equivalent to 31 mg N g^{-1} dry mass. Nodulated P-sufficient soybeans grown on N-free nutrient media attained the following plant nitrogen concentrations: 40–60 mg N g^{-1} dry mass for leaves; <20 mg N g^{-1} dry mass for stems; ±15 mg N g^{-1} for roots and 60–80 mg N g^{-1} dry mass for nodules (Sa and Israel 1995); averaged over the total plant biomass the mean tissue nitrogen concentration is in the region of 30 mg N g^{-1} dry mass. Reported values for nodule acetylene reduction activity in soybean range from 100 to 300 μmol C_2H_4 g^{-1} nodule dry mass h^{-1} (Parsons et al. 1992; Ribet and Drevon 1995; Sa and Israel 1995). Predictions of N_2 fixation rates based on acetylene reduction assays need to be treated with caution. A major problem in the use of the acetylene reduction based nitrogenase assay to estimate absolute rates of N_2 is the reliability of the C_2H_2 to N_2 calibration. The C_2H_2/N_2 ratio can range from 2.66 to 4.33 (Minchin et al. 1983). For an average C/N ratio value of 13 which corresponds to a tissue nitrogen concentration of approximately 30 mg N g^{-1} dry mass for a plant with a C content of 40%, a specific growth rate of 0.115 g g^{-1} day^{-1} can be calculated for soybean plants having a specific nitrogenase activity of 300 μmol C_2H_4 g^{-1} dry mass h^{-1} and an average nodule:plant mass ratio of 0.06 (given a theoretical average C_2H_2/N_2 ratio of 3.5). The calculated specific growth rate gives a fairly accurate estimation of the actual measured mean RGR, 0.120 g g^{-1} day^{-1} (data for this comparison were derived from Sa and Israel 1991; 1995).

10 Rhizobium Nodule Phosphorus Requirements

Phosphorus deficiency is a major limiting factor for N_2 fixation. Specific nitrogenase activity decreases with the onset of P-deficiency (Sa and Israel 1991). Several physiological and metabolic properties were associated with lower specific nitrogenase activity in nodules of P-deficient plants: Bacteroid mass per unit nodule mass, bacteroid N concentrations, plant cell ATP concentrations and energy charge were significantly lower in nodules of P-deficient plants (Sa and Israel 1991). Because nitrogenase is localized in the bacteroids, lower bacteroid mass per unit nodule mass and N concentration could account for decreased specific nitrogenase activity under P-deficiency (Sa and Israel 1991). Legumes dependent on symbiotic N_2 fixation have a higher internal P requirement for optimum nitrogen assimilation compared to plants dependent on combined inorganic nitrogen in the form of nitrate and ammonium (Israel 1987). Soybean grown with limiting P supply showed a reduction in nodule numbers; nodule mass; individual nodule mass and plant mass (Israel 1993). The growing nodule is a major sink for P in legumes and in soybean the total nodule P concentration is threefold higher than in the rest of the plant (Sa and Israel 1991). Phosphorus deficiency resulted in decreases of *Rhizobium* bacteroid dry mass per unit nodule dry mass by an average of 20% relative to that of P-sufficient controls in soybean; P and N concentrations in bacteroids from P-deficient plants averaged 9 and 95 mg g^{-1} dry mass bacteroid respectively (Sa and Israel 1991). These P and N concentrations were 25 and 17% lower, respectively, than the P and N concentrations in bacteroids from P-sufficient plants. Nodule nitrogenase activity is decreased by plant P deficiency independently of the effectiveness of the rhizobium strain with nonlimiting P supply (Singleton et al. 1985). The concentration of ATP remained constant in whole nodules of P-sufficient and P-deficient plants with the ATP concentration being three to fourfold greater in P-sufficient plant nodules (Sa and Israel 1991). The magnitude of the specific nitrogenase activity is well correlated with legume tissue phosphorus concentration (Sa and Israel 1995). This empirical correlation provides phenomenological grounds for deriving a functional relationship between specific nitrogenase activity and legume P content.

11 Interrelations Between the C:P and C:N Exchange Dynamics

Exchanges of photosynthate, N and P between the symbiotic systems and the legume have important consequences for the overall plant carbon economy. The source capacity of leaves measured as daily gross photosynthetic output will be dependent on the nitrogen and phosphorus concentrations in the leaf tissues. Leaf nitrogen and phosphorus concentrations in legumes are influenced by the degree of nodulation and intensity of VA mycorrhiza fungal infection and the extent to which the external mycorrhizal mycelium explores the soil volume surrounding the root. The ratio of mycorrhizal P-supply and nodule N-supply to the respective carbon demand can give a quantitative index of host carbon expenditure necessary for the acquisition of nitrogen and phosphorus via these two symbiotic associations (Paul and Clark 1989; Twary and Heichel 1991; Bethlenfalvay 1992).

Photosynthetic rates in the host plant of the *Glycine–Glomus–Rhizobium* symbiont system increased linearly with increasing leaf P or N concentration (Brown and Bethlenfalvay 1988). Brown and Bethlenfalvay (1988) showed that the rate of photosynthetic CO_2 assimilation per unit leaf N or P was significantly greater in symbiotic than in nonsymbiotic plants. The experimental results of Brown and Bethlenfalvay (1988) provide sufficient evidence to counter the argument that the enhancement of photosynthetic nutrient-use efficiencies (N and P) in plants with microsymbionts can be explained as being exclusively due to increased stomatal conductance resulting from VA mycorrhizal colonization of the roots as suggested by Koide (1985). They found that soybean plants which had an association with one

(*Glomus* or *Rhizobium*) or both microsymbionts always had greater photosynthetic rates per unit leaf N or P than nonsymbiotic plants with similar leaf N or P concentrations. Also efficient nutrient utilization by the N- and P-deficient symbiotic plants relative to the N- and P-sufficient nonsymbiotic plants is shown by higher CO_2 assimilation rates in the former. The thresholds of N- and P-deficiency for youngest mature soybean leaves have been defined as < 40 mg N and <1.5 mg P per gram of leaf dry mass (De Mooy et al. 1973). Brown and Bethlenfalvay (1988) reported that most of the leaves of their symbiotic plants were below these values. The mechanism responsible for the greater photosynthetic N- and P-utilization efficiencies that have been observed in N-and P-deficient symbiotic plants remains obscure.

The magnitude or degree of *benefits* for the host measured in terms of enhanced growth potential or elevated photosynthetic rates resulting from the exchange of C for N supplied by the N_2-fixing nodules, and P supplied by the mycorrhizal system depends on two sets of efficiencies: (1) host nutrient utilization efficiencies which include the plant's nitrogen utilization efficiency and phosphate utilization efficiency and (2) microsymbiont nutrient acquisition efficiencies which include the nodules' N_2-fixation efficiency, to be called in this context the nitrogen acquisition efficiency and the mycorrhiza's phosphate acquisition efficiency. The nitrogen use efficiency of the legume is given by the derivative, $\mathrm{d}W_{\mathrm{plt}}/\mathrm{d}W_{\mathrm{N}}$, which defines the increment of biomass production per unit of N supply, and the legume's phosphorus utilization efficiency can be similarly expressed as $\mathrm{d}W_{\mathrm{plt}}/\mathrm{d}W_{\mathrm{P}}$, where W_{plt} is plant host's structural dry mass, W_{N} is mass of N in plant tissue and W_{P} is mass of P in plant tissue.

A formal-analytical approach may help elucidate the fundamental conceptual issues underlying plant nutrient-utilization efficiencies. One approach for achieving this involves the *decoupling* of photosynthesis and growth; photosynthesis involves nonstructural carbon or photosynthate (starch and sucrose) production from growth as structural biomass production. Nutrient use efficiencies, for N and P, can be construed as follows: N-utilization efficiency (NUE) is the differential of substrate carbon production (W_{C}) per unit nitrogen ($\mathrm{d}W_{\mathrm{N}}$)

$$\mathrm{NUE} = \mathrm{d}W_{\mathrm{C}}/\mathrm{d}W_{\mathrm{N}} \tag{1}$$

and P-utilization efficiency (PUE) which is the differential of substrate carbon utilization for biomass production per unit phosphate ($\mathrm{d}W_{\mathrm{P}}$)

$$\mathrm{PUE} = \mathrm{d}W_{\mathrm{C}}/\mathrm{d}W_{\mathrm{P}} \tag{2}$$

Using the above equations the following two differential equations for photosynthate production and growth can be derived:

$$\mathrm{d}W_{\mathrm{C}}/\mathrm{d}t = (\mathrm{d}W_{\mathrm{C}}/\mathrm{d}W_{\mathrm{N}})(\mathrm{d}W_{\mathrm{N}}/\mathrm{d}t) - \mathrm{d}W_{\mathrm{plt}}/\mathrm{d}t \tag{3}$$

or

$$dW_{plt}/dt = (dW_C/dW_P)(dW_P/dt) \quad (4)$$

In the above equations, the rate of W_C production is given as the product of the photosynthetic NUE and the rate of N acquisition uptake. In second equation, the rate of structural biomass (W_{plt}) is given as product of PUE and the rate of P acquisition.

12 Nitrogen Utilization in Legume Biomass Production

Under any given ambient CO_2 partial pressure, the upper limit of light-saturated photosynthetic rates is fixed by a relatively small set of physical and biochemical factors, for example, stomatal conductance, concentration of activated rubisco catalytical sites, steady-state concentrations of sugar–phosphate intermediates of the Calvin cycle and chlorophastic orthophosphate concentration. Ågren (1985a, b) provides a useful summary of the potential rates of biomass production theoretically achievable when the rate of CO_2 assimilation is calculated as a function of irradiance, water supply, ambient CO_2 or nitrogen. He concluded that nitrogen tissue concentration sets the upper limit to plant productivity. In order to develop a meaningful analysis of how nitrogen contributes to plant productivity, it is useful to begin with a quantitative analysis of the physical characteristics of the leaf photosynthetic system.

The results of a number of quantitative analyses have been used to construct the physical features characterizing the photosynthetic system of an "average" C_3 leaf. For an idealized leaf the following physical estimates or vital statistics have been derived (Lawlor 1987): 5.5×10^{11} chloroplasts m^{-2} leaf; total chloroplast volume per m^2 of leaf is 1.8×10^{-5} m^3; total volume of thylakoid lumen m^{-2} leaf is 1.2×10^{-6} m^3; 16.8 cm^3 stromal volume per m^2 of leaf; concentration of rubisco catalytic sites is 4 mM; inorganic phosphate concentration is 100 mM and ribulose 1,5-bisphosphate concentration is 0.1–2 mM; these values are from the data for spinach, tobacco and wheat given by different workers (Esua 1958; Heath 1969; Nobel 1974; Mühlethaler 1977). Rubisco accounts for up to 50% of the soluble leaf protein in C_3 plants (Schmitt and Edwards 1981).The molecular weight of rubisco is 557,000 Da with each molecule of rubisco containing eight potential catalytic sites. In one mole of rubisco, there is 83,550 g of N (assuming that the average fraction of N in the 20 different amino acids is approximately 0.15) and if each mole of rubisco contains 8× Avogadro's number (6.022×10^{23}) catalytic sites, then the ratio of catalytic sites per gram N of rubisco is 5.77×10^{19} rubisco catalytic sites per gram N of rubisco or 9.6×10^{-2}mmoles rubisco catalytic sites per gram N of rubisco.

The general response of light saturated rates of photosynthesis to increasing leaf nitrogen (g N m^{-2}) is generally curvilinear with two critical response regions: (1) a lower limit for photosynthetic rates corresponding to a certain threshold of leaf nitrogen concentration below which photosynthetic rates approach zero and (2) an upper limit for photosynthetic rates which corresponds to threshold leaf nitrogen concentrations above which no increase in photosynthetic rate occurs (Sinclair and Horie 1989). In soybeans the rate of photosynthesis is zero at 1.0 g N m^{-2} and increases linearly for leaf nitrogen concentration between 1.0 and 2.4 g N m^{-2}; above 2.4 g N m^{-2} the response is curvilinear reaching a maximum of 1.6 mg CO_2 m^{-2} s^{-1} (Sinclair and Horie 1989). Using the above data, an average hypothetical soybean or C_3 plant with a leaf nitrogen content of 2.4 g N m^{-2} (rubisco accounting for 50% of the leaf N) should have approximately 6.92×10^{19} or 0.115 mmol of rubisco catalytic sites per m^2 leaf. In order to compare the calculated A_{max} with the measured light saturated values, let the Vc_{max} of the rubisco carboxylation reaction be equal to the product kcat$[R_{sites}]^a$, where kcat is the catalytic turnover time for rubisco (2 s^{-1}) and $[R_{sites}]^a$ is the concentration of activated catalytic sites (catalytic sites g^{-1} rubisco N) while a stands for activated sites. The rate of RuBP carboxylation under light saturated conditions when rubisco catalytic sites are saturated with RuBP depends on the concentrations of CO_2 and O_2 in the stroma and A_{max} can be taken as being equal to Farguhar and von Caemmerer's (1982) expression:

$$A_{max} = (\text{kcat}[R_{sites}]^a)(\text{Ca} - \ni)/(\text{Ca} + K_c(1 + O/K_o)) - R_d \quad (5)$$

where, Ca is the ambient CO_2 concentration; $\ni$ is the CO_2-light compensation value (4–5 Pa CO_2 in 21 kPa O_2) and is equal to $(0.5Vo_{max}K_cO/Vc_{max}K_c)$, and Vo_{max} is the maximum velociy of the RuBP oxygenase reaction; K_c and K_o are Michaelis–Menten constants for CO_2 (12–20 μM for C_3 plants) and O_2 (250 μM), respectively; R_d is "dark" respiration in the light.

At standard atmospheric pressure (101,325 Pa) and with a leaf temperature of 25°C, the ambient CO_2 and O_2 concentrations are 13.9 mmol m^{-3} and 8.6 mol m^{-3} respectively. An average light saturated CO_2 compensation $\ni$ value for C_3 plants is approximately 1.6 mmol CO_2 m^{-3}. On the basis of assumptions concerning the proportion of rubisco N with respect to total leaf N values, A_{max} can be calculated for C_3 leaves using the above equation. An average hypothetical C_3 leaf consists of 16.8 cm^3 chloroplast stromal volume per m^2 of leaf tissue. A leaf with an N concentration of 2.4 g N m^{-2} will have a rubisco catalytic site concentration in the region of 6.85 mM if 50% of leaf N is partitioned into rubisco. All the data necessary to generate A_{max} with (5) are summarized in Table 1. Substitution of Table 1 data into (5) gives a net A_{max} value of 55.8 μmol CO_2 m^{-2} s^{-1} or 2.46 mg CO_2 m^{-2} s^{-1} for a hypothetical soybean leaf with a leaf N concentration of 2.4 g N m^{-2}.

The above calculation assumes that ribulose 1,5 bisphosphate is at substrate saturating levels in light saturated leaves and that rubisco is 100% activated. The calculated A_{max} value for the theoretical soybean leaf gives an over estimation of 1.54 times the measured value. If only 65% of rubisco is in the activated state under light saturating conditions, then the calculated and the measured rates are the same.

Table 1 Gas and kinetic data for calculation of A_{max} using (5)

Variable or parameter	Units
Kcat	$2\ s^{-1}$
$[R_{sites}]$[a]	$0.115\ mmol\ m^{-2}$
Ca	13.9×10^{-3} mM
∋	1.6×10^{-3} mM
O	8.6 mM
K_c	1.0×10^{-3} mM
K_o	0.250 mM
Specific leaf mass	$22\ g\ m^{-2}$
Maintenance respiratory coefficient[a]	3.18×10^{-6} g $CO_2\ g^{-1}DM\ s^{-1}$

[a]Amthor (1986)

Therefore, it is reasonable to assume that the factor that contributes greatest to the calculated overestimation of A_{max} is the estimated percentage activation state of rubisco in leaves. The percentage of activated sites is dependent on the size of the carbon substrate storage pool, N supply and P supply (Fig. 1).

13 Phosphorus Utilization in Legume Biomass Production

Plant nutrient sufficiency levels for phosphorus in alfalfa on a percent dry mass basis ranges from 0.26 to 0.7% (Miller and Donahue 1990 pp 370–371). Higher P tissue concentrations result in toxic effects. An interesting consideration is the relationship or correlation between the legume's RGR and the capacity of the external hyphal system of the AM fungi to acquire and transfer phosphorus from the surrounding soil volume into the plant root system. It has been proposed that only 20% of the total root mass is involved in nutrient acquisition (Robinson 1986), and it is most probable that the AM fungal–root association is restricted to this fraction of the total root mass. The fraction of root dry mass attributable to mycorrhizal fungi ranges from 12 to 14% for AMF colonized legume roots (Sieverding 1991). The specific activity of the AMF can be expressed as mg P cm^{-1} hypha day^{-1}. Functional relationships should exist between a legume's specific growth rates and the corresponding uptake rate per unit root mass of any mineral nutrient (Garnier et al. 1989). When plants are grown at steady-state nutritional supply rate, it has been shown for P (Eissenstat et al. 1993) and for N (Cromer and Jarvis 1990; Ingestad and Ågren 1991) that the slope of the RGR vs plant nutrient concentration is linear until a maximum *RGR* is reached. With steady-state conditions of phosphate supply during exponential growth, tissue phosphorus concentration remains constant, i.e., $dP_c/dt = 0$, where $P_c = W_p/W_{plt}$ is the tissue phosphorus concentration, W_p equals plant phosphorus mass and W_{plt} is the plant dry mass.

For sustained steady-state exponential growth, a dynamic functional equilibrium must exist between the size and the specific activities of the microbial symbiotic associations and the plant. Maintenance of this dynamic functional equilibrium involves the regulatory control by the plant of the proportion of carbon lost in exchange for the quantity of phosphate necessary for sustaining exponential growth. Therefore, with regard to phosphate-limited growth, the maintenance of a constant tissue phosphorus concentration above a P-deficiency threshold is a necessary condition for sustained steady-state exponential growth. The constraint that plant tissue P-concentration imposes on plant growth is mediated via the effects that P has on key growth driving metabolic and physiological processes such as the Calvin cycle and phloem loading. Low-P treatment (plants grown on 0.05 mM phosphate) has a greater impact on plant biomass production (60% reduction compared to P-sufficient plants) than on the rate of photosynthesis; low-P treatment effected photosynthesis much less at low irradiances than at high radiances relative to P-sufficient plants (plants grown on 1.0 mM phosphate), light saturated rates in leaves of P-deficient plants were decreased by 35% (Rao and Terry 1989). P-deficiency may decrease light saturated rates of photosynthesis by decreasing RuBP regeneration capacity and/or decreasing the concentration of activated rubisco (Brooks 1986; Brooks et al. 1988). Under low-P conditions, RuBP concentrations declined to half the rubisco binding site concentration (Rao et al. 1989a, b). The activities of the phosphatases, FBPase and SBPase increased with P-deficiency in leaves, possibly to promote more rapid *recycling* of phosphate, and the activities of the enzymes associated with RuBP regeneration, i.e., PGA kinase, NADP-G3P dehydrogenase and Ru5P kinase, declined with P-deficiency in leaves (Rao and Terry 1989). The concentrations of Calvin cycle intermediates declined in P-limited leaves while starch increased to levels higher than in control leaves (Rao et al. 1990). Starch, sucrose and glucose concentrations increased in P-deficient leaves, resulting in a higher accumulation of leaf photosynthate compared to P-sufficient plants (Rao et al. 1990; Rao and Terry 1995). Low-P treatment decreases leaf ATP concentrations considerably (Rao et al. 1989a, b) and it is well known that the loading of sugars into the phloem has a large requirement for ATP (Giaguinta 1983), and therefore, there may be insufficient ATP in P-deficient leaves to maintain the rates of photosynthate export necessary to support high growth rates (Rao et al. 1990).

Maintenance of optimal rates of photosynthetic carbon assimilation, photosynthate translocation and nitrogen fixation during exponential growth of legumes requires the maintenance of an optimal constant concentration of phosphate in the leaf tissue and root nodules. Put differently, in the legume symbiotic system, phosphorus uptake rates must satisfy the metabolic phosphate demands of the Calvin cycle, phloem phosphate-loading system and the N_2-fixing nodule system, which are the necessary P-demands for sustaining optimal RGRs. Consistent with the premises of the model depicted in Fig. 1, tight coupling between the RGR and nitrogen supply rates has been shown to exist (Hirose 1986, 1988; Hirose et al. 1988; MacDuff et al. 1993); also the same tight coupling applies between growth rates and phosphate supply rates.

A definition for phosphate demand and utilization efficiency as a function of optimal inherent growth capacity can be developed as follows: Under steady-state exponential growth, given mass flow and mass conservation, the following equalities hold: rate of phloem loading = rate of phloem unloading = rate of biomass production in sink organ. In this case, the sink organ can be conceptualized as a biomass growth sink such as growing leaves, elongating nodes and growing root systems. Given this scenario of events, it is proposed that phloem loading is the rate limiting step for biomass production. Phloem loading is an energetically uphill process requiring the consumption of ATP and if ATP is treated as a P equivalent, then phloem loading can expressed as a function of tissue phosphate concentration. Phloem loading can be treated as a two substrate enzyme reaction involving sucrose and ATP. Sucrose is usually present at the phloem loading site at substrate saturating levels and ATP can be treated as the rate limiting substrate for the phloem loading process. Metabolic demand for phosphate is linked to its role in ATP synthesis for phloem loading. The maximum rate of phloem loading is proportional to product of the concentration of phloem loading sites, (sites g^{-1} dry mass) and the turnover time of the loading site, (sites per second). In the model depicted in Fig. 1, the close coupling between P supply and C mobilization into the metabolic pool for biomass synthesis could be linked to its role as the "high energy phosphate ester" in ATP driven phloem loading. In addition, the close coupling between P supply and C utilization in the metabolic pool for biomass synthesis could also be linked to its role, either direct or indirect, as the "high energy phosphate ester" in ATP driven polymerization reactions (protein synthesis and polysaccharide synthesis).

14 Conclusion

In this chapter, an analysis of how legume growth is influenced by C:N:P stoichiometries has been put forward. The supply of N and P by the two microsymbionts plays a major role in the control of photosynthetic CO_2 assimilation and in the control of conversion of photosynthate into plant biomass (Fig. 1). Resource (C, N and P) allocation dynamics have been investigated in terms of (1) C:P exchange dynamics with respect to the AMF symbiotic partner, (2) C:N exchange dynamics with respect to the *Rhizobium* symbiotic partner and (3) how the C:N:P supply ratio controls photosynthetic rates and the rate of conversion of photosynthate into plant biomass. The existence and maintenance of balanced exponential growth or steady-state exponential growth of all plant components and microbial symbiotic associations in the tripartite complex have been an underlying functional assumption in this investigation. Furthermore it has been assumed that resource allocation strategies within the tripartite complex ensure the maintenance of functional equilibrium between plant components and microbial symbiotic associations. The intensities of metabolic activities of different plant components and microbial associations represent the resource acquisition or resource utilization capacities of the plant components and the microbial associations. If during balanced exponential growth of the tripartite complex a functional equilibrium has been maintained between the steady-state rates of resource acquisitions (C, N, P), then it logically follows that the concentrations of various substrates (C, N, P) within the plant tissues will remain constant and the RGRs of the various components and associations can be expressed as a function of one or more of these substrate concentrations. This will greatly help in simplifying any effort to model the resource allocation and growth dynamics of the tripartite complex. The internal concentration values of the different resources (C, N and P) within the tripartite complex will at any given time be determined by the rate of resource acquisition from the external environment. Rate of acquisition will depend on the available supplies of the resources. Given the realities of global climate change arising as a consequence of increasing levels of atmospheric CO_2, it has become relevant to formulate plant–microbe resource allocation optimization problems in terms of C:N:P stoichiometries. For example with respect to the legume tripartite association, what would be the optimal microbial symbiotic to plant mass ratio that ensures maximum specific legume growth rates for a given supply of soil nitrogen and phosphate at elevated levels of ambient CO_2 concentration? How does N and P availability affect plant productivity at elevated levels of ambient CO_2 concentration?

References

Ågren GI (1985a) Limits to plant production. J Theor Biol 113:89–92

Ågren GI (1985b) Theory for growth of plants derived from then nitrogen productivity concept. Physiol Plant 64:17–28

Ågren GI (2004) The C:N:P stoichiometry of autotrophs – theory and observations. Ecol Lett 7:185–191

Allen MF, Smith WK, Moore TS Jr, Christensen M (1981) Comparative water relations and photosynthesis of mycorrhizal and non-mycorrhizal *Bouteloua gracilis* H.B.K. lag ex Steud. New Phytol 88:683–693

Amthor JS (1986) Evolution and applicability of a whole plant respiration model. J Theor Biol 122:473–490

Atkins CA (1984) Efficiencies in the legume/*Rhizobium* symbiosis. A review. Plant Soil 82:273–284

Azcón De Aguilar C, Azcón R, Barea JM (1979) Endomycorrhizal fungi and *Rhizobium* as biological fertilizers for *Medicago sativa* in normal cultivation. Nature 279:325–327

Azcón R, Ocampo JA (1981) Factors affecting the vesicular-arbuscular infection and mycorrhizal dependency of thirteen wheat cultivars. New Phytol 87:677–685

Azcón R, Ocampo JA (1984) Effect of root exudation on VA mycorrhizal at early stages of plant growth. Plant Soil 82:133–138

Azcón R, Barea JM, Elatrach F (1988) Influence of mycorrhiza vs soluble phosphate on growth, nodulation, and N_2 fixation (N-15) in alfalfa under different levels of water potential. Biol Fertil Soils 7:28–31

Azcón R, Gomez M, Tobar R (1992) Effects of nitrogen source on growth, nutrition, photosynthetic rate and nitrogen metabolism of mycorrhizal and phosphorus-fertilized plants of *Lactuca sativa* L. New Phytol 121:227–234

Barea JM, Azcón-Aguilar C (1983) Mycorrhizas and their significance in nodulating nitrogen-fixing plants. Adv Agron 36:1–34

Barea JM, Azcon R, Azcon-Aguilar C (1992) Vesicular-arbuscular mycorrhizal fungi in nitrogen-fixing systems. Methods Microbiol 24:391–416

Bayne HG, Brown MS, Bethlenfalvay GJ (1984) Defoliation effects on mycorrhizal colonization, nitrogen fixation and photosynthesis in the Glycine-*Glomus*- symbiosis. Physiol Plant 62:576–580

Bergersen FJ (1971) The biochemistry of nitrogen fixation in legumes. Annu Rev Plant Physiol 22:121–140

Bethlenfalvay GJ (1992) Mycorrhizae and crop productivity. In: Bethlenfalvay GJ, Linderman RG (eds) Mycorrhizae in sustainable agriculture. ASA Special Publication Number 54. American Society of Agronomy, Inc. Crop Science Society of America, Inc. Soil Science Society of America, Inc. Madison, WI, USA, pp 1–27

Bethlenfalvay GJ, Brown MS (1985) The *Glycine-Glomus-Rhizobium* symbiosis II. Antagonistic effects between mycorrhizal colonization and nodulation. Plant Physiol 79:1054–1058

Bethlenfalvay GJ, Newton WE (1991) Agro-ecological aspects of the mycorrhizal, nitrogen-fixing legume symbiosis. In: Keister DL, Cregan P (eds) The rhizosphere and plant growth. Kluwer, Dordrecht, The Netherlands, pp 349–354

Bethlenfalvay GJ, Brown MS, Pacovsky RS (1982) Relationships between host and endophyte development in mycorrhizal soybeans. New Phytol 90:537–543

Betlenfalvay GJ, Yoder JF (1981) The Glycine-Glomus-Rhizobium symbiosis I. Phosphorus effect on nitrogen fixation and mycorrhizal infection. Physiol Plant 52:141–145

Black KG, Mitchell DT, Osborne BA (2000) Effect of mycorrhizal-enhanced phosphate status on carbon partitioning, translocation and photosynthesis in cucumber. Plant Cell Environ 23:797–809

Bould C, Hewitt EJ, Needham P (1983) Diagnosis of mineral disorders in plants, Chapter 3. In: Robinson JBD (ed) Principles, vol 1. Her Majesty's Stationery Office, London, p 133

Brooks A (1986) Effects of phosphorus nutrition on ribulose-1, 5-bisphosphate carboxylase activation, photosynthetic quantum yield and amounts of some Calvin-cycle metabolites in spinach leaves. Aust J Plant Physiol 13:221–237

Brooks A, Woo KC, Wong SC (1988) Effects of phosphorus nutrition on the response of photosynthesis to CO_2 and O_2 activation of ribulose bisphosphate carboxylase and amounts of ribulose bisphosphate and 3-phosphoglycerate in spinach leaves. Photosynth Res 15: 133–141

Brown MS, Bethlenfalvay GJ (1986) The *Glycine-Glomus-Rhizobium* symbiosis. VI. Endophyte effects on leaf carbon, nitrogen and phosphorus nutrition. J Plant Nutr 9:1199–1212

Brown MS, Bethlenfalvay GJ (1988) The *Glycine-Glomus-Rhizobium* symbiosis. VII. Photosynthetic nutrient-use efficiency in nodulated, mycorrhizal soybeans. Plant Physiol 86:1292–1297

Brown R, Broadbent D (1950) The development of cells in the growing zones of the root. J Exp Bot 1:249

Buwalda JG, Goh KM (1982) Host-fungus competition for carbon a cause of growth depressions in vesicular-arbuscular mycorrhizal ryegrass. Soil Biol Biochem 14:103–106

Cihacek LJ (1993) Phosphorus source effects on alfalfa yield, total nitrogen content, and soil test phosphorus. Commun Soil Sci Plant Anal 24:2043–2057

Cockburn W, Baldry CW, Walker DA (1967a) Some effects of inorganic phosphate on O_2 evolution by isolated chloroplasts. Biochim Biophys Acta 143:614–624

Cockburn W, Baldry CW, Walker DA (1967b) Oxygen evolution by isolated chloroplasts with carbon dioxide as the hydrogen acceptor. A requirement for orthophosphate or pyrophosphate. Biochim Biophys Acta 131:594–596

Cox G, Sanders FE, Tinker PB, Wild JA (1975) Ultrastructural evidence relating to host-endophyte transfer in a vesicular-arbuscular mycorrhiza. In: Sanders FE, Mosse B, Tinker PB (eds) Endomycorrhizas. Academic, London, p 297

Cralle HT, Heichel GH (1986) Photosynthate and dry matter distribution of effectively and ineffectively nodulated alfalfa. Crop Sci 26:117–121

Cralle HT, Heichel GH, Barnes DK (1987) Photosynthate partitioning in plants of alfalfa population selected for high and low nodule mass. Crop Sci 27:96–100
Cromer RN, Jarvis PG (1990) Growth and biomass partitioning in *Eucalyptus grandis* seedlings in response to nitrogen supply. Aust J Plant Physiol 17:503–515
Daft MJ (1978) Nitrogen fixation in nodulated and mycorrhizal crop plants. Ann Appl Biol 86:461–462
De Mooy CJ, Pesek E, Spaldon E (1973) Mineral nutrition. In: Caldwell BE (ed) Soybeans: improvement, production and uses. American Society of Agronomy, Madison, WI, pp 267–352
Diedrichs C (1983a) Influence of light on the efficacy of vesicular-arbuscular mycorrhiza in tropical and subtropical plants. II. Effect of light intensity under growth chamber conditions. Angew Bot 57:45–53
Diedrichs C (1983b) Influence of light on the efficacy of vesicular-arbuscular mycorrhiza in tropical and subtropical plants. III. The effect of day length. Angew Bot 57:55–67
Douds DD Jr, Johnson CR, Koch KE (1988) Carbon cost of the fungal symbiont relative to net leaf P accumulation in a split-root VA mycorrhizal symbiosis. Plant Physiol 86:491–496
Eissenstat DM, Graham JH, Syvertsen JP, Drouillard DL (1993) Carbon economy of sour orange in relation to mycorrhizal colonization. Ann Bot 71:1–10
Ericsson T, Ingestad T (1988) Nutrition and growth of birch seedlings at varied relative phosphorus addition rates. Physiol Plant 72:227–235
Esua K (1958) Plant anatomy. Wiley/Chapman and Hall, New York
Farquhar GD, von Caemmerer S (1982) Modelling of photosynthetic response to environmental conditions. In: Lange O, Nobel PS, Osmond CB, Ziegler H (eds) Encyclopedia of plant physiology (N.S.), vol. 12B, vol II, Physiological plant ecology. Springer, Berlin, pp 549–587
Fay P, Mitchell DT, Osborne BA (1996) Photosynthesis and nutrient-use efficiency in barley in response to low arbuscular mycorrhizal colonization and addition of phosphorus. New Phytol 132:425–433
Fitter AH (1991) Costs and benefits of mycorrhizas: implications for functioning under natural conditions. Experientia 47:350–355
Fredeen AL, Terry N (1988) Influence of vesicular-arbuscular mycorrhizal infection and soil phosphorus level on growth and metabolism of soybean. Can J Bot 66:2311–2316
Garnier E, Koch GW, Roy J, Mooney HA (1989) Responses of wild plants to nitrate availability: relationships between growth rate and nitrate uptake parameters, a case study with two *Bromus* species, and a survey. Oecologia 79:542–550
Giaguinta RT (1983) Phloem loading of sucrose. Annu Rev Plant Physiol 34:347–387
Gordon AJ, Ryle GJA, Mitchell DF, Powell CE (1985) The flux of 14C labelled photosynthate through soybean root nodules during N_2 fixation. J Exp Bot 36:756–759
Graham JH, Leonard RT, Menge JA (1981) Membrane-mediated decrease in root exudation responsible for phosphorus inhibition of vesicular-arbuscular mycorrhiza formation. Plant Physiol 68:548–552
Gray VM (1996) Alfalfa. In: Zamski E, Schaffer AA (eds) Photoassimilate distribution in plants and crops: source–sink relationships. Marcel Dekker, New York, pp 759–779
Gray VM (2000) A comparison of two approaches for modelling (*Manihot esculenta* Crantz.) crop growth. Ann Bot 85:77–90
Greenwood DJ, Gastal F, Lemaire G, Draycott A, Millard P, Neeteson JJ (1991) Growth rate and % N of field grown crops: theory and experiments. Ann Bot 67:181–190
Grünweig J, Korner C (2003) Differential phosphorus and nitrogen effects drive species and community responses to elevated CO_2 in semi-arid grassland. Funct Ecol 17:766–777
Haaker H (1988) Biochemistry and physiology of nitrogen fixation. Bioessays 9:112–117
Harris D, Paul EA (1987) Carbon requirements of vesicular-arbuscular mycorrhizae. In: Safir GR (ed) Ecophysiology of mycorrhizal plants. Florida, CRC, pp 93–105
Harris RS, Pacovsky RS, Paul EA (1985) Carbon economy of soybean-*Rhizobium-Glomus* associations. New Phytol 101:427–440

Heath OVS (1969) The physiological aspects of photosynthesis. Stanford University Press, Stanford, CA
Hedin LO (2004) Global organization of terrestrial nutrient interactions. Proc Natl Acad Sci USA 101:10849–10850
Heytler PG, Hardy RWF (1984) Calorimetry of nitrogenase-mediated reductions in detached soybean nodules. Plant Physiol 75:304–310
Hirose T (1986) Nitrogen uptake and plant growth. II. An empirical model of vegetative growth and partitioning. Ann Bot 58:487–496
Hirose T (1988) Modelling the relative growth rate as a function of plant nitrogen concentration. Physiol Plant 72:185–189
Hirose T, Freusen AHJ, Lambers H (1988) Modelling the responses to nitrogen availability of two *Plantago* species grown at a range of exponential nutrient addition rates. Plant Cell Environ 11:827–834
Hostak MS, Henson CA, Duke SH, Vandenbosch KA (1987) Starch granule distribution between cell types of alfalfa nodules as affected by symbiotic development. Can J Bot 65:1108–1115
Ingestad T, Ågren GI (1991) The influence of plant nutrition on biomass allocation. Ecol Appl 1:168–174
Irving DE, Silsbury JH (1987) A comparison of the rate of maintenance respiration in some crop legumes and tobacco determined by three methods. Ann Bot 59:257–264
Israel DW (1987) Investigation of the role of phosphorus in symbiotic dinitrogen fixation. Plant Physiol 84:835–840
Israel DW (1993) Symbiotic dinitrogen fixation and host-plant growth during development of and recovery of phosphorus deficiency. Physiol Plant 88:294–300
Jakobsen I, Rosendahl L (1990) Carbon flow into soil and external hyphae from roots of mycorrhizal cucumber plant. New Phytol 115:77–83
Jayachandran K, Schwab AP, Hetrick BAD (1992) Mineralization of organic phosphorus by vesicular-arbuscular mycorrhizal fungi. Soil Biol Biochem 24:897–903
Jia Y, Gray VM (2004a) Interrelationships between nitrogen supply and photosynthetic parameters in *Vicia faba* L. Photosynthetica 41:605–610
Jia Y, Gray VM (2004b) Influence of phosphorus nitrogen on photosynthetic parameters and growth in *Vicia faba* L. Photosynthetica 42:535–542
Jia Y, Gray VM (2007) The influence N and P supply on the short-term responses of elevated CO_2 in faba bean (*Vicia faba* L.). S Afr J Bot 73:466–470
Jia Y, Gray VM (2008) Growth yield of *Vicia faba* L in response to microbial symbiotic associations. S Afr J Bot 74:25–32
Jia Y, Gray VM, Straker CJ (2004) The influence of *Rhizobium* and arbuscular mycorrhizal fungi on nitrogen and phosphorus accumulation by *Vicia faba*. Ann Bot 94:251–258
Kaschuk G, Kuyper TW, Leffelaar PA, Hungria M, Giller KE (2009) Are the rates of photosynthesis stimulated by carbon sink strength of rhizobial and arbuscular mycorrhizal symbioses? Soil Biol Biochem 41:1233–1244
Koch KE, Johnson CR (1984) Photosynthate partitioning in split-root citrus seedlings with mycorrhizal and nonmycorrhizal root systems. Plant Physiol 75:26–30
Koide R (1985) The effect of VA mycorrhizal infection and phosphorus status on sunflower hydraulic and stomatal properties. J Exp Bot 36:1087–1089
Kouchi H, Nakaji K, Yoneyama T, Ishizuka J (1985) Dynamics of carbon photosynthetically assimilated in nodulated soybean plants under steady state conditions. 3. Time course of study 14C incorporation into soluble metabolites and respiratory evolution of CO_2 from roots and nodules. Ann Bot 56:333–346
Kucey RMN, Paul EA (1982a) Biomass of mycorrhizal fungi associated with bean roots. Soil Biol Biochem 14:413–414
Kucey RMN, Paul EA (1982b) Carbon flow, photosynthesis and N_2 fixation in mycorrhizal and nodulated faba beans (*Vicia faba* L.). Soil Biol Biochem 14:407–412

Lawlor DW (1987) Photosynthesis: metabolism, control, and physiology. Longman Scientific & Technical, New York
MacDuff JH, Jarvis SC, Larsson CM, Oscarson P (1993) Plant growth in relation to the supply and uptake of NO_3^-: A comparison between relative addition rate and external concentration as driving variables. J Exp Bot 44:1475–1484
Mächler F, Schnyder H, Nösberger J (1984) Influence of inorganic photosynthesis of wheat chloroplasts. J Exp Bot 35:481–487
Mahon JD (1983) Energy relationships, Chapter 8. In: Broughton WJ (ed) Nitrogen fixation, vol 3. Clarendon, Oxford, UK
Marcus Y, Gurevitz M (2000) Activation of cyanobacterial RuBP-carboxylase/oxygenase is facilitated by inorganic phosphate via two independent mechanisms. Eur J Biochem 267:5995–6003
McCree KJ, Silsbury JH (1978) Growth and maintenance requirements of subterranean clover. Crop Sci 18:13–18
Miller RW, Donahue RL (1990) Soils: an introduction to soils and plant growth, 6th edn. Prentice Hall, Englewood Cliffs, NJ
Miller RM, Miller SP, Jastrow JD, Rivetta CB (2002) Mycorrhizal mediated feedbacks influence net carbon gain and nutrient uptake in *Andropogon gerardii*. New Phytol 155:149–162
Minchin FR, Witty JF (2005) Respiratory/carbon costs of symbiotic nitrogen fixation in legumes. In: Lambers H, Ribas-Carbo M (eds) Plant respiration. Springer, Dordrecht, pp 195–205
Minchin FR, Summerfield RJ, Hadley P, Roberts EH, Rawsthorne S (1981) Carbon and nitrogen nutrition of nodulated roots of grain legumes. Plant Cell Environ 4:5–26
Minchin FR, Witty JF, Sheehy JE, Müller M (1983) A major error in the acetylene reduction assay: decreases in nodular nitrogenase activity under assay conditions. J Exp Bot 34:641–649
Mühlethaler K (1977) Introduction to structure and function of the photosynthesis apparatus. In: Tresbst A, Avon M (eds) Encyclopedia of pant physiology (N.S.), Vol 5, vol I, Photosynthesis. Springer, Berlin, pp 503–521
Nobel PS (1974) Biophysical plant physiology. WH Freeman, San Francisco
Parsons R, Raven JA, Sprent JL (1992) A simple open flow system used to measure acetylene reduction in *Sesbania rostrata* stem and root nodules. J Exp Bot 43:595–604
Paul EA, Clark FE (1989) Soil microbiology and biochemistry. Academic, San Diego
Paul EA, Kucey RMN (1981) Carbon flow in plant microbial associations. Science 213:473–474
Payne WJ (1970) Energy and growth of heterotrophs. Annu Rev Microbiol 24:17–52
Pearson JN, Jakobsen I (1993) Symbiotic exchange of carbon and phosphorus between cucumber and three arbuscular mycorrhizal fungi. New Phytol 124:481–494
Philippot S, Allirand JM, Chartier M, Gosse G (1991) The role of different daily irradiations on shoot growth and root/shoot ratio in Lucerne (*Medicago sativa* L.). Ann Bot 68:329–335
Phillips DA (1980) Efficiency of symbiotic nitrogen fixation in legumes. Annu Rev Plant Physiol 31:29–49
Piccini D, Ocampo JA, Bedmar EJ (1988) Possible influence of *Rhizobium* on VA mycorrhiza metabolic-activity in double symbiosis of alfalfa plants (*Medicago sativa* L.). Biol Fertil Soils 6:65–67
Pieters A, Paul MJ, Lawlor DW (2001) Low sink demand limits photosynthesis under Pi deficiency. J Exp Bot 52:1083–1091
Rainbird RM, William DH, Hardy RWF (1984) Experimental determination of the respiration associated with Soybean/*Rhizobium* nitrogenase function, nodule maintenance, and total nodule nitrogen fixation. Plant Physiol 75:49–53
Rao IM, Terry N (1989) Leaf phosphate status, photosynthesis, and carbon partitioning in sugar beet. I. Changes in growth, gas exchange, and Calvin cycle enzymes. Plant Physiol 90:814–819
Rao IM, Terry N (1995) Leaf phosphate status, photosynthesis, and carbon partitioning in sugar beet. IV. Changes with time following increased supply of phosphate to low-phosphate plants. Plant Physiol 107:1313–1321

Rao IM, Aruanantham AR, Terry N (1989a) Leaf phosphate status, photosynthesis and carbon partitioning in sugar beet. II. Diurnal change in sugar phosphates, adenylates, and nicotinamide nucleotides. Plant Physiol 90:820–826

Rao IM, Aruanantham AR, Terry N (1989b) Diurnal change in sugar phosphates, adenylates, and nicotinamide nucleotides in sugar beet leaves. Photosynth Res 23:205–212

Rao IM, Fredeen AL, Terry N (1990) Leaf phosphate status, photosynthesis, and carbon partitioning in sugar beet. III. Diurnal changes in carbon partitioning and carbon export. Plant Physiol 92:29–36

Rawsthorne S, Minchin FR, Summerfield RJ, Cookson C, Coombs J (1980) Carbon and nitrogen metabolism in legume root nodule. Phytochemistry 19:341–355

Ribet J, Drevon JJ (1995) Increase in permeability to oxygen and in oxygen uptake of soybean nodules under limiting phosphorus nutrition. Physiol Plant 94:298–304

Robinson D (1986) Compensatory changes in the partitioning of dry matter in relation to nitrogen uptake and optimal variations in growth. Ann Bot 58:841–848

Ryle GJ, Arnott RA, Powell CE, Gordon AJ (1984) N2 fixation and the respiratory costs of nodules, nitrogenase activity, and nodule growth and maintenance in Fiskeby soybean. J Exp Bot 35:1156–1165

Sa TM, Israel DW (1991) Energy status and functioning of phosphorus-deficient soybean nodules. Plant Physiol 97:928–935

Sa TM, Israel DW (1995) Nitrogen assimilation in nitrogen-fixing soybean plants during phosphorus deficiency. Crop Sci 35:814–820

Sadras VO (2006) The N:P stoichiometry of cereal, grain legume and oilseed crops. Field Crops Res 95:13–29

Salsac L, Drevon J, Zengbe M, Cleyet Mariel JC, Obaton M (1984) Energy requirements of symbiotic nitrogen fixation. Physiol Veg 22:509–521

Same BI, Robson AD, Abbott LK (1983) Phosphorus, soluble carbohydrates and endomycorrhizal infection. Soil Biol Biochem 15:593–597

Scaife A, Turner M, Wood P (1983) Vegetables: diagnosis of mineral disorders in plants. In: Robinson JBD (ed) Vol 2. Her Majesty's Stationery Office, London.

Scannnerini S, Bonfante-Fasolo P (1983) Comparative ultrastructural analysis of mycorrhizal associations. Can J Bot 61:917–943

Schmitt MR, Edwards GE (1981) Photosynthetic capacity and nitrogen use efficiency of maize, wheat, and rice: A comparison between C_3 and C_4 photosynthesis. J Exp Bot 32:459–466

Schubert KR, Ryle JA (1980) The energy requirement for nitrogen fixation in nodulated legumes. In: Summerfield RJ, Bunting AH (eds) Advances in legume science. Royal Botanic Gardens, Kew, pp 85–96

Schubert KR, Wolk CP (1982) The energetic of biological nitrogen fixation. Workshop summaries. American Society of Plant Physiology, Rockville, MD, USA

Schwab SM, Menge JA, Leonard RT (1983) Comparison of stages of vesicular-arbuscular mycorrhiza formation in Sudan grass grown under two levels of phosphorus nutrition. Am J Bot 70:1225–1231

Schwab SM, Menge JA, Tinker PB (1991) Regulation of nutrient transfer between host and fungus in vesicular-arbuscular mycorrhizas. New Phytol 117:387–398

Sieverding E (1991) Vesicular-arbuscular mycorrhiza management in tropical agrosystems. Deutsche Gesellschaft fur TechnischeZusammernarbeit (GTZ), GmbH, Eschborn

Sinclair TR, Horie T (1989) Leaf nitrogen, photosynthesis, and crop radiation use efficiency: a review. Crop Sci 29:90–98

Singleton PW, Abdel-Magid HM, Tavares JW (1985) Effect of phosphorus on the effectiveness of strains of *Rhizobium japonicum*. Soil Sci Soc Am J 49:613–616

Skot L, Hirsch PR, Witty JF (1986) Genetic factors in *Rhizobium* affecting the symbiotic carbon costs of N_2 fixation and host plant biomass production. J Appl Bacteriol 61:239–246

Snellgrove RC, Splittstoesser WE, Stribley DP, Tinker PB (1982) The distribution of carbon and the demand of the fungal symbiont in leek plants with vesicular-arbuscular mycorrhizas. New Phytol 92:75–87

Son CL, Smith SE (1988) Mycorrhizal growth responses:interactions between photon irradiance and phosphorus nutrition. New Phytol 108:305–314

Sterner RW, Elser JJ (2002) Ecological stoichiometry: the biology of elements from molecular to the biosphere. Princeton University Press, Princeton

Stribley KG, Tinker PB, Rayner JH (1980) Relation of internal phosphorus concentration and plant weight in plants infected by vesicular-arbuscular mycorrhizas. New Phytol 86:261–266

Tester M, Smith SE, Walker NA (1986) Effects of photon irradiance on the growth of shoots and roots, on the rate of initiation of mycorrhizal infection and on the growth of infection units in *Trifolium subterraneum* L. New Phytol 103:375–390

Toth R, Miller RM (1984) Dynamics of arbuscule development and degeneration in a *Zea mays* mycorrhiza. Am J Bot 71:449–460

Toth R, Doane C, Bennett E, Alexander T (1990) Correlation between host-fungal surface areas and percent colonization in VA mycorrhizae. Mycologia 82:519–522

Twary S, Heichel GH (1991) Carbon costs of dinitrogen fixation associated with dry matter accumulation alfalfa. Crop Sci 31:985–992

Usuda H, Edward GE (1982) Influence of varying CO_2 and orthophosphate concentrations on rates of photosynthesis, and synthesis of glycolate and dihydroxyacetone phosphate by wheat chloroplasts. Plant Physiol 69:469–473

Usuda H, Shimogawara K (1991) Phosphate deficiency in maize. I. Leaf phosphate status, growth, photosynthesis and carbon partitioning. Plant Cell Physiol 32:497–504

Vance CP, Heichel GH (1991) Carbon in N2 fixation: limitation or exquisite adaptation. Annu Rev Plant Physiol Plant Mol Biol 42:373–390

Walsh KB, Vessey JK, Layzell DB (1987) Carbohydrate supply and N_2 fixation in soybean. Plant Physiol 85:135–144

Wang GM, Coleman DC, Freckman DW, Dyer MI, McNaughton SJ, Acra MA, Goeschl JD (1989) Carbon partitioning patterns of mycorrhizal versus non-mycorrhizal plants: real-time dynamic measuring using $^{11}CO_2$. New Phytol 112:489–493

Warren CR, Adams MA (2001) Distribution of N, Rubisco and photosynthesis in *Pinus pinaster* and acclimation to light. Plant Cell Environ 24:598–609

Wilcox HE (1991) Mycorrhizae. In: Waisel Y, Eschel A, Kafkafi U (eds) Plant roots: the hidden half. Marcell Dekker, New York, pp 731–765

Wright DP, Scholes JD, Read DJ (1998a) Effects of VA mycorrhizal colonization on photosynthesis and biomass production of *Trifolium repens* L. Plant Cell Environ 21:209–216

Wright DP, Read DJ, Scholes JD (1998b) Mycorrhizal sink strength influences whole plant carbon balance of *Trifolium repens* L. Plant Cell Environ 21:881–891

Regulation of Nitrogen Assimilation in Foliar Fed Legume Plants at Insufficient Molybdenum Supply

Marieta Hristozkova, Maria Geneva, and Ira Stancheva

Contents

1 Introduction 418
2 Foliar Application of Nutrient Elements: Connection with Function of Symbiotic Systems and Nitrogen Assimilation 419
3 Molybdenum Importance for Plant Nitrogen Metabolism 421
4 Plant Biomass Accumulation, Number of Nodules and Nitrogen Fixing Activity 423
5 Free Amino Acid Composition 425
6 Conclusion 428
References 429

Abstract Formation and function of N_2-fixing systems between bacteria from *Rhizobiaceae* family and legume plants from *Fabaceae* family are especially sensitive to molybdenum (Mo) deficiency. The hypothesis of the present work was that nitrogen fixation and assimilation in Mo deficient pea and alfalfa plants are enhanced when the nutrients were supplied through the foliage. It was established that foliar fertilization resulted in the increase of nitrogen fixation and biomass accumulation in the absence of Mo. The positive effect of foliar fertilization at insufficient Mo supply on the nitrogen uptake is better expressed in garden pea than in alfalfa. Otherwise, alfalfa was more sensitive to Mo starvation than the pea plants. Insufficient Mo supply leads to significant reduction in plant Mo content and nitrogen fixing activity, while stress induced free amino acids increased repeatedly. The negative effect of Mo exclusion from the nutrient media on nitrogen assimilation and biomass accumulation diminished through the foliar absorbed nutrients.

M. Hristozkova, M. Geneva, and I. Stancheva (✉)
Acad. M. Popov, Institute of Plant Physiology, Bulgarian Academy of Sciences, Acad. G. Bonchev str., block 21, Sofia 1113, Bulgaria

D.K. Maheshwari (ed.), *Plant Growth and Health Promoting Bacteria*, Microbiology Monographs 18, DOI 10.1007/978-3-642-13612-2_18,

1 Introduction

Molybdenum (Mo) is among the micronutrients that are very essential for the plant growth and is required in small amounts. The symptoms associated with Mo deficiency are closely related to nitrogen metabolism (Gupta and Lipsett 1981), and nitrogen assimilatory processes in plants tissues strongly depend on the plant Mo levels (Kaiser et al. 2005). Mo is an important constituent of several enzymes catalyzing different chains of nitrogen metabolism: nitrate reductase (EC 1.6.6.1), nitrogenase (EC 1.18.6.1), xanthine dehydrogenase (EC 1.1.1.204), etc. (Mendel and Haensch 2002). Loss of Mo-dependent activity (directly or indirectly through low internal Mo levels) impacts nitrogen metabolism and eventually plant development.

Amino acids in plant tissues are determined by complex interplay of factors, which may be influenced by nutrition, developmental stage, and environmental conditions (Foyer and Noctor 2002). In legumes, ammonia can be formed by direct fixation of atmospheric dinitrogen atoms within root nodules (Lam et al. 1996). It is established that the major pathway for ammonia incorporation into nontoxic organic compounds occurs through glutamine synthetase-glutamate synthase (GS-GOGAT) cycle (Ireland and Lea 1999). The GS-GOGAT cycle converts ammonia by the combined activity of the two enzymes to produce two molecules of glutamate. Amino groups are then transferred out of the GS-GOGAT cycle, predominantly via glutamate to other amino acids, such as aspartate and alanine and a range of transamination products are formed. Leaf amino acid content increased with enhanced supply of nitrogen during growth. In legumes, fluctuations in the amino acid proportion might reflect changes in the source of nitrogen for growth (Peoples et al. 1987) or in the effectiveness of symbiosis (Rosendahl and Jakobsen 1987).

According to Streeter (1981), mineral nitrogen availability in the legumes rhizosphere is a limiting factor for nodule formation and nitrogen fixation. Foliar application of nutrients, including N, allows avoiding the harmful direct action of inorganic nitrogen on symbiotic processes (Marschner 1995). Foliar uptake of N is not only supplementary but can influence the N assimilation of the whole plant. The significance of foliar fertilizer application may lie in the localization and regulation of the enzyme systems involved in nitrogen assimilation. Although the effect of foliar application of several nitrogenous fertilizers has already been investigated, information on the effect of foliar application on the content of amino acids has been limited.

There are not enough data about the localization and regulation of nitrogen assimilatory enzymes and amino acid accumulation in case of changing the site of primary N assimilation through the foliar feeding especially when plants are inoculated with the respective nitrogen fixing bacteria strain under conditions of insufficient Mo supply.

Experiments on the effects of additional foliar feeding on the nitrogen fixation and free amino acid accumulation in pea and alfalfa plants (temperate legumes with amide compounds transport), inoculated with *Rhizobium leguminosarum bv. Vicae* and *Sinorhizobium meliloti*, respectively, were carried out under the influence of Mo deficiency.

2 Foliar Application of Nutrient Elements: Connection with Function of Symbiotic Systems and Nitrogen Assimilation

Symbiotic systems *Medicago sativa* L. – *S. meliloti* and *Pisum sativum* L. – *R. leguminosarum* bv. *viciae*, form indeterminated nodules with apical permanently functioning meristem. In these plants from temperate latitudes, the fixed and reduced nitrogen is transported toward the xylem as amide compounds – asparagine and glutamine (Atkins 1991). It was known that nitrates present in the nutrient media depressed the root nodule formation and symbiotic nitrogen fixation (Schulze et al. 1998). One of the ways to diminish such negative influence is to change the place of uptake and assimilation of exogenous supplied mineral nitrogen through the foliar application (Boote et al. 1978; Poole et al. 1983). Foliar fertilization (or foliar feeding) entails the application via spraying of nutrients (minor and major nutrients, plant hormones, stimulants and other beneficial substances). Observed effects of foliar fertilization have included yield increases, resistance to diseases and insect pests, improved drought tolerance, and enhanced crop quality (Kuepper 2003).

Foliar fertilization caused the plant to pump more sugars and other exudates from its roots into the rhizosphere. Beneficial microbial populations in the root zone are stimulated by the increased availability of these exudates. On the other hand, absorption of water and nutrients by the roots toward xylem increased (Alexander 1986). The uptake of nutrients through the foliage is affected by a number of interacting factors (Kuepper 2003; Wójcik 2004) of which a few have been known recently (Table 1).

Table 1 Influences determining the efficacy of foliar nutrient sprays (Clarkson 1985)

Plant	Environment	Spray solution
• Epicuticular wax	• Temperature (max. 30°C)	• Concentration
• Cuticular wax	• Light	• Application rate
• Age of the leaf	• Photo period	• Application technique
• Stomata and guard cells	• Air movement	• pH (5.5–8.5)
• Leaf hairs	• Humidity	• Polarity
• Adaxial and abaxial leaf sides	• Time of day	• Hygroscopicity
• Leaf turgor	• Nutrient ratio	• Sticking ability
• Surface moisture		• Carriers, penetrates
• Cation exchange capacity		• Chelates
• Root osmotic potential		
• Nutritional status		
• Species and cultivar		

Two related plant processes are associated with the effectiveness of foliar application of fertilizers (Doring and Gericke 1986). The first relates to the rates of absorption for each of the nutrients. The second is concerned with the mobility or the extent to which each foliar absorbed nutrient is translocated out of the leaves to the other parts of the plant including the stem, roots, flowers, and seeds. The structure of epicuticular wax helps limit penetration of water molecules and ions across the membrane (Marschner 1995).

The flow of cations through the cuticular membrane is much easier than that of anions. It is estimated that cation ability to penetrate the cuticular membrane is 1,000 times higher than for the anions (Mengel 2002). Generally, the movement of low-molecular-weight solutes (e.g., ions, organic acids, amino acids) from the leaf surface to the epidermal cell wall is a nonmetabolic process driven by diffusion and electrochemical potential formed by a negative charge increase across the cuticular membrane (Kannan 1980; Tyree et al. 1992). Selective transport of nutrients across the plasma membrane requires energy, specific carriers, permeases and channels and may also be a passive process driven by diffusion.

Much interest in foliar feeding has centered on the use of nitrogen. Plants respond to foliar sprays of nitrates, ammonium compounds and urea (Wójcik 2004). Generally, it seems that the ability of leaves to absorb different N forms depends considerably on plant species. The leafy absorbed nitrogen has to be metabolized rapidly before it is involved into high molecular compounds. Andrews (1986) reported that the parts of the plants above the ground are the main place of nitrogen assimilation. It was established (Peuke et al. 1998) that under nitrate nutrition the highest incorporation of N was found in the roots of *Ricinus communis* L., but when ammonium was sprayed, it was in the growing leaves. It was not surprising that N uptake was significantly higher in ammonium-sprayed leaves than in nitrate-sprayed leaves (Garten and Hanson 1990). It seems that the positively charged ammonium was more easily transported across the cuticle than was the negatively charged nitrate.

Many authors showed a beneficial effect of foliar fertilization on the nitrogen fixing plants on the different developmental stages (Wojcieska and Kocon 1997; Palta et al. 2005). Hanway (1979) reported about a significant increase of soybean seed yields that received nitrogen through the leaves as urea. Application of urea resulted in avoiding early fall of leaves and senescence. Similar research with soybean showed that foliar fertilization with macronutrients at early vegetative stages enhanced nutrient uptake through the roots, photosynthetic rates and seed yields (Haq and Mallarino 2000). The authors suggested that low input of N, P, and K could stimulate growth without inhibiting nodulation.

Da Silva et al. (1993) have concluded that bean plants (*Phaseoulus vulgaris* L.) are capable for nitrogen fixation in the case of the exogenous nitrogen supplied through the leaves which resulted in the increase of seed yields with high content of nitrogen.

It is likely that foliar fertilization alleviated problems with early nutrient uptake, which sometimes occur even in high-testing soils. Positive effect of foliar fertilization on the legume plants growth increased in the case of application of combined foliar fertilizers containing macro- and micronutrients. Complex foliar fertilizers are absorbed easily with high efficiency (Garcia and Hanway 1976; Schon and Blevins 1990; Fenn et al. 1995; Haq and Mallarino 2000).

Despite many studies carried out on mineral nutrient absorption by leaf tissues many aspects of foliar fertilization are still unknown. At present, it is believed that such fertilization of plants is a valuable complement to the application of nutrients to the soil. It is proposed that this treatment should be recommended in the integrated plant production because it is environment friendly and increases productivity and yield quality (Wójcik 2004).

3 Molybdenum Importance for Plant Nitrogen Metabolism

Among the micronutrients that are essential for plant growth, Mo is required in the smallest amounts. Plant species significantly differ in their requirements for Mo. Grasses from nonlegume plants contained less Mo (0.2–1.0 ppm) than legume plants (0.5–20 ppm) per gram dry weight (Gupta and Lipsett 1981). Its mobility is proved by the translocation of foliar supplied molybdenum. The form in which Mo is translocated is unknown, but its chemical properties indicated that is most likely transported as MoO_4^{2-}, rather than a complex ion (Marschner 1995).

Arnon and Stout (1939) first reported about Mo deficiency symptoms in tomato plants. The visual Mo deficiency symptoms were shown in cauliflower (Davies 1945; Mitchell 1945) alfalfa (Anderson 1942), clover (Anderson 1946), and grape (Williams et al. 2004). The symptoms associated with the deficiency of Mo are closely related to metabolism of nitrogen. In plants only a few enzymes have been found to contain molybdenum as cofactor. These include enzymes that catalyzed different chains of nitrogen metabolism. The molybdenum requirements of higher plants, therefore, depend on the mode of nitrogen supply. Under Mo deficiency conditions, plant molybdoenzymes can be broken down to those involved in nitrogen reduction and assimilation (Kaiser et al. 2005). Harper and Paulsen (1969) observed significant accumulation of nitrates in wheat seedlings starved for Mo and a negative correlation between nitrate content and nitrate reductase activity. On the other hand, it could be expected that under molybdenum-deficiency conditions, application of nitrogen in a form different from nitrate-N, for example NH_4^+-N do not influence so strongly, nitrogen assimilation and plant development (Notton 1983).

The other notable influence of Mo on plant nitrogen metabolism is in nitrogen-fixing legumes. The symbiotic bacterial enzyme nitrogenase is comprised of two subunits one of which is the MoFe protein directly involved in the reduction of N_2 to NH_3. What is known, with respect to molybdenum and legume nitrogen fixation, is that Mo availability is closely correlated with nodule development and Mo requirement of nitrate reductase is lower than for nitrogenase (Anderson 1956; Kaiser et al. 2005). Depending on the plant species, the critical deficiency levels of molybdenum vary between 0.1 and 1.0 ppm leaf dry weight (Gupta and Lipsett

1981). Frame et al. (1998) showed that critical Mo levels for alfalfa are 0.5–0.9 ppm in plant dry weight.

Anderson (1956) clarified the association of deficiency symptoms with seed reserves and pointed it out as a probable reason for the relative absence in early experience of deficiency of large-seeded legumes. Jongruaysup et al. (1997) suggested that in case of high Mo seed reserves deficiency symptoms were unlikely even on low Mo soils. Hagstrom and Berger (1965) observed that large-seeded crops, such as peas (*P. sativum* L.,) responded to soil application of Mo when the seeds contained less than 0.2 ppm Mo, but not when they contained enough Mo (0.5–0.7 ppm) to supply the Mo needs of the crop. Most frequently occurred Mo deficiency symptoms are chlorosis, golden-yellow coloration of older leaves along the apex and the apical leaf margins as well as necrotic areas extending back along the apex and the apical leaf margins (Agarwala et al. 1979; Gupta 1997). Plants had short internodes and reduced foliage (Gupta and Lipsett 1981). It would appear that nodules accumulate significantly more Mo than what is required in order to support bacterial nitrogenase activity and symbiotic nitrogen fixation (Kaiser et al. 2005).

Mo distribution among the plant organs strongly depended on the Mo soil reserves. In case of poor Mo supply, Mo transport is directed to the roots and nodules while in the case of Mo excess Mo is accumulated mainly in the leaves (Becking 1961; Franco and Munns 1981).

Mo content in plant samples was determined with inductively coupled plasma spectrometry. The seeds, roots and shoots of nitrogen fixing pea and alfalfa plants were analyzed (Table 2) (Hristozkova et al. 2009). As the alfalfa plants are more sensitive than pea to Mo contents in the nutrient media, Mo reserves in alfalfa seeds were ten times higher than those in pea seeds. The content of Mo in Mo deficient organs was obviously lower than the content in normally supplied plants and this trend appeared more clearly in alfalfa. Thus, Mo contents in plants grown in the absence of Mo significantly decreased – with 99% in the shoots and 98% in the roots in comparison with relevant Mo adequate plants. In this connection, typical Mo deficiency visual symptoms expressed as chlorosis of the young mature leaves as in alfalfa plants only (Fig. 1). Higher amounts of Mo were accumulated into the shoots in case of pea and in roots of Alfalfa. This trend was observed both in deficient plants and those supplied with Mo. Higher levels of Mo in the pea shoots coincided with higher nitrogen content in shoots (data not shown).

Total Mo content (Table 2) (Hristozkova et al. 2009) in the alfalfa Mo deficient plants (roots and shoots) is relevant to the initial Mo level in the seeds and is much

Table 2 Molybdenum content in plant organs (ppm) (Hristozkova et al. 2009)

Plant organs	Pea	Alfalfa
Seeds	0.17	1.73
Shoots (+Mo)	4.94	23.8
Roots (+Mo)	0.72	64.5
Shoots (−Mo)	0.69	0.31
Roots (−Mo)	0.17	1.20

Fig. 1 Alfalfa plants grown at reduced Mo supply

higher in roots. Jongruaysup et al. (1997) showed that in Mo deficient nitrogen fixing plants roots are the main storage organ for Mo accumulation, necessary for the forming seeds. According to Brodrick and Giller (1991) when plants suffer from Mo shortage, Mo become more mobile and its transport is orientated from the shoots toward roots and nodules in order to support nitrogen fixing activity.

4 Plant Biomass Accumulation, Number of Nodules and Nitrogen Fixing Activity

Significant changes in plant metabolism related to nitrogen assimilation and biomass accumulation of pea and alfalfa, grown under conditions of Mo deficiency were observed. Under conditions of optimal Mo supply, the favorable effect of foliar fertilization on root and shoot dry biomass accumulation was expressed and more distinct in pea plants. Foliar fertilization resulted to increase of shoot dry biomass with 67% and root dry biomass with 10% in comparison with the control. In alfalfa plants, only shoot biomass increased with 33% (Table 3).

Under Mo deficiency conditions, additional foliar nutrition also enhanced shoot and root dry biomass of both plant species as compared with the root nutrition. According to the obtained results excluding Mo from the nutrient media resulted to a less reduction of plant biomass in case of additional nutrient supply though the foliage. Therefore, plant dry biomass accumulation of Mo deficient foliar fed plants were close to plant biomass with root nutrition and sufficient Mo supply.

Table 3 Effect of foliar feeding on the dry biomass accumulation of nitrogen fixing 35-days old pea and alfalfa plants grown at different Mo supply

Variants	Dry biomass (g plant^{-1})		Dry biomass (g plant^{-1})	
	Shoots	Roots	Shoots	Roots
	Pea		Alfalfa	
F1+Mo-control	0.600 ± 0.019^{fb*}	0.480 ± 0.021^{c}	0.03 ± 0.002^{b}	0.022 ± 0.0015^{c}
F2+Mo	1.000 ± 0.044^{d}	0.530 ± 0.016^{e}	0.04 ± 0.002^{c}	0.019 ± 0.0014^{b}
F1−Mo	0.501 ± 0.015^{f}	0.420 ± 0.017^{b}	0.02 ± 0.0011^{f}	0.013 ± 0.0011^{f}
F2−Mo	0.559 ± 0.022^{a}	0.494 ± 0.015^{c}	0.029 ± 0.0012^{a}	0.015 ± 0.0013^{a}

F1 – root nutrition; F2 – combined root and foliar nutrition

*Different letters indicate significant differences assessed by Fisher LSD test ($P \leq 0.05$) after performing ANOVA multifactor analysis

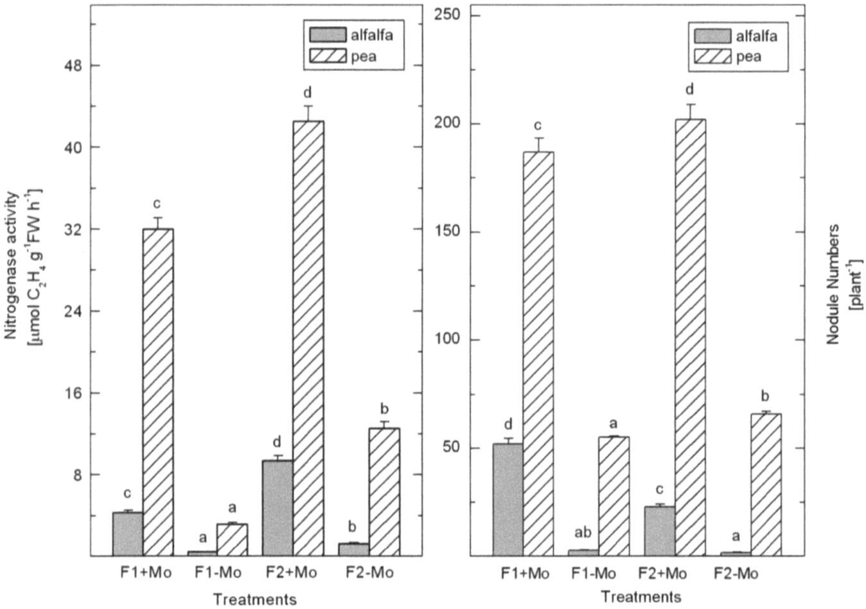

Fig. 2 Nitrogenase activity and nodule number in pea (Hristozkova et al. 2007a) and alfalfa (Hristozkova et al. 2009) plants grown at different Mo supply: (F1+Mo)-control Mo supplied plants with root nutrition; (F2+Mo)-Mo supplied plants with root and foliar nutrition; (F1−Mo)-Mo deficient plants with root nutrition; (F2−Mo)-Mo deficient plants with root and foliar nutrition. Different letters indicate significant differences assessed by Fisher LSD test ($P \leq 0.05$) after performing ANOVA multifactor analysis

The number of nodules in the pea plants (Hristozkova et al. 2007a) grown without Mo declined and this reduction is expressed to a less extent in foliar fed (F2−Mo) plants (Fig. 2). The exclusion of Mo from the nutrient media resulted in the reduction of nodule numbers with 63% for the plants with root nutrition and 15% for those with root and foliar nutrition. A depression of nitrogenase activity

(NG) with 98% was observed in Mo deficient pea plants with root nutrition in comparison with the relevant Mo supplied plants (Fig 2) (Hristozkova et al. 2007a). Nitrogenase activity in Mo deficient pea plants with root and foliar nutrition decreased to a less extent (with about 68%) in comparison with relevant Mo adequate plants.

However, the number of nodules in alfalfa plants with additional foliar feeding declined in comparison with the plants with root nutrition both in the presence and absence of Mo (Fig. 2). Nitrogenase activity in foliar fed plants – treatments F2+Mo and F2−Mo was higher than the activity in root fed plants – treatments F1+Mo and F1−Mo (Hristozkova et al. 2009). Therefore, the number of nodules was not relevant to their NG activity (Fig. 2). The lack of correspondence between the nodule number and nitrogen fixing activity was also suggested by Puppo et al. (2005).

5 Free Amino Acid Composition

Connection between Mo deficiency and nitrogen metabolism strongly affected protein synthesis. Low molecular nitrogen compounds accumulated (amino acids, amides) as a result of high rubo nuclease and low aminotransferase activities (Marschner 1995). In temperate legumes with amide compounds transport, both fixed and supplied inorganic nitrogen are assimilated into amino acids glutamine (Gln), glutamate (Glu), asparagine (Asn) and aspartate (Asp), which serve as important nitrogen carriers in plants (Ta et al. 1984). When the pea plants were supplied with normal Mo concentration the highest content of Asp/Asn was found in the roots, especially in the plants with foliar fertilization (F2+Mo) – 35% of total amino acid content (Fig. 3). Relatively high content of proline (Pro) in root fed plants (F1+Mo) and in foliar fed plants (F2+Mo) was observed both in the shoots and roots followed by the content of alanine (Ala). The content of Asp/Asn and Glu/Gln in the shoots of Mo supplied plants (F1+Mo, F2+Mo) was lower than in the roots (Fig. 3). Ta et al. (1984) suggested that in legumes with amide compounds transport, Asn is a major nitrogen transport compound. Rosendahl and Jakobsen (1987) have studied the concentration of major amino acids and amides in the root xylem sap of *P. sativum* in relation to the efficiency of various strains of *Rhizobium*. The authors concluded that Asn contents were clearly higher than the Gln in the most efficient symbiosis. According to Fougére et al. (1991) in alfalfa roots, Glu and γ-aminobutyrate (GABA) are predominantly higher and represented 25% and 18% of the total amino acid fraction, respectively. We observed high levels of GABA in the shoots (Fig. 3) of Mo supplied pea plants (F1+Mo, F2+Mo). GABA amino acid is found in plants as a significant component of the free amino acid pool. In higher plants GABA is synthesized primarily from L-glutamate (Bown and Shelp 1997). This is in correspondence with lower Glu/Gln levels in the shoots (Fig. 3). Some authors suggested that GABA might play a role in signaling (Kathiresan et al. 1997).

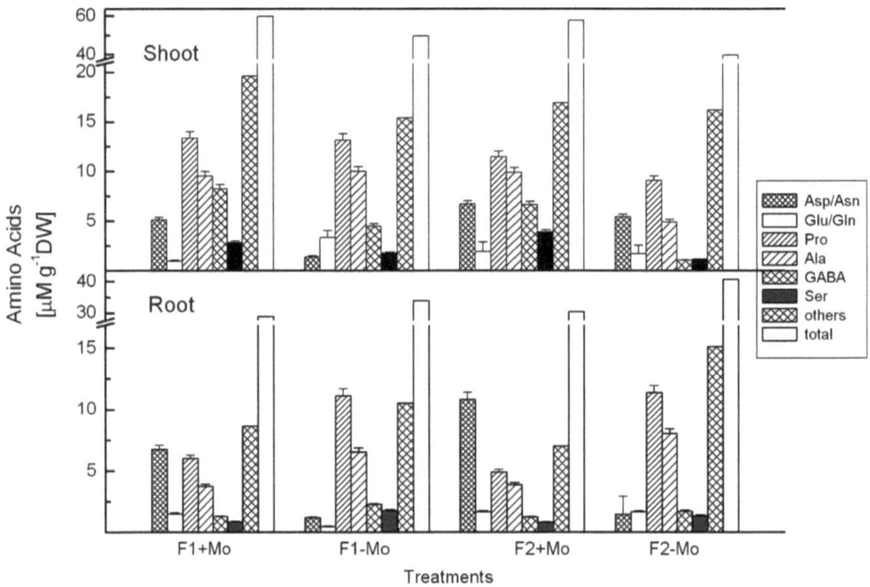

Fig. 3 Shoot and root amino acids content in pea plants (Hristozkova et al. 2007b). Treatments: F1+Mo; F1−Mo; F2+Mo; F2−Mo are described in detail at Fig. 2. Bars represent the standard error of the mean. Reported values were averaged from three independent extractions ($n = 3$)

In the variants without Mo (F1−Mo, F2−Mo) the highest Pro content in the free amino acid pool was measured both in pea shoots and roots, and varied from 22 to 32% of the total amino acid fraction (Fig. 3). The addition of foliar nutrients resulted in lower proline content in the pea shoots. The proline content in the roots of Mo deficient plants was higher than the Mo adequate pea plants, while in the shoots significant changes were not observed. Marked increase in free proline occurs in many plants during moderate or severe water and salt stress; this accumulation, mainly as a result of increased proline biosynthesis, is usually the most outstanding change among the free amino acids (Fougére et al. 1991). Hence, Mo exclusion from the nutrient media could be considered as stress factor for the pea plants. The content of Asp/Asn in Mo deficient plants is low with the exception of the value in the shoots of foliar supplied F2−Mo pea plants (Fig. 3). The Asp/Asn concentration in F2−Mo treatment is close to the value in the shoots of Mo supplied plants with root nutrition (F1+Mo). Therefore, the transport of major nitrogen compound Asp/Asn toward the leaves is not suppressed in Mo deficient plants in case of foliar nutrient application. High content of alanine (Ala) of Mo deficient pea shoots and roots were accounted for and reached values between 12 and 20% of the total amino acid pool (Fig. 3). Ta and Joy (1986) pointed out that Ala has a major involvement in photorespiration and those other amino acids and amides such as Asn are also involved although to a lesser extent. The synthesis of alanine may occur at the expense of the acidic amino acids, glutamate and aspartate (Stewart and Larher 1980), and occurs concomitantly with the accumulation of GABA (Wallace et al. 1984; Ratcliffe 1995). We observed similar correlation regarding the shoots of Mo supplied pea plants (F1+Mo, F2+Mo). In plants grown in Mo restrictive media such correlation was not found.

The content of total free amino acids (Hristozkova et al. 2009) in alfalfa roots was significantly higher in the plants grown in Mo absence compared with the relevant treatments when Mo was supplied (Fig. 4). The lowest free amino acids content was established in F2+Mo treatment. High content of alanine (Ala) was found in the all treatments. In the roots of F1−Mo plants, the main nitrogen carriers' aspartate/asparagine (Asp/Asn) and glutamate/glutamine (Glu/Gln) content decreased in comparison with F1+Mo treatments.

The levels of Asp/Asn and Glu/Gln in the roots of F2−Mo alfalfa plants were higher than in F2+Mo. The highest content in the roots of Ala, γ-aminobutyrate (GABA), proline (Pro), threonine (Thr) and serine (Ser) was observed in F1−Mo treatment. In the shoots of Mo deficient plants, the total content of free amino acids exceeded three times than that of Mo supplied – F1+Mo and F2+Mo plants (Fig. 4). Additional foliar nutrition did not significantly change the total amount of free amino acids independently on Mo supply. In the shoots of F1−Mo and F2−Mo treatments, the level of stress induced amino acids Ala, GABA, Pro, Thr, and Ser mainly increased in comparison with the controls (F1+Mo). Enhanced Asp/Asn and Glu/Gln levels were observed in Mo deficient plants with root and foliar nutrition in comparison with Mo supplied treatments.

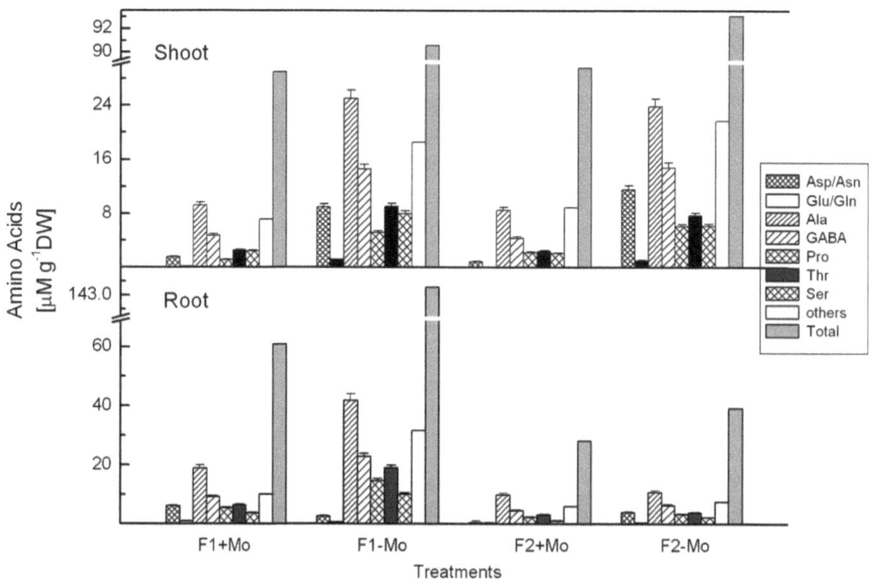

Fig. 4 Shoot and root amino acids content in alfalfa plants (Hristozkova et al. 2009). Treatments: F1+Mo; F1−Mo; F2+Mo; F2−Mo are described in detail at Fig. 2. Bars represent the standard error of the mean. Reported values were averaged from three independent extractions ($n = 3$)

Molybdenum starvation resulted in an increase of total content of free amino acids especially in the alfalfa roots in comparison with the Mo supplied plants. Influence of foliar feeding was appeared in free amino acids accumulation mainly in the roots especially in Mo deficient plants. The increase of free amino acids is a result of enhanced flow of assimilates toward the roots in foliar supplied plants.

Increased levels of stress induced amino acids such as Ala, GABA, Trh, Pro, and Ser in the roots and shoots of Mo deficient plants indicated that alfalfa plants are very sensitive to insufficient Mo supply (Fig. 4). The main nitrogen transport compounds Asp/Asn increased at insufficient Mo supply especially in the shoots. In the all treatments, Asp/Asn content in the shoots and roots was higher than the Glu/Gln content, and its values varied in dependence on Mo presence in the nutrient media as well as additional foliar nutrition. On the other hand, Lea et al. (2007) observed accumulation of Asn in plant tissues during the periods of suppressed protein synthesis. Foliar fertilization resulted in lowering of total free amino acid content predominantly in the roots both in the Mo supplied and Mo deficient plants compared to the plants with root nutrition. It was also observed that the foliar fertilization reduced the inhibitory effect of Mo shortage on the aspartate/asparagine content in the pea shoots.

6 Conclusion

Biological nitrogen fixation provides a large proportion of plant nitrogen requirements and contributes to agricultural sustainability. Symbiotic nitrogen fixation by the legume–*Rhizobium* symbiosis is a finely regulated process. Mechanisms for the regulation of symbiotic N_2 fixation under conditions of nutrient deficiency may be very different in the different symbiotic types. Molybdenum is now known to be essential in the fixation of nitrogen by the symbiotic bacteria associated with leguminous plants. It was established that Mo shortage in the nutrient media resulted in nodule number and biomass reduction, lowered nitrogenase activity and suppressed plant biomass accumulation of pea and alfalfa plants. The composition of free amino acids and amides changed under Mo deficiency conditions – the main nitrogen transport compounds aspartate/asparagines and glutamate/glutamine decreased, while stress induced amino acids as alanine, GABA, threonine, proline, and serine accumulated. It was shown that alfalfa was more sensitive to Mo starvation than the pea plants.

The results connected with nitrogen assimilation and biomass accumulation of pea and alfalfa showed that efficiency of nitrogen fixation and assimilation could be improved through the application of Agroleaf® in 0.3% concentration. The positive effect of foliar nutrition on nitrogen assimilation was better expressed in pea plants under Mo deficiency conditions. In addition it was observed that the negative effect of Mo deficiency on the nitrogen fixation and assimilation in pea and alfalfa plants (both temperate legumes with amides nitrogen transport) was lowered through the foliar absorbed nutrients.

Acknowledgments This study was financially supported by The Project "Progress in plant investigations for the improvement of sustainability of agriculture (PISA-INI 14/01.09.2005)" Bulgarian Ministry of Education and Sciences.

References

Agarwala S, Chatterjee C, Sharma P, Sharma C, Nautiyal N (1979) Pollen development in maize plants subjected to molybdenum deficiency. Can J Bot 57:1946–1950

Alexander A (1986) Optimum timing of foliar nutrient sprays. In: Alexander A (ed) Foliar fertilization. Kluwer, Dordrecht, The Netherlands, pp 44–60

Anderson A (1942) Molybdenum deficiency on a South Australian ironstone soil. J Aust Inst Agr Sci 8:73–75

Anderson A (1946) Molybdenum in relation to pasture improvement in South Australia. J Counc Scient Indust Res 19:1–15

Anderson A (1956) Molybdenum deficiencies in legumes in Australia. Soil Sci 81:173–192

Andrews M (1986) The partitioning of nitrate assimilation between root and shoot of higher plants. Plant Cell Environ 9:511–519

Arnon D, Stout P (1939) Molybdenum as an essential element for higher plants. Plant Physiol 14:599–602

Atkins CA (1991) Ammonia assimilation and export of nitrogen from the legume nodule. In: Dilworth MJ, Glen AR (eds) Biology and biochemistry of nitrogen fixation. Elsevier, Amsterdam, pp 293–319

Becking JH (1961) A requirement of molybdenum for the symbiotic nitrogen fixation in alder (*Alnus glutinosa* Gaertn.). Plant Soil 15:217–227

Boote K, Gallagher R, Robertson W, Hinson K, Hammond LC (1978) Effect of foliar fertilization on photosynthesis, leaf nutrition and yield of soybean. Agron J 70:787–791

Bown A, Shelp B (1997) The metabolism and functions of gamma-aminobutyric acid. Plant Physiol 115:1–5

Brodrick S, Giller K (1991) Root nodules of phaseolus: efficient scavengers of molybdenum for N_2-fixation. J Exp Bot 42:679–686

Clarkson DT (1985) Factors affecting mineral nutrient acquisition by plants. Annu Rev Plant Physio 36:77–115

Da Silva P, Tsai S, Bonetti R (1993) Response to inoculation and N fertilization for increased yield and biological nitrogen fixation of common bean *(Phaseolus vulgaris* L.). Plant Soil 152:123–130

Davies E (1945) A case of molybdenum deficiency in New Zealand. Nature 156:392–393

Doring H, Gericke R (1986) The efficiency of foliar fertilization in arid and semi-arid regions. In: Alexander A (ed) Foliar fertilization. Kluwer, Dordrecht, The Netherlands, pp 29–35

Fenn L, Hasanein B, Burks C (1995) Calcium-ammonium effects on growth and yield of small grains. Agron J 87:1041–1046

Fougére F, Le Rudulier D, Streeter J (1991) Effects of salt stress on amino acid, organic acid, and carbohydrate composition of roots, bacteroids, and cytosol of alfalfa (*Medicago sativa L.*). Plant Physiol 96:1228–1236

Foyer CH, Noctor G (2002) Nitrogen assimilation and photosynthesis: energetics, interaction and control. In: Foyer CH, Noctor G (eds) Photosynthetic nitrogen assimilation and associated carbon and respiratory metabolism. Advances in photosynthesis research, vol 12. Kluwer, The Netherlands, pp 1–22

Frame J, Charlton J, Laidlaw A (1998) Temperate forage legumes. CAB International, Wallingford

Franco A, Munns D (1981) Response of *Phaseolus vulgaris* L. to molybdenum under acid conditions. Soil Sci Soc Am J 45:1144–1148

Garcia R, Hanway J (1976) Foliar fertilization of soybean during the seed-filling period. Agron J 68:653–657

Garten CT, Hanson PJ (1990) Foliar retention of ^{15}N-nitrate and ^{15}N-ammonium by red maple (*Acer rubrum*) and white oak (*Quercus alba*) leaves from simulated rain. Environ Exp Bot 30:333–342

Gupta U (1997) Symptoms of molybdenum deficiency and toxicity in crops. In: Gupta UC (ed) Molybdenum in agriculture. Cambridge University Press, New York, pp 160–170

Gupta U, Lipsett J (1981) Molybdenum in soil, plants, and animals. Adv Agron 34:73–115

Hagstrom G, Berger K (1965) Molybdenum deficiencies of Wisconsin soils. Soil Sci 100:52–56

Hanway J (1979) Foliar fertilization of soybeans during seed-filling. In: Corbin FT (ed) World soybean research conference proceeding, vol 2, pp 409–416

Haq M, Mallarino A (2000) Soybean yield and nutrient composition as affected by early season foliar fertilization. Agron J 92:16–24

Harper J, Paulsen G (1969) Nitrogen assimilation and protein synthesis in wheat seedlings as affected by mineral nutrition. II. Micronutrients. Plant Physiol 44:636–640

Hristozkova M, Geneva M, Stancheva I, Georgiev G (2007a) Response of inoculated foliar fed pea plants (*Pisum sativum* L.) to reduced Mo supply. Acta Biol Hung 58:87–92

Hristozkova M, Geneva M, Stancheva I, Georgiev G (2007b) Nitrogen assimilatory enzymes and amino acid content in inoculated foliar fertilized pea plants grown at reduced molybdenum concentration. J Plant Nutr 30:1409–1419

Hristozkova M, Geneva M, Stancheva I (2009) Effect of foliar feeding on nitrogen assimilation in alfalfa plants at insufficient molybdenum supply. Acta Biol Hung 60:211–219

Ireland R, Lea P (1999) The enzymes of glutamine, glutamate, asparagine, and aspartate metabolism. In: Singh BK (ed) Plant amino acids: biochemistry and biotechnology. Marcel Dekker, New York, pp 49–109

Jongruaysup S, Dell B, Bell RW, O'Hara GW, Bradley JS (1997) Effect of molybdenum and inorganic nitrogen on molybdenum redistribution in Black Gram (*Vigna mungo L. Hepper*) with particular reference to seed fill. Ann Bot (London) 79:67–74

Kaiser B, Gridley K, Brady J, Phillips T, Tyerman S (2005) The role of Molybdenum in agricultural plant production. Ann Bot (London) 96:745–754

Kannan S (1980) Mechanisms of foliar uptake of plant nutrients: accomplishments and prospects. J Plant Nutr 2:717–735

Kathiresan A, Tung P, Chinnappa C, Reid D (1997) Gamma-aminobutyric acid stimulates ethylene biosynthesis in sunflower. Plant Physiol 115:129–135

Kuepper G (2003) Foliar fertilization. ATTRA, National Sustainable Agriculture Information Service, Fayetteville, AR

Lam H, Coschigano K, Oliveira I, Melo-Oliveira R, Coruzzi G (1996) The molecular-genetics of nitrogen assimilation into amino acids in higher plants. Annu Rev Plant Phys 47: 569–593

Lea P, Sodek L, Parry M, Shewry P, Halford N (2007) Asparagine in plants. Ann Appl Biol 150:1–26

Marschner H (1995) Mineral nutrition of higher plants. Academic, London

Mendel R, Haensch R (2002) Molybdoenzymes and molybdenum cofactor in plants. J Exp Bot 375:1689–1698

Mengel K (2002) Alternative or complementary role of foliar supply in mineral nutrition. Acta Hortic 594:33–48

Mitchell K (1945) Preliminary note on the use of ammonium molybdate to control whiptail in cauliflower and broccoli crops. New Zeal J Sci 27:287–293

Notton B (1983) Micronutrients and nitrate reductase. In: Robb DA, Pierpoint WS (eds) Metals and micronutrients: uptake and utilization by plants. Academic, Bath, pp 219–240

Palta JA, Nandwal AS, Kumari S, Turner NC (2005) Foliar nitrogen application increase the seed yield and protein content in chickpea (*Cicer arietinum* L.) subject to terminal drought. Aust J Agr Res 56:105–112

Peoples M, Sudin M, Herridge D (1987) Translocation of nitrogenous compounds in symbiotic and nitrate-fed amide-exporting legumes. J Exp Bot 38:567–579

Peuke A, Jeschke W, Dietz K, Schreiber L, Hartung W (1998) Foliar application of nitrate or ammonium as sole nitrogen supply in *Ricinus communis* I. Carbon and nitrogen uptake and inflows. New Phytol 138:675–687

Poole W, Randall G, Ham G (1983) Foliar fertilisation of soybean. I. Effect of fertiliser sources, rates and frequency of application. Agron J 75:195–200

Puppo A, Groten K, Bastian F, Carzaniga R, Soussi M, Lucas M, Rosario de Felipe M, Harrison J, Vanacker H, Foyer C (2005) Legume nodule senescence: roles for redox and hormone signalling in the orchestration of the natural aging process. New Phytol 165:683–701

Ratcliffe R (1995) Metabolic aspects of the anoxic response in plant tissue. In: Smirnoff N (ed) Environment and plant metabolism: flexibility and acclimation. Bios Scientific, Oxford, pp 111–127

Rosendahl L, Jakobsen I (1987) *Rhizobium* strain effects on pea: the relation between nitrogen accumulation, phosphoenolpyrovate carboxylase activity in nodules and asparagine in root bleeding sap. Physiol Plant 71:281–286

Schon M, Blevins D (1990) Foliar boron applications increase the final number of branches and pods on branches of field-grown soybeans. Plant Physiol 92:607–609

Schulze J, Shi L, Blumenthal J, Samac DA, Gantt JS, Vance CP (1998) Inhibition of alfalfa root nodule phosphoenolpyruvate carboxylase through an antisense strategy impacts nitrogen fixation and plant growth. Phytochemistry 49:341–346

Stewart G, Larher F (1980) Accumulation of amino acids and related compounds in relation to environmental stress. In: Miflin BJ (ed) The biochemistry of plants, vol 5. Academic, New York, pp 609–635

Streeter JG (1981) Effect of nitrate in the rooting medium on carbohydrate composition of soybean nodules. Plant Physiol 69:1429–1434

Ta T, Joy K (1986) Metabolism of some amino acids in relation to the photorespiratory nitrogen cycle of pea leaves. Planta 169:117–122

Ta T, Joy K, Ireland R (1984) Amino acid metabolism in pea leaves. Plant Physiol 74:822–826

Tyree M, Wescot C, Tabor C, Morse A (1992) Diffusion and electric mobility of KCl within isolated cuticles of *Citrus aurantium*. Plant Physiol 99:1057–1061

Wallace W, Secor J, Schrader L (1984) Rapid accumulation of gamma-aminobutyric acid and alanine in soybean leaves in response to an abrupt transfer to lower temperature, darkness, or mechanical manipulation. Plant Physiol 75:170–175

Williams C, Maier N, Bartlett L (2004) Effect of molybdenum foliar sprays on yield, berry size, seed formation, and petiolar nutrient composition of 'Merlot' grapevines. J Plant Nutr 27:1891–1916

Wojcieska U, Kocon A (1997) Reaction of faba been plants to soil and foliar N application and K nutrition. Acta Physiol Plant 19:23–28

Wójcik P (2004) Uptake of mineral nutrients from foliar fertilization. J Fruit Ornam Plant Res 12:201–218, Special edition

Index

A
Abiotic factors, 101
Abiotic stresses, 2, 4, 8, 9, 368, 372, 380
Abscisic acid, 345
ACC. *see* 1-Aminocyclopropane-1-carboxylate
Acetobacter diazotropicus, 51, 52
Acetobacter peroxydans, 304
Acetoin, 345, 348
Acidobacteria, 105, 253
Acidovorax, 59
Acid rain, 9
Acinetobacter, 46, 60, 98, 335
Acinetobacter rhizosphaerae, 288
Actinobacteria, 253
Actinomycete, 161
Actinoplanes, 236
Acyclobacillaceae, 335–336
Acyclobacillus, 336, 337
Acylated homoserine lactones (AHLs), 84, 85
Adenocallyma alliaceum, 242
Adjuvants, 219
AdoMet. *See* S-adenosyl L-methionine
Aerenchyma, 322
Aerial diseases, 202
Aeromonas hydrophila, 258
AFPs. *See* Anti freeze proteins
Agrobacterium, 4, 23, 31, 51, 59, 66, 98, 199, 236, 334, 344
  A. radiobacter, 303–304
  A. rhizogenes, 176
  A. rubi, 50, 54–57
  A. rubi A-1, A-16 and A-18, 50, 54, 55, 57
  A. rubi strain A16, 54
  A. tumefaciens, 50, 345–346
Agrochemicals, 199
Agroecosystems, 199, 204, 276, 282–283, 290
AHLs. *See* Acylated homoserine lactones
Alanine (Ala), 418, 425–428
Alanine racemase, 370
Alcaligenes, 46, 51, 55, 66, 98, 236
  A. faecalis, 320
  A. xylosoxydans, 238–239
Alfalfa, 286
Alginate, 259–260, 266–267
Allium sativum, 242
Alnus glutinosa, 52, 344
Alpine soil, 275
Alternaria brassicae, 346
Alternaria mali, 351
American Phytopatology Society (APS), 35
AMF. *See* Arbuscular mycorrhizal fungi
AM fungi. *See* Arbuscular mycorrhizal fungi
Amino acids, 418–420, 425–428
1-Aminocyclopropane-1-carboxylate acid (ACC)
  deaminase, 22, 24, 47, 51–53, 256, 266, 284, 366
  oxidase, 368, 371, 375
  synthase, 368, 370–371, 375, 377
Ammonia, 24, 33, 366, 369–372
Amorphosporangium, 236
Ampelomyces, 98
Amphisin, 160
Amplified ribosomal DNA restriction analysis (ARDRA), 27
Amylases, 119, 128
Andropogon gerardii, 392–393
Anoxic, 253
Antarctic bacterium, 282
Anthocyanin, 289, 368
Anthracnose, 171, 176
Anthurium andraeanum, 87

Anthyllis cytisoides, 256–257
Antibiosis, 159, 199, 214, 216, 220
Antibiotics, 6, 82, 84, 90, 118–120, 129, 130, 137, 138, 143–148, 199, 200, 237, 239, 275
Antibodies, 322
Anti freeze proteins (AFPs), 276, 281, 289
Antifungals, 199, 200, 202
Antimicrobial compounds, 82, 90
API kits, 25
APS. *See* American Phytopatology Society
Arabidopsis thaliana, 90, 177, 289–290, 345, 349–350, 355
Arachis hypogaea, 12
Arbuscular mycorrhizal fungi (AMF), 9, 64, 105, 387–410
ARDRA. *See* Amplified ribosomal DNA restriction analysis
Arthrobacter, 3, 46, 49, 98, 236, 334
Arthrobacter aureofaciens, 49
Asparagines (Asn), 419, 425, 426, 427, 428
Asparagus adscendens, 242
Aspartate (Asp), 418, 425–428
Aspartate aminotransferase, 370
Aspergillus, 200
  A. flavus, 240
  A. niger, 223, 240, 341, 352, 356
  A. oryzae, 352
Astragalus/Oxytropis spp., 286
Asymbiotic, 274, 286
Asymbiotic N_2 fixation, 34
ATPase, 56
Autotrophs, 275
Auxins, 46, 50, 88, 283–284, 319, 325
Avena sativa, 49
Azoarcus spp., 3, 46, 67, 105, 304, 305, 318, 319, 322, 323, 328
Azorhizobium, 4
Azospirillum spp., 3, 34, 36, 46, 54, 56, 86, 98, 118, 303, 304, 322–325, 334
  A. brasilense, 5, 9, 11–13, 48, 50–52, 60, 62, 300, 302, 305, 307, 308, 319
  A. brasilense Sp7, 13
  A. brasilense sp246, 48
  A. brasiliense, 265
  A. irakense, 5
  A. lipoferum, 52, 300, 302, 308, 309, 319
Azotobacter spp., 46, 48, 62, 67, 86, 98, 236, 305, 307, 334
  A. chroococcum, 9, 13, 303
  A. chroococcum strain C2, 54
Azotobacter sp. S8, strain E11, 48
Azotospirillum brasilense, 48, 50–52, 60, 62

B
Bacillaceae, 27, 335–336
Bacillales, 335–336
Bacilli, 334–337, 340, 343
Bacillibactin, 344
Bacillomycin F, 351
Bacillopeptins, 119, 122, 351
Bacillus BA16, RC03, RC23, 55
Bacillus M3 and/or OSU-142, 49, 53
Bacillus OSU-142, 48, 53–54, 57
Bacillus RC03, 50, 51, 55
Bacillus RC23, 50, 51, 55
Bacillus spp., 3, 8, 23, 26, 28–32, 34–36, 46–57, 60–62, 66, 67, 86, 98, 118–121, 138, 143, 145, 162–164, 169–173, 175, 179, 199–205, 215, 218, 221, 223, 224, 236–243, 255, 256, 265–267, 320, 366–367, 373
  B. alvei, 337
  B. amyloliquefaciens, 65, 336–337, 343, 345–346, 348, 351
  B. amyloliquefaciens 937a, 63
  B. amyloliquefaciens BNM122, 14, 200, 203, 204
  B. amyloliquefaciens IN937a, 55, 59, 63
  B. amyloliquefaciens RC-2, 345
  B. amyloliquifaciens, 221
  B. amylolyticus, 337
  B. anthracis, 335
  B. azotofixans, 337
  B. brevis, 239, 337, 340
  B. cepacia, 222
  B. cereus, 24, 107, 119, 126–128, 149, 161, 164–165, 167, 174, 216, 217, 222, 280, 336–340, 344, 345, 347, 348, 351
  B. cereus MJ-1, 51, 52
  B. circulans, 107, 217, 222, 337, 340, 343, 345
  B. coagulans, 31, 343
  B. firmus, 322, 337, 340
  B. globiosporus, 375–376
  B. gordonae, 337
  B. ircalmous, 343
  B. larvae, 337
  B. lautus, 337
  B. lentimorbus, 337
  B. lentus, 337
  B. licheniformis, 7, 49, 52, 217, 222, 336, 340, 344, 345, 351
  B. licheniformis (strain-2042), 240
  B. licheniformis BC 98, 340, 344
  B. licheniformis RC08, 49
  B. macerans, 337

B. macquariensis, 337
B. macroides CJ-29, 51, 52
B. megaterium, 49–51, 55, 57, 61, 119, 335, 338, 340, 343–345, 351, 353
B. megaterium M3, 49, 57
B. megaterium RC01, 50, 51
B. megaterium RC07, 49
B. megaterium strain M3, 55
B. megatorium M3 and RC01, 49–51, 55, 57
B. mojavensis, 336
B. mycoides, 337, 338, 344, 345, 348, 349
B. mycoides Bm J, 202
B. pabuli, 337
B. pasteurii T4, 55
B. polymyxa, 337–338, 340, 343, 351, 353
B. popilliae, 337
B. pseudomycoides, 337
B. psychrophilus, 337, 338
B. pulvifaciens, 337
B. pumilus, 7, 51, 52, 66, 336, 340, 344, 345, 348, 349, 353
B. pumilus CJ-69, 51, 52
B. pumilus INR-7, 55, 60
B. pumilus SE34, 55, 63
B. pumilus SE34, INR7 and T4, 63
B. pumilus strain SE34, 349
B. pumilus T-4, 55
B. simplex, 50, 51, 55
B. simplex RC19, 50, 51, 55
B. sorensis, 336–337
B. sphaericus, 205, 345, 347
B. sporothermodurans, 337
B. stearothermophilus, 337, 338
B. subtilis, 7, 24, 28, 31, 34–35, 55, 62, 64–66, 90, 240, 241, 276, 277, 279, 280, 335–340, 343–348, 351, 354, 356, 375–376
B. subtilis (Quadra 136 and 137), 59
B. subtilis AF 1, 239
B. subtilis 937b, 63
B. subtilis GB03, 201, 202
B. subtilis GB03 and IN937b, 63
B. subtilis GBO3, 55, 59, 60, 62
B. subtilis MFD-5, 54, 57
B. subtilis OSU142, 48–50, 54, 55, 57
B. subtilis strain, 119, 143
B. subtilis strain BA142, 55
B. subtilis strain PRBS-1 and AP-3, 346
B. subtilis subsp. *spizizenii,* 336
B. subtilus IN937b, 55, 63
B. thuringiensis, 7, 205, 257, 260, 337–339, 344, 347, 354–355
B. validus, 337
B. vallismortis, 336–337
B. weihenstephanensis, 337
Bacilysocin, 120
Bacteria, 271–281, 283–289
  biopesticides, 199
  communities, 254, 255, 258–266
  diversity, 264–265
  inoculation, 252
  populations, 254
  tolerance, 259
Bacterial wilt of tomato, 161
Bactericidal, 324
Bacteriostatic, 324
Bacterization, 98
Bacteroids, 378, 402–403
Bagasse, 241, 308
Banksia grandis, 161
Barley (*Hordeum vulgare* L.), 301, 303, 304
BBs. *See* Biofilmed biofertilizers
BCA. *See* Biocontrol agents
Bean (*Phaseolus vulgaris* L.), 8
Bean yellow mosaic potyvirus (BYMV), 63
Beijerinckia spp., 46, 86
Bergey's Manual of Determinative Bacteriology, 335
Biochemical tests, 25
Biocidal, 6
Biocontrol, 157–182, 211–226, 256
Biocontrol agents (BCA), 82, 85, 86, 89–91, 103, 107–109, 235–243
Biocontrol efficacy, 102, 107, 212, 218, 219, 225, 226
Biodiversity, 258, 261
Biofertilizers, 23, 48, 82, 85–88, 91, 108, 109, 213, 215, 236, 256–258
Biofilmed biofertilizers (BBs), 82, 86–88
Biofilm-forming cells, 81–91
Biofilms, 106, 255, 259–260, 335, 341, 352–356
  networks, 84
  PGPR, 82–91
Bioformulations, 97–112
Bioinoculants, 98, 103, 111, 112
Biolog assay, 26
Biological buffer, 259–260
Biological control, 198–205, 212, 213, 214, 224, 225
Biological nitrogen fixation (BNF), 47–48, 67, 298, 303–305, 308
Biolog plates, 25
Biopesticides, 7, 203–204, 213, 214
Biopreparations, 63

Biopriming, 30
Bioprotectants, 23
Bio-solubilisation, 88
Biosphere, 274
Biostimulants, 23
Biosurfactants, 120, 127, 199–200, 214
Biotic factors, 101
Biotic stresses, 8, 380
Biotrophism, 105
Blast and blight of rice, 103
BLAST, software package, 28
Blue-green aphids, 65
BNF. *See* Biological nitrogen fixation
Botrytis cinerea, 346
Bradyrhizobium, 4, 5, 12, 23–24, 48, 98, 334, 378
Bradyrhizobium japonicum, 5, 12–13, 31, 54, 62, 240
Bradyrhizobium SEMIA6144, 12
Bradyrhizobium spp., 86, 378
Bradyrhizobium spp. strains S62 and S63, 48
Brassica juncea, 374
Brevibacillus, 336
Brevibacillus brevis, 65
Brevibacillus brevis IPC11, 55, 59, 62
Brevundimonas aurantiaca, 320, 324
Brevundimonas fusiformis, 322
Brevundimonas pumilus, 322
Brine channels, 275
Bromeliads, 261
Burkholderia, 3, 23, 46, 48, 59, 67, 118, 119, 129–141, 144–148, 236, 239, 241, 319, 334
  B. cepacia, 10, 35
  B. gladii BA7, 54–57
  B. gladii OSU 7, 60
  B. gladii strain BA7, 54
  B. gladioli, 237
  B. phytofirmans, 288
  B. unamae, 10
Burkholderia OSU-7, 57
Burkholderia sp. MBf21 and MBf15, 61
Burkholderia sp. MBp1, 61
Burkholderia sp. MSSP, 239, 241
2, 3-Butanediol, 345, 348
Butler, E.J., 233
Butyrolactones, 160
BYMV. *See* Bean yellow mosaic potyvirus
Byssochlamys fulva H25, 351

C
Caatinga, 252, 253, 261–263
Cacti, 261
Caenorhabditis elegans, 106
Cajanin, 235–236
Cajanol, 235–236
Cajanus cajan (Pigeon pea), 223, 231–243
Callose, 177
Camellia sinensis, 50, 55, 86, 87
Cameroon grass, 304
Candida, 98
Candida albicans, 352
Canola (Brassica napus), 375
Caps. *See* Cold acclimation proteins
Capsicum annum L, 60
α-Carbanionic intermediate, 370
Cardinal growth temperatures, 274–275
Carrier, 99, 100, 107
Catechol, 175–176
Catharanthus roseus, 265
Caulobacter crescentus, 322
Cellulase (CMCase), 119, 128, 222, 345, 353
Cellulolytic enzymes, 323
Cellulomonas, 236
Cell wall degrading enzymes, 214, 222–223
Cephalosporium sacchari, 242
Cercospora leaf spot, 202, 203
Cereals, 213, 215–221
Channels, 420
Chaperones, 278, 279
Charcoal, 241
Chemical fertilizers, 8–9, 47
Chemical pesticides, 214, 219
Chickpea, 48–49, 66
Chitin, 107
Chitin (a homopolymer of β–1, 4 linked *N*-acetylglucosamine), 238
Chitinases, 33, 35, 119, 120, 128, 129, 133, 161, 177, 222, 223, 238–239, 256, 339, 341, 346, 349, 350
Chitinolytic, 107, 161
Chlamydospore, 239
Chlorosis, 422
Choline oxidase, 290
Chryseobacterium balustinum Aur9, 8
Chryseobacterium indologenes (ISE14), 61
Cicer arietinum, 14, 343
cis-11-methyl-2-dodecanoic acid, 84
Clavibacter, 59
Clavibacter michiganensis ssp., 59, 60
Clostridia, 320, 324
Clostridiaceae, 335
Clostridiales, 335
Clostridium, 46, 335
Clostridium botulinum, 277
CLUSTAL X, 28

CNR. *See* Critical nutrient range
Cochliobolus sativum, 221
Cocoa peat, 241
Co-inoculation, 88, 267
Cold acclimation proteins (Caps), 276, 278–279
Cold-adapted enzymes, 276, 281–282
Cold-adapted microorganisms, 274, 290
Cold box, 279
Cold shock domain (CSD), 275, 276, 278–280
Cold shock proteins (Csps), 276, 278–280, 286
Cold-tolerant mesophiles, 275, 276, 278
Colletotrichum
C. demantium, 345
C. lindemuthianum, 223
C. orbiculare, 32, 57
C. truncatum, 346
Colonization, 319–323, 325, 326, 328, 329
Comamonas, 103
C. acidovorans RC41, 50, 51, 55
Competition, 199
Coniothyrium, 98
Consortium, 23, 320, 324
Cortex, 322, 323
Cotton, 233
Critical nutrient range (CNR), 401–402
Crown gall, 176
Crown rot, 215, 218, 219
Cryoprotectants, 276, 278, 289
CSD. *See* Cold shock domain
Csps. *See* Cold shock proteins
Cucumber, 161, 162, 167–168, 171, 173, 175–177, 180
Culturable, 104–105
Curcuma longa, 242
Curtobacterium, 59, 60
Curvularia lunata, 345–346
Cyanide, 256
Cyanogas, 237
Cyclic lipopeptide, 160
Cytochrome oxidase, 237–238
Cytokinins, 46, 50–52, 88

D
D-amino acid aminotransferase, 370
Damping-off, 203, 204
DAPG. *See* 2, 4-Diacetylphloroglucinol
DDT, 205
Deciduous, 261
Degradation, 299, 307
Delftia sp., 49
D. terephthalate, 103
D. tsuruhatensis, 103
D. tsuruhatensis HR4, 236
Desulfococcus, 265
Desulfosarcina, 265
Desulfotomaculum, 265
Desulfovibrio, 265
Detoxication, 6
Diabrotica undecimpunctata howardi Barber, 65
2, 4-Diacetylphloroglucinol (DAPG), 129, 131, 143, 160–161, 179, 220, 225
Diadzein, 325
Diazotrophs, 286–287, 317–329
Difeconazole, 202
Difenoconazole, 31
Difficidin, 120, 126, 143, 145, 149
Diffusion, 420
Dinitrogen (N_2) fixation, 393, 400
Disease management, 202, 206
Disease suppression, 203
DNA hybridization, 24, 27–28
DNA microarray, 27

E
Ecological diversity, 275
Ecological niches, 258
Ecological significance, 329
Economic significance, 329
Ecosystems, 252, 253, 258–266
Ectomycorrhizal fungi, 3
Endophytic, 254, 255, 304, 305, 306
bacteria, 29, 30, 103
colonization, 90
Endoribonuclease, 282
Endosymbionts, 343, 388–389, 393
Enterobacter, 3, 46, 48, 51, 59, 60, 67, 118, 137, 236
E. agglomerans, 49, 161, 167, 222
E. cloacae, 319
E. cloacae CAL2 and UW4, 367
E. dissolvens, 320, 324
Enterobacteriaceae, 320
Enterococcus, 336
Environmental microbiology, 320
Environmental stresses, 274, 284
Epidermis, 322
Epiphytic, 254, 255
EPS. *See* Extracellular polymeric substances
Erwinia, 46, 49, 59, 67, 236
E. caratovora, 349
E. carotovora var. *carotovora,* 120
E. herbicola, 50, 60
Escherichia coli, 161, 276, 278–282
Ethyl-cellulose, 107

Ethylene, 22, 24, 32, 47, 50, 52, 58, 177, 265–266, 366–380
Eubacteria, 58
Euphorbiaceae, 261
Exiguobacterium acetylicum, 285
Exogenous, 278
Exopolysaccharides, 259–260
Exoproteases, 90
Exoribonuclease, 282
Extracellular polymeric substances (EPS), 82, 87

F
Facultative endophyte, 319
Fatty acid methyl ester (FAME), 25, 26
Fatty acids, 276–278, 280
Fengycin (plipastatin), 7, 119, 122, 138, 145, 148, 351
Fertilizers, 2, 8–9, 11
Filamentous, 275
Flavobacterium sp., 46, 236, 374
Flavobacterium sp. (GSE09), 61
Flavonoids, 5, 8, 12, 325
Fluorescent pseudomonads, 160, 163, 164, 166–170, 172, 173, 176, 180, 237, 242
Flux, 388, 390, 391, 393, 398, 399
Foliar formulations, 203
Formulation, 197, 199, 203, 204
Foxtail millet *(Setaria italica),* 11
Frankia, 98, 334
Fructose, 278
Fungicides, 108
  bio, 213
  chemical, 212, 214, 219
Fungicides, *Bacillus* sp. L-324–92, 202
Fusarial wilt *(Fusarium udum),* 61
Fusaricidin, 341, 352
Fusaricidin B, 352
Fusarium spp., 169, 171, 200, 203, 231–243
  F. culmorum, 168, 176
  F. graminearum, 221
  F. oxysporium, 285
  F. oxysporum, 29, 120, 215, 216, 218, 219, 221, 222, 223, 341, 346, 351, 352
  F. oxysporum f. sp. *dianthi,* 57
  F. udam, 353
  F. udum, 233–235, 237–243
  F. vasinfectum, 233
  F. verticillioides M1, 61
  F. verticilloides, 216, 221
Fusarium wilt, 29–30, 108, 169, 171, 176, 202, 215, 223, 231–243
Fusarium wilt disease, 346
Fusarium wilt of cucumber, 108
Fusarium wilt of tomato, 176

G
Gaeumanno graminis var. *tritici,* 129
Gaeumannomyces graminis var. *tritici,* 120, 346
Gaeumanomyces graminis var. *tritici,* 220, 225
γ-Aminobutyrate acid (GABA), 425–428
Gamma-butyrolactones, 84
Garden pea, 285
Gavaserin, 351
Genistein, 235–236
Genomisotopic, 106
Geobacillus, 336
GFP. *See* Green fluorescent protein
Gibberellins, 34, 46, 51, 88
Gliocladium virens, 240
Gln. *See* Glutamine
Global warming, 9
Glomerella cingulata, 29
Glomus fasciculatum, 9, 240, 398–399
Glomus intraradices, 260, 343
Glomus mosseae, 9, 351–352
Glu. *See* Glutamate
β–1,3-Glucanase, 161, 222, 242, 256, 349
Glucanases, 33, 35, 119, 350
β–1,3-, β–1,4-and β–1,6-glucanases, 161
Gluconacetobacter, 3, 46, 67, 236
Gluconacetobacter diazotrophicus, 304–306, 322, 325
Gluconic acid, 287, 343
Glucose, 278, 282, 287
β-Glucosidase, 323
Glutamate (Glu), 418, 425–428
Glutamine (Gln), 418, 419, 425, 427, 428
Glutamine synthetase-glutamate synthase (GS-GOGAT), 418
Glycerol, 278
Glycine betaine, 278, 290
Glycine max, 256, 353
Glycine soja, 353
Glycolic, 343
Gnotobiotic cultures, 322
Grapevine, 288–289
Green fluorescent protein (GFP), 224, 241
GS-GOGAT. *See* Glutamine synthetase-glutamate synthase
Guanidine hydrochloride, 278–279

H
Hafnia alvei strain P3, 54
Hansenula satrunus, 366

HCN. *See* Hydrogen cyanide
Helianthus annus, 260
Herbaspirillum spp., 3, 300, 304, 318, 319, 328, 329
Herbospirillum seropedicae, 51, 52
Heterodea, 201
Heterodera cajani, 232
Heterotrophs, 275
HGT. *See* Horizontal gene transfer
Homology, 279–280
Horizontal gene transfer (HGT), 84
Hydrogen cyanide (HCN), 160, 237–238
Hydrolytic enzymes, 118, 120, 201
Hydropathy plots, 281
Hydroxygenistein, 235–236
2-Hydroxymethyl-chroman-4-one, 239

I

IAM. *See* Indole-3-acetamide pathway
Ice minus, 281, 285
Ice nucleation proteins (INP), 281
Ice nucleators, 276, 281
Ice plus, 281
Indigenous indole acetic acid, 256, 257, 267
Indole-3-acetamide (IAM) pathway, 50
Indole-3-acetic acid (IAA), 22, 34, 47, 50–51, 283–284, 344–345, 353, 370–372, 377
Indole-3-pyruvic acid (IPyA) pathway, 50
Induced systemic resistance, 199
Induced systemic resistance (ISR), 161, 167, 171, 172, 175, 177, 178, 214, 217, 223–224, 345, 348–350
Induced systemic tolerance (IST), 4, 8, 265
Induction of systemic resistance, 283
Inoculant, 99–102, 107, 110–112
Inorganic fertilizer, 298–300, 304–306, 310, 311
INP. *See* Ice nucleation proteins
Insecticides, 108
Integrated disease management, 201, 206
Integrated management, 108–110
Integrated Pest Management (IPM), 198, 201, 202, 205
Intersymbiont, 390
IPM. *See* Integrated Pest Management
IPyA. *See* Indole-3-pyruvic acid pathway
Isoflavonoids, 5
Isogenic, 287
Isonovalol, 103
Iso-pentenyladenine, 344–345
ISR. *See* Induced systemic resistance
IST. *see* Induced systemic tolerance
Iturin, 7, 119, 123, 138, 143, 145, 149, 202, 351

J

Jasmonic acid, 32, 175, 177, 373–374

K

Kallar grass (*Leptochloa fusca* (L.) kunth), 304, 319, 322, 328
Kanosamine, 160
α-Ketobutyrate, 24, 366, 369–372
2-Ketogluconic acid, 287, 343
Kings-B medium, 222
Klebsiella, 3, 46, 48, 67, 319
Kluyvera ascorbata SUD165, 367
Kluyvera cryocrescens IN114, 63
Kocuria oxytoca, 319
Kocuria palustris, 322

L

Lactobacillus, 336
Lathryrus, 286
Lathyrus japonicus, 280
Lathyrus pratensis, 280
LCO. *See* Lipochitooligosaccharides
Leghemoglobin, 12
Legume, 84, 86, 87, 387–410
Leguminosae, 261
Leptochloa fusca, 319
Leptosphaeria maculans, 346, 352
Lichenysins, 351
Lignin, 177
Lindane, 103
Lipochitooligosaccharides (LCOs), 8
Lipopolysaccharide, 177, 179
Listeria, 336
Listeria monocytogenes, 277, 278, 280
Listonella, 265
Lotus japonicas, 378
LuxI-type protein, 85
LuxR-type protein, 85
Lytic enzymes, 6, 34

M

Macroconidia, 233
Macrophomina phaseolina, 346, 353
Magnoporthe grisea, 346
Maize *(Zea mays),* 215, 301, 303–309
Malathion, 205
MALDI-TOF. *See* Matrix-assisted laser desorption/ionization time-of-flight
Malonic, 343
Mangroves, 252, 253, 258, 259, 263–266
Manintol, 278
Matacin (Polymyxin M), 351

Matrix-assisted laser desorption/ionization time-of-flight (MALDI-TOF), 26
MBCAs. *See* Microbial biocontrol agents
Medicago, 65
Medicago sativa L., 419
Mefoxam, 202
Meloidogyne spp., 201
M. incognita, 65–66, 239
M. javanica, 64–66, 161, 164, 165, 347
Mesophilic temperatures, 275, 279, 286, 287
Mesorhizobium, 4, 286
Mesorhizobium mediterraneum strain PECA21, 49
Metagenomics, 104–106, 111
Methacrylic copolymer carrier, 107
Methylobacterium fujisawaense, 322
Methylobacterium radiotolerans, 322
MHB. *See* Mycorrhization-helper bacteria
Michaelis–Menton rate equation, 369–370
Microbacterium FS01, 53
Microbial adaptability, 258, 260
Microbial biocontrol agents (MBCAs), 199, 206
Microbial biodiversity, 199
Microbial biomass, 199
Microbial cells, 253
Microbial community, 252–253, 261–263
Microbial inoculants, 9, 82, 83, 85, 88, 89, 91, 256
Microbial interactions, 263–265
Microbial metabolism, 259
Microbial population, 258
Microbial species, 252, 253, 259
Microbiostasis, 101
Micrococcus cryophilus, 277
Micrococcus luteus, 322
Microcolonies, 82
Micromonospora, 236
Micromonospora carbonacea, 161
Micronutrients, 418, 420–421
Microscale (pore), 110
Microsymbionts, 390, 391, 395, 400, 401, 403–404, 409
Mimicking, 281
Mimosa pudica, 378
Molecular diversity, 325
Molybdenum (Mo), 417–428
Molybdoenzymes, 421
Moniliana laxa Ehr., 60
Monocultures, 85, 86, 88, 89, 320
Mycorhizospheric, 159
Mycorrhization-helper bacteria (MHB), 30–31, 120
Mycorrhizosphere, 82, 378–379
Mycosubtilin, 201
Myzus persicae, 65

N
β–1,4-N-acetylglucosamine, 222
N-acyl-homoserine lactones (AHL), 106
Naringenin, 8
Necrotrophs, 59
Nematodes, 158–159, 161–166, 201
Niche, 252, 253, 255, 256, 258, 263–265, 267, 275
Nitrate reductase, 418, 421–422
Nitrogen, 46–50, 56, 67, 256–258, 265–267
Nitrogenase *(nifH),* 35, 86, 325–328, 340, 342, 353, 400–403, 418, 421–422, 424–425, 428
Nitrogen-fixing bacteria, 3, 4, 10
Nodulation, 286–287
Nodule occupancy, 109
Non-legumes, 86, 87
Non-spore formers, 275
Nutrient, 301–303, 305–311
mobilization, 282–283
solubilization, 282–283

O
Obligate endophyte, 319
Ochrobactrum sp., 103, 318
Oil seed rape *(Brassica napus),* 346
Oligomycin A, 160
Oligopeptides, 84
Onobrychis viciifolia, 286
Oomycin A, 160
Organic acids, 48–49, 179
Oryza sativa, 86, 318, 320, 322
Oxalic, 343
Oxychlororaphin, 239
Oxydifficidin, 120, 126, 143

P
Paenibacillus sp., 3, 8, 14, 24, 26, 28, 103, 108, 265, 333–356
P. aeruginosa, 222
P. agglomerans, 318, 320, 324, 325
P. alvei, 215, 219
P. ananatis, 322
P. azotofixans, 26, 337, 339, 340, 342
P. brasilensis, 31, 340
P. brasilensis PB177, 221
P. graminis, 340
P. lentimorbus, 337, 347, 352
P. lentimorbus WJ5, 352

P. macerans, 31, 337, 340, 345
P. odorifer, 340–342, 353
P. peoriae, 340, 351
P. peoriae strain NRRL BD-62, 351–352
P. polymyxa, 8, 14, 51, 66, 85, 90, 163–165, 179, 221, 337, 340–342, 344–347, 352, 355, 356
P. polymyxa HKA-15, 341, 346, 355–356
P. polymyxa PKB1, 352
P. polymyxa RC05, 48, 50, 51, 54, 55
P. polymyxa RC14, 48
P. polymyxa strain HKA-15, 352
P. popilliae, 337, 347
P. rhizosphaerae sp., 14
PAGE. *See* Polyacrylamide gel electrophoresis
PAL. *See* Phenylalanine ammonia lyase
Paneer, 241
Pantoea, 59, 98, 318
Pantoea (Enterobacter) agglomerans, 161
Pantoea ananas, 278–279
Pantoea dispersa, 240–241, 284
Parasitism, 199
Parenchymatic, 255
Pasteuria penetrans, 105
Pathogenesis related proteins, 217, 223
Pathogenic microorganisms, 252, 254, 256
PCA. *See* Phenazine-1-carboxylic acid
PCR. *See* Polymerase chain reaction
PCR–DGGE, 323–324, 326, 327
Pea (*Pisum sativum* L.), 374–375
Penicillium spp., 86, 88, 200
P. citrinum, 240, 367, 373
P. thomii, 352
Pentose phosphate pathway, 282, 284
Perlite, 308
Permafrost soil, 275
Permeases, 420
Peroxidases (POX), 33, 177, 223, 242, 349
Pesticide, 299, 310
PFGE. *See* Pulsed field gel electrophoresis
PGPB. *See* Plant growth promoting bacteria
PGPR. *See* Plant growth promoting rhizobacteria
Phaseolus vulgaris, 8, 12, 267
Phaseoulus vulgaris L., 420
PHB. *See* Poly-β-hydroxybutyrate
Phenazine, 239
Phenazine-1-carboxylic acid (PCA), 160, 220, 221, 239
Phenazines, 90, 220
Phenolics, 176–177, 289
Phenylalanine ammonia lyase (PAL), 33, 177, 223, 242
Phomopsis spp., 346
Phosphatases, 9, 48, 343
Phosphate solubilizing bacteria (PSB), 48–49, 342–344
Phospholipids, 201, 277
Phosphorous (P), 46, 48–49, 56, 266, 267
Phosphorylase, 282
Photinia, 13
Photosynthate, 389–394, 398–401, 403, 404, 408, 409
Photosynthesis, 84
Phyllobacterium, 257, 265
Phyllobacterium myrsinacearum, 49
Phylloplane, 107
Phyllosphere, 7, 308
Phylogenetic, 263–264
Phythium, 285
Phytoalexins, 33, 168, 177, 235–236, 238
Phytohormones, 22–23, 34, 50–52, 56, 67, 265, 266, 283, 300–303, 305–306, 308, 366–367, 376
Phytopathogenic fungi, 118, 144, 145
Phytopathogens, 29, 33, 34, 108, 109, 159, 161, 198, 201, 239, 259, 267, 318
Phytophthora, 203, 204
P. capsici, 61, 62
P. cinnamomi, 161
P. fragariae var. *rubi,* 161
P. palmivora, 341, 349–350
P. sojae, 5
Phytophthora blight, 60–61
Phytophthora blight of pepper, 61
Phytoremediation, 374
Phytostimulators, 82, 236, 256
Pigeon pea, 61, 223
Pinus pinaster, 394
Pistacia vera, 50
Pisum sativum L., 419
Planktonic, 82, 83, 90
Planococcaceae, 336
Plant growth promoting bacteria (PGPB), 23
Plant growth promoting rhizobacteria (PGPR), 3–13, 22–36, 45–67, 82–91, 97–112, 158–182, 211–226, 236–242, 251–268, 297–311, 365–380
Plant growth promotion, 274, 282–285, 287, 290
Plant growth regulators, 46–47, 50–53, 82, 88, 89, 91
Plant hormones, 50
Plant-microbe interactions, 82–85, 212, 225
Plant pathogens, 212, 215, 216, 217, 221, 224
Pleurotus ostreatus, 88, 90

Pln. *See* Pyrrolnitrin
PLP. *See* Pyridoxal phosphate
Plt. *See* Pyoluteorin
Pochonia chlamydospora, 108
Polyacrylamide gel electrophoresis (PAGE), 25, 26
Poly-β-hydroxybutyrate (PHB), 276
Polymerase chain reaction (PCR), 288
Polymethyl galacturonase, 242
Polymyxin B, 351, 352
Polynucleotide kinase, 289
Polypeptides, 282
Polyphenol oxidase (PPO), 33, 177, 242
Polysaccharides, 259
Polyunsaturated fatty acids, 277
Pomegranate (*Punica granatum* L.), 9
Potato (*Solanum tuberosum* L.), 6
POX. *See* Peroxidases
PPO. *See* Polyphenol oxidase
Prn. *See* Pyrrolnitrin
Productivity, 257, 266, 267
Proline (Pro), 284, 289, 425, 426, 427, 428
Proteases, 201, 222, 341, 350
Protectant, 278
Proteobacteria, 84, 103, 253, 326
Prototrophy, 179
Prunus cerasus, 50
PSB. *See* Phosphate solubilizing bacteria
Pseudobactins, 222, 285
Pseudomonads, 160, 163, 164, 166–170, 172, 173, 175–176, 180, 334
Pseudomonas BA-8, 53–57
Pseudomonas sp., 3, 6, 7, 23, 26, 32, 34, 36, 46, 47, 55, 59, 67, 86, 98, 107, 109, 118–120, 130–131, 134, 135, 138, 143–147, 160–174, 176, 199, 203, 220, 222, 224, 236–241, 243, 255–257, 259–260, 267, 320, 324, 344, 366–367
  P. aeruginosa, 64, 83, 108, 129, 132, 133, 137, 146–148, 162–165, 168–169, 171, 176
  P. aeruginosa (N39), 51, 52
  P. aeruginosa PNA1, 239
  P. aureofaciens, 160, 166, 167, 220
  P. aureofaciens strain, 160
  P. brassicacearum, 51, 53
  P. cepacia, 161, 170
  P. chlororaphis, 49, 66
  P. corrugata (CCR04 and CCR80), 61
  P. corrugata, 288
  P. fluorescens, 7–8, 62, 63, 65, 66, 84, 85, 90, 103, 216, 217, 220, 222, 223, 225, 301, 305, 307, 308, 367
  P. fluorescens biotype G (N-3), 56
  P. fluorescens FB11, 63
  P. fluorescens LPK2, 241
  P. fluorescens MPp4, 61
  P. fluorescens strain, 160, 162, 167, 172, 177, 179
  P. fluorescens WCS417r, 173, 177
  P. fragi, 280
  P. jessenii, 288
  P. kilonensis, 288
  P. monteilii, 31
  P. moraviensis, 288
  P. putida, 5, 13, 48, 49, 51–54, 60, 64, 66, 84, 163–167, 169–171, 173, 176, 301, 305, 308
  P. putida BA-8, 54
  P. putida N21, 51, 52
  P. putida PCL1444, 103
  P. putida RC06, 48, 56
  P. putida strain BA8, 54
  P. putida strain HS-2, 53
  P. putida UW, 367, 373
  P. rhodesiae, 103
  P. striata, 343
  P. syringae, 50, 60, 282, 285
  P. syringae pv. glycineae, 347
Pseudomonas sp. CDB 35, 49
Pseudomonas strain Pa324, 237–238
Pseudomonas strain PMZ2, 54
Pseudomonas strains Pf1, TDK1 and PY15, 61
Pseudonodules, 87
Pulsed field gel electrophoresis (PFGE), 27
Pyochelins, 222
Pyoluteorin (Plt), 120, 129, 131, 137, 141, 144, 145, 147, 160–161, 179
Pyoverdine, 176
Pyoverdins, 222
Pyricularia oryzae, 8, 103, 222, 345
Pyridoxal phosphate (PLP), 366
Pyrrolnitrin (Pln), 160
Pyrrolnitrin (Prn), 120, 129–131, 137, 141, 144–148
Pythium root rot, 216, 222
Pythium sp., 167, 176, 238
  P. aphanidermatum, 349–350
  P. irregulare, 216, 221
  P. ultimum, 120, 131, 161, 167, 215–218, 221, 222

Q

Quercetin, 8, 325
Quorum sensing, 83–85, 90, 91

R
Radiolabelling, 279
Ralstonia, 59, 60, 68
  R. solanacearum, 373–374
Raman spectroscopy, 26–27
r-and K-strategists agents, 203
Randomly amplified polymorphic DNA (RAPD), 27
Rathayibacter, 59
Resistance management, 205
Restriction fragment length polymorphism (RFLP), 27
Retama sphaerocarpa, 260
Reverse transcriptase, 328
Rheinheimera sp., 320, 323
Rhizobacteria, 81–91, 211–226
Rhizobia, 1–14, 236–238, 242, 277–278, 282, 286–287, 290, 323, 343, 378
Rhizobium spp., 4–5, 11–13, 46, 50, 51, 67, 84, 86, 98, 238, 239, 240, 241, 256, 258, 267, 334, 340, 343, 347, 373, 378–379, 387–410, 425, 428
  Arbuscular mycorrhizal fungi–legume, 387–388
  R. etli ISP42, 12
  R. leguminosarum, 31, 51, 277, 280, 286
  R. leguminosarum bv. Vicae, 418–419
  R. leguminosarum E11, 48
  R. leguminosarum FBG05, 63
  R. meliloti, 401
  R. phaseoli, 52
  R. radiobacter, 304
  R. tropici, 8, 12
  R. tropici CIAT899, 12
Rhizoctonia bactaticola, 238
Rhizoctonia bataticola, 103, 341, 346, 352, 353
Rhizoctonia solani, 14, 103, 120, 121, 161, 167–170, 216, 221, 222, 223, 285, 307, 346, 373–374
Rhizodeposition, 4, 14
Rhizoplane, 3, 213, 233
Rhizoremediation, 10
Rhizosphere, 3–7, 10, 14, 46–49, 58, 64, 98, 102, 103, 105–111, 158–161, 168, 176, 178–181, 213, 254–257, 259, 334–335, 338–344, 346, 351–354, 356
Rhizosphere colonization, 178–179, 236
Rhodobacter capsulatus RC04, 54
Rhodococcus sp., 374
  R. erythropolis, 49
  R. rodochrous, 259
Rhodopseudomonas sp BL6, 54
Rhodopseudomonas sp. KL9, 54
Rhodosporidium diobovatum (S33), 59
Ribulose-1, 5-bisphosphate carboxylase oxygenase (RubisCO), 394
Rice *(Oryza sativa),* 103, 215, 300, 303–308
Rice husk, 241
Ricinus communis L., 420
RNA degradosomes, 276, 282
Rock phosphate, 82, 88, 241
Root nodulation, 266
Root rot, 215, 216, 218, 219, 221, 222
Root rot of soybean, 204–205
Rosa canina, 50
Rosa dumalis, 50
Rosellina necatrix, 345
Rotylenchulus, 201
RubisCO. *See* Ribulose-1, 5-bisphosphate carboxylase oxygenase

S
SA. *See* Salicylic acid
Saccharomyces cerevisiae, 352
S-adenosyl L-methionine (AdoMet), 368, 371
Sainfoin, 286
Salicylic acid (SA), 175, 177, 348, 349, 373–374
Salmonella spp., 277
Saltavalin, 351
SAR. *See* Systemic acquired resistance
Sarocladium oryzae, 61
Sawdust, 241
Sclerotinia homeocarpa, 160
Sclerotinia sclerotium, 346
Sclerotium rolfsi, 222, 223
Sclerotium rolfsii, 161, 171, 285, 353
Secondary metabolites, 104, 106
Seed coating/pelleting, 102, 107, 108
Semideciduos, 261
Serine (Ser), 427, 428
Serratia, 3, 46, 67, 98, 236, 256, 325, 326, 334
  S. liquefaciens, 258
  S. marcescens, 162, 175–176, 217, 222, 223, 284, 319
  S. marcescens EB 67, 49
  S. proteamaculans, 258
  S. proteamaculans M35, 51, 52
Sesamum, 233
Shelf Life, 100–101, 111
Siderophores, 6, 118, 124, 129, 130, 132, 138, 142, 145, 146, 159, 160, 161, 173, 175–177, 214, 216, 217, 221–222, 237–238, 256, 267, 283–285, 366–367, 374, 379

Silver thiosulphate, 376
Sinorhizobium, 4, 13
  S. fredii KCC5, 241
  S. fredii SMH12, 8
  S. meliloti, 109, 238, 241, 418–419
  S. meliloti 3DOh13, 13
Soilborne diseases, 3, 202, 203, 211–226
Soil inoculants, 212, 213, 215
Sorbitol, 278
Soybean, 48, 49, 54–55, 60, 62, 66, 256, 265, 266, 285, 286
Sphingomonas, 103
Spore-formers, 275
Sporosarcina, 336
Staphylococcus, 336
  S. aureus, 352
Starch, 288
Stele, 320, 322
Stenotrophomonas, 48, 98
  S. maltophila, 161
Streptococcus, 336
Streptomyces spp., 23, 31, 55, 59, 84, 98, 108, 222, 236, 285
  S. clavuligerus, 279
  S. scabies, 6
Succinic acid, 343
Sugar cane (*Saccharum* spp.), 304–306
Sulfur-oxidizing bacteria, 23
Sunflower, 260
Surfactin, 7, 90, 119, 127, 138, 145, 148, 200, 201, 351, 354
Sustainable agriculture, 22, 34, 36, 82, 318
Swaminathania salitolerans, 304
Swimming planktonic cells, 83
Symbioses, 274, 277, 286, 287, 318, 323, 325
Symbiotic nitrogen fixation, 419, 422, 428
Synthetic pesticides, 198, 299
Systemic acquired resistance (SAR), 3, 29, 32, 177
Systemic resistance, 159, 161, 167, 168, 171–173, 175

T
Take-all disease, 85, 220–221, 225
Take-all of wheat, 160, 161, 346
Tannin, 85
Temperature gradient, 275
Tensin, 160
Terrestrial ecosystems, 82
Theophrastus, 3
Thielaviopsis bassicola, 129, 140
Thiobacillus, 98, 334
Threonine (Thr), 427, 428
Tn5 mutants (FM29 and FM13), 239
Tobacco mosaic virus (TMV), 32, 62
Tomato *(Lycopersicon esculentum),* 10, 90, 285
Tomato mottle virus (ToMoV), 63, 348
Tomato spotted wilt virus (TSWV), 63
Transcription, 279, 280
Translation, 280
Trehalose, 108, 278, 289
Tri-calcium, 267
Trichoderma, 98
  T. harzianum, 240–241
  T. harzianum (R), 59
  T. viride, 240
Trifolium subterraneum L., 396
Tripartite, 387–410
Triticum aestivum, 87
Tropical, 251–268
Tropolone, 160
Tryptophan synthase, 50, 370
TSWV. *See* Tomato spotted wilt virus

U
Unculturable fraction, 104–106

V
Variovorax paradox, 373, 374
Variovorax paradoxus 5C-2, 51, 52
Verrucomicrobia, 105
Verticillum longisporum, 346
Vessels, 319, 320, 322
Vibrio, 265
Vibrio cholera, 276
Vibrio harvey, 85
Vicia faba L, 389
Vigna radiata, 285
Viscosinamide, 160
VITEK cards, 25
Vitis vinifera, 288
von Leeuwenhoek, A., 3

W
Wheat, 85–87, 284, 286, 288, 289
Wheat (*Triticum aestivum* L), 215, 300, 301, 303–309
Wheat bran, 241–242
Whey, 241
Whole-cell proteins, 26
Wilting index, 239–240

X
Xanthobacter, 265
Xanthine dehydrogenase, 418
Xanthobaccin, 160
Xanthomonas, 59, 60, 84, 236, 281
Xanthomonas campestris, 29, 347
Xanthomonas campestris pv. campestris, 345, 347
Xanthomonas oryzae, 29, 103
Xanthoparmelia mexicana, 88
Xylem vessel, 319, 322, 419, 425

Y
Yeasts, 201, 366, 367
Yield increasing bacteria (YIB), 28

Z
Zea mays, 266
Zingiber officinale, 242
Zwittermycin A, 160, 351
Zygosaccharomyces bailii, 277